Lecture Notes in Computer Scie

T0238592

Commenced Publication in 1973
Founding and Former Series Editors:
Gerhard Goos, Juris Hartmanis, and Jan van Leeuwen

Editorial Board

Stephen W. Liddle   Klaus-Dieter Schewe
A Min Tjoa   Xiaofang Zhou (Eds.)

Database and Expert Systems Applications

23rd International Conference, DEXA 2012
Vienna, Austria, September 3-6, 2012
Proceedings, Part I

 Springer

Volume Editors

Stephen W. Liddle
Brigham Young University, Marriott School
784 TNRB, Provo, UT 84602, USA
E-mail: liddle@byu.edu

Klaus-Dieter Schewe
Software Competence Center Hagenberg
Softwarepark 21, 4232 Hagenberg, Austria
E-mail: kd.schewe@scch.at

A Min Tjoa
Vienna University of Technology, Institute of Software Technology
Favoritenstraße 9-11/188, 1040 Wien, Austria
E-mail: amin@ifs.tuwien.ac.at

Xiaofang Zhou
University of Queensland
School of Information Technology and Electrical Engineering
Brisbane, QLD 4072, Australia
E-mail: zxf@uq.edu.au

ISSN 0302-9743   e-ISSN 1611-3349
ISBN 978-3-642-32599-1   e-ISBN 978-3-642-32600-4
DOI 10.1007/978-3-642-32600-4
Springer Heidelberg Dordrecht London New York

Library of Congress Control Number: 2012943836

CR Subject Classification (1998): H.2.3-4, H.2.7-8, H.2, H.3.3-5, H.4.1, H.5.3, I.2.1,
I.2.4, I.2.6, J.1, C.2

LNCS Sublibrary: SL 3 – Information Systems and Application, incl. Internet/Web
and HCI

Typesetting: Camera-ready by author, data conversion by Scientific Publishing Services, Chennai, India

Printed on acid-free paper

Springer is part of Springer Science+Business Media (www.springer.com)

Preface

This volume includes invited papers, research papers, and short papers presented at DEXA 2012, the 23rd International Conference on Database and Expert Systems Applications, held in Vienna, Austria. DEXA 2012 continued the long and successful DEXA tradition begun in 1990, bringing together a large collection of bright researchers, scientists, and practitioners from around the world to share new results in the areas of database, intelligent systems, and related advanced applications.

The call for papers resulted in the submission of 179 papers, of which 49 were accepted as regular research papers, and 37 were accepted as short papers. The authors of these papers come from 43 different countries. The papers discuss a range of topics including:

- Database query processing, in particular XML queries
- Labeling of XML documents
- Computational efficiency
- Data extraction
- Personalization, preferences, and ranking
- Security and privacy
- Database schema evaluation and evolution
- Semantic Web
- Privacy and provenance
- Data mining
- Data streaming
- Distributed systems
- Searching and query answering
- Structuring, compression and optimization
- Failure, fault analysis, and uncertainty
- Predication, extraction, and annotation
- Ranking and personalization
- Database partitioning and performance measurement
- Recommendation and prediction systems
- Business processes
- Social networking

In addition to the papers selected by the Program Committee two internationally recognized scholars delivered keynote speeches:

Georg Gottlob: DIADEM: Domains to Databases

Yamie Aït-Ameur: Stepwise Development of Formal Models for Web Services Compositions – Modelling and Property Verification

In addition to the main conference track, DEXA 2012 also included seven workshops that explored the conference theme within the context of life sciences, specific application areas, and theoretical underpinnings.

We are grateful to the hundreds of authors who submitted papers to DEXA 2012 and to our large Program Committee for the many hours they spent carefully reading and reviewing these papers. The Program Committee was also assisted by a number of external referees, and we appreciate their contributions and detailed comments.

We are thankful for the Institute of Software Technology at Vienna University of Technology for organizing DEXA 2012, and for the excellent working atmosphere provided. In particular, we recognize the efforts of the conference Organizing Committee led by the DEXA 2012 General Chair A Min Tjoa. We are gratefull to the Workshop Chairs Abdelkader Hameurlain, A Min Tjoa, and Roland R. Wagner.

Finally, we are especially grateful to Gabriela Wagner, whose professional attention to detail and skillful handling of all aspects of the Program Committee management and proceedings preparation was most helpful.

September 2012   Stephen W. Liddle
                                                   Klaus-Dieter Schewe
                                                   Xiaofang Zhou

Organization

Honorary Chair

Makoto Takizawa   Seikei University, Japan

General Chair

A Min Tjoa   Technical University of Vienna, Austria

Conference Program Chair

Stephen Liddle	Brigham Young University, USA
Klaus-Dieter Schewe	Software Competence Center Hagenberg and Johannes Kepler University Linz, Austria
Xiaofang Zhou	University of Queensland, Australia

Publication Chair

Vladimir Marik   Czech Technical University, Czech Republic

Program Committee

Witold Abramowicz	The Poznan University of Economics, Poland
Rafael Accorsi	University of Freiburg, Germany
Hamideh Afsarmanesh	University of Amsterdam, The Netherlands
Riccardo Albertoni	OEG, Universidad Politécnica de Madrid, Spain
Rachid Anane	Coventry University, UK
Annalisa Appice	Università degli Studi di Bari, Italy
Mustafa Atay	Winston-Salem State University, USA
James Bailey	University of Melbourne, Australia
Spiridon Bakiras	City University of New York, USA
Zhifeng Bao	National University of Singapore, Singapore
Ladjel Bellatreche	ENSMA, France
Morad Benyoucef	University of Ottawa, Canada
Catherine Berrut	Grenoble University, France
Debmalya Biswas	Nokia Research, Germany
Athman Bouguettaya	RMIT, Australia
Danielle Boulanger	MODEME,University of Lyon, France
Omar Boussaid	University of Lyon, France
Stephane Bressan	National University of Singapore, Singapore
Patrick Brezillon	University of Paris VI (UPMC), France
Yiwei Cao	RWTH Aachen University, Germany
Silvana Castano	Università degli Studi di Milano, Italy

Michal Krátký	VSB-Technical University of Ostrava, Czech Republic
Arun Kumar	IBM Research, India
Ashish Kundu	IBM T.J. Watson Research Center, Hawthorne, USA
Josef Küng	University of Linz, Austria
Kwok-Wa Lam	University of Hong Kong, Hong Kong
Nadira Lammari	CNAM, France
Gianfranco Lamperti	University of Brescia, Italy
Mong Li Lee	National University of Singapore, Singapore
Alain Toinon Leger	Orange - France Telecom R&D, France
Daniel Lemire	LICEF Research Center, Canada
Lenka Lhotska	Czech Technical University, Czech Republic
Wenxin Liang	Dalian University of Technology, China
Lipyeow Lim	University of Hawai at Manoa, USA
Tok Wang Ling	National University of Singapore, Singapore
Sebastian Link	University of Auckland, New Zealand
Volker Linnemann	University of Lübeck, Germany
Chengfei Liu	Swinburne University of Technology, Australia
Chuan-Ming Liu	National Taipei University of Technology, Taiwan
Fuyu Liu	Microsoft Corporation, USA
Hong-Cheu Liu	University of South Australia, Australia
Jorge Lloret Gazo	University of Zaragoza, Spain
Miguel Ángel López Carmona	University of Alcalá de Henares, Spain
Jiaheng Lu	Renmin University, China
Jianguo Lu	University of Windsor, Canada
Alessandra Lumini	University of Bologna, Italy
Hui Ma	Victoria University of Wellington, New Zealand
Qiang Ma	Kyoto University, Japan
Stéphane Maag	TELECOM SudParis, France
Nikos Mamoulis	University of Hong Kong, Hong Kong
Elio Masciari	ICAR-CNR, Università della Calabria, Italy
Norman May	SAP AG, Germany
Jose-Norberto Mazón	University of Alicante, Spain
Dennis McLeod	University of Southern California, USA
Brahim Medjahed	University of Michigan - Dearborn, USA
Harekrishna Misra	Institute of Rural Management Anand, India
Jose Mocito	INESC-ID/FCUL, Portugal
Riad Mokadem	IRIT, Paul Sabatier University, France
Lars Mönch	FernUniversität in Hagen, Germany
Yang-Sae Moon	Kangwon National University, Korea
Reagan Moore	University of North Carolina at Chapel Hill, USA

Franck Morvan   IRIT, Paul Sabatier University, Toulouse,
                                France
Mirco Musolesi   University of Birmingham, UK
Ismael Navas-Delgado   University of Málaga, Spain
Wilfred Ng   University of Science and Technology,
                                Hong Kong
Javier Nieves Acedo   Deusto University, Spain
Mourad Oussalah   University of Nantes, France
Gultekin Ozsoyoglu   Case Western Reserve University, USA
George Pallis   University of Cyprus, Cyprus
Christos Papatheodorou   Ionian University and "Athena" Research
                                Centre, Greece
Marcin Paprzycki   Polish Academy of Sciences, Warsaw
                                Management Academy, Poland
Oscar Pastor Lopez   Universidad Politecnica de Valencia, Spain
Jovan Pehcevski   European University, Macedonia
Reinhard Pichler   Technische Universität Wien, Austria
Clara Pizzuti   ICAR-CNR, Italy
Jaroslav Pokorny   Charles University in Prague, Czech Republic
Elaheh Pourabbas   National Research Council, Italy
Fausto Rabitti   ISTI, CNR Pisa, Italy
Claudia Raibulet   Università degli Studi di Milano-Bicocca, Italy
Isidro Ramos   Technical University of Valencia, Spain
Praveen Rao   University of Missour-KaNSAS City, USA
Rodolfo F. Resende   Federal University of Minas Gerais, Brazil
Claudia Roncancio   Grenoble University / LIG, France
Edna Ruckhaus   Universidad Simon Bolivar, Venezuela
Massimo Ruffolo   ICAR-CNR, Italy
Igor Ruiz Agúndez   Deusto University, Spain
Giovanni Maria Sacco   University of Turin, Italy
Shazia Sadiq   University of Queensland, Australia
Simonas Saltenis   Aalborg University, Denmark
Carlo Sansone   Università di Napoli "Federico II", Italy
Igor Santos Grueiro   Deusto University, Spain
N.L. Sarda   I.I.T. Bombay, India
Marinette Savonnet   University of Burgundy, France
Raimondo Schettini   Università degli Studi di Milano-Bicocca, Italy
Erich Schweighofer   University of Vienna, Austria
Florence Sedes   IRIT, Paul Sabatier University, Toulouse,
                                France
Nazha Selmaoui   University of New Caledonia, France
Patrick Siarry   Université Paris 12 (LiSSi), France
Gheorghe Cosmin Silaghi   Babes-Bolyai University of Cluj-Napoca,
                                Romania
Leonid Sokolinsky   South Ural State University, Russia

Bala Srinivasan   Monash University, Australia
Umberto Straccia   Italian National Research Council, Italy
Darijus Strasunskas   Strasunskas Forskning, Norway
Lena Stromback   Swedish Meteorological and Hydrological
                                    Institute, Sweden
Aixin Sun   Nanyang Technological University, Singapore
David Taniar   Monash University, Australia
Cui Tao   Mayo Clinic, USA
Maguelonne Teisseire   Irstea - TETIS, France
Sergio Tessaris   Free University of Bozen-Bolzano, Italy
Olivier Teste   IRIT, University of Toulouse, France
Stephanie Teufel   University of Fribourg, Switzerland
Jukka Teuhola   University of Turku, Finland
Taro Tezuka   University of Tsukuba, Japan
Bernhard Thalheim   Christian Albrechts Universität Kiel, Germany
J.M. Thevenin   University of Toulouse I Capitole, France
Helmut Thoma   Thoma SW-Engineering, Basel, Switzerland
A Min Tjoa   Technical University of Vienna, Austria
Vicenc Torra   IIIA-CSIC, Spain
Traian Truta   Northern Kentucky University, USA
Theodoros Tzouramanis   University of the Aegean, Greece
Marco Vieira   University of Coimbra, Portugal
Jianyong Wang   Tsinghua University, China
Junhu Wang   Griffith University, Brisbane, Australia
Qing Wang   The Australian National University, Australia
Wei Wang   University of New South Wales, Sydney,
                                    Australia
Wendy Hui Wang   Stevens Institute of Technology, USA
Andreas Wombacher   University Twente, The Netherlands
Lai Xu   Bournemouth University, UK
Ming Hour Yang   Chung Yuan Christian University, Taiwan
Xiaochun Yang   Northeastern University, China
Haruo Yokota   Tokyo Institute of Technology, Japan
Zhiwen Yu   Northwestern Polytechnical University, China
Xiao-Jun Zeng   University of Manchester, UK
Zhigang Zeng   Huazhong University of Science and
                                    Technology, China
Xiuzhen (Jenny) Zhang   RMIT University Australia, Australia
Yanchang Zhao   RDataMining.com, Australia
Yu Zheng   Microsoft Research Asia, China
Qiang Zhu   The University of Michigan, USA
Yan Zhu   Southwest Jiaotong University, Chengdu,
                                    China

External Reviewers

Hadjali Allel	ENSSAT, France
Toshiyuki Amagasa	Tsukuba University, Japan
Flora Amato	University of Naples Federico II, Italy
Abdelkrim Amirat	University of Nantes, France
Zahoua Aoussat	University of Algiers, Algeria
Radim Bača	Technical University of Ostrava, Czech Republic
Dinesh Barenkala	University of Missouri-Kansas City, USA
Riad Belkhatir	University of Nantes, France
Yiklun Cai	University of Hong Kong
Nafisa Afrin Chowdhury	University of Oregon, USA
Shumo Chu	Nanyang Technological University, Singapore
Ercument Cicek	Case Western Reserve University, USA
Camelia Constantin	UPMC, France
Ryadh Dahimene	CNAM, France
Matthew Damigos	NTUA, Greece
Franca Debole	ISTI-CNR, Italy
Saulo Domingos de Souza Pedro	Federal University of Sao Carlos, Brazil
Laurence Rodrigues do Amaral	Federal University of Uberlandia, Brazil
Andrea Esuli	ISTI-CNR, Italy
Qiong Fang	University of Science and Technology, Hong Kong
Nikolaos Fousteris	Ionian University, Greece
Filippo Furfaro	DEIS, University of Calabria, Italy
Jose Manuel Gimenez	Universidad de Alcala, Spain
Reginaldo Gotardo	Federal University of Sao Carlos, Brazil
Fernando Gutierrez	University of Oregon, USA
Zeinab Hmedeh	CNAM, France
Hai Huang	KAUST, Saudi Arabia
Lili Jiang	Lanzhou University, China
Shangpu Jiang	University of Oregon, USA
Hideyuki Kawashima	University of Tsukuba, Japan
Selma Khouri	LIAS/ENSMA, France
Christian Koncilia	University of Klagenfurt, Austria
Cyril Labbe	Université Joseph Fourier, Grenoble, France
Thuy Ngoc Le	National University of Singapore, Singapore
Fabio Leuzzi	University of Bari, Italy
Luochen Li	National University of Singapore, Singapore
Jing Li	University of Hong Kong
Xumin Liu	Rochester Institute of Technology, USA
Yifei Lu	University of New South Wales, Australia
Jia-Ning Luo	Ming Chuan University, Taiwan

Table of Contents – Part I

Keynote Talks

XML Queries and Labeling I

Computational Efficiency

XML Queries

Privacy and Provenance

XML Queries and Labeling II

Data Streams

Structuring, Compression and Optimization

Table of Contents – Part II

Ranking and Personalization

Searching I

Database Partitioning and Performance

Semantic Web

Data Mining II

Distributed Systems

Web Searching and Query Answering

Recommendation and Prediction Systems

Query Processing II

Query Processing III

Searching II

Business Processes and Social Networking

Data Security, Privacy, and Organization

DIADEM: Domains to Databases*

Tim Furche, Georg Gottlob, and Christian Schallhart

Department of Computer Science, Oxford University,
Wolfson Building, Parks Road, Oxford OX1 3QD
firstname.lastname@cs.ox.ac.uk

Abstract. What if you could turn all websites of an entire domain into
a single database? Imagine all real estate offers, all airline flights, or
all your local restaurants' menus automatically collected from hundreds
or thousands of agencies, travel agencies, or restaurants, presented as a
single homogeneous dataset.

Historically, this has required tremendous effort by the data providers
and whoever is collecting the data: Vertical search engines aggregate
offers through specific interfaces which provide suitably structured data.
The semantic web vision replaces the specific interfaces with a single one,
but still requires providers to publish structured data.

Attempts to turn human-oriented HTML interfaces back into their
underlying databases have largely failed due to the variability of web
sources. In this paper, we demonstrate that this is about to change: The
availability of comprehensive entity recognition together with advances
in ontology reasoning have made possible a new generation of knowledge-
driven, domain-specific data extraction approaches. To that end, we introduce DIADEM, the first automated data extraction system that can
turn nearly any website of a domain into structured data, working fully
automatically, and present some preliminary evaluation results.

1 Introduction

Most websites with offers on books, real estate, flights, or any number of other
products are generated from some database. However, meant for human
consumption, they make the data accessible only through, increasingly sophisticated, search and browse interfaces. Unfortunately, this poses a significant challenge in automatically processing these offers, e.g., for price comparison, market
analysis, or improved search interfaces. To obtain the data driving such applications, we have to explore human-oriented HTML interfaces and extract the data
made accessible through them, without requiring any human involvment.

Automated data extraction has long been a dream of the web community,
whether to improve search engines, to "model every object on the planet"[1], or to

---

* The research leading to these results has received funding from the European Research Council under the European Community's Seventh Framework Programme
(FP7/2007–2013) / ERC grant agreement DIADEM, no. 246858.

[1] Bing's new aim, http://tinyurl.com/77jjqz6.

S.W. Liddle et al. (Eds.): DEXA 2012, Part I, LNCS 7446, pp. 1–8, 2012.
© Springer-Verlag Berlin Heidelberg 2012

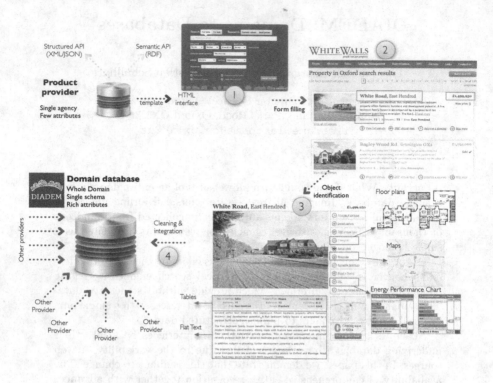

Fig. 1. Data extraction with DIADEM

bootstrap the semantic web vision. Web extraction comes roughly in two shapes, namely web *information extraction* (IE), extracting facts from flat text at very large scale, and web *data extraction* (DE), extracting complex objects based on text, but also layout, page and template structure, etc. Data extraction often uses some techniques from information extraction such as entity and relationship recognition, but not vice versa. Historically, IE systems are domain-independent and web-scale [15,12], but at a rather low recall. DE systems fall into two categories: domain-independent, low accuracy systems [3,14,13] based on discovering the repeated structure of HTML templates common to a set of pages, and highly accurate, but site-specific systems [16,4] based on machine learning.

In this paper, we argue that a **new trade-off** is necessary to make *highly accurate, fully automated web extraction possible at a large scale*. We trade off scope for accuracy and automation: By limiting ourselves to a specific domain where we can provide substantial knowledge about that domain and the representation of its objects on web sites, automated data extraction becomes possible at high accuracy. Though not fully web-scale, one domain often covers thousands or even tens of thousands of web sites: To achieve a coverage above 80% for typical attributes in common domains, it does not suffice to extract only from large, popular web sites. Rather, we need to include objects from thousands of small, long-tail sources, as shown in [5] for a number of domains and attributes.

Figure 1 illustrates the principle of *fully automated data extraction at domain-scale*. The input is a website, typically generated by populating HTML templates from a provider's database. Unfortunately, this human-focused HTML interface is usually the only way to access this data. For instance, of the nearly 50 real estate agencies that operate in the Oxford area, not a single one provides their data in structured format. Thus data extraction systems need to explore and understand the interface designed for humans: A system needs to automatically navigate the search or browse interface (1), typically forms, provided by the site to get to result pages. On the result pages (2), it automatically identifies and separates the individual objects and aligns them with their attributes. The attribute alignment may then be refined on the details pages (3), i.e., pages that provide comprehensive information about a single entity. This involves some of the most challenging analysis, e.g., to find and extract attribute-value pairs from tables, to enrich the information about the object from the flat text description, e.g., with relations to known points-of-interest, or to understand non-textual artefacts such as floor plans, maps, or energy performance charts. All that information is cleaned and integrated (4) with previously extracted information to establish a large database of all objects extracted from websites in that domain. If fed with a sufficient portion of the websites of a domain, this database provides a comprehensive picture of all objects of the domain.

That *domain knowledge* is the solution to high-accuracy data extraction at scale is not entirely new. Indeed, recently there has been a flurry of approaches focused on this idea. Specifically, domain-specific approaches use background knowledge in form of ontologies or instance databases to replace the role of the human in supervised, site-specific approaches. Domain knowledge comes in two fashions, either as instance knowledge (that "Georg" is a person and lives in the town "Oxford") or as schema or ontology knowledge (that "town" is a type of "location" and that "persons" can "live" in "locations"). Roughly, existing approaches can be distinguished by the amount of schema knowledge they use and whether instances are recognised through annotators or through redundancy. One of the dominant issues when dealing with automated annotators is that *text annotators have low accuracy*. Therefore, [6] suggests the use of a top-k search strategy on subsets of the annotations provided by the annotators. For each subset a separate wrapper is generated and ranked using, among others, schema knowledge. Other approaches exploit *content redundancy*, i.e., the fact that there is some overlapping (at least on the level of attribute values) between web sites of the same domain. This approach is used in [11] and an enumeration of possible attribute alignments (reminiscent of [6]). Also [2] exploits content redundancy, but focuses on redundancy on entity level rather than attribute level only.

Unfortunately, all of these approaches are only half-hearted: They add a bit of domain knowledge here or there, but fail to exploit it in other places. Unsurprisingly, they remain stuck at accuracies around $90 - 94\%$. There is also no single system that covers the whole data extraction process, from forms over result pages to details pages, but rather most either focus on forms, result or details pages only.

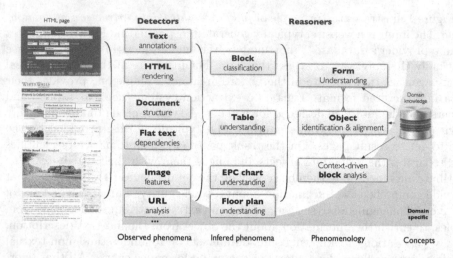

Fig. 2. DIADEM knowledge

To address these shortcomings, we introduce the DIADEM **engine** which demonstrates that through domain-specific knowledge in all stages of data extraction we can indeed achieve high accuracy extraction for entire domain. Specifically, DIADEM implements the full data extraction pipeline from Figure 1 integrating form, result, and details page understanding. We discuss DIADEM, the way it uses domain knowledge (Section 2) and performs an integrated analysis (Section 3) of a web site of a domain in the rest of this paper, concluding with a set of preliminary results (Section 4).

2   DIADEM Knowledge

DIADEM is organised around knowledge of three types, see Figure 2:

1. *What to detect?* The first type of knowledge is all about detecting instances, whether instances of domain entities or their attributes, or instances of a technical concept such as a table, a strongly highlighted text, or an advertisement. We call such instances **phenomena** and distinguish phenomena into those that can be directly *observed* on a page, e.g., by means of text annotators or visual saliency algorithms, and those *inferred* from directly observed ones, e.g., that similar values aligned in columns, each emphasising its first value, constitute a table with a header row.

2. *How to interpret?* However, phenomena alone are fairly useless: They are rather noisy with accuracy in the $70 - 80\%$ range even with state of the art techniques. Furthermore, they are not what we are interested in: We are interested in structured objects and their attributes. How we assemble these objects and assign their attributes is described in the **phenomenology** that is used by a set of reasoners to derive structured instances of domain concepts from the phenomena. Thus a table phenomenon may be used together with price and location

annotations on some cell values and the fact that there is a price refinement form to recognise that the table represents a list of real estate offers for sale. Similarly, we assemble phenomena into instances of high-level interaction concepts such as real-estate forms or floor plans, e.g., to get the rooms and room dimensions from edge information and label annotations of a PDF floor plan.

3. *How to structure?* Finally, the domain knowledge guides the way we structure the final data and resolve conflicts between different interpretations of the phenomena (e.g., if we have one interpretation that a flat has two bedrooms and one that it has 13 bedrooms, yet the price is rather low, it is more likely a two bedroom flat).

For all three layers, the necessary knowledge can be divided into domain-specific and domain-independent. For quick adaptability of DIADEM to new domains, we formulate as much knowledge as possible in general, domain independent ways, either as reusable components, sharing knowledge, e.g., on the UK locations between domains, or as domain independent templates which are instantiated with domain specific parameters. Thus, to adapt DIADEM to a given domain, one needs to select the relevant knowledge, instantiate suitable templates, and sometimes provide additional, truly domain specific knowledge.

Where phenomena (usually only in the form of textual annotators) and ontological knowledge are fairly common, though never applied to this extent in data extraction, DIADEM is unique in the use of explicit knowledge for the mapping between phenomena. These mappings (or phenomenology) are described in Datalog$^{\pm}$, $\neg$ rules and fall, roughly, into three types that illustrate three of the most profligate techniques used in the DIADEM engine:

1. *Finding repetition.* Fortunately, most database-backed websites use templates that can be identified with fair accuracy. Exploiting this fact is, indeed, the primary reason why DE systems are so much more accurate that IE that do not use this information. However, previous approaches are often limited by their inability to distinguish noise from actual data in the repetition analysis (and thus get, e.g., confused by different record types or irregular advertisements). Both is addressed in DIADEM by focusing the search for repetition carrying relevant phenomena (such as instances of domain attributes).

2. *Identifying object instances through context.* However, for details pages not enough repetition may be available and thus we also need to be able to identify *singular* object occurrences. Here, we exploit context information, e.g., from the search form or from the result page through which a details page is reached.

3. *Corroboration of disparate phenomena.* Finally, individual results obtained from annotations and patterns must be corroborated into a coherent model, building not only a consistent model of individual pages but of an entire site.

3   DIADEM Engine

All this knowledge is used in the DIADEM engine to analyse a web site. It is evident that this analysis process is rather involved and thus not feasible for every single page on a web site. Fortunately, we can once again profit from the

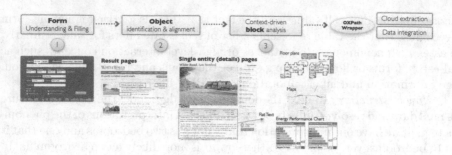

Fig. 3. DIADEM pipeline

template structure of such sites: First, DIADEM analyzes a small fraction of a web site to generate a wrapper, and second, DIADEM executes these wrappers to extract all relevant data from the analyzed sites at high speed and low cost. Figure 3 gives an overview of the high-level architecture of DIADEM. On the left, we show the analysis, on the right the execution stage. In practice, there are far more dependencies and feedback mechanisms, but for space reasons we limit ourselves to a sequential model.

In the first stage, with a sample from the pages of a web site, DIADEM generates *fully automatically* wrappers (i.e., extraction program). This **analysis** is based on the knowledge from Section 2, while the extraction phase does not require any further domain knowledge. The result of the analysis is a wrapper program, i.e., a specification how to extract all the data from the website without further analysis. Conceptually, the analysis is divided into three major phases, though these are closely interwoven in the actual system:

(1) *Exploration:* DIADEM automatically explores a site to locate relevant objects. The major challenge here are *web forms*: DIADEM needs to understand such forms sufficiently to fill them for sampling, but also to generate exhaustive queries for the extraction stage, such that *all* the relevant data is extracted (see [1]). DIADEM's form understanding engine OPAL [8] uses an phenomenology of relevant domain forms for these tasks.

(2) *Identification:* The exploration unearths those web pages that contain actual objects. But DIADEM still needs to identify the precise boundaries of these objects as well as their attributes. To that end, DIADEM's result page analysis AMBER [9] analyses the repeated structure within and among pages. It exploits the domain knowledge to distinguish noise from relevant data and is thus far more robust than existing data extraction approaches.

(3) *Block analysis:* Most attributes that a human would identify as structured, textual attributes (as opposed to images or flat text) are already identified and aligned in the previous phase. But DIADEM can also identify and extract attributes that are not of that type by analysing the flat text as well as specific, attribute-rich image artefacts such as energy performance charts or floor plans. Finally, we also aim to associate "unknown" attributes with extracted objects, if these attributes are associated to suitable labels and appear with many objects of the same type,

At the end of this process, we obtain a sample of instance objects with rich attributes that we use to generate an OXPath wrapper for extraction. Some of the attributes (such as floor plan room numbers) may require post-processing also at run-time and specific data cleaning and linking instructions are provided with the wrapper.

The wrapper generated by the analysis stage can be **executed** independently. We have developed a new wrapper language, called OXPath [10], the first of its kind for large scale, repeated data extraction. OXPath is powerful enough to express nearly any extraction task, yet as a careful extension of XPath maintains the low data and combined complexity. In fact, it is so efficient, that page retrieval and rendering time by far dominate the execution. For large scale execution, the aim is thus to minimize page rendering and retrieval by storing pages that are possibly needed for further processing. At the same time, memory should be independent from the number of pages visited, as otherwise large-scale or continuous extraction tasks become impossible. With OXPath we obtain all these characteristics, as shown in Section 4.

4   DIADEM Results

To give an impression of the DIADEM engine we briefly summarise results on three components of DIADEM: its form understanding system, OPAL; its result page analysis, AMBER; and the OXPath extraction language.

Figures 4a and 4b report on the quality of form understanding and result page analysis in DIADEM's first prototype. Figure 4a [8] shows that OPAL is able to identify about 99% of all form fields in the UK real estate and used car domain correctly. We also show the results on the ICQ and Tel-8 form benchmarks, where OPAL achieves > 96% accuracy (in contrast recent approaches achieve at best 92% [7]). The latter result is without use of domain knowledge. With domain knowledge we could easily achieve close to 99% accuracy as well. Figure 4b [9] shows the results for data area, record, and attribute identification on result pages for AMBER in the UK real estate domain. We report each attribute separately. AMBER achieves on average 98% accuracy for all these tasks, with a tendency to perform worse on attributes that occur less frequently (such as the number of reception rooms). AMBER is unique in achieving this accuracy even in presence of significant noise in the underlying annotations: Even if we introduce an error rate of over 50%, accuracy only drops by 1 or 2%.

For an extensive evaluation on OXPath, please see [10] . It easily outperforms existing data extraction systems, often by a wide margin. Its high performance execution leaves page retrieval and rendering to dominate execution (> 85%) and thus makes avoiding page rendering imperative. We minimize page rendering by buffering any page that may still be needed in further processing, yet manage to keep memory consumption constant in nearly all cases including extraction tasks of millions of records from hundreds of thousands of pages.

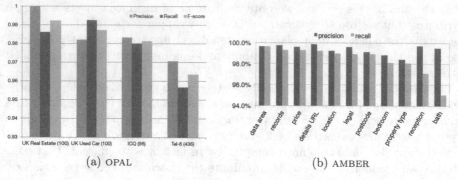

(a) OPAL   (b) AMBER

Fig. 4. DIADEM results

References

1. Benedikt, M., Gottlob, G., Senellart, P.: Determining relevance of accesses at runtime. In: PODS (2011)
2. Blanco, L., Bronzi, M., Crescenzi, V., Merialdo, P., Papotti, P.: Exploiting Information Redundancy to Wring Out Structured Data from the Web. In: WWW (2010)
3. Crescenzi, V., Mecca, G.: Automatic information extraction from large websites. J. ACM 51(5), 731–779 (2004)
4. Dalvi, N., Bohannon, P., Sha, F.: Robust web extraction: an approach based on a probabilistic tree-edit model. In: SIGMOD (2009)
5. Dalvi, N., Machanavajjhala, A., Pang, B.: An analysis of structured data on the web. In: VLDB (2012)
6. Dalvi, N.N., Kumar, R., Soliman, M.A.: Automatic wrappers for large scale web extraction. In: VLDB (2011)
7. Dragut, E.C., Kabisch, T., Yu, C., Leser, U.: A hierarchical approach to model web query interfaces for web source integration. In: VLDB (2009)
8. Furche, T., Gottlob, G., Grasso, G., Guo, X., Orsi, G., Schallhart, C.: Opal: automated form understanding for the deep web. In: WWW (2012)
9. Furche, T., Gottlob, G., Grasso, G., Orsi, G., Schallhart, C., Wang, C.: Little Knowledge Rules the Web: Domain-Centric Result Page Extraction. In: Rudolph, S., Gutierrez, C. (eds.) RR 2011. LNCS, vol. 6902, pp. 61–76. Springer, Heidelberg (2011)
10. Furche, T., Gottlob, G., Grasso, G., Schallhart, C., Sellers, A.: Oxpath: A language for scalable, memory-efficient data extraction from web applications. In: VLDB (2011)
11. Gulhane, P., Rastogi, R., Sengamedu, S.H., Tengli, A.: Exploiting content redundancy for web information extraction. In: VLDB (2010)
12. Lin, T., Etzioni, O., Fogarty, J.: Identifying interesting assertions from the web. In: CIKM (2009)
13. Liu, W., Meng, X., Meng, W.: Vide: A vision-based approach for deep web data extraction. TKDE 22, 447–460 (2010)
14. Simon, K., Lausen, G.: Viper: augmenting automatic information extraction with visual perceptions. In: CIKM (2005)
15. Yates, A., Cafarella, M., Banko, M., Etzioni, O., Broadhead, M., Soderland, S.: Textrunner: open information extraction on the web. In: NAACL (2007)
16. Zheng, S., Song, R., Wen, J.R., Giles, C.L.: Efficient record-level wrapper induction. In: CIKM (2009)

Stepwise Development of Formal Models for Web Services Compositions: Modelling and Property Verification

Yamine Ait-Ameur[1] and Idir Ait-Sadoune[2]

[1] IRIT/INPT-ENSEEIHT, 2 Rue Charles Camichel. BP 7122,
31071 TOULOUSE CEDEX 7, France
yamine@enseeiht.fr
[2] E3S/SUPELEC, 3, rue Joliot-Curie,
91192 GIF-SUR-YVETTE CEDEX, France
idir.aitsadoune@supelec.fr

With the development of the web, a huge number of services available on the web have been published. These web services operate in several application domains like concurrent engineering, semantic web, system engineering or electronic commerce. Moreover, due to the ease of use of the web, the idea of composing these web services to build composite ones defining complex workflows arose. Even if several industrial standards providing specification and/or design XML-oriented languages for web services compositions description, like BPEL, CDL, OWL-S, BPMN or XPDL have been proposed, the activity of composing web services remains a syntactically based approach. Due to the lack of formal semantics of these languages, ambiguous interpretations remain possible and the validation of the compositions is left to the testing and deployment phases. From the business point of view, customers do not trust these services nor rely on them. As a consequence, building correct, safe and trustable web services compositions becomes a major challenge.

It is well accepted that the use of formal methods for the development of information systems has increased the quality of such systems. Nowadays, such methods are set up not only for critical systems, but also for the development of various information systems. Their formal semantics and their associated proof system allow the system developer to establish relevant properties of the described information systems.

This talk addresses the formal development of models for services and their composition using a refinement and proof based method, namely the Event B method. The particular case of web services and their composition is illustrated. We will focus on the benefits of the refinement operation and show how such a formalization makes it possible to formalise and prove relevant properties related to composition and adaptation. Moreover, we will also show how implicit semantics carried out by the services can be handled by ontologies and their formalisation in such formal developments. Indeed, once ontologies are formalised as additional domain theories beside the developed formal models, it becomes possible to formalise and prove other properties related to semantic domain heterogeneity.

The case of BPEL web services compositions will be illustrated.

S.W. Liddle et al. (Eds.): DEXA 2012, Part I, LNCS 7446, p. 9, 2012.

A Hybrid Approach
for General XML Query Processing

Huayu Wu[1], Ruiming Tang[2], Tok Wang Ling[2],
Yong Zeng[2], and Stéphane Bressan[2]

[1] Institute for Infocomm Research, Singapore
huwu@i2r.a-star.edu.sg
[2] School of Computing, National University of Singapore
{tangruiming,lingtw,zengyong,steph}@comp.nus.edu.sg

Abstract. The state-of-the-art XML twig pattern query processing algorithms focus on matching a single twig pattern to a document. However, many practical queries are modeled by multiple twig patterns with joins to link them. The output of twig pattern matching is tuples of labels, while the joins between twig patterns are based on values. The inefficiency of integrating label-based structural joins in twig pattern matching and value-based joins to link patterns becomes an obstacle preventing those structural join algorithms in literatures from being adopted in practical XML query processors. In this paper, we propose a hybrid approach to bridge this gap. In particular, we introduce both relational tables and inverted lists to organize values and elements respectively. General XML queries involving several twig patterns are processed by the both data structures. We further analyze join order selection for a general query with both pattern matching and value-based join, which is essential for the generation of a good query plan.

1 Introduction

Twig pattern is considered the core query pattern in most XML query languages (e.g., XPath and XQuery). How to efficiently process twig pattern queries has been well studied in the past decade. One highlight is the transfer from using RDBMS to manage and query XML data, to processing XML queries natively (see the survey [10]). Now the state-of-the-art XML twig pattern query processing techniques are based on structural join between each pair of adjacent query nodes, which are proven more efficient than the traditional approaches using RDBMS for most cases [16]. After Bruno et al. [4] and many subsequent works bringing the idea of holistic twig join into the structural join based algorithms, it seems that the XML twig pattern matching techniques are already quite developed in terms of efficiency. However, one simple question is whether twig pattern matching is the only issue for answering general XML queries. XQuery is powerful to express any complex query. It is quite often that in an XQuery expression there are multiple XPath expressions involved, each of which corresponds to a twig pattern query; and value-based joins are used to link those

S.W. Liddle et al. (Eds.): DEXA 2012, Part I, LNCS 7446, pp. 10–25, 2012.

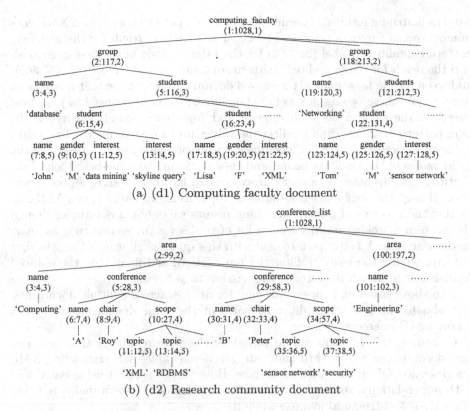

(a) (d1) Computing faculty document

(b) (d2) Research community document

Fig. 1. Two example documents with node labeled

XPath expressions (twig patterns). Matching a twig pattern to a document tree can be efficiently done with current techniques, but how to join different sets of matching results based on values is not as trivial as expected. This limitation also prevents many structural join algorithms in literatures being adopted by practical XQuery processors. In fact, nowadays many popular XQuery processors are still based on relational approach (e.g. MonetDB[15], SQL Server[14]) or navigational approach (e.g. IBM DB2[3], Oracle DB[24], Natix[8]).

Consider two example labeled documents shown in Fig. 1, and a query to find all the computing conferences that accept the same topic as Lisa's research interest. The core idea to express and process this query is to match two twig patterns, as shown in Fig. 2, separately in the two documents, and then join the two sets of results based on the same values under nodes *interest* and *topic*. Twig

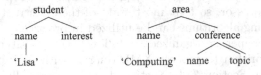

Fig. 2. Two twig patterns to be matched for the example query

pattern matching returns all occurrences of a twig pattern query in an XML document tree, in tuples of positional labels. The matching result for the first twig pattern contains the label (21:22,5) for the *interest* node in the first document, and the result for the second twig pattern contains (11:12,5), (13:14,5), (35:36,5) and so on for the *topic* node in the second document. We can see that it makes no sense to join these two result sets based on labels. Actually we need to join them based on the value under each *interest* and *topic* node found. Unfortunately, twig pattern matching cannot reflect the values under desired nodes. To do this, we have to access the original document (probably with index) to retrieve the child values of the resulting nodes, and then perform the value-based join. Furthermore, in this attempt, a query processor can hardly guarantee a good query plan. If only one conference has the same topic as Lisa's interest, i.e., XML, by joining the two sets of pattern matching results we only get one tuple, though the pattern matching returns quite a lot of tuples for the second twig as intermediate results. A better plan to deal with this query is after matching the first pattern, we use the result to filter the *topic* labels, and finally use the reduced *topic* labels to match the second twig pattern to get result.

The above situation happens not only for queries across multiple documents, but also for queries across different parts of the same document, or queries involving ID references.

Of course, this problem can be solved by the pure relational approach, i.e., transforming the whole XML data into relational tables and translating XML queries into SQL queries with table joins. However, to support value-based join with pure relational approach is not worth the sacrifice of the optimality of many algorithms for structural join (e.g., [4][6]).

In this paper, we propose a hybrid approach to process general XML queries which involve multiple twig patterns linked by value-based joins. "Hybrid" means we adopt both relational tables and inverted lists, which are the core data structures used in the relational approach and the structural join approach for XML query processing. In fact, the idea of hybridising two kinds of approaches is proposed in our previous reports to optimize content search in twig pattern matching [18], and to semantically reduce graph search to tree search for queries with ID references [20]. The contribution of this paper is to use the hybrid approach to process all XML queries that are modeled as multiple twig patterns with value-based joins, possibly across different documents. The two most challenging parts of the hybrid approach are (1) how to smoothly bridge structural join and value-based join, and (2) how to determine a good join order for performance concern. Thus in this paper, we first propose algorithms to link up the two joins and then investigate join order selection, especially when the value-based join can be inner join or outer join. Ideally, our approach can be adopted by an XML query (XQuery) processor, so that most state-of-the-art structural join based twig pattern matching algorithms can be utilized in practice.

The rest of the paper is organized as follows. We revisit related work in Section 2. The algorithm to process XML queries with multiple linking patterns is presented in Section 3. In Section 4, we theoretically discuss how to optimize query processing, with focus on join order selection for queries with inner join and outer join. We present the experimental result in Section 5 and conclude this paper in Section 6.

2   Related Work

In the early stage, many works (e.g., [16][23]) focus on using mature RDBMS to store and query XML data. They generally shred XML documents into relational tables and transform XML queries into SQL statements to query tables. These relational approaches can easily handle the value-based joins between twig patterns, however, the main problem is they are not efficient in twig pattern matching, because too many costly table joins may be involved.

To improve the efficiency in twig pattern matching, many native approaches without using RDBMS are proposed. Structural join based approach is an important native approach attracting most research interests. At the beginning, binary join based algorithms are proposed [2], in which a twig pattern is decomposed into binary relationships, and the binary relationships are matched separately and combined to get final result. Although [22] proposed a structural join order selection strategy for the binary join algorithms, they may still produce a large size of useless intermediate result. Bruno et al. [4] first proposed a holistic twig join algorithm, *TwigStack*, to solve the problem of large intermediate result size. It is also proven to be optimal for queries with only "//"-axis. Later many subsequent works [7][12][11][6] are proposed to either optimize *TwigStack*, or extend it for different problems. However, these approaches only focus on matching a single twig pattern to a document. Since many queries involve multiple twig patterns, how to join the matching results from different twig patterns is also an important issue affecting the overall performance.

We proposed a hybrid approach for twig pattern matching in [18]. The basic idea is to use relational table to store values with the corresponding element labels. Then twig pattern matching is divided into structural search and content search, which are performed separately using inverted lists and relational tables. In [19] we theoretically and experimentally show that this hybrid approach for twig pattern matching is more efficient than pure structural join approach and pure relational approach for most XML cases. Later we extend this approach for queries with ID references, by using tables to capture semantics of ID reference [20]. However, the previous work is not applicable for general XML queries with random value-based joins to link multiple twig patterns.

3   General XML Query Processing

For illustration purpose, we first model a general XML query as a linked twig pattern (*LTP* for short).

3.1   Linked Twig Pattern

Definition 1. *(LTP) A linked twig pattern is* L = ((T, d)*, (u, v)*), *where* T *is a normal twig pattern representation with the same semantics,* d *is document identifier,* (u, v) *is a value link for value-based join between nodes* u *and* v. *Each value link* (u, v) *is associated to a label* l∈{*inner join, left outer join, right outer join, full outer join*}, *to indicate the type of the join based on* u *and* v.

In graphical representation, we use solid edge for the edges in T, *and dotted edge with or without arrow for each* (u, v). *Particularly, a dotted edge without arrow means an inner join; a dotted edge with an arrow at one end means a left/right outer join; and a dotted edge with arrows at both ends means a full outer join. The output nodes in an LTP query are underlined.*

It is possible that there are multiple dotted edges between two twig patterns in an LTP query, which means the two patterns are joined based on multiple conditions. Theoretically, the multiple dotted edges between two twig patterns must be in the same type, as they stand for a unique join between two patterns, just with multiple conditions.

Example 1. Recall the previous query to find all the computing conferences that accept the same topic as Lisa's research interest. Its LTP representation is shown in Fig. 3(a). Another example LTP query containing outer join is shown in Fig. 3(b). This query aims to find the names of all male students in the database group, and for each such student it also optionally outputs the names of all conferences that contain the same topic as his interest, if any.

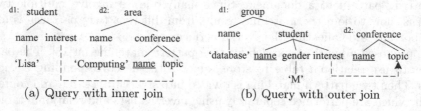

(a) Query with inner join   (b) Query with outer join

Fig. 3. Example LTP queries with inner join and outer join

To process an LTP query, intuitively we need to match each twig pattern in the query to the given documents, and then perform joins (either inner join or outer join, as indicated in the LTP expression) over pattern matching results.

3.2   Algebraic Expression

In this section, we introduce the algebraic operators used to express LTP queries.

Pattern Matching - *PM(T)*: Performs pattern matching for a twig pattern T. When we adopt a structural join based pattern matching algorithm, we also use another notation to express pattern matching, which is $DH_T \bowtie^s DT_T$, where DH_T and DT_T are dummy head and dummy tail of twig pattern T which can be any query node in T, and $\bowtie^s$ indicates a serial of structural joins starting from DH_T, through all other nodes in T and ending at DT_T, to match T to the document. When a pattern matching is after (or before) a value-based join in a query, DH_T (or DT_T) represents the query node in T that is involved in the value-based join. For example, in Fig. 3(a), any node except *interest* in the left hand side twig pattern (temporarily named T_1) can be considered as a dummy head, while *interest* is the dummy tail of T_1 because it is involved in the value-based join with the right hand side twig pattern. Suppose we consider *'Lisa'* as DH_{T_1}, then $DH_{T_1} \bowtie^s DT_{T_1}$ means a series of structural joins between *'Lisa'* and *name*, *name* and *student*, and *student* and *interest*, to finish the pattern matching. The purpose to introduce this notation is for the ease to investigate join order selection for the mixture of structural joins and value-based joins, as shown later.

Value-based join - $\bowtie^l_c$: This operator joins two sets of tuples based on the value condition(s) specified in c. The label $l \in \{null, \leftarrow, \rightarrow, \leftrightarrow\}$ indicates the type of value-based join, where $\{null, \leftarrow, \rightarrow, \leftrightarrow\}$ correspond to inner join, left outer join, right outer join and full outer join respectively.

The LTP query shown in Fig. 3(a) and 3(b) can be expressed by the proposed algebra as follows, where T_1 and T_2 represent the two twig patterns involved in each query:

$$PM(T_1) \bowtie_{T_1.interest=T_2.topic} PM(T_2)$$

and
$$PM(T_1) \bowtie^{\rightarrow}_{T_1.interest=T_2.topic} PM(T_2)$$

Alternatively, we can substitute $PM(T_n)$ by $DH_{T_n} \bowtie^s DT_{T_n}$, which is helpful to explain join order selection, as discussed later.

3.3   LTP Query Processing

To process an LTP query, generally we need to match all twig patterns and join the matching results according to specified conditions. Thus how to efficiently perform the two operators $\bowtie^s$ and $\bowtie^l_c$ is essential to LTP query processing. As reviewed in Section 2, there are many efficient twig pattern matching algorithms proposed. In this part, we focus on how to incorporate value-based join with twig pattern matching .

Joining results from matching different twig patterns is not so trivial as expected. Most twig pattern matching algorithms assign positional labels to document nodes, and perform pattern matching based on node labels. There are many advantages of this attempt, e.g., they do not need to scan the whole XML document to process a query, and they can easily determine the positional relationship between two document nodes from their labels. However, performing pattern matching based on labels also returns labels as matching results. After matching different twig patterns in an LTP query, we have to join them based on relevant values, instead of labels. How to link those pattern matching results by value-based joins becomes a problem for LTP query processing.

We introduce relational table to associate each data value to the label of its parent element node in a document. This data structure is helpful for performing value-based join between sets of labels.

3.3.1   Relational Tables

Relational tables are efficient for maintaining structured data and performing value-based joins. Most structural join based pattern matching algorithms ignore the important role of relational tables in value storage and value-based join, thus they have to pay more on redundant document access to extract values before they can perform value-based join, as well as return final value answers. In our approach, we use a relational table for each type of property node to store the labels of the property and the property values. Thus the relational tables are also referred as *property tables* in this paper.

Definition 2. *(Property Node) A document node which has a child value (text) is called a* property *node (property in short), irrespective of whether it appears as an element type or an attribute type in the XML document.*

Definition 3. *(Property Table) A property table is a relational table built for a particular type of property. It contains two fields: property label and property value. The table name indicates which property type in which document the table is used for.*

In the documents in Fig. 1, name, gender, interest, topic, and so on, are all properties. The property tables for *interest* and *topic* are shown in Fig. 4.

$R_{d1\_interest}$

label	value
(11:12,5)	data mining
(13:14,5)	skyline query
(21:22,5)	XML
...	...
(127:128,5)	sensor network
...	...

R_{d2_topic}

label	value
(11:12,5)	XML
(13:14,5)	RDBMS
(35:36,5)	sensor network
(37:38,5)	security
...	...

Fig. 4. Example property tables for *interest* and *topic* in the two documents

3.3.2   Linking Structural Join and Value-Based Join

When we perform a sequence of joins in relational databases, a query optimizer will choose one join to start based on the generated plan, and the result will be pipelined to the second join and so on. Our approach follows a similar strategy, but the difference is in our LTP algebraic expression pattern matching and value-based join occur alternately. A pair of consecutive value-based joins can be easily performed as they are similar to table joins in relational systems. For a pair of consecutive pattern matchings, existing pattern matching algorithms can be easily extended to perform it, as both the input and the output of a twig pattern matching algorithm are labels. In this section, we investigate how to perform a pair of consecutive pattern matching and value-based join, which needs property tables to bridge the gap between labels and values.

We propose algorithms to handle the two operations: (1) a value-based join follows a pattern matching, i.e. $(DH_T \bowtie^s DT_T) \bowtie^l_c S$ or $S \bowtie^l_c (DH_T \bowtie^s DT_T)$, and (2) a pattern matching follows a value-based join, i.e. $(S \bowtie^l_c DH_T) \bowtie^s DT_T$ or $DH_T \bowtie^s (DT_T \bowtie^l_c S)$ (S is a set of intermediate result from other operations).

Case 1: Pattern matching before value-based join

Because of the commutativity property[1] of value-based join, we only take the expression $(DH_T \bowtie^s DT_T) \bowtie^l_c S$ for illustration. $S \bowtie^l_c (DH_T \bowtie^s DT_T)$ can be processed in a similar way. The algorithm to process the query expression with pattern matching followed by value-based join is shown in Algorithm 1.

Algorithm 1. Pattern Matching First

Input: a query $(DH_T \bowtie^s DT_T) \bowtie^l_{p_1 = p_2} S$ and property tables R_{p_1} and R_{p_2}
Output: a set of resulting tuples
1: identify the output nodes $o_1, o_2, \dots o_m$ in T and S
2: identify the property fields p_1 in T, and p_2 in S
3: load the inverted list streams for all the query nodes in T
4: perform pattern matching for T using inverted lists, to get the result set RS
5: project RS on output nodes and p_1 to get PRS
6: join PRS, R_{p_1}, R_{p_2} and S based on conditions $PRS.p_1 = R_{p_1}.label$, $R_{p_1}.value = R_{p_2}.value$ and $R_{p_2}.label = S.p_2$
7: return the join result with projection on $o_1, o_2, \dots o_m$

The main idea is to use corresponding property tables to bridge the two sets of lable tuples during value-based join. Since each property table has have two fields: label and value, we can effectively join two sets of label tuples based on values using property tables. Also in our illustration, we use only one condition for value-based join. It is possible that two or more conditions are applied.

Example 2. Consider the LTP query shown in Fig. 3(a). Suppose we process this query based on the default order of its algebraic expression:

$$DH_{T_1} \bowtie^s DT_{T_1} \bowtie_{T_1.interest = T_2.topic} DH_{T_2} \bowtie^s DT_{T_2}$$

We first match the pattern T_1 to the document d1, i.e., performing $DH_{T_1} \bowtie^s DT_{T_1}$, and project the result on *interest* as it is used for value-based join. By any twig pattern matching algorithm, we can get the only resulting label (21:22,5). Then we join this result with DH_{T_2}, which corresponds to all labels for *topic* (dummy head of T_2, as it is the property in T_2 for value-based join), through $R_{d1\_interest}$ and R_{d2_topic} (shown in Fig. 4), and get a list of labels (11:12,5), etc.

Case 2: Value-based join before pattern matching

In this case a value-based join will be performed first, and then using the resulting tuples on relevant properties, we perform structural joins for pattern matching. Similarly, due to the commutativity property of $DH_T \bowtie^s DT_T$, we only concern the expression $(S \bowtie^l_c DH_T) \bowtie^s DT_T$, ignoring $DH_T \bowtie^s (DT_T \bowtie^l_c S)$. The algorithm is shown in Algorithm 2.

---

[1] $S_1 \bowtie_c S_2 = S_2 \bowtie_c S_1$, $S_1 \bowtie^{\rightarrow}_c S_2 = S_2 \bowtie^{\leftarrow}_c S_1$, $S_1 \bowtie^{\leftrightarrow}_c S_2 = S_2 \bowtie^{\leftrightarrow}_c S_1$.

Algorithm 2. Value-based Join First

Input: a query $(S\bowtie^t_{p_1=p_2}DH_T)\bowtie^s DT_T$ and property tables R_{p_1} and R_{p_2}
Output: a set of resulting tuples
1: identify the output nodes $o_1, o_2, \dots o_m$ in S and T
2: identify the property fields p_1 in S, and p_2 in T
3: join S, R_{p_1} and R_{p_2} based on the condition $S.p_1 = R_{p1}.label$ and $R_{p1}.value = R_{p2}.value$, to get RS
4: project RS on output nodes and p_2, and sort the labels by p_2 in document order to form I_{p_2}
5: load the inverted list streams for all the query nodes, except p_2 in T
6: perform pattern matching for T using I_{p_2} and other loaded inverted lists, to get the result RS
7: return the matching result with projection on $o_1, o_2, \dots o_m$

The advantage of performing value-based join before pattern matching is that value-based join may reduce the number of labels for the relevant property node in the twig pattern, so that the performance of pattern matching is improved. However, the trade-off is the higher cost for value-based join, compared to the first case where pattern matching is first performed to reduce the labels for value-based join. The details of join order selection are discussed in Section 4.

Example 3. Still consider the query in Fig. 3(a). Suppose $DH_{T_1}\bowtie^s DT_{T_1}$ is already performed and returns a label (21:22,5). Now we join this intermediate result with $R_{d1\_interest}$ to materialize the *interest* values, and then join the values with R_{d2_topic} to get a smaller set of *topic* labels. We use these labels to construct a new inverted list for the *topic* node in T_2, then T_2 can be matched more efficiently with less labels scanned.

Example 2 and 3 take inner join between twig patterns to illustrate the process. Note that the algorithms are also applicable for outer join.

4   Query Optimization

In this section, we discuss how to optimize LTP query processing. We have proposed an approach to optimize twig pattern matching [18]. Generally, it performs content search in each twig pattern query before structural joins, because content search normally leads high selectivity. This is similar to the selection push-ahead in relational database systems. This technique will be inherited when we perform an LTP query with multiple pattern matchings. Now we focus on how to choose a good join order to process LTP queries.

An LTP query expression can be considered as a sequence of mixed structural joins and value-based joins. In particular, any twig pattern matching can be considered as a sequence of structural joins, and to link different twig patterns, we need value-based joins. As in most state-of-the-art twig pattern matching algorithms, the structural joins are performed in a holistic way to reduce useless intermediate result, then it is reasonable to consider a pattern matching wholly as one operation. As mentioned above, we use $DH_T\bowtie^s DT_T$ to express a pattern matching for twig T, in which $\bowtie^s$ means a set of structural joins to holistically find the matched results. When a pattern matching is before or following a value-based join, the dummy tail DT_T or the dummy head DH_T of T represent the sets of properties, based on which the value-based join is performed.

We start with a simple case that the query only involves inner value-based joins between twig patterns. After that we discuss how to handle outer joins.

4.1 Query with Inner Join Only

The rationale behind join order selection in a relational optimizer is the commutativity and the associativity of inner table join. During a pattern matching, how to select the order of structural joins does not affect the matching result [22], so we consider a pattern matching is commutative, i.e. $DH_T\bowtie^s DT_T \equiv DT_T\bowtie^s DH_T$. An inner value-based join has no difference from an inner table join, so it is also commutative. Now we discuss the associativity of a join sequence with both pattern matching and inner value-based joins.

Proposition 1. *An algebraic expression with both pattern matchings ($\bowtie^s$) and inner value-based joins ($\bowtie_c$) operators is reorderable, i.e., $(DH_T\bowtie^s DT_T)\bowtie_c S \equiv DH_T\bowtie^s(DT_T\bowtie_c S)$, where S is a set of labels.*

Proof. [Sketch] A tuple t is in the resulting set of $(DH_T\bowtie^s DT_T)\bowtie_c S$ if and only if t satisfies two conditions: (1) it corresponds to a subtree that matches T, and (2) the label l of a relevant property in t has the same child value as the corresponding label in S. Now consider $DH_T\bowtie^s(DT_T\bowtie_c S)$. Since value-based join $\bowtie_c$ returns all the tuples satisfying (2), and pattern matching $\bowtie^s$ also soundly and completely return all matches satisfying (1) [4], t also exists in the result of $DH_T\bowtie^s(DT_T\bowtie^l S)$. Similarly, we can prove that any tuple from $DH_T\bowtie^s(DT_T\bowtie^l S)$ also exists in $(DH_T\bowtie^s DT_T)\bowtie^l S$.

Based on Proposition 1, there may be multiple execution plans for a LTP query. Consider the example query in Fig. 3(a):

$$(DH_{T_1}\bowtie^s DT_{T_1})\bowtie_{interest=topic}(DH_{T_2}\bowtie^s DT_{T_2})$$

some possible execution plans are shown in Fig. 5. Note that we ignore the outer set and inner set selection for the value-based join.

With an effective cost model of each operation and an efficient method to estimate result size after each step, we can choose a good query plan to proceed.

Cost Models

Twig Pattern Matching: It is proven in [4] that in the worst case, the I/O and CPU time for the holistic join based twig pattern matching algorithm is linear

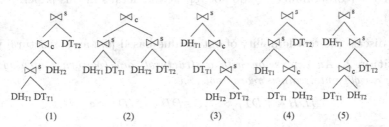

Fig. 5. Different execution plans

to the size of input lists and the size of results. The cost of pattern matching is $f_{in} * sum(|I_1|, ..., |I_n|) + f_{out} * |RS_{p\_out}|$, where f_{in} and f_{out} are the factors for input size and output size, $|I_m|$ is the size of m-th inverted list and $|RS_{p\_out}|$ is the size of output result set.

Value-Based Join: We adopt the cost model for inner table join in relational query optimizer for set join. As relational query optimization has been studied for many years, the details are not repeated here.

Result Size Estimation

There are many approaches to estimate the result size of an inner join in RDBMS. Those approaches can also be used to estimate the result size of a value-based join. Estimating result size of structural joins is also studied in many previous research works, e.g. [21]. We can use them in our approach.

4.2 Handling Outer Value-Based Join

Outer join may appear as a structural join within a twig pattern, or as a value-based join between different twig patterns. The difference is the outer structural join is based on positional labels, while the outer value-based join is based on values. Outer structural join can be handled during pattern matching [6]. In this part, we focus on the outer value-based join between twig patterns.

Generating different query plans relies on the associativity of join operators. Unfortunately, an LTP query expression with outer value-based join may not always be reorderable. In an LTP query execution plan, there are three cases that an operator is adjacent to an outer value-based join: (1) two outer value-based joins are adjacent $(S_1 \bowtie^{\rightarrow/\leftrightarrow}{}_c S_2 \bowtie^{\rightarrow/\leftrightarrow}{}_c S_3)^2$, (2) an inner value-based join and an outer value-based join are adjacent $(S_1 \bowtie^{\rightarrow/\leftrightarrow}{}_c S_2 \bowtie_c S_3$ or $S_1 \bowtie_c S_2 \bowtie^{\rightarrow/\leftrightarrow}{}_c S_3)$, and (3) a structural join and an outer value-based join are adjacent $(S_1 \bowtie^{\rightarrow/\leftrightarrow}{}_c DH_T \bowtie^s DT_T$ or $S_1 \bowtie^s DH_T \bowtie^{\rightarrow/\leftrightarrow}{}_c DT_T)$.

Case (1) and Case (2):

The associativity of inner/outer join between tables (sets) have been well studied [9][13]. Generally, based on a null-rejection constraint, in which an outer join condition is evaluated to false when referring a *null* value, Case (1) expressions can be reordered. For Case (2), some expressions are natively reorderable, e.g. $(S_1 \bowtie_c S_2) \bowtie^{\rightarrow}{}_c S_3 \equiv S_1 \bowtie_c (S_2 \bowtie^{\rightarrow}{}_c S_3)$. For those expressions that are not reorderable, [13] invents a generalized outer join operator to replace the original join, to ensure the reorderability. We do not repeat the details in this paper.

Case (3):

Now we discuss the reorderability of outer value-based join and structural join.

Proposition 2. *An expression with a structural join (pattern matching) followed by a right outer join is reorderable, i.e.*

$$(DH_T \bowtie^s DT_T) \bowtie^{\rightarrow}{}_c S \equiv DH_T \bowtie^s (DT_T \bowtie^{\rightarrow}{}_c S).$$

---

[2] Since $S_1 \bowtie^{\leftarrow}{}_c S_2 \equiv S_2 \bowtie^{\rightarrow}{}_c S_1$, we omit left outer join.

Proof. [Sketch] $(DH_T \bowtie^s DT_T) \bowtie^{\rightarrow}_c S$ preserves all the matches of T in the relevant XML document, because of the right outer join. Also in this expression, every tuple t in S that satisfies c will be bound to the corresponding match of T. Now consider the second expression. $DT_T \bowtie^{\rightarrow}_c S$ preserves all labels of DT_T (relevant query nodes in T, based on which value-based join is performed), so $DH_T \bowtie^s (DT_T \bowtie^{\rightarrow}_c S)$ preserves all the matches of T. Each tuple t in S that satisfies both c and pattern constraint T is also preserved. Thus the two expressions are equivalent.

Example 4. Consider the query in Fig. 3(b). Suppose we match the second twig pattern first and get a result set S, as shown in Fig. 6(a). Now we need to execute the expression $DH_{T_1} \bowtie^s DT_{T_1} \bowtie^{\rightarrow}_{T_1.interest=S.topic} S$ to answer the query. Fig. 6(b) shows the intermediate and final results of two query plan: $(DH_{T_1} \bowtie^s DT_{T_1}) \bowtie^{\rightarrow}_{T_1.interest=S.topic} S$ and $DH_{T_1} \bowtie^s (DT_{T_1} \bowtie^{\rightarrow}_{T_1.interest=S.topic} S)$. We can see that the two expressions return the same result.

S

name	topic
(6:7,4)	(11:12,5)
(6:7,4)	(13:14,5)
...	...
(30:31,4)	(35:36,5)
(30:31,4)	(37:38,5)
...	...

$DH_{T1} \bowtie^s DH_{T1}$

name$_{student}$	interest
(7:8,5)	(11:12,5)
(7:8,5)	(13:14,5)
...	

$(DH_{T1} \bowtie^s DH_{T1}) \bowtie^{\rightarrow}_{interest=topic} S$

name$_{student}$	name$_{conference}$
(7:8,5)	null
...	...

$DH_{T1} \bowtie^{\rightarrow}_{interest=topic} S$

interest	name$_{conference}$
(11:12,5)	null
(13:14,5)	null
(21:22,5)	(6:7,4)
...	...

$DH_{T1} \bowtie^s (DH_{T1} \bowtie^{\rightarrow}_{interest=topic} S)$

name$_{student}$	name$_{conference}$
(7:8,5)	null
...	...

(a) Intermediate set   (b) Join results from different expressions

Fig. 6. Reorderable expression example

However, when a structural join is followed by a left outer join, i.e., $(DH_T \bowtie^s DT_T) \bowtie^{\leftarrow}_c S$ (it is equivalent to a right outer join followed by a structural join, i.e. $S \bowtie^{\rightarrow}_c (DH_T \bowtie^s DT_T)$), then expression is not reorderable.

Example 5. Consider the same query in Fig. 3(b). This time we match the first twig pattern first, to get a result set S shown in Fig. 7(a). Then we execute the expression $S \bowtie^{\rightarrow}_{S.interest=T_2.topic} (DH_{T_2} \bowtie^s DT_{T_2})$. The results are shown in the first two tables in Fig. 7(b). By comparing this result with the execution result of a second expression $(S \bowtie^{\rightarrow}_{S.interest=T_2.topic} DH_{T_2}) \bowtie^s DT_{T_2}$ (shown in the last two tables in Fig. 7(b)), we can see that the expression is not reorderable.

The reason driving the two query plans of an expression $S \bowtie^{\rightarrow}_c DH_T \bowtie^s DT_T$ to have different results is: when we perform the right outer join as the last operation, we can always preserve the tuples in S in the final result; but when we perform the structural join as the last operation, it is not guaranteed to preserve any information (some inverted list contains empty entries for structural join).

$DH_{T2} \bowtie^s DT_{T2}$

name$_{conference}$	topic
(6:7,4)	(11:12,5)
(6:7,4)	(13:14,5)
...	...

$S \bowtie^{\rightarrow}_{interest=topic}(DH_{T2} \bowtie^s DT_{T2})$

name$_{student}$	name$_{conference}$
(7:8,5)	null
...	...

S

name	interest
(7:8,5)	(11:12,5)
(7:8,5)	(13:14,5)
...	...

$S \bowtie^{\rightarrow}_{interest=topic} DH_{T2}$

name$_{student}$	topic
(7:8,5)	null
...	...

$(S \bowtie^{\rightarrow}_{interest=topic} DH_{T2}) \bowtie^s DT_{T2}$

name$_{student}$	name$_{conference}$

(empty set)

(a) Intermediate set   (b) Join results from different expressions

Fig. 7. Unreorderable expression example

To make expressions reorderable in this case, we introduce an extended twig pattern matching operator, $\bowtie^s_A$:

$$DH_T \bowtie^s_A DT_T = (DH_T \bowtie^s DT_T) \bigcup ((\pi_A(DH_T) - \pi_A(DH_T \bowtie^s DT_T)) \times \{null_{[DH_T]+[DT_T]-A}\})$$

where $[R]$ is the schema of R, A is a set of properties, $\pi_A(S)$ is a projection on S based on A, and $null_S$ is a tuple with schema of S and all entries are null values. It is easy to check that $DH_T \bowtie^s_A DT_T$ preserves all values in A.

Example 6. Consider the two lower resulting tables in Fig. 7(b). The expression $S \bowtie^{\rightarrow}_{S.interest=T_2.topic} DH_{T2}$ returns a set of intermediate results, R_i, containing the tuple $((7:8,5), null)$. If we perform $R_i \bowtie^s_{[S]} DT_{T2}$, we will get a set of results containing $((7:8,5), null)$, which is the same as the results from $S \bowtie^{\rightarrow}_{S.interest=T_2.topic} (DH_{T2} \bowtie^s DT_{T2})$ in the second table in Fig. 7(b).

Proposition 3. *The following reordering holds:*

$$S \bowtie^{\rightarrow}_c (DH_T \bowtie^s DT_T) = (S \bowtie^{\rightarrow}_c DH_T) \bowtie^s_{[S]} DT_T.$$

Proof. [Sketch] $S \bowtie^{\rightarrow}_c (DH_T \bowtie^s DT_T)$ preserves all tuples in S, and binds each matching result of T to the corresponding tuple in S (if any). When we perform the right outer join first, as in the second expression, $S \bowtie^{\rightarrow}_c DH_T$ results all tuples in S, plus possible DH_T values. Since $\bowtie^s_{[S]}$ preserves S tuples after pattern matching, the final results also contain all tuples in S, and attach matched patterns to corresponding S tuples.

Proposition 4. *The following reordering holds:*

$$S \bowtie^{\leftrightarrow}_c (DH_T \bowtie^s DT_T) = (S \bowtie^{\leftrightarrow}_c DH_T) \bowtie^s_{[S]} DT_T.$$

The proof is similar to the proof of Proposition 3 and thus omitted. With the help of the new pattern matching operator $\bowtie^s_A$, expressions with both outer join and structural join are reorderable, for good query plan selection.

5   Experiments

In our experiment, we used an 110MB XMark [1] data set and selected seven meaningful complex queries (queries involving multiple twig patterns), in which the first four queries contain two twig patterns in each and the last three contains three twig patterns in each[3]. The experiments were implemented with Java, and performed on a dual-core 2.33GHz processor with 3.5G RAM.

5.1 Effectiveness of Join Order Selection

In this part, based on our cost model we compare the good plan (with lowest estimated cost) and a median plan (with median estimated cost) on estimated cost and real cost. The worst estimated cost of each query is always from the plan that performs value-based join before pattern matching, and the cost is very expensive. Thus we do not compare with the worst plan. Similar to any relational optimizer, the purpose of our query plan selection is to avoid bad plans, rather than finding the best plan to perform a query, which can be hardly guaranteed. The results are shown in Fig. 8. Fig. 8(a) shows the comparison of estimated cost between the selected plan and the median plan for each query, while Fig. 8(b) shows the real costs. From the figures, we can see that our query plan selection mechanism can find a relatively good query processing plan based on the estimated cost for all queries.

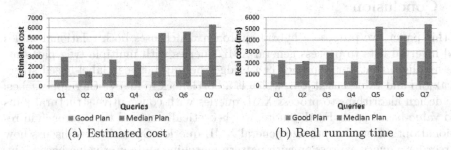

(a) Estimated cost   (b) Real running time

Fig. 8. Query optimization test

5.2 Comparison with Other Approaches

In this part, we compare our hybrid approach with relational approach and structural join approach. We do not compare with commercial XQuery processors because most of them build many extra indexes to speed up query processing during document loading. Thus, we simply compare with the approaches appearing in literatures. For the relational approach, we use the schema-aware relational approach (*SAR* for short) which is first proposed in [16] and refined in [10]. It is generally more efficient than other schemaless approaches [10]. We also compare with a structural join approach. We take the TwigStack [4] algorithm for the implementation in our hybrid approach (*Hybrid* for short) and compared structural join approach (*SJ* for short). Other improved algorithms

---

[3] Due to the space limitation, we do not present the query details in the paper. The meaning, XQuery expression and LTP representation of each query can be found in http://www.comp.nus.edu.sg/~tang1987/queries

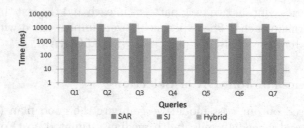

Fig. 9. Different approaches comparison

are also adoptable for both *SJ* and *Hybrid*, but taking TwigStack is enough to reflect the difference between the two approaches. In *SJ*, for each query all twig patterns are matched first, and then results are joined based on values. This is the only way to extend *SJ* for LTP queries.

The result is shown in Fig. 9. Note that the Y-axis is in logscale. *SAR* is the worst approach, because too many table joins seriously affect the performance. *Hybrid* is the most efficient approach for all queries. This proves that the idea of integrating relational tables and inverted lists in *SAR* and *SJ* is promising.

6   Conclusion

In this paper, we propose a hybrid approach which uses both relational tables and inverted lists to process general XML queries with multiple twig patterns. We highlight that our approach can bridge the gap between label-based structural join and value-based join, which is a bottleneck in many other approaches. We design algorithms to process XML queries with consecutive structural joins and value-based joins. Furthermore, we theoretically extend the approach in relational optimizers to optimize general XML queries. Specially, we discuss how to reorder a query expression with pattern matching and outer value-based join. We conduct experiments to validate our query optimization and the performance advantage of our approach over pure relational approach and structural join approach.

References

1. http://www.xml-benchmark.org/
2. Al-Khalifa, S., Jagadish, H.V., Koudas, N., Patel, J.M., Srivastava, D., Wu, Y.: Structural joins: A primitive for efficient XML query pattern matching. In: ICDE, pp. 141–154 (2002)
3. Beyer, K.S., Cochrane, R., Josifovski, V., Kleewein, J., Lapis, G., Lohman, G.M., Lyle, R., Ozcan, F., Pirahesh, H., Seemann, N., Truong, T.C., Van der Linden, B., Vickery, B., Zhang, C.: System RX: One part relational, one part XML. In: SIGMOD, pp. 347–358 (2005)
4. Bruno, N., Koudas, N., Srivastava, D.: Holistic twig joins: optimal XML pattern matching. In: SIGMOD, pp. 310–321 (2002)
5. Chen, L., Gupta, A., Kurul, M.E.: Stack-based algorithms for pattern matching on DAGs. In: VLDB, pp. 493–504 (2005)

6. Chen, S., Li, H.-G., Tatemura, J., Hsiung, W.-P., Agrawal, D., Candan, K.S.: Twig^{2}stack: Bottom-up processing of generalized-tree-pattern queries over XML documents. In: VLDB, pp. 283–294 (2006)
7. Chen, T., Lu, J., Ling, T.W.: On boosting holism in XML twig pattern matching using structural indexing techniques. In: SIGMOD, pp. 455–466 (2005)
8. Fiebig, T., Helmer, S., Kanne, C., Moerkotte, G., Neumann, J., Schiele, R., Westmann, T.: Anatomy of a native XML base management system. VLDB J. 11(4), 292–314 (2002)
9. Galindo-Legaria, C., Rosenthal, A.: Outerjoin simplification and reordering for query optimization. ACM Trans. Database Syst. 22(1), 43–74 (1997)
10. Gou, G., Chirkova, R.: Efficiently querying large XML data repositories: a survey. IEEE Trans. Knowl. Data Eng. 19(10), 1381–1403 (2007)
11. Lu, J., Chen, T., Ling, T.W.: Efficient processing of XML twig patterns with parent child edges: a look-ahead approach. In: CIKM, pp. 533–542 (2004)
12. Lu, J., Ling, T.W., Chan, C.Y., Chen, T.: From region encoding to extended dewey: On efficient processing of XML twig pattern matching. In: VLDB, pp. 193–204 (2005)
13. Rao, J., Pirahesh, H., Zuzarte, C.: Canonical abstraction for outerjoin optimization. In: SIGMOD, pp. 671–682 (2004)
14. Rys, M.: XML and relational database management systems: inside Microsoft SQL Server 2005. In: SIGMOD, pp. 958–962 (2005)
15. Boncz, P.A., Grust, T., van Keulen, M., Manegold, S., Rittinger, J., Teubner, J.: MonetDB/XQuery: a fast XQuery processor powered by a relational engine. In: SIGMOD, pp. 479–490 (2006)
16. Shanmugasundaram, J., Tufte, K., Zhang, C., He, G., DeWitt, D.J., Naughton, J.F.: Relational databases for querying XML documents: Limitations and opportunities. In: VLDB, pp. 302–314 (1999)
17. Wang, H., Li, J., Luo, J., Gao, H.: Hash-based subgraph query processing method for graph structured XML documents. In: VLDB, pp. 478–489 (2008)
18. Wu, H., Ling, T.-W., Chen, B.: VERT: A Semantic Approach for Content Search and Content Extraction in XML Query Processing. In: Parent, C., Schewe, K.-D., Storey, V.C., Thalheim, B. (eds.) ER 2007. LNCS, vol. 4801, pp. 534–549. Springer, Heidelberg (2007)
19. Wu, H., Ling, T.W., Chen, B., Xu, L.: TwigTable: ssing semantics in XML twig pattern query processing. JoDS 15, 102–129 (2011)
20. Wu, H., Ling, T.W., Dobbie, G., Bao, Z., Xu, L.: Reducing Graph Matching to Tree Matching for XML Queries with ID References. In: Bringas, P.G., Hameurlain, A., Quirchmayr, G. (eds.) DEXA 2010. LNCS, vol. 6262, pp. 391–406. Springer, Heidelberg (2010)
21. Wu, Y., Patel, J.M., Jagadish, H.V.: Estimating Answer Sizes for XML Queries. In: Jensen, C.S., Jeffery, K., Pokorný, J., Šaltenis, S., Bertino, E., Böhm, K., Jarke, M. (eds.) EDBT 2002. LNCS, vol. 2287, pp. 590–608. Springer, Heidelberg (2002)
22. Wu, Y., Patel, J.M., Jagadish, H.V.: Structural join order selection for XML query optimization. In: ICDE, pp. 443–454 (2003)
23. Zhang, C., Naughton, J.F., DeWitt, D.J., Luo, Q., Lohman, G.M.: On supporting containment queries in relational database management systems. In: SIGMOD Conference, pp. 425–436 (2001)
24. Zhang, N., Agarwal, N., Chandrasekar, S., Idicula, S., Medi, V., Petride, S., Sthanikam, B.: Binary XML storage and query processing in Oracle 11g. PVLDB 2(2), 1354–1365 (2009)

SCOOTER: A Compact and Scalable Dynamic Labeling Scheme for XML Updates

Martin F. O'Connor and Mark Roantree

Interoperable Systems Group, School of Computing,
Dublin City University, Dublin 9, Ireland
{moconnor,mark}@computing.dcu.ie

Abstract. Although dynamic labeling schemes for XML have been the focus of recent research activity, there are significant challenges still to be overcome. In particular, though there are labeling schemes that ensure a compact label representation when creating an XML document, when the document is subject to repeated and arbitrary deletions and insertions, the labels grow rapidly and consequently have a significant impact on query and update performance. We review the outstanding issues to-date and in this paper we propose SCOOTER - a new dynamic labeling scheme for XML. The new labeling scheme can completely avoid relabeling existing labels. In particular, SCOOTER can handle frequently skewed insertions gracefully. Theoretical analysis and experimental results confirm the scalability, compact representation, efficient growth rate and performance of SCOOTER in comparison to existing dynamic labeling schemes.

1 Introduction

At present, most modern database providers support the storage and querying of XML documents. They also support the updating of XML data at the document level, but provide limited and inefficient support for the more fine-grained (node-based) updates within XML documents. The XML technology stack models an XML document as a tree and the functionality provided by a tree labeling scheme is key in the provision of an efficient and effective update solution. In particular, throughout the lifecycle of an XML document there may be an arbitrary number of node insertions and deletions. In our previous work, we proposed a labeling scheme that fully supported the reuse of deleted node labels under arbitrary insertions. In this paper, we focus on the more pressing problem that affects almost all dynamic labeling schemes to-date, the linear growth rate of the node label size under arbitrary insertions, whether they are single insertions, bulk insertions, or frequently skewed insertions.

1.1 Motivation

The length of a node label is an important criterion in the quality of any dynamic labeling scheme [2] and the larger the label size, the more significant is

S.W. Liddle et al. (Eds.): DEXA 2012, Part I, LNCS 7446, pp. 26–40, 2012.

the negative impact on query and update performance [6]. In the day-to-day management of databases and document repositories, it is a common experience that more information is inserted over time than deleted. Indeed in the context of XML documents and repositories, a common insertion operation is to append new nodes into an existing document (e.g.: heart rate readings in a sensor databases, transaction logging in financial databases) or to perform bulk insertions of nodes at a particular point in a document. These insertion operations are classified as frequently skewed insertions [20]. Over time, such insertions can quickly lead to large label sizes and consequently impact negatively on query and update performance. Furthermore, large label sizes lead to larger storage costs and more expensive IO costs. Finally, large node labels require higher computational processing costs in order to determine the structural relationships (ancestor-descendant, parent child and sibling-order) between node labels. Our objectives are to minimize the update cost of inserting new nodes and while minimizing the label growth rate under any arbitrary combination of node insertions and deletions.

1.2 Contribution

In this paper, we propose a new dynamic labeling scheme for XML called SCOOTER. The name encapsulates the core properties - Scalable, Compact, Ordered, Orthogonal, Trinary Encoded Reusable dynamic labeling scheme. The principle design goal underpinning SCOOTER is to avoid and bypass the congestion and bottleneck caused by large labels when performing fine-grained node-based updates of XML documents. SCOOTER is scalable insofar as it can support an arbitrary number of node label insertions and deletions while completely avoiding the need to relabel nodes. SCOOTER provides compact label sizes by constraining the label size growth rate under various insertions scenarios. Order is maintained between nodes at all times by way of lexicographical comparison. SCOOTER is orthogonal to the encoding technique employed to determine structural relationships between node labels. Specifically, SCOOTER can be deployed using a prefix-based encoding, containment-based encoding or a prime number based encoding. SCOOTER employs the quaternary bit-encoding presented in [9] and uses the ternary base to encode node label values. SCOOTER supports the reuse of shorter deleted node labels when available.

This paper is structured as follows: in §2, we review the state-of-the-art in dynamic labeling schemes for XML, with a particular focus on scalability and compactness. In §3, we present our new dynamic labeling scheme SCOOTER and the properties that underpin it. We describe how node labels are initially assigned and we analyse the growth rate of the label size. In §4, we present our insertion algorithms and our novel compact adaptive growth mechanism which ensures that label sizes grow gracefully and remain compact even in the presence of frequently skewed node insertions. We provide experimental validation for our approach in terms of execution time and total label storage costs by comparing SCOOTER to three dynamic label schemes that provide similar functionality and present the results in §5. Finally in §6, our conclusions are presented.

2   Related Research

There are several surveys that provide an overview and analysis of the principle dynamic labeling schemes for XML proposed to date [15], [4], [16], [12]. In reviewing the state-of-the-art in dynamic labeling schemes for XML, we will consider each labeling scheme in its ability to support the following two core properties: scalability and compactness. By scalable, we mean the labeling scheme can support an arbitrary number of node insertions and deletions while completely avoiding the need to relabel nodes. As the volume of data increases and the size of databases grow from Gigabytes to Terabytes and beyond, the computational cost of relabeling node labels and rebuilding the corresponding indices becomes prohibitive. By compact, we mean the labeling scheme can ensure the label size will have a highly constrained growth rate both at initial document creation and after subsequent and repeated node insertions and deletions. Indeed almost all dynamic labeling schemes to-date [14], [9], [2], [3], [8], [7], [13], [1] are compact only when assigning labels at document creation, but have a linear growth rate for subsequent node label insertions which quickly lead to large labels.

To the best of our knowledge, there is only one published dynamic labeling scheme for XML that is scalable, and offers compact labels at the document initialisation stage, namely the QED labeling scheme [9] (subsequently renamed CDQS in [10]). The Quaternary encoding technique presented in [9] overcomes the limitations present in all other binary encoding approaches. A comprehensive review of the various binary encoding approaches for node labels and their corresponding advantages and limitations is provided in [10]. We now briefly summarise their findings and supplement them with our own analysis.

All node labels must be stored as binary numbers at implementation, which are in turn stored as either fixed length or variable length. It should be clear that all fixed length labels are subject to overflow (and are hence, not scalable) once all the assigned bits have been consumed by the update process and consequently require the relabeling of all existing labels. The problem may be temporarily alleviated by reserving labels for future insertions, as in [11] but this leads to increased storage costs and only delays the relabeling process. There are three types of variable encoding: variable length encoding, multi-byte encoding and bit-string encoding. The V-CDBS labeling scheme [10] employs a variable length encoding. Variable length encodings require the size of the label to be stored in addition to the label itself. Thus, when many nodes are inserted into the XML tree, at some point the original fixed length of bits assigned to store the size of the label will be too small and overflow, requiring all existing labels to be relabeled. This problem has been called the overflow problem in [9]. The Ordpath [14], QRS [1], LSDX [3] and DeweyID [17] labeling schemes all suffer from the overflow problem. In addition, Ordpath is not compact because it only uses odd numbers when assigning labels during document initialisation.

UTF-8 [21] is an example of a multi-byte variable encoding. However UTF-8 encoding is not as compact as the Quaternary encoding. Furthermore, UTF-8 can only encode up to 2^{31} labels and thus, cannot scale beyond 2^{32}. The vector order labeling scheme [19] stores labels using a UTF-8 encoding. Cohen et al.

[2] and EBSL [13] use bitstrings to represent node labels. Although they do not suffer from the overflow problem they have a linear growth rate in label size when assigning labels during document initialisation and also when generating labels during frequently skewed insertions - consequently they are not compact. The Prime number labeling scheme [18] also avoids node relabeling by using simultaneous congruence (SC) values to determine node order. However, order-sensitive updates are only possible by recalculating the SC values based on the new ordering of nodes and in [9], they determined the recalculation costs to be highly prohibitive.

Lastly, we consider the problem of frequently skewed insertions. To the best of our knowledge, there are only two published dynamic labeling schemes that directly address this problem. The first is the vector order labeling scheme [19] which as mentioned previously uses a UTF-8 encoding for node labels which does not scale. We will detail in our experiments section that for a relatively small number of frequently skewed insertions that are repeated a number of times, the vector order labels can quickly grow beyond the valid range permitted by UTF-8 encoding. The second labeling scheme is Dynamic Dewey Encoding (DDE) [20] which has the stated goal of supporting both static and dynamic XML documents. However, although the authors indicate their desire to avoid a dynamic binary encoding representation for their labels due to the overhead of encoding and decoding labels, they do not state the binary representation they employ to store or represent labels and thus, we are not in a position to evaluate their work.

In summary, there does not currently exist a dynamic labeling scheme for XML that is scalable, can completely avoid relabeling node labels and is compact at both document initialisation and during arbitrary node insertions and deletions, including frequently skewed insertions.

3   The SCOOTER Labeling Scheme

In this section, we introduce our new dynamic labeling scheme for XML called SCOOTER, describe how labels are initially assigned at document creation and then, highlight the unique characteristics of our labeling scheme.

SCOOTER adopts the quaternary encoding presented in [9] so we now briefly introduce quaternary codes. A Quaternary code consists of four numbers, "0", "1", "2", "3", and each number is stored with two bits, i.e.: "00", "01", "10", "11". A SCOOTER code is a quaternary code such that the number "0", is reserved as a separator and only "1", "2", "3" are used in the SCOOTER code itself. The SCOOTER labeling scheme inherits many of the properties of the QED labeling scheme, such as being orthogonal to the structural encoding technique employed to represent node order and to determine relationships between nodes. Specifically, node order is based on lexicographical order rather than numerical order and SCOOTER may be deployed as a prefix-based or containment-based labeling scheme. The containment based labeling schemes exploit the properties of tree traversal to maintain document order and to determine the various

Decimal	SCOOTER 2 digits	SCOOTER 3 digits	QED
1	12	112	112
2	13	113	12
3	2	12	122
4	22	122	123
5	23	123	13
6	3	13	132
7	32	132	2
8	33	133	212
9		2	22
10		212	222
11		213	223
12		22	23
13		222	232
14		223	3
15		23	312
16		232	32
17		233	322
18		3	323
19		312	33
20		313	332
Total Size		104	100

The algorithm:

```
input  : k - the kth node to be labelled.
input  : childCount
input  : base - the base to encode in.
output : label - the label of node k.
1  digits = ⌈log_base (childCount + 1)⌉
2  divisor = base^digits
3  quotient = k
4  label = null
5  while (i=1; i < digits; i++) do
6      divisor ⟵ divisor / base
7      code ⟵ ⌊ quotient / divisor ⌋ + 1
8      label ⟵ label ⊕ code
9      remainder ⟵ quotient % divisor
10     if remainder == 0 then
11         return label
12     else
13         quotient ⟵ remainder
14     end
15 end
16 return label ⟵ label ⊕ (quotient + 1)
```

(a) Function AssignNodeK.   (b) SCOOTER & QED labels

Fig. 1

structural relationships between nodes. Two SCOOTER codes are generated to represent the start and end intervals for each node in a containment-based scheme. In a prefix-based labeling scheme, the label of a node in the XML tree consists of the parent's label concatenated with a delimiter (separator) and a positional identifier of the node itself. The positional identifier indicates the position of the node relative to its siblings. In the prefix approach, the SCOOTER code represents the positional identifier of a node, also referred to as the selflabel.

3.1 Assigning Labels

A SCOOTER code must end in a "2" or a "3" in order to maintain lexicographical order in the presence of dynamic insertions due to reasons outlined in [10]. For the purpose of presenting our algorithms, we shall assume a prefix-based labeling scheme in this paper. The QED labeling scheme adopts a recursive divide-and-conquer algorithm to assign initial labels at document creation [9]. The SCOOTER AssignInitialLabels algorithm presents a novel approach described in algorithm 1. The SCOOTER codes presented in Figure 1(b) are examples generated using algorithm 1. The algorithm takes as input a parent node P (the root node is labeled '2'), and $childCount$ - the number of children of P. When $childCount$ is expressed as an positive integer of the base three, Line 1 computes the minimum number of digits required to represent $childCount$ in the base three. The minimum number of digits required will determine the maximum label size generated by our AssignInitialLabels algorithm.

Algorithm 1. Assign Initial Labels.

```
input : P - a parent node
input : childCount - the number of child nodes of P
output: a unique SCOOTER selflabel for each child node.
```

1   maxLabelSize = $\lceil \log_3 (childCount + 1) \rceil$
2   selfLabel[1] = null

```
   /* Compute the SCOOTER selflabel of the first child.                    */
```
3   **while** *(i=1; i < maxLabelSize; i++)* **do**
4 | selfLabel[1] ⟵ selfLabel[1] ⊕ 1 ;   // ⊕ means concatenation.
5   **end**
6   selfLabel[1] ⟵ selfLabel[1] ⊕ 2

```
   /* Now compute the SCOOTER selflabels for all remaining children.       */
```
7   **while** *(i=2; i <= childCount; i++)* **do**
8 | selfLabel[i] ⟵ Increment (selfLabel[i - 1], maxLabelSize)
9   **end**

There are a small number of rules used to determine the assignment of labels in order to ensure a compact label size and to maintain lexicographical order between labels. The first 4 of these rules concern the first label while the remaining 2 rules, determine remaining labels.

- The first label must terminate with "2".
- It must have no other "2" other than the final digit.
- It can never have the digit "3".
- It will always be the maximum allowable length.
- The second and remaining labels can never end in "1".
- The second and remaining labels can be of any allowable length.

As previously stated, the first label will always end with a '2' symbol but as it must be of maximum length, it is preceded by a sequence of '1' symbols. The sequence of '1' symbols may be zero (empty) if the minimum number of digits required to represent *childCount* in the base three is one digit. All subsequent child labels after the first label are generated by incrementing the child label immediately preceding it. Figure 1(b) shows the sequence of labels for both 2- and 3-digit initially assigned labels. The maximum number of labels that may be assigned with D digits is 3^D - 1 labels.

The Increment algorithm (algorithm 2) takes as input a node label and the maxLabelSize and returns a new node label that is the immediate lexicographical increment of the input node. The Increment algorithm will never receive a label longer than maxLabelSize. Furthermore, the Increment algorithm will never receive a label with a length of maxLabelSize **and** consisting of all '3' symbols by virtue of line 1 in algorithm 1. Lastly, although the Increment algorithm will never receive a node label from the AssignInitialLabels algorithm that ends with a '1' symbol, we may pass substrings of labels that end with a '1' symbol to the Increment algorithm when handling dynamic node label insertions and deletions (discussed in the next section).

SCOOTER's AssignInitialLabels algorithm has three distinct properties which make it quite different from the QED labeling scheme.

Algorithm 2. Increment

---

input : N_{left} - a node label; maxLabelSize - maximum number of symbols allowed in label.
output: N_{new} - a new self_label such that $N_{left} \prec N_{new}$
1   $N_{temp} \longleftarrow N_{left}$

2   **if** $Length(N_{temp}) == maxLabelSize$ **then**
3   **if** Last symbol in N_{temp} is '1' **then**
4 | $N_{new} \longleftarrow N_{temp}$ with last symbol changed to '2'
5   **else if** (Last symbol in N_{temp} is '2' **then**
6 | $N_{new} \longleftarrow N_{temp}$ with last symbol changed to '3'
7   **else if** (Last symbol in N_{temp} is '3' **then**
8   **while** last symbol of N_{temp} is '3' **do**
9 | $N_{temp} \longleftarrow N_{temp}$ with last symbol removed.
10   **end**
11   **if** Last symbol in N_{temp} is '1' **then**
12 | $N_{new} \longleftarrow N_{temp}$ with last symbol changed to '2'
13   **else if** (Last symbol in N_{temp} is '2' **then**
14 | $N_{new} \longleftarrow N_{temp}$ with last symbol changed to '3'
15   **end**
16   **end**
17   **else if** $Length(N_{temp}) < maxLabelSize$ **then**
18   **while** $(i = Length(N_{temp}) + 1; i < maxLabelSize; i++)$ **do**
19 | $N_{temp} \longleftarrow N_{temp} \oplus 1$
20   **end**
21      $N_{new} \longleftarrow N_{temp} \oplus 2$
22   **end**
23   **return** N_{new}

---

1. Firstly, each SCOOTER label can be determined solely based on the label of the node to the immediate left (and immediate right but we omit the Decrement algorithm due to lack of space). This is a key property which we will exploit to enable and maintain compact node labels in the presence of an arbitrary number of node insertions and deletions. This property also facilitates the reuse of deleted node labels. In contrast, the QED encoding algorithm employs the mathematical **round** function which introduces an approximation function into the QED assign initial labels process. In order words, the label value of node n is not and cannot be determined solely from the label values of node $n+1$ or node n-1. The QED encoding algorithm can guarantee lexicographical order but cannot guarantee the accurate calculation of the size of a node label n or indeed the label n itself, based solely on the node labels immediately adjacent to node n. It follows that when node n is deleted, the QED labeling scheme cannot guarantee the accurate calculation of the deleted node label n (and its size), and consequently cannot guarantee that the deleted node label n will be reused.

2. One significant limitation arising from the sequential determination of initially assigned node labels is that to generate a label for node n, we must first generate all n-1 node labels. This limitation can be a significant bottleneck when processing very large XML files. Hence, the second distinct property: the SCOOTER initially assigned labels may also be computed independent of each other as illustrated in function `AssignNodeK` in Figure 1(a). Specifically, given n child nodes to be labeled, the function `AssignNodeK` can determine an arbitrary k^{th} child node label without having to compute any other child node label. When parsing very large XML documents, the ability to compute node labels independent of one another opens up the possibility

of parallel processing in a multi-threaded and multi-core environment and may offer sizeable gains in computation time.

3. The third distinct property: the SCOOTER label encoding mechanism is independent of the underlying numeric base, as illustrated by function **Assign-NodeK** in Figure 1(a). Our SCOOTER dynamic labeling scheme and compact adaptive growth mechanism may be applied and implemented using any numeric base greater than or equal to two. In mathematical numeral systems, the base or radix is the number of unique symbols that a positional numeral system uses to represent numbers. For example, the decimal system uses the *base 10*, because it uses the 10 symbols from 0 through 9. The highest symbol usually has the value of one less than the base of that numeric system. In [5], the authors demonstrate the most economical radix for a numbering system is e, the base of the natural logarithms, with a value of approximately 2.718. Economy is measured as the product of the radix and the number of digits needed to express a range of given values. Consequently the economy is also a measure of how compact is the numerical representation of a given radix. In [5], the authors also demonstrate that the integer 3, being the closest integer to e, is almost always the most economical integer radix or base. For this reason, in this paper, we have chosen to use the numeric *base 3* and consequently quaternary codes to represent SCOOTER labels.

4   Compact Adaptive Growth Mechanism

In this section, we present our novel compact adaptive growth mechanism and related node label insertion algorithms which ensure a highly constrained label growth rate irrespective of the quantity of arbitrary and repeated node label insertions and deletions. We will begin with a simple example to provide an overview of the conceptual approach followed by a more detailed analysis of the underlying properties.

Consider an XML tree consisting of a root node R and two child nodes with selflabels '2' and '3'. We insert a sequence of 100 nodes to the right of the rightmost child node. Table 1 in Figure 2(a) illustrates the first 18 insertions. The first two $(3^1 - 1)$ node labels generated consist of a *prefix* string '3' and a *postfix* string that mirrors the labels normally generated for a maxLabelSize of 1 digit (i.e.: '2' and '3'). For the next 8 $(3^2 - 1)$ insertions, from the third to the tenth insertion inclusive, the newly generated labels consist of a *prefix* string '33' and a *postfix* string that mirrors the labels generated for a maxLabelSize of 2 digits (e.g.: '12', '13', '2' and so on). For the next 26 $(3^3 - 1)$ insertions, from the 11th to the 36th insertion inclusive, labels consist of a *prefix* string '3333' and a *postfix* string that mirrors the labels generated for a maxLabelSize of 3 digits (e.g.: '112', '113', '12' and so on). This process is repeated as many times as required.

We now provide an analysis of the underlying properties. Conceptually, we consider a label as comprising of two components: a prefix and a postfix. We define the smallest permissible prefix length to be 1 symbol. When the prefix has length 1, we define the maximum postfix length permissible to be 1 also

Insert after rightmost node	SCOOTER Label
	2
	3
1	32
2	33
3	3312
4	3313
5	332
6	3322
7	3323
8	333
9	3332
10	3333
11	3333112
12	3333113
13	333312
14	3333122
15	3333123
16	333313
17	3333132
18	3333133

Range Start	Range End	Node Start	Node End	Prefix Length	Postfix Length	Maximum SCOOTER SelfLabel Length	SCOOTER SelfLabel Total bits
3^0	3^1-1	1	2	1	1	2	4
3^1	3^2-1	3	10	2	2	4	8
3^2	3^3-1	11	36	4	3	7	14
3^3	3^4-1	37	116	7	4	11	22
3^4	3^5-1	117	358	11	5	16	32
3^5	3^6-1	359	1,086	16	6	22	44
3^6	3^7-1	1,087	3,272	22	7	29	58
3^7	3^8-1	3,273	9,832	29	8	37	74
3^8	3^9-1	9,833	29,514	37	9	46	92
3^9	$3^{10}-1$	29,515	88,562	46	10	56	112
3^{10}	$3^{11}-1$	88,563	265,708	56	11	67	134
3^{11}	$3^{12}-1$	265,709	797,148	67	12	79	158
3^{12}	$3^{13}-1$	797,149	2,391,470	79	13	92	184
3^{13}	$3^{14}-1$	2,391,471	7,174,438	92	14	106	212
3^{14}	$3^{15}-1$	7,174,439	21,523,344	106	15	121	242
3^{15}	$3^{16}-1$	21,523,345	64,570,064	121	16	137	274
3^{16}	$3^{17}-1$	64,570,065	193,710,226	137	17	154	308
3^{17}	$3^{18}-1$	193,710,227	581,130,714	154	18	172	344
3^{18}	$3^{19}-1$	581,130,715	1,743,392,180	172	19	191	382
3^{19}	$3^{20}-1$	1,743,392,181	5,230,176,580	191	20	211	422

(a) Table 1.   (b) Table 2. Label Insertion - Compact Adaptive Growth Rate

Fig. 2

(please refer to Table 2 in Figure 2(b)). The maximum label length will always equal the sum of the prefix length and the maximum postfix length.

- When inserting a new rightmost node label, we extend the length of the prefix *if and only if* the current rightmost label consists of all '3' symbols *and* the length of the current rightmost node label equals the sum of the current prefix length and maximum postfix length. For example, given the current rightmost node label '33' with a prefix length of 1 and a maximum postfix length of 1; in order to insert a new node after node '33', we must extend the prefix and postfix lengths.
- The rule governing the adaptive growth rate of the prefix and postfix lengths is simple: the new prefix length is assigned the value of the previous maximum label length; the new maximum postfix length is assigned the value of the previous maximum postfix length plus 1. This rule is codified in lines 6–13 in algorithm 3.

All bit-string dynamic labeling schemes (including QED) have a label growth rate of one bit per node insertion. Therefore, after one thousand insertions and one million insertions, the largest selflabel sizes are 1,000 and 1,000,000 bits respectively. In contrast, after one thousand insertions and one million insertions, the largest SCOOTER selflabels are 44 bits and 184 bits respectively. Thus, SCOOTER labels may be several orders of magnitude smaller than the labels of all existing bit-string labeling schemes when processing frequently skewed insertions. Furthermore, in contrast to all existing dynamic labeling schemes, SCOOTER generates compact labels without requiring advance knowledge of

---

Algorithm 3. Insert New Node After Rightmost Node.

---

    input : left self_label N_{left}, N_{left} is not empty.
    output: New self_label N_{new} such that $N_{left} \prec N_{new}$

 1 **if** *first symbol in N_{left} is '1'* **then**
 2 | $N_{new} \longleftarrow$ '2'
 3 **else if** *first symbol in N_{left} is '2'* **then**
 4 | $N_{new} \longleftarrow$ '3'
 5 **else if** *first symbol in N_{left} is '3'* **then**
 6   numConsecThrees $\longleftarrow$ the number of consecutive '3' symbols at start of N_{left}
 7   prefixLength $\longleftarrow$ postfixLength $\longleftarrow$ 1
 8   labelLength $\longleftarrow$ prefixLength + postfixLength
      /* Compute the prefixLength and postfixLength based on numConsecThrees.   */
 9   **while** *labelLength <= numConsecThrees* **do**
10 | prefixLength $\longleftarrow$ prefixLength + postfixLength
11 | postfixLength $\longleftarrow$ postfixLength + 1
12 | labelLength $\longleftarrow$ prefixLength + postfixLength
13   **end**
14   postfix $\longleftarrow$ substring(N_{left}, 1 + prefixLength, length(N_{left}))
15   **if** *postfix is not empty* **then**
      /* An arbitrary number of nodes may have been deleted, thus the label may be
        longer than the postfixLength. If it is longer, trim it.   */
16 | postfix $\longleftarrow$ substring(postfix, 1, postfixLength)
17 | **if** *last symbol in postfix is '1'* **then**
18 | | postfix $\longleftarrow$ postfix with last symbol changed to '2'
19 | **else**
20 | | postfix $\longleftarrow$ Increment (postfix, postfixLength)
21 | **end**
22   **else if** *postfix is empty* **then**
23 | **while** *i=1; i < postfixLength; i++* **do**
24 | | postfix $\longleftarrow$ postfix $\oplus$ 1
25 | **end**
26 | postfix $\longleftarrow$ postfix $\oplus$ 2
27   **end**
28   prefix = null
29   **while** *i=1; i <= prefixLength; i++* **do**
30 | prefix $\longleftarrow$ prefix $\oplus$ 3
31   **end**
32    $N_{new} \longleftarrow$ prefix $\oplus$ postfix
33 **end**
34 **return** N_{new}

---

the number of nodes to be inserted. The compact adaptive growth mechanism is made possible by virtue of the deterministic property of our AssignInitialLabels algorithm. The compact adaptive growth mechanism may also be applied when inserting new nodes before the leftmost node, however in this case we count the number of consecutive '1' symbols to determine the length of the prefix.

4.1 Insertion between Two Consecutive Non-empty Node Labels

The most difficult insertion scenario is between two non-empty consecutive node labels. Between any two consecutive nodes, there may have been an arbitrary number of node deletions. The ability to determine whether deletions have occurred must be determined from the information encoded in the label alone. In addition, there are 4 distinct insertion scenarios permitted when inserting a new node between two consecutive node labels:

Algorithm 4. InsertBetweenTwoNodes_LessThan.

input : left self_label N_{left}; right self_label N_{right}; both labels not empty.
output: New self_label N_{new} such that $N_{left} \prec N_{new} \prec N_{right}$.

```
1  if length(N_left) < length(N_right) then
2      if N_left is a prefix of N_right then
3          if symbol at N_right[length(N_left)+1] is a '3' then
4              N_new ⟵ N_left ⊕ 2
5          else if symbol at N_right[length(N_left)+1] is a '2' then
6              N_new ⟵ N_left ⊕ 12
7          else
8              N_temp ⟵ N_left
9              Let P ⟵ length(N_left) + 1
10             while symbol at position P in N_right is '1' do
11                 N_temp ⟵ N_temp ⊕ 1
12                 P ⟵ P + 1
13             end
14             if symbol at N_right[P] is a '3' then
15                 N_new ⟵ N_temp ⊕ 2
16             else
17                 N_new ⟵ N_temp ⊕ 12
18             end
19         end
20     else if N_left is not a prefix of N_right then
21         Let P ⟵ first position of difference between N_left and N_right
22         if P == 1 then
23             N_new ⟵ Increment (first symbol in N_left, 1)
24         else if P > 1 then
25             N_temp ⟵ substring(N_left, 1, P - 1)
26             N_new ⟵ N_temp ⊕ Increment (symbol at position P in N_left, 1)
27         end
28     end
29 end
30 return N_new
```

---

1. The left label is a prefix string of the right label;
2. The left label is shorter than the right label but not a prefix of the right label;
3. The left label is the same length as the right label; or
4. The left label is longer than the right label

The SCOOTER labeling scheme provides the same highly constrained adaptive growth rate when processing node label insertions in all four scenarios. In the remainder of this section, we analyse the four algorithms and highlight some observations.

In algorithm 4, the InsertBetweenTwoNodes_LessThan algorithm processes the first two insertion scenarios. The new label returned will always be shorter than both input labels *if and only if* a shorter deleted node label is available for reuse. By available, we mean a shorter unique and valid SCOOTER code that is lexicographically ordered between the left and right node labels. If no shorter label is available (such as when the left label is a prefix of the right label), the algorithm will still return the smallest valid label lexicographically ordered between the two input labels. When the two labels are the same size, if the position of difference between the two input labels is the last symbol in both labels, then both input labels must be lexicographical neighbours with no deleted node label available between them. Consequently a new label is generated by concatenating a '2' symbol

---

Algorithm 5. InsertBetweenTwoNodes_GreaterThan.

---

input : left self_label N_{left}; right self_label N_{right}; both labels not empty.
output: New self_label N_{new} such that $N_{left} \prec N_{new} \prec N_{right}$.

1 **if** $length(N_{left}) > length(N_{right})$ **then**
2   Let P ⟵ first position of difference between N_{left} and N_{right}
3   **if** $P < length(N_{right})$ **then**
       /* If the position of difference is not the last symbol of N_{right} */
4   N_{temp} ⟵ substring(N_{left}, 1, P)
5   N_{new} ⟵ InsertBetweenTwoNodes_LessThan (N_{temp}, N_{right})
6   **else if** $P == length(N_{right})$ **then**
7   N_{temp} ⟵ substring(N_{left}, 1, P - 1)
8   **if** (symbol at position P in N_{left} is '1') and (symbol at position P in N_{right} is '3')
       then
9   N_{new} ⟵ $N_{temp} \oplus 2$
10   **else**
11   Affix ⟵ substring(N_{left}, 1, P)
12   N_{temp} ⟵ substring(N_{left}, P + 1, length(N_{left}))
13   numConsecThrees ⟵ the number of consecutive '3' symbols at start of N_{temp}
14   **if** numConsecThrees == 0 **then**
15   postfix ⟵ Increment (first symbol of N_{temp}, 1)
16   N_{new} ⟵ Affix $\oplus$ postfix
17   **else if** numConsecThrees > 0 **then**
             /* Replace the PLACEHOLDER with lines 7 through 31 inclusive from
               Algorithm 3, substituting all references to N_{left} with N_{temp}.   */
18   PLACEHOLDER
19   N_{new} ⟵ Affix $\oplus$ prefix $\oplus$ postfix
20   **end**
21   **end**
22 **end**
23 **end**
24 **return** N_{new}

---

to the end of the left input label. Otherwise, the left input label is trimmed and algorithm InsertBetweenTwoNodes_LessThan is invoked which will reuse a shorter deleted node label lexicographically ordered between the two input labels. Finally, in algorithm 5, the InsertBetweenTwoNodes_GreaterThan algorithm processes the fourth insertion scenario: the left input label is longer than the right input label. Algorithm 5 will reclaim and reuse the shortest deleted node label, if one exists. Otherwise, it will ensure that all newly generated labels will be assigned according to our compact adaptive growth rate insertion mechanism.

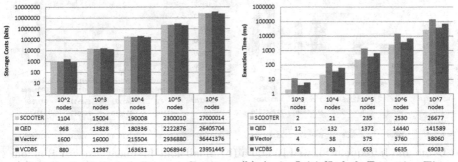

(a) AssignInitialLabels Storage Costs. (b) AssignInitialLabels Execution Time.

Fig. 3

5   Experiments

In this section, we evaluate and compare our SCOOTER labeling scheme with three other dynamic labeling schemes, namely QED [9], Vector [19] and V-CDBS [10]. The three labeling schemes were chosen because they each offer those properties we have encapsulated in SCOOTER - scalability, compactness and the ability to process frequently skewed insertions in an efficient manner. QED is the only dynamic labeling scheme that offers a compact labeling encoding at document initialisation, while overcoming the overflow problem and completely avoiding node relabeling. The SCOOTER labeling scheme inherits these properties by virtue of the quaternary encoding. The Vector labeling scheme is the only dynamic labeling scheme that has as one of its design goals, the ability to process frequently skewed insertions efficiently. SCOOTER has also been designed with this specific property in mind. Lastly, V-CDBS is the most compact dynamic labeling scheme presented to-date, as illustrated in [10]. Although V-CDBS is subject to the overflow problem and thus, cannot avoid relabeling nodes, we want to compare SCOOTER against the most compact dynamic encoding available. All the schemes were implemented in Python and all experiments were carried out on a 2.66Ghz Intel(R) Core(TM)2 Duo CPU and 4GB of RAM. The experiments were performed 10 times and the results averaged.

In Figure 3(a), we illustrate the total label storage cost of 10^2 through 10^6 initially assigned node labels. As expected V-CDBS is the most compact, QED in second place and SCOOTER has marginally larger storage costs than QED. However, SCOOTER has the most efficient processing time, illustrated in Figure 3(b), when generating initially assigned labels, due to the efficient Increment algorithm. Although SCOOTER theoretical scales efficiently under frequently skewed insertions - a result that was validated by experimental analysis - it was necessary to evaluate SCOOTER under multiple frequently skewed insertions and in this process, revealed some interesting results. Figure 4(a) illustrates large skewed node label insertions performed at a randomly chosen position and repeated a small number of times. Figure 4(b) illustrates small skewed node

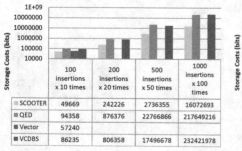

	100 insertions x 10 times	200 insertions x 20 times	500 insertions x 50 times	1000 insertions x 100 times
SCOOTER	49669	242226	2736355	16072693
QED	94358	876376	22766866	217649216
Vector	57240			
VCDBS	86235	806358	17496678	232421978

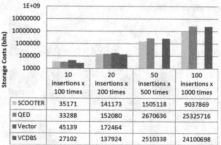

	10 insertions x 100 times	20 insertions x 200 times	50 insertions x 500 times	100 insertions x 1000 times
SCOOTER	35171	141173	1505118	9037869
QED	33288	152080	2670636	25325716
Vector	45139	172464		
VCDBS	27102	137924	2510338	24100698

(a) Large and Infrequent Skewed Random Insertions.   (b) Small and Frequent Skewed Random Insertions.

Fig. 4

label insertions performed at a randomly chosen position and repeated many times. SCOOTER initially performs best in the former case, because our adaptive growth mechanism is designed to give greater savings as the quantity of insertions increase. As the quantity or frequency of insertions scale, SCOOTER offers significant storages benefits in all cases. The vector labeling scheme is absent from some of the results because the label sizes grew beyond the storage capacity permitted by UTF-8 encoding.

6 Conclusions

Updates for XML databases and caches provide ongoing problems for both academic and industrial researchers. One of the primary issues is the labeling scheme that provides uniqueness and retrievability of nodes. In this, there are two major issues for researchers: the length of the label as it may negatively impact performance; and an efficient insertion mechanism to manage updates. In this paper, we introduced the SCOOTER labeling scheme and algorithms for assigning node labels and inserting new labels. We developed new algorithms for assigning labels and a novel highly compact adaptive insertion mechanism that compare favourably to all existing approaches. Our evaluations confirmed these findings through a series of experiments that examined both a very large number of label assignments and similarly large insertion operations.

Although SCOOTER offers compact labels at document initialisation, we are investigating the possibility of an improved AssignInitialLabels algorithm that generates deterministic labels as compact as the labels initially assigned by V-CDBS. Secondly, SCOOTER generates and maintains compact labels under frequently skewed insertions, such as appending a large number of node labels before or after any arbitrary node. However, when a large number of node labels are inserted at a fixed point, the label size grows rapidly. We are investigating a modification to our compact adaptive growth mechanism, such that label sizes will always have a highly constrained growth rate under any insertion scenario. Lastly, we are adapting SCOOTER to work in binary (and not use the quaternary encoding) and we are investigating a new binary encoding to overcome the overflow problem that the quaternary encoding set out to address.

References

1. Amagasa, T., Yoshikawa, M., Uemura, S.: QRS: A Robust Numbering Scheme for XML Documents. In: ICDE, pp. 705–707 (2003)
2. Cohen, E., Kaplan, H., Milo, T.: Labeling Dynamic XML trees. In: PODS, pp. 271–281. ACM, New York (2002)
3. Duong, M., Zhang, Y.: LSDX: A New Labelling Scheme for Dynamically Updating XML Data. In: ADC, pp. 185–193 (2005)
4. Härder, T., Haustein, M.P., Mathis, C., Wagner, M.: Node Labeling Schemes for Dynamic XML Documents Reconsidered. Data Knowl. Eng. 60(1), 126–149 (2007)

5. Hayes, B.: Third Base. American Scientist 89(6), 490–494 (2001)
6. Kay, M.: Ten Reasons Why Saxon XQuery is Fast. IEEE Data Eng. Bull. 31(4), 65–74 (2008)
7. Kobayashi, K., Liang, W., Kobayashi, D., Watanabe, A., Yokota, H.: VLEI code: An Efficient Labeling Method for Handling XML Documents in an RDB. In: ICDE, pp. 386–387 (2005)
8. Li, C., Ling, T.-W.: An Improved Prefix Labeling Scheme: A Binary String Approach for Dynamic Ordered XML. In: Zhou, L.-Z., Ooi, B.-C., Meng, X. (eds.) DASFAA 2005. LNCS, vol. 3453, pp. 125–137. Springer, Heidelberg (2005)
9. Li, C., Ling, T.W.: QED: A Novel Quaternary Encoding to Completely Avoid Re-labeling in XML Updates. In: CIKM, pp. 501–508 (2005)
10. Li, C., Ling, T.W., Hu, M.: Efficient Updates in Dynamic XML Data: from Binary String to Quaternary String. VLDB Journal 17(3), 573–601 (2008)
11. Li, Q., Moon, B.: Indexing and Querying XML Data for Regular Path Expressions. In: VLDB, pp. 361–370 (2001)
12. O'Connor, M.F., Roantree, M.: Desirable Properties for XML Update Mechanisms. In: EDBT/ICDT Workshops (2010)
13. O'Connor, M.F., Roantree, M.: EBSL: Supporting Deleted Node Label Reuse in XML. In: Lee, M.L., Yu, J.X., Bellahsène, Z., Unland, R. (eds.) XSym 2010. LNCS, vol. 6309, pp. 73–87. Springer, Heidelberg (2010)
14. O'Neil, P.E., O'Neil, E.J., Pal, S., Cseri, I., Schaller, G., Westbury, N.: ORDPATHs: Insert-Friendly XML Node Labels. In: SIGMOD Conference, pp. 903–908 (2004)
15. Sans, V., Laurent, D.: Prefix based Numbering Schemes for XML: Techniques, Applications and Performances. PVLDB 1(2), 1564–1573 (2008)
16. Su-Cheng, H., Chien-Sing, L.: Node Labeling Schemes in XML Query Optimization: A Survey and Trends. IETE Technical Review 26, 88–100 (2009)
17. Tatarinov, I., Viglas, S., Beyer, K.S., Shanmugasundaram, J., Shekita, E.J., Zhang, C.: Storing and Querying Ordered XML using a Relational Database System. In: SIGMOD Conference, pp. 204–215 (2002)
18. Wu, X., Lee, M.L., Hsu, W.: A Prime Number Labeling Scheme for Dynamic Ordered XML Trees. In: ICDE, pp. 66–78 (2004)
19. Xu, L., Bao, Z., Ling, T.-W.: A Dynamic Labeling Scheme Using Vectors. In: Wagner, R., Revell, N., Pernul, G. (eds.) DEXA 2007. LNCS, vol. 4653, pp. 130–140. Springer, Heidelberg (2007)
20. Xu, L., Ling, T.W., Wu, H., Bao, Z.: DDE: From Dewey to a Fully Dynamic XML Labeling Scheme. In: SIGMOD Conference, pp. 719–730 (2009)
21. Yergeau, F.: UTF-8, A Transformation Format of ISO 10646, Request for Comments (RFC) 3629 edn. (November 2003)

Reuse the Deleted Labels for Vector
Order-Based Dynamic XML Labeling Schemes

Canwei Zhuang and Shaorong Feng[*]

Department of Computer Science, Xiamen University, 361005 Xiamen, China
cwzhuang0229@163.com, shaorong@xmu.edu.cn

Abstract. Documents obeying XML standard are intrinsically ordered and typically modeled as a tree. Labeling schemes encode both document order and
structural information so that queries can exploit them without accessing the
original XML file. When XML data become dynamic, it is important to design
labeling schemes that can efficiently facilitate updates as well as processing
XML queries. Recently, vector order-based labeling schemes have been proposed to efficiently process updates in dynamic XML data. However the updates are focused on how to process the labels when a node is inserted into the
XML; how to process the deleted labels is not considered in the previous researches. In this paper, we propose new algorithms to generate the labels with
smallest size and therefore reuse all the deleted labels to control the label size
increasing speed; meanwhile the algorithms can completely avoid the re-
labeling also. Extensive experimental results show that the algorithms proposed
in this paper can control the label size increasing speed and enhance the query
performance.

Keywords: XML, Dynamic labeling schemes, reuse the deleted labels.

1   Introduction

XML has become a standard of information exchange and representation on the web.
To query XML data effectiveness and efficiency, the labeling schemes are widely
used to determine the relationships such as the ancestor-descendant between any two
elements. If the XML is static, the current labeling schemes, for examples, containment scheme [1], Dewey scheme [2] and prime scheme [3]can process different queries efficiently. However, when XML data become dynamic, a large amount of nodes
need re-labeling, which is costly and becomes a bottleneck for XML-Database performance. Designing dynamic labeling schemes to avoid the re-labeling of existing
nodes has been recognized as an important research problem.

Different labeling schemes [4-9] have been proposed to support dynamic XML
document. One state-of-the-art approach to design dynamic labeling schemes is assigning labels based on vector order rather than the numerical order. Vector order can
be applied to both range-based and prefix-based labeling schemes. The resulting

---

[*] Corresponding author.

S.W. Liddle et al. (Eds.): DEXA 2012, Part I, LNCS 7446, pp. 41–54, 2012.

labeling schemes including V-containment [7], V-Prefix [8], DDE [9] and CDDE [9] are not only tailored for static documents, but also can completely avoid re-labeling for updates. However, the previous works are focused on how to update the labels when nodes are inserted into the XML. How to process the deleted labels are not discussed previously. We think that the deleted labels should be reused for the benefit of reducing the label size and improving the query performance. Thus the objective of this paper is to propose algorithms to reused the deleted labels and meanwhile to keep the low update cost. We design the algorithm which can find the vector with the smallest size between any two consecutive vectors. It can be applied broadly to different vector order-based labeling schemes to control the label size increasing speed in the update environment with both insertions and deletions.

2   Preliminary and Motivation

In this section, we describe the related labeling schemes and show the limitations of vector order-based labeling schemes which motivate this paper.

2.1   Labeling Tree-Structured Data

Containment Labeling Scheme. In containment labeling scheme[1], every node is assigned three values: *start*, *end* and *level*, where *start* and *end* denote an interval and *level* refers to the level of the node in the XML tree. For any two nodes u and v, u is an ancestor of v if and only if the interval of v is contained in the interval of u. Additionally, with using the *level* of a node, the parent-child relationship can be determined efficiently. Document order can also be deduced well by the comparison of *start* values.

Dewey Labeling Scheme[2] assigns each node a Dewey label which is a concatenation of its parent's label and its local order. The local order of a node is i if it is the ith child of its parent. A Dewey label uniquely identifies a path from the root to an element. Thus structural information can be derived from Dewey labels. Fig.1 shows examples of containment labeling scheme and Dewey labeling scheme.

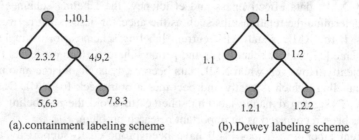

(a).containment labeling scheme   (b).Dewey labeling scheme

Fig. 1. Labeling tree-structured data

While both containment labeling scheme and Dewey labeling scheme work well for static XML documents, an insertion of a node incurs relabeling of large amounts of nodes, which is costly and becomes a performance bottleneck.

2.2 Vector Order and Updates Processing

Vector order[7] is proposed to efficiently process XML updates. We review the vector order and vector order-based schemes in this section.

2.2.1 Vector Order and Updates Processing

Definition 2.1 *(Vector Code). A vector code is a two-tuple of the form (x, y) where x and y are integers with $y \neq 0$.*

Given two vector codes $A:(x_1, y_1)$ and $B:(x_2, y_2)$, *vector preorder* and *vector equivalent* are defined as:

Definition 2.2 *(Vector preorder). A precedes B in vector preorder (denoted as $A \prec_v B$) if and only if and only if $x_1/y_1 < x_2/y_2$.*

Definition 2.3 *(Vector equivalence). A is equivalent to B (denoted as $A \equiv_v B$) if and only if $x_1/y_1 = x_2/y_2$.*

A vector code (x, y) can be interpreted as a fraction x/y, and a fraction x/y can be represented as a vector code (x, y). Given the one-to-one correspondence between vector and fraction, we will use the two terms interchangeably in the rest of the paper.

Let $A=(x_1, y_1)$ and $B=(x_2, y_2)$ be two vector codes. *Addition* of A and B and *Multiplication* of a scalar r and a vector A are defined as:

$$A + B = (x_1 + x_2, y_1 + y_2)$$
$$r \times A = (r \times x, r \times y)$$

Let $A:(x_1, y_1)$ and $B:(x_2, y_2)$ be two vector codes, the following property can be easily deduced:

Lemma 2.1. *Suppose $A \prec_v B$, then $A \prec_v (A + B) \prec_v B$.*

The following theorem guarantees that the vector order-based labeling schemes can process updates without relabeling.

Theorem 2.1. *Given two vector codes V_L and V_R where $V_L \prec_v V_R$, it can be always find V_M which is also a vector code such that $V_L \prec_v V_M \prec_v V_R$.*

V_M can be calculated based on Lemma **2.1**.

2.2.2 V-Containment Labeling Scheme

The ranges in a set of containment labels come from a sequence of integers from 1 to $2n$ for an XML tree with n elements. Linear transformation is defined to transform the integers to vector codes such that the vector codes preserve the order of the original labels. Let Z denote the set of integers and V denote the set of vector codes, Linear transformation $f : Z \rightarrow V$ is defined as: $f(i) = (i, 1)$ *for* $i \in Z$. It is an order-preserving transformation since, given any $i, j \in Z$ such that $i < j$, then $f(i) \prec_v f(j)$.

Appling linear transformation to containment scheme is called V-containment scheme[7]. Based on Theorem 2.1, it can be always inserted a new vector code between two consecutive ones in vector order, and thus V-containment scheme can completely avoids re-labeling for updates in XML document. Fig.2 shows an example of processing insertion with V-containment labeling scheme. Node "*a*" is inserted between two consecutive element nodes. Its *start* and *end* should be between the *end* of its preceding sibling (6,1)and the *start* of its following siblings (7,1). Thus, the *start* of "*a*" should be (13, 2), which is equal to (6,1)+(7,1). and *end* of "*a*" should be (20, 3), which is equal to (13, 2) + (7,1).

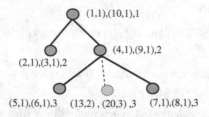

Fig. 2. V-containment labeling scheme

2.2.3   V-Prefix Labeling Scheme

V-Prefix labeling scheme[8] is the straight forward application of vector order to Dewey labeling scheme. It is derived from Dewey labeling scheme by transforming every Dewey label into a sequence of vector codes through linear transformation. V-Prefix label can be seen as a generalized Dewey label where every component is a vector code. Similar to V-containment scheme, V-Prefix scheme can process updates without re-labeling. Fig.3. shows an example of processing insertion with V-Prefix labeling scheme. Node "*a*" is inserted between two consecutive element nodes. Its parent label is the same as its parent's label whereas its local order should fall between the local orders of its two siblings. That is, (3,2)=(1,1)+(2,1).

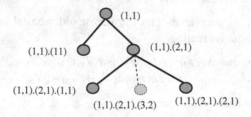

Fig. 3. V-Prefix labeling scheme

2.2.4   DDE and CDDE

DDE[9] and CDDE[9] assign labels based on vector order as well. By viewing every component of a Dewey label as a vector code, DDE directly transforms the static Dewey into a fully dynamic labeling scheme. The initial labeling of DDE labeling scheme is the same as Dewey. However, the semantic meanings of DDE and Dewey are different. A DDE label $x.y_1.y_2... y_m$ is considered as $v_1.v_2...v_m$ where $v_1=(y_1, x)$, $v_2=(y_2, x),..., v_m=(y_m, x)$. That is, the DDE label is interpreted as a sequence of vector codes that share a common Y component. In this way, DDE is dynamic enough to completely avoid re-labeling while introducing minimum additional complexity to static documents. In addition, paper [9] also introduced CDDE which is designed to optimize the performance of DDE for insertions. Both DDE and CDDE can be incorporated into existing systems and applications that are based on Dewey labeling schemes with minimum efforts.

2.3   Motivation

It can be seen that the vector order-based labeling schemes avoid the node re-labeling when a node is inserted into the XML. But they cannot reuse the deleted labels and guarantee that the inserted label has the smallest size when there are deletions as well.

Example 1. When we delete V-containment label "(2,1), (3,1)" (see Fig.2) and want to insert another node at this place, the inserted label will be "(5,2), (9,3)", which is equal to "((1,1)+(4,1)),((5,2)+(4,1))". The deleted label "(2,1), (3,1)" is not reused although it has small size. The original scheme cannot reuse the deleted label and therefore its size increases fast. It is the same for other vector-order based schemes.

It is not good to process the deleted labels in this way. If we want to improve the query performance, all the deleted labels should be reused. Therefore we propose the improved method for vector order-based schemes to reuse the deleted labels and to control the label size increasing speed in the update environment.

3   Reuse the Deleted Labels for Vector-Order Based Schemes

We first introduce a definition **Granularity Sum**[7] used to measure the size of a vector.

Definition 3.1(*Granularity Sum*). *The Granularity Sum of a vector $V:(x, y)$ is defined as $|x|+|y|$(denoted by GS(V)).*

We then propose another definition.

Definition 3.2 (*Middle Vector With the Smallest Granularity Sum, MVWSGS*). *Given two vectors $V_L:(x_1, y_1)$ and $V_R:(x_2, y_2)$ where $V_L \prec V_R$, the Middle Vector With the Smallest Granularity Sum is the vector V_M such that $V_L \prec V_M \prec V_R$ and GS (V_M) is smallest (denoted by MVWSGS(V_L, V_R) or MVWSGS(x_1, y_1, x_2, y_2)).*

Vector-order based labeling schemes are the applications of vector order. When updating takes place, the core operation is to find the middle vector V_M such that $V_L \prec V_M \prec V_R$. The choice is not unique, actually there are infinitely many vectors possible; however, to reuse the deleted labels and slow down the increase rate of label size, we would want to find the vector that has the smallest Granularity Sum, i.e. V_M should be MVWSGS(V_L, V_R). Although we can always use the sum of the two consecutive vectors, the resulting vector may not always yield the minimum Granularity Sum.

3.1   Designing the *MVWSGS*'s Algorithm

We design the algorithm to compute $MVWSGS(x_1, y_1, x_2, y_2)$. With no loss of generality, we consider the case that all the four arguments are nonnegative integers, and satisfy that: (1). $x_1/y_1 < x_2/y_2$; (2). x_1 and y_1 are relatively prime; (3). x_2 and y_2 are relatively prime.

We first propose some elementary properties of the *MVWSGS* function.

Property 3.1. *Let* $(m, n) = MVWSGS(x_1, y_1, x_2, y_2)$, *then* $MVWSGS(y_2, x_2, y_1, x_1) = (n, m)$.

Proof. Suppose to the contrary that $MVWSGS(y_2, x_2, y_1, x_1) = (n', m')$, where $(n', m') \neq (n, m)$. Then we have $y_2/x_2 < n'/m' < y_1/x_1$ and $n' + m' < n + m$, which implies $x_1/y_1 < m'/n' < x_2/y_2$ and $m' + n' < m + n$. That is, we can construct a new middle vector $V' = (m', n')$, such that $(x_1, y_1) \prec V' \prec (x_2, y_2)$; however, $GS(V') = m' + n' < m + n = GS((m, n))$, which is contrast to the fact $MVWSGS(x_1, y_1, x_2, y_2) = (m, n)$. □

Property 3.2. *Let* $(m, n) = MVWSGS(x_1, y_1, x_2, y_2)$ *and t be a integer no larger than* $\lfloor x_1/y_1 \rfloor$, *then* $MVWSGS(x_1 - t\, y_1, y_1, x_2 - t\, y_2, y_2) = (m - t\, n, n)$.

Proof. Suppose to the contrary that $MVWSGS(x_1 - t\, y_1, y_1, x_2 - t\, y_2, y_2) = (m', n')$, where $(m', n') \neq (m - t\, n, n)$. Then we have: $(x_1 - t\, y_1)/y_1 < m'/n' < (x_2 - t\, y_2)/y_2$ and $GS(m', n') < GS(m - t\, n, n)$, i.e. $m' + n' < m - t\, n + n$.

If $n' \geq n$, then the inequality $m' + n' < m - t\, n + n$ implies $m' < m - t\, n + n - n' \leq m - t\, n$. Thus $m'/n < (m - t\, n)/n = m/n - t < x_2/y_2 - t = (x_2 - t\, y_2)/y_2$; and since $n' \geq n$, then $m'/n > m'/n' > (x_1 - y_1)/y_1$; Therefore, $(x_1 - y_1)/y_1 < m'/n < (x_2 - t\, y_2)/y_2$. Consequently, $x_1/y_1 < (m' + t\, n)/n < x_2/y_2$. That is, we can construct a vector $V' = (m' + t\, n, n)$, such that $(x_1, y_1) \prec_v V' \prec_v (x_2, y_2)$; However, $GS(V') = m' + t\, n + n < m - t\, n + t\, n + n = m + n = GS((m, n))$, which is contrast to the fact $MVWSGS(x_1, y_1, x_2, y_2) = (m, n)$.

If $n' < n$, then consider two cases. *Case 1*: $m - t\, n < m'$. Then $(m - t\, n)/n' < m'/n' < (x_2 - t\, y_2)/y_2$; and since $n' < n$, then $(m - t\, n)/n' > (m - t\, n)/n > (x_1 - t\, y_1)/y_1$. Thus, $(x_1 - t\, y_1)/y_1 < (m - t\, n)/n' < (x_2 - t\, y_2)/y_2$. Consequently, $x_1/y_1 < (m - t\, n + t\, n')/n' < x_2/y_2$. That is, we can construct a vector $V' = (m - t\, n + t\, n', n')$, such that $(x_1, y_1) \prec_v V' \prec_v (x_2, y_2)$. However, $GS(V') = m - t\, n + t\, n' + n' < m - t\, n' + t\, n' + n = m + n = GS((m, n))$, which is contrast to the fact $MVWSGS(x_1, y_1, x_2, y_2) = (m, n)$; *Case 2*: $m - t\, n \geq m'$. Since $(x_1 - t\, y_1)/y_1 < m'/n' < (x_2 - t\, y_2)/y_2$, we have $x_1/y_1 < (m' + t\, n')/n' < x_2/y_2$. That is, we can construct a vector $V' = (m' + t\, n', n')$, such that $(x_1, y_1) \prec_v V' \prec_v (x_2, y_2)$. However, $GS(V') = m' + t\, n' + n' < m - t\, n + t\, n + n = m + n = GS((m, n))$, which is contrast to $MVWSGS(x_1, y_1, x_2, y_2) = (m, n)$. For all cases, the property is proved. □

The following theorem is derived by Property 3.1 and Property 3.2 .

Theorem 3.1. *Let* $(m, n) = MVWSGS(x_1, y_1, x_2, y_2)$, *then* $MVWSGS(y_2, x_2 - \lfloor x_1/y_1 \rfloor y_2, y_1, x_1 \bmod y_1) = (n, m - \lfloor x_1/y_1 \rfloor n)$.

Proof. Let t in Property 3.2 be $\lfloor x_1/y_1 \rfloor$, we have $x_1 - t\, y_1 = x_1 - \lfloor x_1/y_1 \rfloor y_1 = x_1 \bmod y_1$. Property 3.2 then implies: $MVWSGS(x_1 \bmod y_1, y_1, x_2 - \lfloor x_1/y_1 \rfloor y_2, y_2) = (m - \lfloor x_1/y_1 \rfloor n, n)$. Then from Property 3.1, we have $MVWSGS(y_2, x_2 - \lfloor x_1/y_1 \rfloor y_2, y_1, x_1 \bmod y_1) = (n, m - \lfloor x_1/y_1 \rfloor n)$. □

To compute *MVWSGS*, we introduce the other theorem.

Theorem 3.2. *If* $\lfloor x_1/y_1 \rfloor + 1 < x_2/y_2$, *then* $MVWSGS(x_1, y_1, x_2, y_2) = (\lfloor x_1/y_1 \rfloor + 1, 1)$.

Proof. Suppose to the contrary that $MVWSGS(x_1, y_1, x_2, y_2) = (m, n)$, where $(m, n) \neq (\lfloor x_1/y_1 \rfloor + 1, 1)$. Then we have: $x_1/y_1 < m/n < x_2/y_2$, and $GS((m, n)) < GS((\lfloor x_1/y_1 \rfloor + 1, 1))$, i.e. $m + n < \lfloor x_1/y_1 \rfloor + 2$. If $n=1$, then $m < \lfloor x_1/y_1 \rfloor + 1$. Thus $m \leq x_1/y_1$ and $m/n \leq x_1/y_1$, which is contrast to the inequality $m/n > x_1/y_1$; If $n>1$, the inequality $m + n < \lfloor x_1/y_1 \rfloor + 2$ implies $m < \lfloor x_1/y_1 \rfloor$. Thus $m/n < \lfloor x_1/y_1 \rfloor /y < \lfloor x_1/y_1 \rfloor /1 \leq x_1/y_1$, which is also contrast to the inequality $m/n > x_1/y$. $\square$

Algorithm. *MVWSGS*'s algorithm is expressed as a recursive program based directly on Theorem 3.1 and Theorem 3.2. The inputs x_1, y_1, x_2, y_2 are nonnegative integers, satisfying that: (1). $x_1/y_1 < x_2/y_2$; (2). x_1 and y_1 are relatively prime; (3). x_2 and y_2 are relatively prime.

Algorithm 1. $MVWSGS(x_1, y_1, x_2, y_2)$ // *MVWSGS* is a recursive algorithm

```
1    if  |x₁/y₁| + 1 < x₂/y₂  then
2        return  (⌊x₁/y₁⌋ + 1, 1);
3    else
4        (m', n') ←MVWSGS ( y₂, x₂ −⌊x₁/y₁⌋ y₂, y₁, x₁ mod y₁);
5        return  ( n' + ⌊x₁/y₁⌋ m', m');
6    end
```

If $\lfloor x_1/y_1 \rfloor + 1 < x_2/y_2$, then Algorithm 1 returns the correct result based on Theorem 3.2. Otherwise, to obtain (m, n) such that $(m, n) = MVWSGS(x_1, y_1, x_2, y_2)$, Algorithm 1 first computes (m', n') such that $(m', n') = MVWSGS(y_2, x_2 - \lfloor x_1/y_1 \rfloor y_2, y_1, x_1 \bmod y_1)$; Based on Theorem 3.1, we have: $m' = n; n' = m - \lfloor x_1/y_1 \rfloor n$. Thus $n = m'$ and $m = n' + \lfloor x_1/y_1 \rfloor m'$, proving the correctness of Algorithm 1.

We then show that the algorithm can always terminate. To obtain $MVWSGS(x_1, y_1, x_2, y_2)$, the algorithm first calls $MVWSGS(y_2, x_2 - \lfloor x_1/y_1 \rfloor y_2, y_1, x_1 \bmod y_1)$; then continuously calls: $MVWSGS (x_1 \bmod y_1, y_2 - t (x_1 \bmod y_1), x_2 - \lfloor x_1/y_1 \rfloor y_2, y_2 - t (x_2 - \lfloor x_1/y_1 \rfloor y_2))$ $(t = \lfloor y_2/(x_2 - \lfloor x_1/y_1 \rfloor y_2) \rfloor)$. We can see that after each two recursive calls, none of the four arguments increases and at least one of the four arguments strictly decreases(except the case that $x_1 < y_1$ and $t=0$, however, this case implies $x_1/y_1 < 1$ and $x_2/y_2 > 1$, and the algorithm terminates without any recursive call). Together with the fact that all the arguments are always nonnegative, the algorithm cannot recurse indefinitely.

$MVWSGS(x_1, y_1, x_2, y_2)$ makes the recursive call $MVWSGS(y_2, x_2 - \lfloor x_1/y_1 \rfloor y_2, y_1, x_1 \bmod y_1)$. Since x_1 and y_1 are relatively prime, then y_1 and $(x_1 \bmod y_1)$ are relatively prime. And since x_2 and y_2 are relatively prime, then $(x_2 - \lfloor x_1/y_1 \rfloor y_2)$ and y_2 are relatively prime. Therefore, the fact that the first argument and the second argument are relatively prime and the fact that the third argument and the fourth argument are relatively prime always hold in each recursive call.

Fig. 4 illustrates the execution of Algorithm 1 with the computation of *MVWSGS* (8, 5, 5,3),which makes four recursive calls.

	x_1	y_1	x_2	y_2	result
Inputs	8	5	5	3	(13, 8)
1th call	3	2	5	3	(8, 5)
2th call	3	2	2	1	(5, 3)
3th call	1	1	2	1	(3, 2)
4th call	1	1	1	0	(2, 1)

Fig. 4. An example of the execution of *MVWSGS'*s algorithm

The running time of *MVWSGS*'s algorithm. We discuss the worst-case running time of the algorithm of $MVWSGS(x_1, y_1, x_2, y_2)$. We assume with no loss of generality that $x_1 > y_1 \geq 1$. This assumption can be justified by the discussion of the following cases: (1). $x_1 = 0$. The procedure terminates at once or after one recursive call; (2). $y_1 = 0$. The case is not permitted since there is no vector larger than (x_1, y_1) where $y_1 = 0$; (3). $x_1 = y_1 > 1$. Since x_1 and y_1 are relatively prime, thus $x_1 = y_1 = 1$ and the procedure terminates at once or after one recursive call; (4). $x_1 < y_1$. If $x_2 > y_2$, then $\lfloor x_1/y_1 \rfloor + 1 < x_2/y_2$ and the procedure terminates at once; If $x_2 = y_2$, then $x_2 = y_2 = 1$ and the procedure terminates after two recursive calls at most. If $y_2 > x_2$, $MVWSGS(x_1, y_1, x_2, y_2)$ immediately makes the recursive call $MVWSGS(y_2, x_2, y_1, x_1)$ with first argument larger than the second. That is, *MVWSGS* spends one recursive call swapping its arguments and then proceeds.

$MVWSGS(x_1, y_1, x_2, y_2)$ calls $MVWSGS(y_2, x_2', y_1, x_1 \bmod y_1)$ (here $x_2' = x_2 - \lfloor x_1/y_1 \rfloor y_2$) recursively. Since $x_1 > y_1 \geq 1$ implies $y_1 > (x_1 \bmod y_1)$, thus $y_1/(x_1 \bmod y_1) > 1$. If $y_2 < x_2'$, then $\lfloor y_2/x_2' \rfloor = 0$ and the procedure terminates as $\lfloor y_2/x_2' \rfloor + 1 < y_1/(x_1 \bmod y_1)$. If $y_2 = x_2'$, then $y_2 = x_2' = 1$ and the procedure terminates after one recursive call at most. Thus for the worst case, $y_2 > x_2'$. That is, the first argument should be still larger than the second one after each recursive call. We therefore make the other assumption that the first argument is always larger than the second one in each recursive call to discuss the worst running time of *MVWSGS*'s algorithm.

The overall running time of *MVWSGS* is proportional to the number of recursive calls it makes. The worst running time analysis makes use of the *EUCID* algorithm [10] designed for computing the greatest common divisor of two integers efficiently.

$EUCID(x, y)$

1 **if** $y=0$ **then** return x;
2 **else return** $EUCID(y, x \bmod y)$;

$EUCID(x, y)$ first calls $EUCID(y, x \bmod y)$, and then continuously calls $EUCID(x \bmod y, y \bmod (x \bmod y))$. Compared *EUCID* with *MVWSGS*, $MVWSGS(x_1, y_1, x_2, y_2)$ first calls $MVWSGS(y_2, x_2 - \lfloor x_1/y_1 \rfloor y_2, y_1, x_1 \bmod y_1)$, and then continuously calls $MVWSGS(x_1 \bmod y_1, y_1 - t(x_1 \bmod y_1), x_2 - \lfloor x_1/y_1 \rfloor y_2, y_2 - t(x_2 - \lfloor x_1/y_1 \rfloor y_2))$ ($t = \lfloor y_2/(x_2 - \lfloor x_1/y_1 \rfloor y_2) \rfloor$). That is, after two recursion, $EUCID(x, y)$ calls $EUCID(x \bmod y, y \bmod (x \bmod y))$ and $MVWSGS(x_1, y_1, x_2, y_2)$ calls $MVWSGS(x_1 \bmod y_1, y_1 - t(x_1 \bmod y_1), x_2', y_2')$. We have make the assumption that the first argument of *MVWSGS* is larger than the second one in each recursive call to discuss the worst running time of *MVWSGS*'s algorithm, i.e. $x_1 \bmod y_1 > y_1 - t(x_1 \bmod y_1)$. Then we have

$t > -1 + y_1/(x_1 \bmod y_1)$, Thus, $y_1 - t \, (\, x_1 \bmod y_1)\,) \leq y_1 \bmod (x_1 \bmod y_1)$. That is to say, after two recursive calls, the first argument of $MVWSGS$ is the same as $EUCID$, and the second argument of $MVWSGS$ is no larger than $EUCID$. [10] has proved that the number of recursive calls in $EUCID(x, y)$ is $O(\lg y)$, and hence the number of recursive calls in $MVWSGS(x_1, y_1, x_2, y_2)$ is $O(\lg y_1)$.

3.2   Optimization

If there are only insertions in updates, the original vector order-based label schemes guarantee that the inserted label has the smallest size; that is, if there is no deleted label with smaller size than its neighbor labels, directly using the sum of the two consecutive vectors to compute the middle vector may yield the minimum Granularity Sum, i.e. $MVWSGS(x_1, y_1, x_2, y_2)=(x_1+x_2, y_1+y_2)$. In this section, we establish the premise to hold the equation and prove it, which therefore improve the running time of $MVWSGS$'s algorithm when no deletion occurs between two vector order-based labels. Our method is based on the *Stem-Brocot* tree[11].

Stem-Brocot tree is one beautiful approach to construct the set of all nonnegative fractions x/y with x and y are relatively prime. Its idea is to start with the two fractions $0/1$ and $1/0$ and then to repeat the following operation as many times as desired: Insert $(x_1+x_2)/(y_1+y_2)$ between two adjacent fractions x_1/y_1 and x_2/y_2.

For example, the first step gets one new entry between $0/1$ and $1/0$: $0/1, 1/1, 1/0$; and the next gives two more: $0/1, 1/2, 1/1, 2/1, 1/0$; and then it will get 4, 8 and so on.

The entire array can be regarded as an infinite binary tree structure whose top levels look like Fig.5.

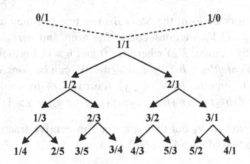

Fig. 5. *Stem-Brocot* tree[11]

In the *Stem-Brocot* tree, each fraction V is derived by the sum of its two special ancestors, defined as **leftV** and **rightV**, i.e. $V =leftV +rightV$, where $leftV$ is the nearest ancestor above and to the left, and $rightV$ is the nearest ancestor above and to the right. Based on the construction, **Consecutive Fractions**[11] is defined as follows.

Definition 3.3 (Consecutive Fractions). *If V_L and V_R are consecutive fractions at any stage of the construction, then V_L and V_R are consecutive fractions.*

The followings are some properties of consecutive fractions.

Lemma 3.1. *If V_L and V_R are consecutive fractions and $V_M = V_L + b \times V_R$, then V_M and V_R are consecutive fractions.*

Proof. Initially, (0,1) and (1,0) are consecutive fractions. The new fraction V is derived by the sum of two consecutive fractions: V_L and V_R. Thus, $(V_L + V_R)$ and V_R are consecutive fractions. $(V_L + 2V_R)$ and V_R are consecutive fractions since $(V_L + 2V_R)$ is derived by the sum of consecutive fractions $(V_L + V_R)$ and V_R. Recursively, $(V_L + b \times V_R)$ and V_R are consecutive fractions.   □

Lemma 3.2. *If V_L and V_R are consecutive fractions and V_M is an another fraction such that $V_L \prec V_M \prec V_R$, then $V_M = a \times V_L + b \times V_R$ where a and b are positive integers.*

Proof. All of the possible fractions can occur in *Stem-Brocot* tree exactly once(the proof can refer to [11]), and *Stem-Brocot* tree is strictly order-preserved (an inorder traversal of the tree visits the fractions in increasing order). Thus, all of the fractions between two consecutive fractions V_L and V_R can be derived by starting with the two fractions V_L and V_R and then repeating the following operation as many times as desired: Insert $(x_1+x_2)/(y_1+y_2)$ between two adjacent fractions x_1/y_1 and x_2/y_2.

That is, the first step gets one new entry between V_L and V_R: V_R, $V_L + V_R$, V_R;
and the next gives two more: V_L, $2V_L + V_R$, $V_L + V_R$, $V_L + 2V_R$, V_R;
and then it will get 4, 8 and so on.

Thus, all of the fractions between V_L and V_R can be expressed as $a \times V_L + b \times V_R$.   □

Lemma 3.3. *If $x_2 y_1 - x_1 y_2 = 1$, then $V_L:(x_1, y_1)$ and $V_R:(x_2, y_2)$ are consecutive fractions.*

Proof. We prove equivalently that if $V_L:(x_1, y_1)$ and $V_R:(x_2, y_2)$ are not consecutive fractions, then $x_2 y_1 - x_1 y_2 > 1$. There are three cases to be considered.

Case 1: V_R is an ancestor of V_L in the *Stem-Brocot* tree but they are not consecutive fractions. Since $leftV_R$ and V_R are consecutive fractions and $leftV_R \prec V_L \prec V_R$, thus $V_L = a \times leftV_R + b \times V_R$(by Lemma 3.2) where $a > 0$ and $b > 0$. Furthermore, a cannot be 1 since if $a=1$, then $V_L = leftV_R + b \times V_R$ and V_L and V_R can be consecutive fractions(by Lemma 3.1).Thus $a > 1$. Suppose $leftV_R = (x, y)$, then $x_1 = ax + bx_2$ and $y_1 = ay + by_2$. We have $x_2 y_1 - x_1 y_2 = x_2(a x + bx_2) - (ax + b x_2) y_2 = a (x_2 y - x y_2) = a > 1$.

Case 2: V_L is an ancestor of V_R but they are not consecutive fractions. This case can be proved similarly to that of case 1.

Case 3: V_L and V_R are not of Ancestor-Descendent relationships. Denote the common ancestor of V_L and V_R as V, then $V_L = a \times leftV + b \times V$ and $V_R = c \times rightV + d \times V$ (a, b, c and d are positive integer). Suppose $V=(m, n)$, $leftV=(m_1, n_1)$ and $rightV = (m_2, n_2)$, then $x_1 = am_1 + bm$, $y_1 = a n_1 + bn$, $x_2 = cm_2 + dm$ and $y_2 = cn_2 + dn$. Therefore, $x_2 y_1 - x_1 y_2 = ac(m_2 n_1 - n_2 m_1) + bc(m_2 n - m n_2) + ad (mn_1 - m_1 n) = ac + bc + ad > 1$.   □

Now we are ready to provide the premise which guarantees the equation $MVWSGS(x_1, y_1, x_2, y_2) = (x_1 + x_2, y_1 + y_2)$.

Theorem 3.3. *If $x_2 y_1 - x_1 y_2 = 1$, then $MVWSGS(x_1, y_1, x_2, y_2) = (x_1 + x_2, y_1 + y_2)$.*

Proof. From Lemma 3.3, $V_L:(x_1, y_1)$ and $V_R:(x_2, y_2)$ are consecutive fractions. Suppose V_M is an another fraction such that $V_L \prec V_M \prec V_R$, then $V_M = a \times V_L + b \times V_R$ where a and b are both positive integer(Lemma 3.2). $GS(V_M)$ is smallest if $a=1$ and $b=1$. Thus, $MVWSGS(x_1, y_1, x_2, y_2)=V_L+V_R=(x_1+x_2, y_1+y_2)$.   □

We can combine Theorem 3.3 with Algorithm 1 for the better performance to compute $MVWSGS$, as we show in Algorithm 2.

Algorithm 2. *Extended-MVWSGS(x_1, y_1, x_2, y_2)*

---
1   **if** $x_2 y_1 - x_1 y_2 = 1$ **then return** (x_1+x_2, y_1+y_2);
2   **else return** $MVWSGS(x_1, y_1, x_2, y_2)$; //Algorithm 1

The Running Time Analysis. If there are only insertions in updates, there is no deleted label to be reused and the running time of *Extended-MVWSGS(x_1, y_1, x_2, y_2)* is $O(1)$ since the cases implies that x_1/y_1 and x_2/y_2 are consecutive fractions and "$x_2 y_1 - x_1 y_2 = 1$". If there are deletions as well, the chance that we have "$\lfloor x_1/y_1 \rfloor + 1 < x_2/y_2$" (see line 1 in Algorithm 1) is actually high especially in the case that large amounts of nodes are deleted. The worst running time of $MVWSGS$ is $O(lg\ y_1)$ as we have shown in Sec.3.1. Considering that it is highly unlikely for y_1 to be very large in practice, the computation of $MVWSGS$ is efficient. This extra computation is worthy because it can guarantee that the inserted label has the smallest size and therefore reduce the label sizes and enhance the query performances. On the other hand, since the original approaches usually have the larger label sizes, their advantage in running time is not noticeable. In most cases, the different updating time between our methods and the original approaches is actually negligible.

4   Experiment and Results

We evaluate our improved methods against the original vector order-based labeling schemes, including V-containment scheme, V-Prefix scheme, DDE and CDDE. We present the results of our Improved-CDDE against the original CDDE only as the others show similar trends. The initial labels of Improved-CDDE is the same as the original CDDE. We evaluate and compare the update performances of different schemes. The test dataset is "Hamlet.xml"[12]. Hamlet has totally 6,636 nodes, and there are 5 "*act*" sub-trees under root. We repeat the following operations ten times(noted that each time is based on the file generated by the last time): the *odd* positions of the 5 "*act*" sub-trees, i.e. the first, the third and the fifth "*act*" sub-trees, are deleted then inserted; then based on the new file, the *even* positions of the five "*act*" sub-trees are deleted and inserted in the same way.

Updating Time. We evaluate the updating time of different schemes and the results are shown in Fig.6. We observe that the updating time of our Improved-CDDE is approximately similar to that of the original CDDE. This result conforms to our previous discussions that the different updating time between our improved methods and the original approaches is actually negligible.

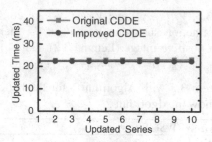

Fig. 6. Updating Time

Label Size. After every update, the new Hamlet is the same as the original; however, the labels are different. We evaluate the average label size of different schemes and show the results in Fig.7. We observe that the label sizes of Improve-CDDE does not increase in all the cases since we produce the optimal size and thus reuse all the deleted labels. On the other hand, the label size of Original-CDDE increases fast. The reason is that Original-CDDE computes the middle vector using the sum of the two consecutive vectors and therefore cannot reused the deleted labels for the inserted nodes. The experimental results confirm that our Improved-CDDE can reuse all the deleted labels, and therefore efficiently control the increasing speed of the label size.

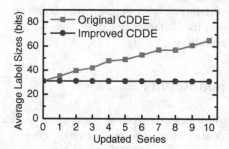

Fig. 7. Label Size

Query Performance. We evaluate the query performance on the updated labels after ten deletions and insertions. We test the performance by computing the most commonly used five relationships: document order, AD, PC, sibling and LCA. We choose all the 6,636 labels and compute all the five relationships for each pair of the labels. As shown in Fig.8, for all the five functions our method outperforms than the previous scheme. The reason is clear from analysis of the label sizes of different schemes. The original CDDE has the large label size and therefore degrade the query performance. Moreover, the original CDDE cannot support the computation of LCA as we discuss before. The results confirm that our method improves the query performance and has a better query support.

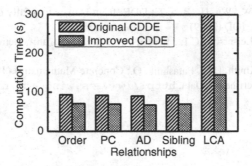

Fig. 8. Query Performance

5   Conclusion

In this paper, we propose the algorithms which can be applied broadly to different vector order-based labeling schemes to reuse all the deleted labels. In this way, we efficiently control the label size increasing speed and thus optimize query performance when XML frequently updates. The experimental results also show that with the algorithms we can greatly decrease the label size when a lot of nodes are deleted and inserted. In summary, our improved methods are more appropriate to process the dynamic XML updates.

References

1. Zhang, C., Naughton, J.F., DeWitt, D.J., Luo, Q., Lohman, G.M.: On Supporting Containment Queries in Relational Database Management Systems. In: SIGMOD (2001)
2. Tatarinov, I., Viglas, S., Beyer, K.S., Shanmugasundaram, J., Shekita, E.J., Zhang, C.: Storing and Querying Ordered XML Using a Relational Database System. In: SIGMOD (2002)
3. Wu, X., Lee, M., Hsu, W.: A prime number labeling scheme for dynamic order XML tree. In: ICDE (2004)
4. O'Neil, P., O'Neil, E., Pal, S., Cseri, I., Schaller, G., Westbury, N.: ORDPATHs: Insert-friendly XML Node Labels. In: SIGMOD (2004)
5. Li, C., Ling, T.W., Hu, M.: Efficient Updates in Dynamic XML Data: from Binary String to Quaternary String. VLDB J. (2008)
6. Li, C., Ling, T.W.: QED: A Novel Quaternary Encoding to Completely Avoid Re-labeling in XML Updates. In: CIKM (2005)
7. Xu, L., Bao, Z., Ling, T.-W.: A Dynamic Labeling Scheme Using Vectors. In: Wagner, R., Revell, N., Pernul, G. (eds.) DEXA 2007. LNCS, vol. 4653, pp. 130–140. Springer, Heidelberg (2007)
8. Xu, L., Ling, T.W., Wu, H.: Labeling Dynamic XML Documents: An Order-Centric Approach. TKDE (2010)

9. Xu, L., Ling, T.W., Wu, H., Bao, Z.: DDE: from Dewey to a fully dynamic XML labeling scheme. In: SIGMOD (2009)
10. Cormen, T.H., Leiserson, C.E., Rivest, R.L., Stein, C.: Introduction to Algorithms, 2nd edn. (2001)
11. Graham, R.L., Knuth, D.E., Patashnik, O.: Concrete Mathematics (1989)
12. NIAGARA Experimental Data, http://www.cs.wisc.edu/niagara/data

Towards an Efficient Flash-Based
Mid-Tier Cache

Yi Ou[1], Jianliang Xu[2], and Theo Härder[1]

[1] University of Kaiserslautern
{ou,haerder}@cs.uni-kl.de
[2] Hong Kong Baptist University
xujl@comp.hkbu.edu.hk

Abstract. Due to high access performance and price-per-byte considerations, flash memory has been recommended for use as a mid-tier cache
in a multi-tier storage system. However, previous studies related to flash-
based mid-tier caching only considered the indirect use of flash memory
via a flash translation layer, which causes expensive flash-based cache
maintenance. This paper identifies the weaknesses of such indirect methods, with a focus on the *cold-page migration* problem. As improvements,
we propose two novel approaches, an indirect approach called LPD (logical page drop) and a native approach called NFA (native flash access).
The basic idea is to drop cold pages proactively so that the garbage collection overhead can be minimized. Our experiments demonstrate that
both approaches, especially the native one, effectively improve the use of
flash memory in the mid-tier cache. NFA reduces the number of garbage
collections and block erasures by up to a factor of five and improves the
mid-tier throughput by up to 66%.

1 Introduction

Despite fast page-oriented random access, storage devices based on flash memory
(flash-based devices), e. g., flash SSDs, are still too expensive to be the prevalent
mass storage solution. In fact, flash memory perfectly bridges the gap between
RAM and magnetic disks (HDDs) in terms of price per capacity and performance[1]. Therefore, using them in the mid-tier of a *three-tier storage hierarchy*
is a much more realistic approach. In such a hierarchy, the top tier incorporates
fast but expensive RAM-based buffer pools, while the bottom tier is based on
slow but cheap storage devices such as HDDs or even low-end flash SSDs. The
middle tier acts as a cache larger but slower than the top-tier buffer pool. Such
a three-tier storage system can be deployed as, e. g., the storage sub-system
of a database (DB) system. Nevertheless, flash memory has some distinguishing
characteristics and limitations that make its efficient use technically challenging.
This paper studies its efficient use for mid-tier caching.

---

[1] We focus on NAND flash memory due to its suitability for storage systems.

S.W. Liddle et al. (Eds.): DEXA 2012, Part I, LNCS 7446, pp. 55–70, 2012.

1.1   Flash Memory and FTL

Flash memory supports three basic operations: *read*, *program*, and *erase*. Read and program (also known as write) operations must be performed in units of *flash pages*, while erase operations have to be done at a larger granularity called *flash block (block)*, which contains a multiple of (e. g., 128) flash pages. Read operations have a very small latency (in the sub-millisecond range). However, program operations are much slower than them, typically by an order of magnitude. Erase operations are even slower than program operations, by another order of magnitude. Let C_{fr}, C_{fp}, and C_{fe} be the costs of read, program, and erase operations, respectively, we have: $C_{fr} < C_{fp} < C_{fe}$.

An erase operation turns a block into a *free block*, and, consequently, each of its flash pages into a *free flash page*, which is a flash page that has never been programed after the block erasure. Only free flash pages can be programed, i. e., a non-free flash page can become programmable again – but only after an erase of the entire block. This limitation, known as *erase-before-write*, implies that an *in-place update* of flash pages would be very expensive [1]. Furthermore, after enduring a limited number of program/erase (P/E) cycles[2], a block becomes highly susceptible to bit errors, i. e., it becomes a bad block.

Mainly due to the aforementioned limitations, flash memory is often accessed via an intermediate layer called *flash translation layer (FTL)*, which supports *logical* read and write operations, i. e., reading and writing of *logical pages*. A flash page normally consists of a *main area* of several KBs for storing user data, and a *spare area* of a few bytes to facilitate the FTL implementation. For brevity, we assume that (the size of) a logical page corresponds to a bottom-tier page (e. g., a DB page) and they are of the same size of the main area, and we use the term *page* to refer to a *logical page*, which is to be distinguished from a *physical page*, i. e., a flash page.

To avoid the expensive in-place updates, FTL follows an *out-of-place update* scheme, i. e., each logical page update is served using a free flash page prepared in advance. Consequently, multiple writes to a logical page can result in multiple page versions co-existing in flash memory. The term *valid page* refers to the latest version, while *invalid pages* refer to the older versions. Similarly, a *valid flash page* is the physical page where the latest version resides. When free flash pages are in short supply, some space taken by invalid pages has to be reclaimed. This is done by a procedure called *garbage collection (GC)*, which reduces the number of invalid pages and increases the number of free flash pages.

To keep track of the valid flash page of a *(logical) page*, an address mapping $m_{\text{FTL}} : A_F \mapsto A_f$ is maintained, where A_F represents the set of FTL logical addresses (FLAs), i. e., logical page numbers supported by the FTL, and A_f the set of FTL physical addresses (FPAs), i. e., flash page addresses available on the device. Depending on the map-entry granule of the m_{FTL} implementation, FTL algorithms can be classified into three categories: page-level mapping [2,3], block-level mapping [4,5], and hybrid mapping [6,7]. Among them, page-level mapping

---

[2] The number of cycles depends on density, vendor, and flash memory type. SLC NAND flash memory is typically rated for ~100,000 P/E cycles.

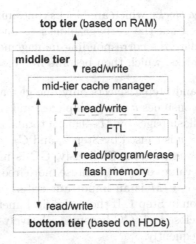

Fig. 1. Three-tier storage system with indirect use of flash memory by the mid-tier cache. FTL makes the native interface of the flash memory transparent to the mid-tier cache manager. The two components of a flash-based device, surrounded by the dashed line, appear as a "black box" and act together as a persistent array of logical pages that can be read and overwritten.

has the greatest performance potential, but it also has the highest resource requirements, mainly due to the mapping table size. However, recent studies have shown that the resource problem of page-level mapping can be effectively dealt with using methods such as demand paging of the mapping table [8] or new hardware such as PCM (phase-change memory) as the mapping table storage media [9]. Therefore, in this paper, we focus on page-level mapping in favor of the performance potential, although the problems studied and the basic ideas leading to our solutions are not specific to any FTL implementation.

1.2 Problem

Previous studies on flash-based mid-tier caching only considered the *indirect* use of flash memory, i. e., the use of flash memory via an FTL, as shown in Fig. 1. Although simplifying the use of flash memory, the indirect approach has some fundamental problems. FTL implementations are usually vendor-specific and proprietary [1,10]. The proprietary FTL logic makes it impossible to accurately model or predict the performance of flash-based devices. This is not acceptable for performance-critical applications, because their optimization is often based on the cost model of the underlying storage devices. Furthermore, without direct control over potentially expensive procedures such as GC, the response time becomes indeterministic for the application. It has been reported that GC can take up to 40 seconds[11], which is not only an issue for applications with real-time requirements, but also intolerable for normal use cases.

For flash-based mid-tier caching, the indirect approach has an even more serious problem related to GC. This problem is explained in the following with the help of a simplified GC procedure, which involves three steps:

1. Select a set of *garbage blocks*, which are blocks containing some invalid pages.
2. Move all valid pages from the garbage blocks to another set of (typically free) blocks, and update the corresponding management information.
3. Erase the garbage blocks, which then become free blocks.

If a block has M pages and Step 1 selects only one garbage block, which has v valid pages, then Step 2 consumes v free flash pages, and the procedure increases the total number of free flash pages by $M - v$, at a total cost of $(C_{fr} + C_{fp}) \times v + C_{fe}$, where $(C_{fr} + C_{fp}) \times v$ is caused by Step 2, and C_{fe} caused by Step 3. The ratio v/M is called *block utilization*. Obviously, GC is more effective and also more efficient for smaller values of v/M, because more free flash pages are gained at a lower cost. Therefore, v/M is an important criterion to be considered for the garbage block selection in Step 1. If the entire flash memory is *highly utilized*, i. e., v/M is statistically close to 1, GC becomes relatively expensive, ineffective, and has to be invoked frequently.

Although for a cache, only hot pages should be kept and cold pages should be evicted, FTL must guarantee each valid page is accessible no matter the page is cold or hot. This means that, during GC processing, cold pages have to be moved along with hot pages (Step 2), while the cold ones, which make v/M unnecessarily high, could actually be discarded from the cache manager perspective. We call this problem the *cold-page migration (CPM)* problem.

More specifically, the CPM problem negatively impacts mid-tier performance in two aspects: 1. The cost of GC, due to the (unnecessary) CPM; 2. The frequency of GC, because, if cold pages are regarded valid, fewer pages can be freed by one invocation of GC, and, as a result, the subsequent GCs have to be invoked earlier. Furthermore, the GC frequency is proportional to the number of block erases, which is inversely proportional to the device life time due to the endurance limitation.

A similar problem exists when flash SSDs are used as the external storage under a file system. File deletion is a frequent operation, but the information about deleted files is normally kept in OS and not available to the SSD. The latter has to keep even the deleted data valid, at a potentially high operational cost. As solution, a *Trim* attribute for the Data Set Management command has been recently proposed and become available in the ATA8-ACS-2 specification [12]. This attribute enables disk drives to be informed about deleted data so that their maintenance can be avoided.

However, no sufficient attention has been paid to the CPM problem, which actually impacts the performance in a more serious way. First, when used in the mid-tier cache, flash-based devices experience a much heavier write traffic than that of file systems, because pages are more frequently loaded into and evicted from the cache. To flash-based devices, heavy write traffic means frequent GCs. Second, the capacity utilization of a mid-tier cache is always full, which makes GC expensive and ineffective (especially for heavy write workloads). In contrast, the GC issue is less critical to file systems, because typically a large portion of their capacity is unused.

1.3   Solution

To solve the CPM problem, we develop two approaches, which share the same basic idea: drop cold pages proactively and ignore them during GCs.

1. The first approach, LPD (logical page drop), accesses flash memory *indirectly* via an *extended FTL*, which can be informed about proactively evicted cold pages, and ignore them during GCs.
2. The second approach, NFA (native flash access), manages flash memory in a *native* way, i.e., it implements the out-of-place update scheme and handles GC by the cache manager, without using an FTL.

According to our experiments, both approaches significantly outperform the normal indirect approach, by improving the GC effectiveness and reducing its frequency. For example, NFA reduces the GC frequency by a factor of five, which not only contributes to the data access performance, but also implies a greatly extended device life time. In terms of overall performance (IOPS), NFA achieves an improvement ranging from 15% to 66%, depending on the workload.

1.4   Contribution

To the best of our knowledge, our work is the first that identifies the CPM problem. Our work is also the first that considers managing flash memory natively in the mid-tier cache. Our further major contributions are:

- We propose two novel approaches for flash-based mid-tier caching: LPD and NFA, both of them effectively deal with the CPM problem.
- Our study shows that, for a flash-based mid-tier cache, our native approach significantly improves the storage system performance while reducing the resource requirements at the same time.
- More importantly, the results of our study urge the reconsideration of the architectural problem of optimally using flash memory in a DB storage system, i.e., whether it should be managed natively by the DBMS or indirectly via the proprietary FTL implementations.

1.5   Organization

The remainder of this paper is organized as follows: Section 2 discusses related works. Section 3 presents and discusses our approaches. Section 4 reports our experiments for the evaluation of both approaches. The concluding remarks are given in Section 5.

2   Related Work

Before flash memory became a prevalent, disruptive storage technology, many studies, e.g., [13,14,15,16], addressed the problem of multi-level caching in the context of client-server storage systems, where the first-level cache is located at the client side and the second-level (mid-tier) cache is based on RAM in storage server. However, these studies did not consider the specific problems of a *flash-based* mid-tier cache. Our proposals are orthogonal to and can be combined with their approaches, because their primary goal is to reduce the disk I/O of the storage server, while our approaches primarily focus on the operational costs of the middle tier.

In one of the pioneer works on flash-aware multi-level caching [17], Koltsidas et al. studied the relationships between page sets of the top tier and the mid-tier caches, and proposed flash-specific cost models for three-tier storage systems. In contrast, a detailed three-tier storage system implementation and performance study was presented in [18]. Their empirical study has demonstrated that, for certain spectrum of applications, system performance and energy efficiency can be both improved at the same time, by reducing the amount of energy-hungry RAM-based memory in the top tier and using a much larger amount of flash memory in the middle tier.

Not only academia, but also industry has shown great interest in flash-based mid-tier caching. Canim et al. [19] proposed a temperature-aware replacement policy for managing an SSD-based mid-tier, based on access statistics of disk regions. In [20], the authors studied three design alternatives of an SSD-based mid-tier, which mainly differ in the way how to deal with dirty pages evicted from the first-tier, e. g., write through or write back.

Although flash-specific cost models and their difference to those of traditional storage devices have been taken into account by previous works on flash-based mid-tier caching [17,18,19,20], they commonly only consider the indirect approach, while hardly any efforts have been made to examine the internals of flash-based devices when used as a mid-tier cache. Such efforts fundamentally distinguish our work from the previous ones.

3   Our Approaches

As introduced in Section 1.3, our basic idea is to drop cold pages proactively and ignore them during GCs. A question critical to the success is to what extent valid but cold pages are dropped. Note, if we drop valid pages too greedily, the benefit will not be covered by the cost of increased accesses to the bottom tier.

Which pages are cold and can be dropped is the decision of the cache manager, while the decision, when and how to do GC, is typically made by the FTL – if we follow the architecture of Fig. 1. Therefore, another important question is how to bring these two pieces of information together.

3.1   LPD

The LPD approach is basically an indirect approach, which follows the architecture shown in Fig. 1. However, to make the basic idea working, we propose, as an extension to the FTL interface, a *delete* operation, in addition to the read and

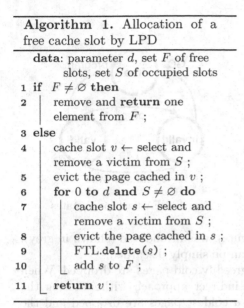

Algorithm 1. Allocation of a free cache slot by LPD

data: parameter d, set F of free slots, set S of occupied slots
1 **if** $F \neq \varnothing$ **then**
2   remove and **return** one element from F ;
3 **else**
4   cache slot $v \leftarrow$ select and remove a victim from S ;
5   evict the page cached in v ;
6   **for** 0 to d **and** $S \neq \varnothing$ **do**
7   cache slot $s \leftarrow$ select and remove a victim from S ;
8   evict the page cached in s ;
9   FTL.delete(s) ;
10   add s to F ;
11   **return** v ;

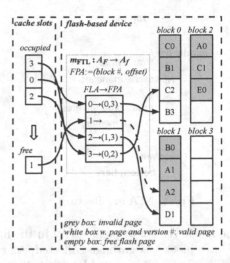

Fig. 2. Example of logical page drop. Note m_{LPD} is not shown in the figure.

write operations. Similar to the read and write operations, the delete operation is also a logical operation. Upon such a delete request, FTL should mark the corresponding flash page invalid (and update other related management information properly) so that it can be discarded by subsequent GCs.

LPD has some typical cache manager data structures. To tell whether and where a page is cached, it maintains an address mapping table $m_{\text{LPD}} : A_b \mapsto A_F$, where A_b denotes the set of bottom-tier addresses (BTAs) and A_F the set of FLAs. A *cache slot* is a volatile data structure corresponding to exactly one FLA. In addition to the FLA, the cache slot uses one bit to represent the clean/dirty state of the cached page. A *dirty page* contains updates not yet propagated to the bottom tier. Therefore, evicting such a page involves writing it back to the bottom tier. A *free*[3] cache slot is a cache slot ready to cache a new page. Such a slot is needed when a read or write cache miss occurs, so that the missing page can be stored at the corresponding FLA. Storing the page turns a free cache slot into an *occupied* slot, which becomes free again when the page is evicted.

For the mid-tier cache manager to make use of the extended FTL, the procedure of allocating a free cache slot has to be enhanced by some additional code as shown in Algorithm 1. The piece of code (Line 6 to 10) evicts up to the d coldest pages and instructs FTL to delete them, i.e., *dropping a page* involves evicting it from the cache and deleting it logically via the extended FTL. Page dropping happens after the standard logic of cache replacement (Line 4 to 5), which is only required when there is no free cache slot available.

An example of LPD is shown in Fig. 2, where the cache slot with FLA = 1 was just dropped and became free. The corresponding flash page, although containing

---

[3] There is no connection between free cache slot and free flash page, although both concepts use the word "free" by convention.

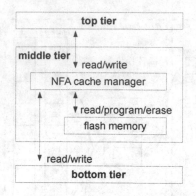

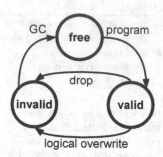

Fig. 3. NFA architecture   **Fig. 4.** NFA flash page states

the latest version of page A (A2 in the figure), was invalidated (shown in grey). If later block 1 is garbage-collected, A2 can be simply discarded.

The tuning parameter d controls how greedily cold pages are dropped. When $d = 0$, LPD degenerates to the normal indirect approach without using the extension. In contrast, when $d > 0$, the d coldest pages are dropped and the same number of cache slots are turned into free slots, ready to be used for the subsequent allocations of free cache slots (Line 1 to 2).

The LPD approach is orthogonal to the cache replacement policy responsible for the victim selection (Line 4 and Line 7), which shall identify the coldest page as per its own definition. In other words, LPD is compatible with other cache management techniques, which can be used to further improve the hit ratio.

3.2 NFA

In contrast to the indirect approaches, NFA does not require an FTL. Instead, it manages flash memory natively. As shown in Fig. 3, the operations available to the NFA cache manager are read and program of *flash pages*, and erase of *blocks*. Besides the common cache management functionality, NFA has to provide the implementation of an out-of-place update scheme and GC.

For the cache management functionality, NFA maintains a mapping table $m_{\mathrm{NFA}} : A_b \mapsto A_f$, where A_b denotes the set of BTAs and A_f the set of FPAs. Note in LPD (and other indirect approaches), two mapping tables are required: m_{LPD} for cache management, and m_{FTL} maintained by FTL (see Section 1.1).

A volatile data structure, *block management structure (BMS)* represents the state of a block. BMS contains two bit vectors, *validity* and *cleanness*, which mark the valid/invalid and clean/dirty states for each flash page in the block. Validity is used by the GC processing, while cleanness is checked when dropping a page. Furthermore, BMS stores, for each of its valid flash pages, the corresponding BTA to speed up reverse lookups, and the corresponding last access time, which is used by the page-dropping logic. The memory consumption of BMS is very low, e. g., using 4 bytes per BTA and another 4 bytes per access time, for 8 KB pages, the memory overhead of BMS is 0.1% at maximum.

Algorithm 2. Allocation of a free flash page by NFA	**Algorithm 3.** NFA GC
data: pointer wp, set F of free blocks, watermarks w_l, w_h	**data:** page-dropping threshold t

1 **if** *current block is fully written* **then**	1 block $b \leftarrow$ select a garbage block ;
2 $\quad$ $wp \leftarrow$ the first flash page of a free block ;	2 **if** *all pages in b are valid* **then**
3 $\quad$ **if** $\|F\| \leq w_l$ **then**	3 $\quad$ $b \leftarrow$ select a victim block ;
4 $\quad$ $\lfloor$ **while** $\|F\| < w_h$ **do** GC;	4 $\quad$ $t \leftarrow$ the last access time of b ;
5 $\quad$ **return** wp ;	5 **foreach** *page $p \in b$* **do**
6 **else**	6 $\quad$ **if** *last access time of $p \leq t$* **then**
7 $\quad$ $\lfloor$ **return** $wp \leftarrow wp + 1$;	7 $\quad\quad$ $\lfloor$ drop(p) ;
	8 $\quad$ **else**
	9 $\quad\quad$ $\lfloor$ move p to a free flash page ;
	10 erase b and mark it a free block ;

Following the out-of-place update scheme, both serving a write request and caching a (not yet cached) page consume a free flash page, which is allocated according to Algorithm 2. The algorithm maintains a write pointer wp, which always points to the next free flash page to be programed. After the program operation, wp moves to the next free flash page in the same block, until the block is fully written – in that case, wp moves to the begin of a new free block.

Because GC is a relatively expensive procedure, it is typically processed by a separate thread. NFA uses a low watermark w_l and a high watermark w_h to control when to start and stop the GC processing. GC is triggered when the number of free blocks is below or equal to w_l, and stops when it reaches w_h, so that multiple garbage blocks can be processed in one batch. The available number of blocks and the high watermark determine the logical capacity of the cache. If we have K blocks with M pages per block, the logical capacity of the cache is: $(K - w_h) \times M$. We say that w_h blocks are *reserved* for GC processing.

Note that the out-of-place update scheme and the use of reserved blocks for GC processing shown in Algorithm 2 are common FTL techniques. They are presented here for comprehension and completeness, because they are now integral to the NFA approach.

The NFA GC procedure (shown in Algorithm 3) is similar to that of a typical FTL in some steps (Line 1, 9, and 10 roughly correspond to Step 1, 2, and 3 of the simplified GC discussed in Section 1.2). The difference is due to the dropping of *victim blocks* and cold pages. Victim blocks are selected by a *victim-selection policy* based on the temporal locality of block accesses. In contrast, garbage blocks are selected by a *garbage-selection policy*, for which block utilization is typically the most important selection criterion. Except for these basic assumptions, the NFA approach is neither dependent on any particular garbage selection policy (Line 1) nor on any particular victim selection policy (Line 3).

Dropping of a victim block happens when the selected garbage block is fully utilized, i.e., all its pages are valid. Garbage-collecting such a block would not gain any free flash page. Furthermore, such a garbage block signals that the overall flash memory utilization is full or close to full (otherwise the garbage-selection policy would return a block with lower block utilization). Therefore, instead of processing the garbage block, a *victim* block is selected by the victim-selection policy (Line 3). The last access time of the block is used to update the page-dropping threshold t. This has the effect that all pages of the victim block are then dropped immediately (Line 6 to 7). The dynamically updated threshold t is passed on to subsequent GC invocations, where the threshold makes sure that valid pages accessed earlier than t are dropped as well.

A flash page managed by NFA has the same set of possible states (shown in Fig. 4) as those managed by a FTL: free, valid, and invalid. However, NFA has a different set of possible state transitions, e. g., a read or write page miss in an NFA cache can trigger a program operation (for storing the missing page) which changes the state of a free flash page into valid, while for FTL, serving a read request does not require a program operation. Obviously, the drop transition is not present in any FTL, either. The semantic of NFA page dropping is similar to that of LPD: the page is evicted (removing the corresponding entry from m_{NFA}, and, if the page is dirty, it is written back to the bottom tier), and then the corresponding flash page is marked invalid.

From the NFA cache manager perspective, the free flash pages for storing pages newly fetched from the bottom tier (due to page faults) are completely provided by the GC procedure in units of blocks. Therefore, NFA does not require page-level victim selection and eviction, which are common in classical caching.

3.3   Discussion

Although sharing the same basic idea, the presented approaches, LPD and NFA, are quite different from each other. While NFA directly integrates the drop logic into the GC processing, LPD can only select the drop candidates and delete them logically. LPD can not erase a block due to the indirection of FTL – the intermediate layer required by an indirect approach. Therefore, contiguously dropped logical pages may be physically scattered over the flash memory and LPD has no control when these pages will be garbage-collected, which is again the responsibility of FTL. Such dropped pages can neither contribute to the mid-tier cache hit ratio nor contribute to the reduction of GC cost, until the space taken by them is eventually reclaimed by some GC run. In contrast to the LPD approach, the pages dropped by NFA immediately become free flash pages.

To control how greedily pages are dropped, LPD depends on the parameter d, for which an optimal value is difficult to find, while the "greediness" of NFA is limited to M pages (one block at maximum). However, due to the victim selection based on block-level temporal statistics, the NFA hit ratio could be slightly compromised and a few more accesses to the bottom tier would be required.

4   Experiments

To evaluate our approaches, we implemented a three-tier storage system simulator supporting both architectures depicted in Fig. 1 and Fig. 3. The simulated

Table 1. Size ratios of the top tier and mid-tier relative to the DB size

trace	top tier	middle tier	DB size (max. page number)
TPC-C	2.216%	13.079%	451,166
TPC-H	0.951%	5.611%	1,051,590
TPC-E	0.002%	0.013%	441,138,522

flash memory and HDD modules used in our experiments were identical for both architectures.

The workloads used in our experiments originate from three buffer traces, which contain the logical page requests received by DB buffer managers under the TPC-C, TPC-H, and TPC-E benchmark workloads. The TPC-C and TPC-H buffer traces were recorded by ourselves, while the TPC-E trace was provided by courtesy of IBM. Therefore, the buffer traces represent typical, strongly varying workloads to the top tier and our results are expected to be indicative for a broad spectrum of applications.

The logical page requests recorded in the buffer traces were sent to the top tier to generate the *mid-tier traces* running the experiments. The top tier, which had a buffer pool of 10,000 pages managed under an LRU replacement policy, served the requests directly from the buffer pool whenever possible. In cases of buffer faults or eviction of dirty pages, it had to read pages from and write pages to the middle tier. The sequences of read and write requests received by the middle tier were recorded and served as the mid tier traces used in the experiments. We used them to stress the systems containing the middle and bottom tiers. As a result, the access statistics to the flash memory and HDD modules were collected for the performance study.

Three approaches were under comparison: NFA, LPD (with $d = 1024$ unless otherwise specified), and a baseline (BL), which is a mid-tier cache with indirect flash access (but without the delete extension). Our FTL implementation uses page-level mapping, which is the ideal case for the indirect approaches LPD and BL. For all three approaches, the LRU replacement policy was used for selecting victim cache pages (LPD and BL) or victim blocks (NFA), and the *greedy policy* [21,22] is used for selecting garbage blocks, which always selects the block having the least number of valid pages.

For each approach, the flash memory module was configured to have 512 blocks of 128 pages. Similar to [23] and [24], the low and high watermarks for GC were set to 5% and 10%, respectively. Due to this setting, the logical size of the mid-tier cache is 59,008 pages ((512 − 51) × 128) for all approaches. In Table 1, we list the ratios of the top-tier buffer pool size and the logical size of the mid-tier cache relative to the DB size (using the maximum page number as an estimate[4]).

---

[4] For the TPC-E trace, the DB size estimation is coarse because the trace was converted from a proprietary format addressing more than 20 DB files whose sizes and utilization were unavailable to us.

4.1 Overall Performance

We use the throughput of the mid-tier, i. e., the throughput seen by the top tier, as the overall performance metric, which is defined as: throughput $= N/t_v$, where N is the number of page requests in a trace and t_v its execution time, which is further defined as:

$$t_v = t_m + t_b \tag{1}$$

t_m represents the total operational cost in the flash memory, and t_b the total disk I/O cost. t_m is defined as $t_m = n_{fr} \times C_{fr} + n_{fp} \times C_{fp} + n_{fe} \times C_{fe}$, where n_{fr}, n_{fp}, and n_{fe} are the numbers of flash read, program, and erase operations. t_b is similarly defined as $t_b = n_H \times C_H$, with C_H being the cost of a disk access and n_H the number of disk accesses. Therefore, the trace execution time t_v is the weighted sum of all media access operations performed in the middle tier and bottom tier while running the trace.

For the costs of flash operations, C_{fr}, C_{fp}, and C_{fe}, we used the corresponding performance metrics of a typical SLC NAND flash memory of a leading manufacturer, while the disk access cost corresponds to the average latency of a WD1500HLFS HDD [25]. These costs are listed in Table 2.

Fig. 5 compares the overall performance of the three approaches under the TPC-C, TPC-H, and TPC-E workloads. Our two approaches, NFA and LPD, significantly outperformed BL, and NFA had a clear performance advantage over LPD. For the TPC-C workload, NFA achieved an improvement of 43% and 66% compared with LPD and BL respectively.

The performance improvement of our approaches can be entirely credited to the cost reduction in the mid-tier, because both of our approaches do not focus on minimizing disk accesses. In fact, they even had a slightly higher number of disk accesses due to proactive page dropping. It is expected that a small fraction of the dropped pages are re-requested shortly after the dropping, which increases disk accesses. However, this is the small price we have to pay in order to achieve the overall performance gain.

Fig. 6 confirms our expectation, where we provide a breakdown of the execution times according to (1). The mid-tier cost t_m is further broken down into two fractions: the fraction caused by GCs, denoted as t_g, and the fraction caused by

Table 2. Operation costs

operation	cost (ms)
C_{fr}	0.035
C_{fp}	0.350
C_{fe}	1.500
C_H	5.500

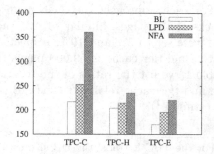

Fig. 5. Throughput (IOPS)

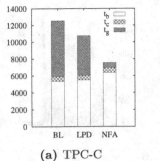
(a) TPC-C

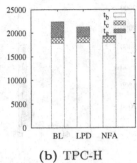
(b) TPC-H

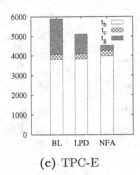
(c) TPC-E

Fig. 6. Breakdown of the trace execution time (seconds) into the fractions of GC t_g, cache overhead t_c, and disk accesses t_b

normal caching operations (e. g., read operations due to cache hits and program operations due to cache replacements), denoted as t_c, such that

$$t_v = t_m + t_b = (t_g + t_c) + t_b$$

As clearly shown in Fig. 6, both our approaches effectively improved the GC fraction, without significantly increasing the cost of other two fractions.

The remainder of this section is a detailed analysis of the experimental results. Due to space constraints, we only focus on the performance metrics collected under the TPC-C workload and omit those of the TPC-H and TPC-E workloads, from which similar observations were made.

4.2   Detailed Analysis

To further understand why our approaches improved the GC efficiency and reduced the number of its invocations, we plotted, in Fig. 7, the distribution of the number of valid pages in garbage-collected blocks. The majority of blocks garbage-collected in the BL configuration had a number of valid pages very close to 128, which resulted in a poor efficiency of GC. Compared with Fig. 7a, the dense region in Fig. 7b is located slightly farther to the left, meaning fewer valid pages in the garbage blocks. For NFA, the majority of garbage-collected blocks had less than 96 valid pages per block, i. e., more than 32 pages could be freed for each garbage block.

Interestingly, in Fig. 7c, the region between 96 and 127 is very sparse. This is the filtering effect (Line 6 to 7 of Algorithm 3). The valid pages in a block either become invalidated due to logical overwrites or are filtered out when they become cold. Therefore, the probability that a block has full or close-to-full utilization is artificially reduced.

For LPD, we ran the trace multiple times scaling d from 0 up to 65,536, which controls how greedily pages are dropped from the cache. For $d = 0$, LPD is equivalent to BL, which does not use the extended FTL and does not drop any pages. For $d = 65536$, it drops all pages from the cache whenever a cache replacement occurs (Line 5 to 11 of Algorithm 1).

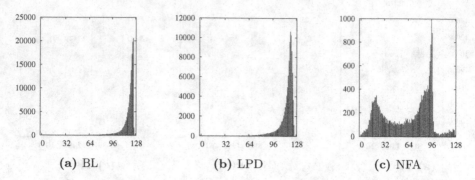

Fig. 7. Distribution of the number of valid pages in garbage-collected blocks. A bar of height y at position x on the x-axis means that it happened y times that a block being garbage-collected contains x valid pages. Note the different scales of the y-axis.

Under the same workload (independent of d), NFA processed 22,106 GCs and achieved a hit ratio of 0.7438. Relative to these values, Fig. 8 plots the number of GCs and the hit ratio of LPD, with d scaled from 0 to 65,536. For $d = 0$, LPD (and BL, due to equivalence) obtained a slightly higher hit ratio than NFA (by 5.84%), however, its number of GCs was much higher than that of NFA (by a factor of five). For $d = 65536$, although LPD's number of GCs was greatly reduced (still higher than that of NFA by 21%), its hit ratio drastically dropped and became only 63.1% of the NFA hit ratio. Note, we could not find a value for $d \in [0, 65536]$ for LPD, such that the number of GCs is lower and the hit ratio is higher than those of NFA at the same time.

4.3 Wear Leveling

So far, we have not discussed other aspects of flash memory management such as wear leveling and bad block management, which are not the focus of our current work, because they can be dealt with using standard techniques proposed in previous works related to FTL. However, fortunately, our approaches seem to

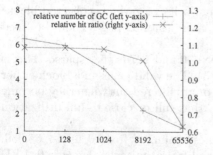

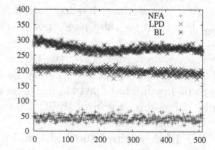

Fig. 8. Number of GCs and hit ratio of LPD relative to NFA, when d is scaled from 0 to 65,536

Fig. 9. Number of erases for each block. Each position on the x-axis refers to a block.

have automatically distributed the erases uniformly to the blocks, as shown in Fig. 9, where the number of erases for each of the 512 blocks is plotted for all three approaches under comparison.

5   Conclusion

In this paper, we studied the problem of efficiently using flash memory for a mid-tier cache in a three-tier storage system. We identified the problems of using flash memory indirectly, which is the common approach taken by previous works. Among these problems, the most important one is the CPM problem, which not only greatly impacts performance, but also shortens the life time of flash-based devices used in the cache. Our basic idea to solve this problem is to drop cold pages proactively and ignore them during GCs. Based on this basic idea, we proposed two approaches, an indirect one and a native one, that effectively handle the problem, as shown by our experiments. The experiments also demonstrated the gravity of the CPM problem, which is ignored so far by typical indirect approaches represented by the baseline. The cache-specific knowledge (e. g., which pages can be dropped) and the direct control over the flash memory (e. g., when is the GC to be started) is the key to the significant performance gain achieved by NFA, the native approach.

We believe that the optimal use of flash memory in a mid-tier cache can only be achieved when the flash memory is managed natively by the cache management software. For similar reasons, system designers should seriously consider how to natively support flash memory in the database software.

Acknowledgement. Yi Ou's work is partially supported by the German Research Foundation and the Carl Zeiss Foundation. Jianliang Xu's work is partially supported by Research Grants Council (RGC) of Hong Kong under grant nos. HKBU211510 and G_HK018/11. The authors are grateful to German Academic Exchange Service (DAAD) for supporting their cooperation. They are also grateful to anonymous referees for valuable comments.

References

1. Gal, E., Toledo, S.: Algorithms and data structures for flash memories. ACM Computing Surveys 37(2), 138–163 (2005)
2. Ban, A.: Flash file system, US Patent 5,404,485 (April 1995)
3. Birrell, A., Isard, M., Thacker, C., Wobber, T.: A design for high-performance flash disks. SIGOPS Oper. Syst. Rev. 41(2), 88–93 (2007)
4. Ban, A.: Flash file system optimized for page-mode flash technologies. US Patent 5,937,425 (October 1999)
5. Estakhri, P., Iman, B.: Moving sequential sectors within a block of information in a flash memory mass storage architecture. US Patent 5,930,815 (July 1999)
6. Kim, J., Kim, J.M., et al.: A space-efficient flash translation layer for CompactFlash systems. IEEE Trans. on Consumer Electronics 48(2), 366–375 (2002)

7. Lee, S.W., Park, D.J., et al.: A log buffer-based flash translation layer using fully-associative sector translation. ACM Trans. Embed. Comput. Syst. 6(3) (July 2007)
8. Gupta, A., Kim, Y., Urgaonkar, B.: DFTL: a flash translation layer employing demand-based selective caching of page-level address mappings. In: Proc. of ASPLOS 2009, pp. 229–240. ACM, New York (2009)
9. Kim, J.K., Lee, H.G., et al.: A PRAM and NAND flash hybrid architecture for high-performance embedded storage subsystems. In: Proc. of EMSOFT 2008, pp. 31–40. ACM, New York (2008)
10. Chung, T., Park, D., Park, S., Lee, D., Lee, S., Song, H.: A survey of flash translation layer. Journal of Systems Architecture 55(5), 332–343 (2009)
11. Chang, L.P., Kuo, T.W., Lo, S.W.: Real-time garbage collection for flash-memory storage systems of real-time embedded systems. ACM Trans. Embed. Comput. Syst. 3(4), 837–863 (2004)
12. INCITS T13: Data Set Management commands proposal for ATA8-ACS2 (revision 6) (2007), http://t13.org/Documents/UploadedDocuments/docs2008/e07154r6-Data_Set_Management_Proposal_for_ATA-ACS2.doc
13. Zhou, Y., Chen, Z., et al.: Second-level buffer cache management. IEEE Trans. on Parallel and Distributed Systems 15(6), 505–519 (2004)
14. Chen, Z., Zhang, Y., et al.: Empirical evaluation of multi-level buffer cache collaboration for storage systems. In: Proc. of SIGMETRICS 2005, pp. 145–156. ACM (2005)
15. Jiang, S., Davis, K., et al.: Coordinated multilevel buffer cache management with consistent access locality quantification. IEEE Trans. on Computers, 95–108 (2007)
16. Gill, B.: On multi-level exclusive caching: offline optimality and why promotions are better than demotions. In: Proc. of FAST 2008, pp. 1–17. USENIX Association (2008)
17. Koltsidas, I., Viglas, S.D.: The case for flash-aware multi-level caching. Technical report, University of Edinburgh (2009)
18. Ou, Y., Härder, T.: Trading Memory for Performance and Energy. In: Xu, J., Yu, G., Zhou, S., Unland, R. (eds.) DASFAA Workshops 2011. LNCS, vol. 6637, pp. 241–253. Springer, Heidelberg (2011)
19. Canim, M., Mihaila, G., et al.: SSD bufferpool extensions for database systems. In: Proc. of VLDB 2010, pp. 1435–1446 (2010)
20. Do, J., DeWitt, D., Zhang, D., Naughton, J., et al.: Turbocharging DBMS buffer pool using SSDs. In: Proc. of SIGMOD 2011, pp. 1113–1124. ACM (2011)
21. Rosenblum, M., Ousterhout, J.K.: The design and implementation of a log-structured file system. ACM Trans. Comput. Syst. 10, 26–52 (1992)
22. Kawaguchi, A., Nishioka, S., Motoda, H.: A flash-memory based file system. In: Proc. of TCON 1995. USENIX Association, Berkeley (1995)
23. On, S.T., Xu, J., et al.: Flag Commit: Supporting efficient transaction recovery on flash-based DBMSs. IEEE Trans. on Knowledge and Data Engineering 99 (2011)
24. Prabhakaran, V., Rodeheffer, T.L., Zhou, L.: Transactional flash. In: Proc. of OSDI 2008, pp. 147–160. USENIX Association, Berkeley (2008)
25. Western Digital Corp.: Specifications for the 150 GB SATA 3.0 Gb/s VelociRaptor drive, model WD1500HLFS (2011), http://wdc.custhelp.com/app/answers/detail/search/1/a_id/2716

Evacuation Planning of Large Buildings
Using Ladders

Alka Bhushan, Nandlal L. Sarda, and P.V. Rami Reddy

GISE Lab., Department of Computer Science and Engineering,
Indian Institute of Technology Bombay,
Mumbai, India
{abhushan,nls,pvrreddy}@cse.iitb.ac.in
http://www.cse.iitb.ac.in

Abstract. Evacuation planning of a building in case of an emergency
has been widely discussed in literature. Most of the existing approaches
consider a building as a static graph with fixed, predefined exits. However, in severe disaster situations, it is desirable to create additional
exits for evacuation purposes. A simple and practical way of creating
additional exits is to place ladders at those locations that can reduce
evacuation time effectively. For large buildings, finding optimal locations
for a limited number of available ladders to utilize them effectively is not
possible without using any systematic approach.

In this paper, we first show that the problem of finding optimal locations for a given number of ladders among all feasible locations to
minimize either the average evacuation time or evacuation egress time is
NP-hard. The feasible locations for ladders are referred as dynamic exit
points and the exits created by ladders are referred as "dynamic exits".
Next, we propose a heuristic for placing a given number of ladders at
dynamic exit points based on the distribution of evacuees in the building. We extend the existing capacity constrained route planner (CCRP)
algorithm for dynamic exit based evacuation planning.

Our model is illustrated by performing a set of experiments and comparing it with existing optimization models that minimize either the
average evacuation time or the evacuation egress time. The results show
that the proposed heuristic produces close to the optimal solution and
has significantly less computational cost as compared to the optimization
models.

Keywords: Evacuation planning, Routing and scheduling, Dynamic
Exits, Building network.

1 Introduction

Building evacuation planning is necessary to move occupants safely from danger
areas to safe areas in case of emergency events such as fire, terrorist attack etc.

Over the years, the evacuation planning problem has been studied as an
optimization problem by various research communities. Typical aims of such

S.W. Liddle et al. (Eds.): DEXA 2012, Part I, LNCS 7446, pp. 71–85, 2012.
© Springer-Verlag Berlin Heidelberg 2012

approaches have been to minimize average evacuation time, minimize evacuation egress time or maximize flow of evacuees. Some of these optimization based models are presented in [2,3,6,10]. These optimization based models produce optimal solutions but are suitable only for small sized buildings due to high computational cost.

Recently, a heuristic, namely capacity constrained route planner (CCRP) has been widely discussed in the literature as a spatio temporal database approach [9,11,12,14,15]. The aim of this approach is to reduce computational time while minimizing evacuation time. The proposed heuristic works efficiently and produces high quality solution.

Most of these approaches consider a building as a static graph with fixed, predefined exits. However, in the event of severe disasters, it is desirable to create additional exits, to reduce the evacuation time. A simple and effective means of creating such additional exits is to create exits by placing ladders. A building potentially may have large number of places such as window, balcony etc., where ladders can be placed. An interesting problem then is to identify the places to use for limited number of available ladders to reduce the evacuation time to an acceptable value.

Usage of ladders in building evacuation planning was first suggested in [4]. Recently, optimization based models have been proposed for various scenarios to incorporate dynamic exits in the evacuation planning [1]. The proposed models are formulated as integer linear programming problems and produces optimal solutions. These models require large memory and have high computational complexity. Also, they need an estimated upper bound on evacuation egress time. If estimated upper bound is less than the actual evacuation egress time then solution becomes infeasible.

The main objectives of this paper are as follows: (1) show that the problem of finding optimal locations for a given number of ladders among all feasible locations to minimize either the average evacuation time or the evacuation egress time is NP-hard (2) present a heuristic approach for placing a given number of ladders at dynamic exit points based on the distribution of evacuees in the building, and (3) extend the existing CCRP for dynamic exit based evacuation planning.

To the best of our knowledge, this paper presents the first efficient heuristic approach for finding an evacuation plan using ladders.

The paper is structured as follows: in Section 2, we briefly describe modeling of a building and a ladder as considered in this paper. In Section 3, we present a formal statement of the problem considered in this paper. In Section 4, we prove that the problem of finding optimal locations for a given number of ladders among all feasible locations to minimize either the average evacuation time or evacuation egress time is NP-hard. In Section 5, we describe integer linear programming formulations for solving the evacuation planning problem using dynamic exits. The heuristics proposed for evacuation planning using ladders are described in Section 6. In Section 7, we present results for various experiments before concluding the paper in Section 8.

2   Modeling of A Building with Ladders

For evacuation planning purposes, generally a building is modeled as a directed graph [2,3,6,10,11], where nodes represent the corridors, lobbies, rooms and intersection points while edges represent stairways, hallways or connections between two nodes. Each node is associated with a *maximum capacity* equal to the maximum number of people that can be accommodated there. Each edge has an *in-take capacity* equal to the maximum number of people that can be admitted per unit of time. Direction of each edge is given by predetermining possible movements of evacuees towards exits, which is possible only if exits are known beforehand. In case of adding additional exits during evacuation process, it is unreasonable to determine again the direction of every edge for a large building.

We suggest the following modifications in the modeling of a building, and the modeling of a ladder for our purposes. A detailed discussion can be found in [1].

1. We consider the edges to be bidirectional unless an edge is specifically restricted to be unidirectional. Since the in-take capacity of an edge is determined by the width of the corresponding edge, an edge can be used in both direction using the same in-take capacity with an additional constraint that the total number of evacuees present on it at any point of time should not be more than its maximum in-take capacity.

2. A ladder can be placed at wide window, balcony etc. Since these places are associated with a room therefore a corresponding node is marked as a *dynamic exit point*.

3. A ladder is modeled as an edge (ladder edge) which connects a dynamic exit point to the ground. Without loss of generality, distinct destination nodes corresponding to the ground are considered for each ladder. These destination nodes are called as *"dynamic exits"*. The following characteristics are associated with each ladder edge.
   - In-take capacity of each ladder is taken to be 1.
   - Travel time of a ladder is a function of length of the ladder which is nothing but a function of height of the corresponding dynamic exit point.
   - Unlike other conventional edges of a building graph, a ladder may not be strong enough to bear load equal to its travel time multiplied by its in-take capacity. Hence, a new parameter denoted as *maximum load* is introduced for a ladder which represents the maximum number of evacuees that can be present on the ladder at any point of time.
   - A ladder may not be set up in zero time and need some time to place at the desired location. Further, each ladder need not require same amount of placement time since it would depend on the position and height of the location. Thus, *placement time* is associated with each ladder edge and a ladder edge becomes active after its placement time.
   - A dynamic exit point may have wide space to place more than one ladder. Thus more than one ladder can be placed at the same location. Further, different types of ladders can be placed at the same location and hence maximum load, placement time and travel time would differ for each

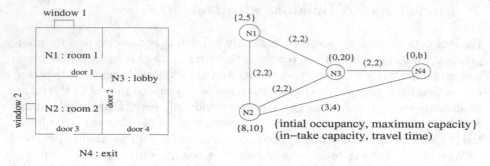

Fig. 1. Building and the corresponding graph for the example

type of ladder. To keep our model simple, we assume that only one type of ladder can be placed at one dynamic exit point. Further, we assume that only one ladder can be placed at a dynamic exit point.

Thus, a ladder edge at a dynamic exit point has the following characteristics:

p_i : ladder placement time
σ_i : maximum load
b_i : ladder in-take capacity $(= 1)$
λ_i : travel time

Example: The example presented in Figure 1 is adapted from [7]. Initial number of evacuees and node capacity is listed next to each node. For example, node 2 has 8 evacuees initially and can accommodate at most 10 evacuees. Further, each edge is labeled with its in-take capacity and travel time. For example, edge (2,4) has in-take capacity 3 and travel time equal to 4. Nodes 1 and 2 are marked as possible dynamic exit points. The characteristics of these dynamic exit points are shown in Table 1.

Table 1. Characteristics of dynamic exit points 1 and 2

Parameters	Node 1	Node 2
ladder placement time (p)	0	1
maximum load (σ)	1	1
ladder in-take capacity (b)	1	1
travel time (λ)	2	2

This example is used later to explain the results of our heuristic algorithm.

3   Problem Statement

Let an undirected graph $G = (N, E)$ and L number of ladders be given for a building, where N is the node set and E is the edge set.

- Each node i is associated with maximum capacity (a_i) and initial occupancy (q_i).
- Each edge (i, j) is associated with in-take capacity (b_{ij}) and travel time (λ_{ij}).
- Each dynamic exit point i is associated with ladder placement time (p_i), maximum load (σ_i), ladder in-take capacity (b_i) and travel time on a ladder (λ_i).

All nodes of non-zero initial occupancy are included in source node set S. Similarly, all destination nodes and dynamic exit points are included in destination node set D and dynamic exit point set P, respectively.

Problem: Find an evacuation plan that finds positions for placing L number of available ladders, a set of source-destination routes, and a schedule of evacuees on each route.

Objective: minimize either the evacuation egress time which is the maximum evacuation time taken by an evacuee or the average evacuation time which is the average time taken by any evacuee.

4   Complexity

We denote the problem of finding optimal dynamic exit points for placing a given L number of ladders to minimize the evacuation egress time as SDEP. A problem can be shown an NP hard problem by reducing a known NP hard problem into the problem. Here, we reduce the k-center problem, an NP-hard problem into the SDEP problem and show that an algorithm which solves SDEP in polynomial time can also solve k-center problem in polynomial time. The k-center problem is shown to be an NP-hard problem in [5].

k-center problem: Given an undirected complete graph $G' = (N', E')$ with node set N', edge set E', and weight on each edge that satisfies triangle inequality, and an integer k, find a set $C \subseteq N'$ such that $|C| \leq k$ and it minimizes the following function

$$\max_{i \in N'} \min_{j \in C} w_{ij}$$

where w_{ij} is the weight of edge (i, j).

Lemma 1. *SDEP is NP-hard.*

Proof. We reduce the k-center problem into SDEP by converting the graph G' into a new graph G'_{new} as follows:

1. Assign one evacuee to each node $i \in N'$.
2. Make each node $i \in N'$ as a dynamic exit point and assign 0 value to placement time and value 1 to travel time, maximum load and in-take capacity.
3. Assign ∞ value to the maximum capacity of each node and in-take capacity of each edge.

Weight of each edge corresponds to its travel time. Since weights of edges satisfy triangle inequality, the edge (i, j) is always the shortest path between the corresponding nodes i and j. Since ladder travel time is 1 unit of time for each dynamic exit point, it will not change the output. Hence, output of SDEP for graph G'_{new} must be the optimal output of k-center problem for graph G'.

Similarly, it can be shown that the problem of selecting dynamic exit points for a given L number of ladders to minimize the average evacuation time is NP-hard by showing a reduction from the k-median problem to the problem. The k-median problem is also a well known NP-hard problem [5].

5   Integer Linear Programming Formulation

Evacuation planning problem has been formulated as a linear programming problem whose objective is to minimize the average evacuation time of each evacuee [2,7]. Triple optimization theorem presented in [8] states that a flow which minimizes average evacuation time also minimizes the evacuation egress time. But, optimal locations for a given number of ladders to minimize the average evacuation time may not be the optimal locations for the given number of ladders to minimize the evacuation egress time. Thus, separate algorithms are required to optimize the each objective. We have presented integer linear programming formulations for each scenarios [1]. We also showed that the formulation which minimizes the average evacuation time for a given number of ladders does not necessarily minimizes the evacuation egress time and vice versa. Our integer linear programming formulations have used the constraints used in the linear programming formulation given in [7] in addition to the new constraints needed for our problem.

6   A Heuristic for Evacuation Planning Using Ladders

We propose an algorithm for placing a given number of ladders at dynamic exit points based on the distribution of evacuees in the building. We need to modify CCRP since the edges representing ladders need to be handled differently from other edges in the building graph. A modified CCRP, namely DCCRP in presented in Section 6.2. We next present an algorithm for selecting dynamic exit points.

6.1   An Algorithm for Selecting Dynamic Exit Points

Our approach is based on the following two observations: (i) high density areas would need more exits, and (ii) a ladder placed at a dynamic exit point would be mainly used by the evacuees located in nearby locations. One simple method to calculate the density for each dynamic exit point is to compute the total number of evacuees located in l-neighborhood. The l-neighborhood of a node i can be defined as

$$l\text{-neighborhood}(i) = \{j | d(i, j) \le l\}$$

Algorithm 1. Calculate density of each dynamic exit point

---

densityFactor(i,j) ← 0 /* for all source nodes i and all dynamic exit points j */
density(j) ← 0 /* for all dynamic exit points j */
sum(i) ← 0 /* for all source nodes i */
for each source node $i \in S$ **do**
   for each dynamic exit point $j \in P$ **do**
      compute shortest travel time travel-time(i,j) from node i to dynamic exit point j
      densityFactor(i,j) = $\frac{1}{\text{travel-time}(i,j)}$
      sum(i) = sum(i) + densityFactor(i,j)
   end for
end for
for each dynamic exit point $j \in P$ **do**
   for each source node $i \in S$ **do**
      density(j) = density(j) + $\frac{q_i \times \text{densityFactor}(i,j)}{\text{sum}(i)}$
   end for
end for

---

where $d(i, j)$ is the shortest path from node i to node j.

But, this method has the following drawbacks.

- For a small value of l, a source node containing large number of evacuees may not be in l-neighborhood of any dynamic exit point and thus all dynamic exit points near it may not be selected. If we increase l then initial occupancy of a source node may be added in large number of dynamic exit points. In both the cases, selection of a dynamic exit point based on l-neighborhood function may not lead to a good selection of dynamic exit points.
- Each source node in l-neighborhood of a dynamic exit point is contributing equally for selection. However, a source node nearer to a dynamic exit point should contribute more than a source node farther from it.
- There can be more than one dynamic exit point who have common source nodes in l-neighborhood and can be selected for placing ladders due to large density which may not be a good selection.

To overcome these problems, we propose a new approach to calculate the density of each dynamic exit point such that each source node contribute a fraction of its initial occupancy to each dynamic exit point based on its distance.

The pseudo code for calculating the density of each dynamic exit point is given in Algorithm 1.

Once density for each dynamic exit point is calculated, L dynamic exit points are selected whose density is largest among all the dynamic exit points. Ladders can be used at these dynamic exit points until all evacuees evacuate the building.

6.2   DCCRP: Modified Capacity Constrained Route Planner

In this section, first we briefly summarize CCRP given in [11]. The CCRP considers an input graph as a directed graph with maximum capacity and initial occupancy on every node, in-take capacity and travel time on every edge, and fixed destination nodes. A supersource node is added and connected to each source node by an edge of infinite capacity and zero travel time. The CCRP iterates over the following steps until no evacuees are left to assign a route:

1. compute a shortest path from the supersource node to a destination node such that minimum available capacity of the path is nonzero and the path is smallest among all shortest paths of non-zero capacity between supersource node and all destination nodes.
2. compute the minimum of maximum available capacity of the path (the actual number of evacuees who will traverse through the path) and the remaining number of evacuees.
3. reserve the capacity of nodes and edges of the selected path for those evacuees.

For our purposes, we modify the CCRP in the following two ways:

1. Since we consider each undirected edge as bidirectional until it is restricted to be bidirectional, the sum of flows on both direction should not be more than the in-take capacity of an edge at any point time. We add this constraint in the CCRP.
2. For a ladder edge, we need to ensure that at any point of time total number of evacuees on it should not be more than its maximum load. This constraint is added by creating a time window of length equal to its travel time which is open when an evacuee enters into the ladder edge. It remains open until its length and allows to enter evacuees on the ladder edge until number of evacuees is less than its load.

Pseudocode of our evacuation planner using ladders is given in Algorithm 2.

Algorithm 2. Evacuation Planner using Ladders

---

   for each dynamic exit point i **do**
      compute density(i) using algorithm 1
   end for
   select L dynamic exit points whose density is largest among all dynamic exit points
   for each selected dynamic exit point i **do**
      create a new exit l_i of infinite capacity
      create a new ladder edge (i, l_i) and set in-take capacity equal to b_i, travel time equal to λ_i, maximum load equal to σ_i
      set start time for the ladder edge to placement time p_i
   end for
   run DCCRP

Consider Example given in Figure 1. When we run our planner for 0 ladder (which is equivalent to CCRP), it gives 6 units of evacuation egress time. When we run our planner for 1 ladder, it reduces evacuation egress time from 6 time units to 5 time units and selects node 2 for placing the ladder. The corresponding evacuation plans are presented in Table 2. Values in parentheses give the time when an evacuee reaches the node written adjacent to it.

Table 2. Evacuation Schedule of the Example 1 for 0 ladder and 1 ladder

	0 Ladder					1 Ladder	
Source	No. of Evac.	Route with Schedule	Dest. Time	Source	No. of Evac.	Route with Schedule	Dest. Time
1	2	1(0.0)–3(2.0)–4(4.0)	4.0	1	2	1(0.0)–3(2.0)–4(4.0)	4.0
2	3	2(0.0)–4(4.0)	4.0	2	1	2(1.0)–ladder-exit(3.0)	3.0
2	3	2(1.0) – 4(5.0)	5.0	2	3	2(0.0)–4(4.0)	4.0
2	2	2(2.0)–4(6.0)	6.0	2	1	2(3.0)–ladder-exit(5.0)	5.0
				2	3	2(1.0) – 4(5.0)	5.0

6.3 Computational Complexity

Our proposed evacuation planner involves the following main steps: (i) computing density of each dynamic exit point and (ii) running DCCRP. Algorithm 1 is used in computing the density for each dynamic exit point. In the algorithm, dominating step is to compute shortest path from each source node to each dynamic exit point which can be done by using single source shortest path algorithm from each source node. If m number of evacuees are present then at most m source nodes will be in the graph. Since single source shortest path algorithm requires $O(|N| \log |E|)$ time, the upper bound on running time for computing the density for all dynamic exit points is $O(m|N| \log |E|)$.

The upper bound of CCRP is $O(m|N| \log |E|)$ which remains the same for the DCCRP.

Thus, the upper bound on the running time of the evacuation planner using ladders is still $O(m|N| \log |E|)$.

7 Experimental Results

We consider our department building named as Kanwal Rekhi (KR) Building to generate test data for our experiments. KR is a five floor building and the corresponding graph contains 342 nodes and 362 edges. The nodes representing rooms with windows are marked as dynamic exit points. Two other buildings are also used from the literature namely Building 101 and Building 42 [2,13]. The Building 101 is a 11 floor building and the corresponding graph contains 56 nodes and 66 edges. The Building 42 is a six floor building and the corresponding graph contains 111 nodes and 140 edges. We have taken the same input data for nodes and edges of these buildings as taken in [1]. We have taken travel time as 1 for each edge whose travel time is 0.

The experiments are performed on a dual-CPU AMD Athlon(tm) II X2 250 processor with 3.8 GB of memory running Linux (Ubuntu 10.04). The language used is java. We use ILOG CPLEX® solver in java to solve the optimization model [1].

We next compare the results of each experiment with the optimization models DEEP-1 and DEEP-2 proposed in [1]. The objective of DEEP-1 is to minimize the average evacuation time while the objective of DEEP-2 is to minimize the evacuation egress time.

7.1 Experiment 1: Varying the Number of Ladders

For this experiment, KR building is used. For the given number of evacuees at each source node, number of ladders are varied from 0 to 50 to compute average evacuation time and evacuation egress time. Total number of evacuees are equal to 300 and number of source nodes are equal to 49. For models DEEP-1 and DEEP-2, we have taken 39 time units as an upper bound on evacuation egress time.

Figure 2 contains two plots depicting the variation in average evacuation time as a function of the number of ladders. While plot A shows average evacuation time obtained by our proposed planner, plot B shows optimal average evacuation time obtained by DEEP-1. From these plots we find that the use of ladders significantly reduces the average evacuation time. For example, in plot B the average evacuation time obtained with 50 ladders is 53 percent lower than the evacuation time obtained without using any ladder. Further we observe that our planner gives results close to the optimal for all the cases with the maximum and average deviation from the optimal times being 10.44 and 4.86 percent.

The variation in evacuation egress time with the number of ladders is displayed in Figure 3. Plot A shows evacuation egress time obtained by our proposed planner while plot C shows optimal evacuation egress time obtained by DEEP-2. Similar to the result in Figure 2, we find that the use of ladders reduces the evacuation egress time by upto 53 percent. Further, compared to the optimal solutions, the maximum deviation in the evacuation egress times computed by our proposed approach is 16.66 percent with the average deviation being 4.97 percent.

Figure 4 shows three plots depicting the running time of each model. Plots A, B, C show the running times obtained by our model, DEEP-1 and DEEP-2

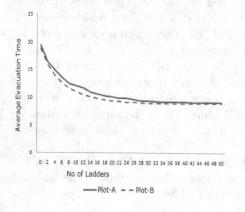

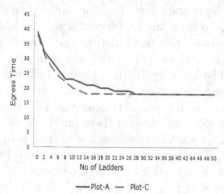

Fig. 2. Average evacuation time vs number of ladders (experiment 1)

Fig. 3. Evacuation egress time vs number of ladders (experiment 1)

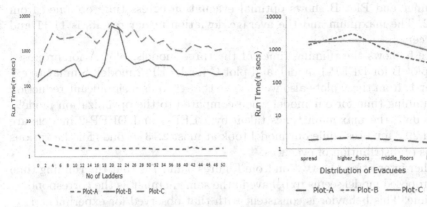

Fig. 4. Running time for varying number of ladders (experiment 1)

Fig. 5. Running time for varying distribution of evacuees (experiment 2)

respectively. From these plots we observe that there is a significant reduction in the running times for our model compared to the optimization models. To illustrate, the maximum running time required by DEEP-1 and DEEP-2 is 80 and 99 minutes respectively while the maximum run time for our model is only 4.02 seconds. Further, from plot A it is interesting to note that the running time of the proposed approach decreases with increasing number of ladders which in turn corresponds to decreasing evacuation egress time (Figure 3).

This experiment shows that our planner gives better results for average evacuation time than evacuation egress time and improves running time significantly as compared to DEEP-1 and DEEP-2.

7.2   Experiment 2: Varying Distribution of Evacuees

For this experiment, KR building is used. For a given number of evacuees and number of ladders, distribution of evacuees is varied and the average evacuation times and the evacuation egress times are computed. In all, 300 evacuees and 20 ladders are taken for the experiment. The upper bound on evacuation egress time for DEEP-1 and DEEP-2 is taken to be 30 time units. Following scenarios are considered for different distributions of 300 evacuees: (i) all evacuees are on lower floors, (ii) all evacuees are on middle floors, (iii) all evacuees are on higher floors, and (iv) the evacuees are spread across all floors. The results are presented in Figures 5-7.

Figure 6 shows the variation in average evacuation time for different distribution of evacuees. In particular, plot A depicts this variation for our proposed planner while plot B is for DEEP-1 model. For all the cases, our heuristic solution is close to the optimal solution, with maximum deviation being 8.19 percent and average deviation being 5.47 percent.

Figure 7 shows two plots on variation in evacuation egress time for different distribution of evacuees. Plot A shows evacuation egress time obtained from our planner and Plot B shows optimal evacuation egress time obtained from DEEP-2. The maximum and the average deviation in our results is 11.11 and 8.33 percent.

Figure 5 shows the running times of the three models: plot A for proposed model, plot B for DEEP-1 model, and plot C for DEEP-2 model. Similar to experiment 1, from these plots also we observe that there is a significant reduction in the running time for our model when compared to the optimization models. In particular, the maximum times taken by DEEP-1 and DEEP-2 models are 49.5 and 29.8 minutes while our model took at most 2.13 seconds for the various scenarios of distribution of evacuees.

Another interesting observation from Figures 5 and 7 is that the running time of our proposed model seems to behave in the same manner as the corresponding egress time. This behavior is consistent with that observed for experiment 1.

7.3   Experiment 3: Varying Number of Nodes

For this experiment, building 101, building 42 and KR building are used. These buildings contain 56, 111 and 342 nodes, and 323, 678 and 504 evacuees respectively. Number of ladders used for all buildings is 4. The upper bound on evacuation egress time for DEEP-1 and DEEP-2 models is 47 time units. The results are presented in Figures 8-10.

Figure 8 shows the variation in average evacuation times obtained by the proposed (plot A) and DEEP-1 (plot B) models. Compared to the optimal times, the maximum and average deviations in the times computed by our proposed approach are 6.62 and 3.07 percent.

Figure 9 shows the variation in evacuation egress time for varying number nodes. As before plot A shows evacuation egress times obtained by our proposed model while plot C shows optimal evacuation egress times obtained by DEEP-2.

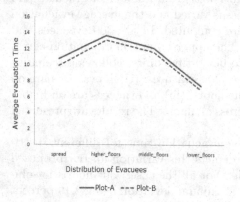

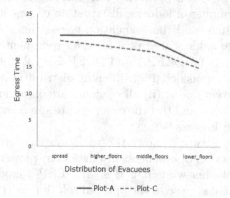

Fig. 6. Average evacuation time vs different distribution of evacuees (experiment 2)

Fig. 7. Evacuation egress time vs different distribution of evacuees (experiment 2)

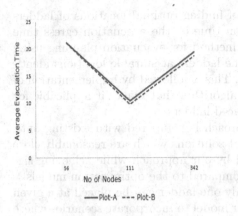

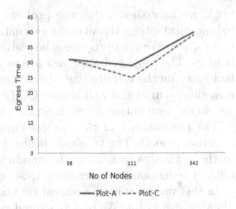

Fig. 8. Average evacuation time vs number of nodes (experiment 3)

Fig. 9. Evacuation egress time vs number of nodes (Experiment 3)

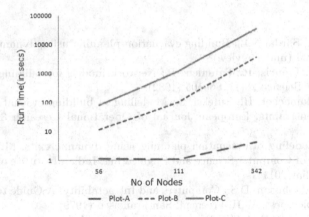

Fig. 10. Run time for varying number of nodes

For this scenario, the maximum and the average deviations in our results, when compared to the optimal time, are 16 and 5.26 percent.

Figure 10 shows the running times for the three models: plot A for the proposed model, and plots B and C for DEEP-1 and DEEP-2 models. Once again, the running times required by our proposed approach are much lower than those required by the optimization models. Further, similar to Experiments 1 and 2, the running time for our proposed model behaves in the same manner as the evacuation egress times.

8   Conclusions

An effective and practical way of creating additional exits in a building to reduce evacuation times is to place ladders at optimally selected locations. In this

work, we have shown that the problem of finding optimal locations of ladders
to minimize either the average evacuation time or the evacuation egress time
is NP-hard. We also propose a heuristic method for evacuation planning using
ladders. The proposed approach places the ladders at suitable locations identified based on the distribution of evacuees. This is followed by implementation of
a modification of the well known CCRP algorithm that makes it applicable for
developing evacuation plans using the placed ladders.

The performance of the proposed approach is compared with existing optimization models. The results indicate that solutions which are reasonably close
to the optimal solutions can be obtained by our approach while achieving significant reduction in the running times compared to the optimization models.

In this work, we have assumed that only one ladder can be placed at a given
dynamic exit point. We plan to extend our model to incorporate scenarios where
multiple ladders can be placed at a dynamic exit point. We also plan to test our
model by applying it to more complex buildings.

References

1. Bhushan, A., Sarda, N.L.: Building evacuation planning using dynamic exits. Fire
   Safety Journal (under review)
2. Chalmet, L., Francis, R., Saunders, P.: Network models for building evacuation.
   Management Science 28(1), 86–105 (1982)
3. Choi, W., Hamacher, H., Tufekci, S.: Modelling of building evacuation problems
   with side constraints. European Journal of Operational Research 35(1), 98–110
   (1988)
4. Garg, P.: Modeling of evacuation planning using dynamic exits, M.Tech Thesis,
   Department of Computer Science and Engineering, Indian Institute of Technology
   Bombay, India (2011)
5. Garey, M.R., Johnson, D.S.: Computers and Intractability: A Guide to the Theory
   of NP-Completeness. W.H. Freeman, San Francisco (1979)
6. Gupta, A., Yadav, P.: SAFE-R a new model to study the evacuation profile of a
   building. Fire Safety Journal 39(7), 539–556 (2004)
7. Hamacher, H., Tjandra, S.: Mathematical modelling of evacuation problems - a
   state of the art. In: Pedestrian and Evacuation Dynamics, pp. 227–226. Springer
   (2002)
8. Jarvis, J.J., Ratliff, H.D.: Some equivalent objectives for dynamic network flow
   problems. Management Science 28(1), 106–109 (1982)
9. Kim, S., George, B., Shekhar, S.: Evacuation route planning: Scalable heuristics.
   In: Proceedings of the 15th Annual ACM International Symposium on Advances
   in Geographic Information Systems, p. 20. ACM (2007)
10. Kisko, T., Francis, R.: Evacnet+: a computer program to determine optimal build-
    ing evacuation plans. Fire Safety Journal 9(2), 211–220 (1985)
11. Lu, Q., George, B., Shekhar, S.: Capacity Constrained Routing Algorithms for
    Evacuation Planning: A Summary of Results. In: Medeiros, C.B., Egenhofer, M.,
    Bertino, E. (eds.) SSTD 2005. LNCS, vol. 3633, pp. 291–307. Springer, Heidelberg
    (2005)

12. Lu, Q., Huang, Y., Shekhar, S.: Evacuation Planning: A Capacity Constrained Routing Approach. In: Chen, H., Miranda, R., Zeng, D.D., Demchak, C.C., Schroeder, J., Madhusudan, T. (eds.) ISI 2003. LNCS, vol. 2665, pp. 111–125. Springer, Heidelberg (2003)
13. Tjandra, S.: Dynamic network optimization with application to the evacuation problem. PhD thesis, Department of Mathematics, University of Kaiserslautern, Kaiserslautern, Germany (2003)
14. Zeng, M., Wang, C.: Evacuation Route Planning Algorithm: Longer Route Preferential. In: Yu, W., He, H., Zhang, N. (eds.) ISNN 2009. LNCS, vol. 5551, pp. 1062–1071. Springer, Heidelberg (2009)
15. Zhou, X., George, B., Kim, S., Wolff, J., Lu, Q., Shekhar, S.: Evacuation planning: A spatial network database approach. IEEE Data Engineering Bulletin (Special Issue on Spatio-temporal databases) 33(2), 26–31 (2010)

A Write Efficient PCM-Aware Sort

Meduri Venkata Vamsikrishna*, Zhan Su, and Kian-Lee Tan

School of Computing, National University of Singapore
meduri@cwi.nl,
{suzhan,tankl}@comp.nus.edu.sg

Abstract. There is an increasing interest in developing Phase Change Memory (PCM) based main memory systems. In order to retain the latency benefits of DRAM, such systems typically have a small DRAM buffer as a part of the main memory. However, for these systems to be widely adopted, limitations of PCM such as low write endurance and expensive writes need to be addressed. In this paper, we propose PCM-aware sorting algorithms that can mitigate writes on PCM by efficient use of the small DRAM buffer. Our performance evaluation shows that the proposed schemes can significantly outperform existing schemes that are oblivious of the PCM.

1 Introduction

Design of database algorithms for modern hardware has always been a prominent research area. Phase Change Memory (PCM) is a latest addition to the list of modern hardware demanding the design of PCM-friendly database algorithms. With PCM being better than flash memory (SSD) in terms of write endurance, read and write latency, focus has shifted to exploring the possibilities of exploiting PCM for databases. Because of the high density and lower power consumption of PCM as compared to DRAM, it is evident that PCM might be an alternative choice for main memory [2,3]. However, as PCM has a low write endurance and high write latency as compared to DRAM, it is essential to design algorithms such that they do not incur too many writes on PCM and thus prevent the hardware from getting worn out soon.

In-memory sorting is a write-intensive operation as it involves huge movement of data in the process of ordering it. Quick sort is the most expensive in terms of data movement though it is the best in time complexity. Selection sort involves fewest data movement but it has quadratic time complexity and also incurs a lot of scans on the data. In this paper, we design write-aware in-memory sorting algorithms on PCM. Like [3], we also use a small DRAM buffer to alleviate PCM writes using efficient data structures. We assume that the data movement between DRAM and PCM can happen seamlessly. This can be achieved through a hardware driven page placement policy that migrates pages between DRAM and PCM [6].

---

* Please note that this work was done while the author was at the National University of Singapore.

S.W. Liddle et al. (Eds.): DEXA 2012, Part I, LNCS 7446, pp. 86–100, 2012.
© Springer-Verlag Berlin Heidelberg 2012

Our first sort algorithm constructs a histogram that allows us to bucketize the in-memory data such that either the depth or width of each bucket is DRAM-size bound. This is an important heuristic that we introduce to make our algorithm write-efficient as well as run-time efficient. Quick sort is employed on buckets that are depth bound and counting sort is used to sort the buckets that are width bound such that minimum writes are incurred on PCM. We further minimize PCM writes aggressively with an improved version of our algorithm. In this variant, we construct the histogram even before the data is read into PCM by sampling data directly from the disk.

If the unsorted data on disk sorted doesn't entirely fit into the PCM at one go, external sort is performed which involves getting the data in chunks to main memory one by one, sorting the chunk, creating the runs on disk and finally merging those runs. We show in the experiments that our algorithm performs well even in the scenario where the entire data is not memory resident.

There have been few works in adapting database algorithms for PCM. In [1], a B+ tree index structure that incurs fewer writes to PCM (which is used as main memory) is proposed. Though the idea to reduce memory writes is beneficial, the B+ tree nodes (at least the leaves) are kept unsorted to obtain fewer modifications during insertion and deletion of data to the index. This leads to more expensive reads. An algorithm to adapt hash join is also proposed in [1] which minimizes the writes from the partitioning phase on PCM by storing the record ids of the tuples (or the differences between consecutive record ids) in the hash partitions. The records are in-place accessed during the join phase using the record ids. The algorithm also aims at achieving fewer cache misses.

PCM-aware sorting is of significance in the context of databases as it is used in many query processing and indexing algorithms. Our work to produce a PCM aware and efficient sorting algorithm can help alleviate the heavy read exchange the existing B+ tree index algorithm in [1] does. It can also be extended to obtain a PCM-aware sort merge join algorithm. To our knowledge, this is the first report work on sorting algorithms in memory-based PCM.

We present our basic and advanced PCM sorting algorithms in Sections 2.3 and 2.4 respectively. We report results of an extensive performance study in Section 3.

2   PCM-Aware Sorting Algorithm

Our goal is to design efficient sorting algorithms that incur as few writes on PCM as possible. As context, we consider external multi-way merge sort that comprises two main phases: (a) generating sorted runs, and (b) merging the runs into a single sorted file. Our key contribution is in the first phase where PCM-aware in-memory sorting is proposed.

For our hybrid architecture, we divide the main memory into three sections. PCM is divided into two partitions - a large chunk to hold the incoming unsorted data and a small chunk for use by a histogram. We refer to the first partition as sort-chunk, and the second partition as hist-chunk. The DRAM forms the third partition.

In the following subsection, we describe the run generation phase. We first look at a naive strategy, afterwhich we present our two proposed solutions.

2.1 A Naive Approach

Under the naive strategy, we fill the sort-chunk with as many tuples from disk as possible. Next, we adopt the scheme in [4] to generate an equi-depth histogram (in hist-chunk) using a single scan of the data in the sort-chunk.

The depth of each bucket in our equi depth histogram will be no greater than DRAM size. Once the histogram is constructed, the input data in the PCM buffers is shuffled according to the bucket ranges in the histogram. That means, the data belonging to each bucket range is brought together such that data belonging to bucket 0 is placed first, bucket 1 next and so on. So the input data is sequentially arranged according to the bucket intervals, still unsorted within each interval. Now each bucket is sent to the DRAM and a quick sort is done upon the bucket data in DRAM. The sorted data will be written to the sorted run on disk directly (assuming DRAM has direct access to the disk). Figure 1 illustrates the manner in which bucket transfer is done from PCM to DRAM.

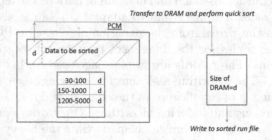

Fig. 1. PCM-aware sorting scheme (naive)

The PCM writes in this scheme will not exceed $2n + sizeof(Histogram)$ (including the disk fetches) where "n" denotes the total size of the input block of data fetched into PCM. This is because, histogram storage on PCM and shuffling of unsorted data according to bucket range contribute to writes.

2.2 Refining the Naive Sorting Scheme

We observe that the naive scheme described above is insensitive to data skew. Where data is skewed, using quicksort is an overkill (and costly). Instead, we can potentially improve the performance by adopting the Counting Sort. Instead of swapping the elements to be sorted, it focusses on the count of each element. If the data range is small enough to fit in memory, a count array is constructed with array index starting from minimum element of the input data to the maximum element. The count array determines the number of elements less than or equal to each data element. Thus it determines the final positions of each data element in the sorted array. Please refer to [10] for complete description of the algorithm.

One of the key disadvantages of Counting Sort is that it is effective only for small data ranges. However, this seemingly disadvantageous trait indeed benefits our scheme. For skewed data, our naive scheme fails to identify the skewedness and constructs several equi depth buckets. Whereas, if we can identify that the frequency of values in a certain range is higher and if the range manages to fit into DRAM, we can replace quick sort with counting sort for the elements of that range.

- The bucket data need not fit into DRAM. Instead if the count array is smaller than DRAM size, it can help us sort many elements in one go, thus saving time.
- This makes our histogram hybrid and compact: either its width or the depth is DRAM-bound.

In the example shown in Figure 2, DRAM size is 50 elements, and PCM holds an unsorted data block of 600 elements. These 600 data elements are distributed over three non-overlapping ranges as 50, 500 and 50 elements. It can be observed that the ranges of the first and third buckets are large spanning over 1000 element width, but the depth is just 50. On the other hand, the middle bucket which has just a range of 50 elements holds 500 elements because of several duplicates. Not paying attention to the skewed range of a small interval holding a whopping 500 values, an equi-depth histogram is constructed by our naive scheme with 12 buckets of 50 elements each since 50 is the available DRAM size. But if we can pay attention to the width of each bucket and strive to make it DRAM-contained, we can achieve a compact histogram with just 3 buckets. This is because the second bucket's width is 50 ($<=$ size of DRAM) and hence a DRAM-contained count array can be constructed for the data in the second bucket. Buckets 1 and 3 are depth-bound, so quick sort is employed to sort them by transferring the bucket data to DRAM.

That means, while constructing a bucket we need to check whether the depth or width of a bucket is hitting the DRAM limit first as input (unsorted) data keeps getting added to a bucket. Whichever parameter hits the limit later is the one we use to decide the bucket boundaries. For example, if the data is sparsely distributed, bucket width crosses DRAM limit easily but it will take a while for its depth to hit the DRAM limit, in that case the bucket is depth-bound.

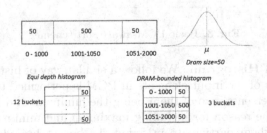

Fig. 2. Compact Histogram achieved using Counting Sort for normal data distribution

In case of densely distributed data, it is the other way round and the bucket becomes width bound. So our histogram is robust to both sparsely and densely distributed data.

One more advantage in having a compact histogram is that, the amount of data that has to be transferred to DRAM is minimal. Whenever there is an opportunity to avoid data transfer, we grab it by employing counting sort. In the worst case where the entire data is uniformly distributed, our hybrid histogram gracefully reduces to an equi-depth histogram since the need to use counting sort arrives only in the case of non-uniformly distributed data. Perhaps, a constraint in using counting sort is that it is basically applicable to sorting integers.

2.3   Algorithm 1: Basic PCM-Aware Sorting

Our scheme consists of the following steps:

  – Construction of the hybrid DRAM-bounded width / depth histogram
  – Shuffling the input data block according to bucket range
  – Since the histogram constructed is not 100% accurate, efforts should be made to combat it
  – Transfer of data / range to DRAM to perform quick / counting sort

Once the hybrid histogram is constructed (Refer Figure 3), the algorithm transfers the bucket data to DRAM and performs quick sort if the depth ≤ DRAM size. On the contrary, if the width ≤ DRAM size, the count array is built in DRAM and accordingly the data is moved within PCM to its final places by looking up to the count array. It should be noted that we do not use a separate output array for counting sort, rather the data on PCM is moved in-place to reach the sorted order.

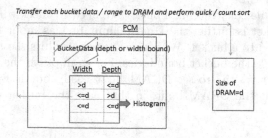

Fig. 3. Basic PCM-aware sort scheme

Construction of Histogram. We follow a similar way of histogram construction as [5]. A scan of the input elements in PCM is performed to uniformly pick up "d" random elements, two of them being the minimum and maximum of the scanned block. The reason for including maximum and minimum values in the sample is to let them participate in fixing the bucket boundaries of the first and the last buckets respectively. Else, we would miss some potential values we need to sort. With "*d*" being the DRAM size, the sampled elements are sorted inside DRAM using quicksort. Now the sorted sample array present in DRAM is scanned to construct the required histogram.

The constructed histogram is for the sample array and it needs to be scaled up for the original PCM resident data. It should be understood that the width (range) of each bucket in the histogram is appropriate because we made sure we did not even miss including the original data's minimum and maximum values while fixing the bucket ranges. The only difference is that, since we are operating on the sampled array, the maximum allowed depth of each bucket (in cases where the width has already exceeded the DRAM limit) is $\frac{d^2}{sizeof|inputelements|}$ as against the regular value of d. Here *inputelements* refers to the original unsorted data residing in the PCM. Since the histogram is constructed on PCM, the depth values of all the buckets are scaled up before storing them in the histogram.

Firstly, PCM writes are incurred by the histogram after its construction (because it is stored on PCM). The number of writes is equal to the size of the histogram.

Kolmogorovs statistic gives a theoretical support to sampling. It fixes a bound of 0.05 error for sample size of 1024 tuples and 740 tuples with confidence 99% and 95% respectively([5]).

Shuffling of Data According to Histogram Buckets. The unsorted data held in the PCM buffer is re-arranged such that all the elements belonging to the same bucket are brought to contiguous locations though unsorted. This is used in our naive scheme as well (refer section 2.1). The easiest way to do this is to create a new array and to transfer the elements from the original array to the new array according to the requirement. But we choose to do in-place shuffling to be memory conscious.

If the total number of PCM elements are "*n*", at max, there can only be "*n*" moves and thus "*n*" integer writes, i.e, $n * sizeof(int)$ byte writes. On the other hand if we are sorting tuples, the number of writes would rather be $n * sizeof(tuple)$.

Appropriating the Hybrid Histogram. We make use of Kolmogorovs statistic to reduce the error in bucket depths, followed by a rigorous bucket correction procedure.

We introduce a tightening factor Δ into the bucket depth to reduce the possibility of errors. As errors in the histogram are induced because of miscalculating bucket depths, we apply the tightening factor to depth bound and not width bound. Instead of DRAM size "*d*", the depth bound is fixed to $(1 - \Delta) * d$, where Δ could range from 0.05 to 0.1 meaning 5% to 10% error is accommodated. Even though this bound tightening is applied, bucket depths can still cross the DRAM size with a small probability. To correct it we apply iterative splitting and shuffling on the erring buckets till we reach a correct hybrid histogram.

The extra writes incurred during histogram correction are dependant on the number of buckets which are corrected. Each time a bucket is split, it incurs a new bucket write on the histogram and the PCM data belonging to that bucket alone is shuffled incurring PCM writes. If "*new*" is the number of new buckets that are created because of additional splits, on an average it leads to $new * (hist\_bucket + PCM\_bucket\_data)$ writes on PCM because of histogram manipulation and bucket shuffling. These writes are few, because on an average, very few buckets are corrected and no bucket from our experiments encounters more than two rounds of additional splits.

Writes During Quick / Counting Sort. As mentioned in section 2.1, quick sort is done in DRAM for those buckets whose depth is DRAM bounded and the sorted elements can be written directly to the disk provided DRAM in the hybrid architecture has direct access to the disk incurring zero PCM writes. In cases where the bucket width is DRAM bounded, the count array is computed in the DRAM. Out of memory consideration, we perform in-place movement of data on PCM looking up to the aggregated count array present in DRAM. So the writes are again at max "n" provided there are "n" data elements. It is clear that because of counting sort, to get the sorted order a linear number of writes happen on PCM. But these writes are minimum and worthwhile given the speed and compact histogram we achieve because of counting sort using our hybrid scheme. Moreover if the data is uniformly distributed, our algorithm automatically reduces to our naive scheme.

Because we use counting sort for width-bound buckets, we achieve some linearity in time complexity. Since counting sort is used only in the case of non-uniform data distribution, suppose "r" buckets are depth bound and "$n - r$" buckets are width bound, the computation goes as $\sum_{i=0}^{r-1} O(n_i \log n_i) + \sum_{i=r}^{n-1} O(n_i + k_i)$ where k_i indicates the width of "i"th bucket as against a relatively expensive $\sum_{i=0}^{n-1} O(n_i \log n_i)$ incurred by a non-skew aware scheme.

2.4   Algorithm 2: Advanced PCM-Aware Sorting

The best running time is achieved by quick sort and the least memory writes are achieved by selection sort. But quick sort is worse in terms of writes because of numerous swaps and selection sort is bad in terms of the huge reads and long sorting time. We try to achieve an algorithm which is close to both the ideal attributes of least writes and running time in our algorithm 1.

In this section, we design an algorithm that improves over the basic PCM aware sort scheme by aggressively reducing the PCM writes. The writes incurred in main memory because of comparisons and swaps have already been reduced using the basic PCM aware algorithm. The aspect that incurs any extra writes (other than disk fetches to PCM) is histogram creation and shuffling. We already ensure that the histogram is always compact. But shuffling is write-intensive. Because our histogram is constructed after fetching the data from disk to memory, we need to shuffle the data in the memory to ensure the unsorted data belonging to each bucket gets collected together. This causes a number of writes directly proportional to the data size that is memory resident, because in the worst case each tuple (assuming that we are sorting tuples) has to move to get to its intended bucket.

Our idea is to construct the histogram even before the data is fetched into main memory (please refer Figure 4(a)), so that once the actual data from disk starts arriving, it can directly go to its respective bucket and thus avoid a shuffle. This can avoid PCM writes totally except for the disk fetch. The efficiency of this approach still depends on the accuracy of the histogram. And the accuracy of the histogram depends on the extent to which sampling helps us. But this incurs a lot of random reads from the disk.

So we adopt two steps to enhance the accuracy of our histogram while keeping the random reads from disk within limit.

– While fetching the data from the disk to DRAM to construct the sampling array for sorting, we drop all attributes in the tuples other than the sorting attribute.
– At the same time, we do not fill the entire DRAM with the sampling array. We just use a small fraction of the DRAM to construct the sample array. So this avoids the overhead of sorting too many tuples beforehand and the overhead of random reads from the disk.

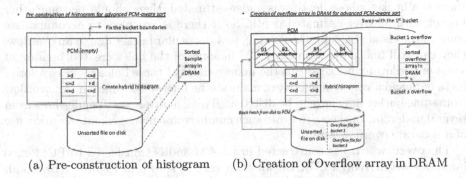

(a) Pre-construction of histogram   (b) Creation of Overflow array in DRAM

Fig. 4. The Advanced PCM-Aware Sort

It is important to note that in this advanced scheme, we do not need to re-scan data and correct the histogram depths. Correction is done as the data is fetched from disk to PCM.

Errors and Correction during Memory Fetch. As the tuples are fetched into PCM, they go into their respective bucket boundaries which are computed from the histogram widths. But in this case, both widths and depths can be at fault unlike algorithm 1 which needs to correct wrong depths. Since the histogram is being pre-constructed, the minimum and maximum elements of the fetched block are not known. So the first bucket and the last bucket in our histogram can be erroneous with respect to the minimum boundary of the first bucket and the maximum range boundary of the last bucket. To accommodate the wrong estimate, we keep extending the border buckets' boundaries as elements come in. If a tuple with its sort attribute value ≤ minimum-boundary of bucket 1, the new tuple will now belong to bucket 1 and the minimum boundary is updated. The same applies to elements arriving beyond the last bucket as well. But there are two problems with this.

- The new elements can cause the depth to go beyond the estimate. If the depths cross DRAM size and if the bucket is depth-bound, sorting the bucket using DRAM is difficult.
- The new element can cause the width to go beyond the estimated boundary. If the bucket is width bound and if counting sort was planned to be employed, the bucket can no longer be sorted using DRAM as the count array is too big to fit into DRAM.

The width error is solved using a slack (Δ) 5% to 10% (same as the tightening factor in Section 2.3) which is sufficient for uniformly distributed data. But in the case of non-uniformly distributed data, the global maximum and minimum which we know from pre-computed database statistics are always included. This is because if the skew in data distribution is extremely high, it is impossible for slack to control width errors.

To correct depth errors, we introduce the construction of an overflow array. If there is a bucket whose depth was under-estimated, there should be some other bucket with an over-estimated depth. So if there is no space to accommodate some of the elements (tuples) belonging to a particular bucket due to an overflow, they can still find a place on the PCM in some of the holes created because of some underflowing buckets. But the management of these holes requires a lot of meta data and a very tedious way of maintaining bucket information. An overflow from some bucket may have to be distributed over multiple small underflows from several underflow buckets. To avoid such cumbersome management, we make use of a separate overflow array.

The overflow array is constructed in DRAM and is transferred to PCM once buckets start arriving to DRAM for their sort as shown in Figure 4(b). So all the tuples which overflow will at first be stored in DRAM in the overflow array. The number of overflow tuples is expected to be under DRAM size, given the accuracy of the histogram is not so bad. Otherwise, the overflow elements spill over to the disk in the individual overflow files, one for each bucket.

There are two cases that need to be handled during the actual sort of the buckets on PCM. The first of them is when the overflow is restricted to main memory alone, which means the overflow array doesn't exceed the DRAM size. In such a case, the overflow array is sorted in DRAM first according to bucket id's and next according to the sort attribute, before transferring it to the PCM. Before the first PCM-resident bucket data moves over to DRAM for sorting, the sorted overflow is also prepared for movement to PCM in the void space. This swap of memory blocks takes place using some vacant input or output buffer as the intermediate swap media. The crucial condition for this to happen is that the first bucket is always depth-bound. Otherwise, if the bucket is width-bound, the count array needs to be constructed in DRAM and the overflow elements have no place to go. It is always theoretically possible to make a bucket depth bound by enumerating elements during histogram construction, than to make it width bound as the values of the elements and distribution are beyond our control. Once the overflow array is transferred to PCM, it will stay there till the sort of all the PCM-resident buckets finishes.

Sorting during in-memory Overflow. If the bucket is width bound, a counting array is constructed over the data present in PCM buffers as well as the overflow array and an in-place counting sort is performed on this aggregate array. In case of a depth bound bucket, the bucket data in DRAM is first quick sorted and merged with the already sorted overflow data belonging to the same bucket residing in PCM (See Figure 5(a)).

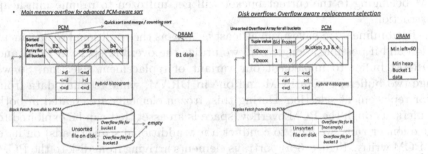

(a) Handling main memory overflow for a bucket.

(b) Overflow aware replacement selection.

Fig. 5. Advanced PCM sort-aware scheme

Sorting for Disk Overflow. For buckets whose overflow elements are on disk, neither quick sort nor counting sort is applicable. So, a variant of replacement selection is applied. Our algorithm is different from the conventional replacement selection as we have two main memory resident data structures that need to be maintained. One is the overflow array and another one is the minimum-heap constructed over the bucket data present in DRAM. But sorting the overflow array by constructing a heap on it is avoided as it incurs additional PCM writes. It is important to note that the overflow array was previously sorted when it was formed inside DRAM. Now it is no longer sorted as new elements keep entering it from the disk. And moreover we have overflow elements belonging to several buckets keep arriving into the PCM-resident overflow array.

So in addition to bucket id (*bid*), we maintain one more field called "*frozen*" which accepts a boolean value (see Figure 5(b)). If an overflow element belonging to any bucket has left the overflow array, it makes space for the disk resident overflow tuples belonging to the current bucket. Though these elements arrive in unsorted order, they belong to the current run as along as they are $\geq$ the minimum value (or root) that has just left the min-heap. Else the element is marked as frozen and belongs to the next run.

Once there is no more space for any more disk elements to enter the PCM overflow array, the bucket elements present in the min-heap of DRAM will start initiating the replacement selection. The root of the min-heap will scan the PCM-resident overflow array to know if any elements ≤ itself belonging to the same bucket are present. If yes, those elements are sent out to the sorted run before the current root. Once the current root leaves the min-heap, another element from the overflow array is brought into the DRAM for min-heap reconstruction and the fetch of overflow elements from disk into the overflow array is repeated provided it has space. If the overflow array is filled with other bucket values and there no more elements in the overflow array belonging to the current bucket, our overflow based replacement selection reduces to the conventional replacement selection. And the elements from the disk are fetched straight into the DRAM heap. Finally the frozen elements in the overflow array and in the DRAM belonging to the current bucket will get unfrozen to resume min-heap reconstruction.

The bottomline is, to have a run at least as long as the traditional replacement selection and if possible a longer run, we utilize the overflow space in PCM and the DRAM buffer to implement our variant of replacement selection. So we manage two buffers, one in PCM and one in DRAM, while fetching data from disk for replacement selection. During this, frozen elements are created in both the buffers. Though the PCM overflow space is an asset that can be exploited to produce longer runs, we have to conduct a few additional scans (reads) on it to avoid PCM writes. Figure 6(a) portrays elements arriving from disk to the PCM buffer and being assigned a frozen value depending on their comparison with the min-heap root present in DRAM. Likewise, Figure 6(b) shows elements from disk directly arriving into the DRAM heap following the absence of unfrozen current bucket elements and space in the PCM buffer.

3   Performance Study

Our PCM simulator is actually DRAM based and uses the measures from [1] to simulate the PCM write latency on DRAM. By default our hybrid memory architecture reserves 3% of the simulators' main memory obtained from actual DRAM for simulated DRAM and 97% behaves like PCM.

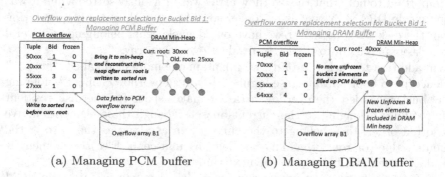

(a) Managing PCM buffer   (b) Managing DRAM buffer

Fig. 6. Managing buffer for replacement selection

The experiments were run on a PC with Intel(R) Xeon(R) 2.33GHz CPU. All the experiments are performed with a default simulator memory size of 1,000,000 tuples with each tuple being 100 bytes wide. The experiments are conducted on data with uniform and non-uniform distribution. The default file size is kept at 1 million tuples.

Our basic and advanced PCM aware sort schemes were compared against quick sort, selection sort and counting sort. Our comparison is with the sort having best running time at one end and other existing sorts which can potentially provide few PCM writes at the other end of the spectrum. For fair comparison we allocate the total PCM + DRAM main memory to all other schemes we compare with. Due to space constraint, we only present representative results here.

3.1 Uniform Distribution

Figure 7(a), 7(b) and 7(c) compare the various schemes in terms of their efficiency (time), total number of PCM writes (after the final merge) and total number of PCM reads (after the final merge) respectively. The data size varies from 1 to 5 million tuples, and the data are uniformly distributed data. Selection sort takes the longest sorting time as against quick sort which is the fastest. Beyond an unsorted disk file size of 5,000,000 tuples, selection sort takes longer than 4 hours of sort time. As expected, quick sort is the weakest in write endurance and incurs lot of writes and reads (during and after merge). Selection sort, though good in writes, performs poorly with respect to PCM reads. It is

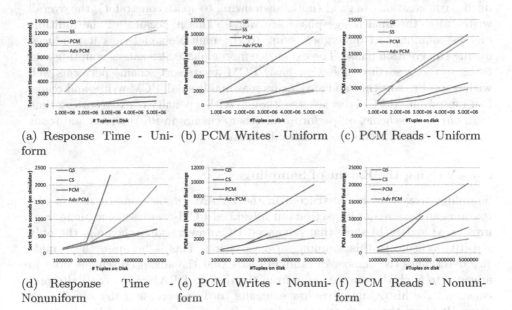

(a) Response Time - Uniform

(b) PCM Writes - Uniform

(c) PCM Reads - Uniform

(d) Response Time - Nonuniform

(e) PCM Writes - Nonuniform

(f) PCM Reads - Nonuniform

Fig. 7. Comparison among various sorting schemes

interesting to note that there are no readings for Counting Sort in Figure 7 owing to its inability to deal with uniform distribution. Counting sort creates numerous runs with tiny run size that it takes too long to finish run generation and merge. The runs are small because, the incoming block has its range too wide in the case of a uniform distribution. So counting sort ends up fetching small blocks to let their count array fit into the main memory.

PCM-aware sort(basic) comes close to quick sort in running time, and also good in reads but it is worse than selection sort with respect to PCM writes. This can be attributed to the writes expended in shuffling of data after the construction of histogram. Since advanced PCM sort was designed with an aim to get rid of those writes, it performs as well as selection sort in terms of writes. It also incurs the least PCM reads.

3.2   Non-uniform Distribution

For non-uniform distribution, we use the normal distribution. The skew is set to a default value by fixing the values of mean and variance. While mean is set to $(min + max)/2$, standard deviation is set to $(max - min)/k$ by basic definitions. In our default settings, $k = 100$. Figure 7(d), 7(e) and 7(f) shows the response time, PCM write counts and PCM read counts for the various schemes. The values of selection sort are not plotted in Figure 7 because the performance of selection sort remains similar to that of the uniform distribution. Counting sort is applicable here as it is sensitive to the distribution of elements. However, in our experiments, counting sort sustains upto an unsorted disk file size of 3 million as it can be seen from the figures. Though the counting sort PCM writes during run creation are good (not shown owing to space constraint), the overall writes after the final merge phase are worse. This is because of the multiple merges counting sort undergoes owing to the multiple small runs it produced during run creation phase. The expensive merge phase also causes counting sort to have a long sorting time. Our basic PCM aware sort scheme performs well with respect to sort time and also incurs reasonably small PCM writes and reads. But our advanced PCM sort scheme, though with a penalty of extra sorting time than our basic scheme, outperforms all other schemes in PCM writes and reads by aggressively reducing them to a minimum.

3.3   Varying the Extent of Sampling

Sampling array that is constructed in the DRAM for the construction of histogram can have major impact on our advanced PCM-aware scheme alone. Figures 8(a) and 8(b) show that advanced PCM aware sort scheme is the only sensitive scheme to this sampling size variation. This is because, the sample size determines the number of random reads that the advanced scheme does in advance to fetch the sample array from the disk to DRAM and eventually preconstruct the histogram. Overflow elements to disk increase if the sampling is poor. But here the overhead in random disk reads to construct a large sample array outweigh the savings obtained by accuracy from it. We can see the decline

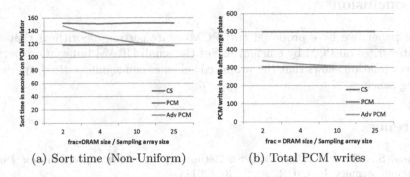

(a) Sort time (Non-Uniform)   (b) Total PCM writes

Fig. 8. Effect of sampling

in the sorting time and PCM writes count of the advanced PCM sorting scheme as the sample array gets smaller. Counting sort is plotted just for reference.

3.4   Varying DRAM Size

DRAM size influences every scheme because it helps alleviate PCM writes in all the schemes. Figure 9 shows the results. We do not present the results for the selection sort because of its long running time. As we know counting sort can perform well in special cases when there is a large DRAM to fit the count array of the entire memory block. As shown in the result, counting sort is faster when a DRAM as large as 20% of the main memory is available. Beyond 10% of DRAM size, the sample array becomes large demanding more random reads from advanced PCM sort. This is unrealistic as such a large DRAM buffer is not good for a hybrid PCM architecture because it defeats the purpose of using PCM as main memory. Basic PCM sort performs well with respect to time, PCM reads and writes. As usual, though advanced PCM sort takes longer to sort, it aggressively reduces PCM reads and writes. Counting sort performs poorly in PCM reads. Realistically, if we consider the interval of 3% to 10% for DRAM buffer size, advanced PCM sort emerges the overall winner.

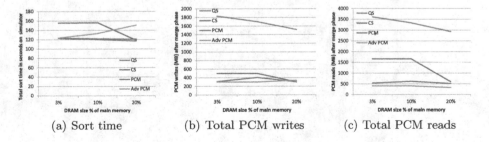

(a) Sort time   (b) Total PCM writes   (c) Total PCM reads

Fig. 9. Effect of DRAM size (non-uniform distribution)

4   Conclusion

In this paper, we have proposed two PCM-aware sorting algorithms that can mitigate writes on PCM by efficient use of the small DRAM buffer. Our performance evaluation shows that the proposed schemes can significantly outperform existing schemes.

References

1. Chen, S., Gibbons, P.B., Nath, S.: Rethinking Database Algorithms for Phase Change Memory. In: CIDR, pp. 21–31 (2011)
2. Lee, B.C., Ipek, E., Mutlu, O., Burger, D.: Architecting phase change memory as a scalable dram alternative. In: ISCA, pp. 2–13 (2009)
3. Qureshi, M.K., Srinivasan, V., Rivers, J.A.: Scalable high performance main memory system using phase-change memory technology. In: International Symposium on Computer Architecture (2009)
4. Mousavi, H., Zaniolo, C.: Fast and accurate computation of equi-depth histograms over data streams. In: EDBT, pp. 69–80 (2011)
5. Piatetsky-Shapiro, G., Connell, C.: Accurate Estimation of the Number of Tuples Satisfying a Condition. In: SIGMOD Conference, pp. 256–276 (1984)
6. Ramos, L.E., Gorbatov, E., Bianchini, R.: Page placement in hybrid memory systems. In: ICSC, pp. 85–95 (2011)
7. Goetz, M.A.: Internal and tape sorting using the replacement-selection technique. Commun. ACM, 201–206 (1963)
8. Graefe, G.: Implementing sorting in database systems. ACM Computing Surveys (CSUR) 38(3) (2006)
9. Knuth, D.: The Art of Computer Programming, 2nd edn. Sorting and Searching, vol. 3. Addison-Wesley (1998)
10. Cormen, T.H., Leiserson, C.E., Rivest, R.L., Stein, C.: Introduction to Algorithms, 2nd edn. (2001)

Performance Analysis of Algorithms to Reason about XML Keys*

Flavio Ferrarotti[1], Sven Hartmann[2], Sebastian Link[3], Mauricio Marin[4], and Emir Muñoz[4,5,**]

[1] Victoria University of Wellington
[2] Clausthal University of Technology
[3] The University of Auckland
[4] Yahoo! Research
[5] University of Santiago de Chile
Flavio.Ferrarotti@vuw.ac.nz

Abstract. Keys are fundamental for database management, independently of the particular data model used. In particular, several notions of XML keys have been proposed over the last decade, and their expressiveness and computational properties have been analyzed in theory. In practice, however, expressive notions of XML keys with good reasoning capabilities have been widely ignored. In this paper we present an efficient implementation of an algorithm that decides the implication problem for a tractable and expressive class of XML keys. We also evaluate the performance of the proposed algorithm, demonstrating that reasoning about expressive notions of XML keys can be done efficiently in practice and scales well. Our work indicates that XML keys as those studied here have great potential for diverse areas such as schema design, query optimization, storage and updates, data exchange and integration. To exemplify this potential, we use the algorithm to calculate non-redundant covers for sets of XML keys, and show that these covers can significantly reduce the number of XML keys against which XML documents must be validated. This can result in enormous time savings.

1 Introduction

The increasing popularity of XML for persistent data storage and data processing has triggered the demand for efficient algorithms to manage XML data. Both industry and academia have long since recognized the importance of keys in XML data management. Over the last decade, several notions of XML keys have been proposed and discussed in the database community. The most influential proposal is due to Buneman et al. [3,4] who defined keys on the basis of an XML

---

* This research is supported by the Marsden Fund Council from Government funding, administered by the Royal Society of New Zealand.
** The contribution of this author was based on his Master's thesis, which was supported by grants from the University of Santiago de Chile and Yahoo! Labs.

S.W. Liddle et al. (Eds.): DEXA 2012, Part I, LNCS 7446, pp. 101–115, 2012.
© Springer-Verlag Berlin Heidelberg 2012

tree model similar to the one suggested by DOM [1] and XPath [6]. While Buneman et al. studied keys as a concept orthogonal to schema specification (such
as DTD or XSD), their proposal has been adopted by the W3C for the XML
Schema standard [15] subject to some minor, though essential modifications (see
[2] for a discussion). Today, all major XML-enabled DBMS, XML parsers and
editors (such as XMLSpy) support keys.

Example 1. Figure 1 shows an XML tree, in which nodes are annotated by their
type: E for element nodes, A for attribute nodes, and S for text nodes. For
the data represented in Figure 1 we have the following keys: (a) A *project* node
is identified by *pname*, no matter where the *project* node appears in the document. (b) A *team* node can be identified by *tname* relatively to a *project* node.
(c) Within any given subtree rooted at *team*, an *employee* node is identified by
name. The first key is an example of an *absolute key* since it must hold globally
throughout the entire tree. The last two are examples of *relative keys* since they
hold locally within some subtrees. Note that a given team of employees can work
on several projects and thus a *team* node cannot be identified in the entire tree
by its *tname*. However, it holds locally within each subtree rooted at a *project*
node. Similarly, a given employee can work on different teams and thus cannot
be identified in the entire tree by its *name*.   □

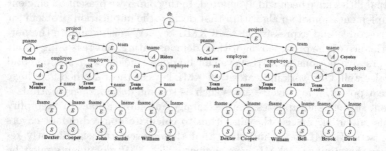

Fig. 1. An XML tree representing an XML document

For relational data, keys have been widely used to improve the performance of
many perennial tasks in database management, ranging from consistency checking to query answering. The hope is that keys will turn out to be equally beneficial for XML. One of the most fundamental questions on keys is that of logical
implication, that is, deciding if a new key holds given a set of known keys. Among
other things, this is important for minimizing the cost of validating that an XML
document satisfies a set of keys gathered as business rules during requirements
engineering.

Example 2. Suppose, the database designer has already specified the keys (a),
(b) and (c). Now she considers a further key (d) which expresses that a *project*
node can be identified by its child nodes *pname* and *team*. By key (a) one already
knows that a *project* node can also be identified by just its *pname*. It is easy

to see that (a) actually implies (d), in the sense that every XML tree that satisfies (a) also satisfies (d). Thus, instead of checking whether an XML tree T satisfies (a) and (d), we could just check whether T satisfies (a). We would like to emphasize that *project* nodes have complex content. Thus, checking whether two *project* nodes in T violate (d) is quite costly in terms of time, since it involves testing whether the subtrees rooted at their *team* nodes are isomorphic to one another with the identity on string values. In contrast, checking whether two *project* nodes violate (a) only involves checking equality on text of their respective *pname* attribute nodes.   □

Example 3. No less important, the implication of XML keys is of interest for semantically rich data exchange. Suppose, the company wants to share part of their project data with a business partner. For that they generate a view over the XML tree T but skip the *lname* nodes for the sake of privacy. Thus, key (c) is no longer meaningful. To provide the business partner with relevant semantic information it should be checked whether the specified keys allow one to conclude a further key stating that an *employee* node is identified by their remaining descendant nodes *fname* and *role* within any given *team* subtree.   □

The definition of keys adopted by the W3C for XML Schema [15] is currently the industry standard for specifying keys. However, Arenas et al. [2] have shown the computational intractability of the associated consistency problem, i.e., the question whether there exists an XML document that conforms to a given XSD and satisfies the specified keys. A further issue pointed out by Buneman et al. [3] is the fact that XML Schema restricts value equality to string-valued data items. But there are cases in which keys are not so restricted (see Section 7.1 of [3] for discussion). In particular, keys (c) and (d) in our examples utilize a less restricted notion of equality, since they require to test equality between *name* nodes and *team* nodes, respectively, none of which are string-valued. On the other hand, the expressiveness and computational properties of XML keys with good reasoning capabilities have been deeply studied from a theoretical perspective [3,4,9]. In practice, however, expressive yet tractable notions of XML keys have been ignored so far.

Aiming to fill this gap between theory and practice, we initiate in this work an empirical study of an expressive XML key fragment, namely the fragment of XML keys with nonempty sets of simple key paths. As shown in [4,9], automated reasoning about this XML key fragment can be done efficiently, in theoretical terms. Our work confirms this fact in practice. Incidentally, note that all the examples of XML keys described above belong to this fragment.

In this paper, we describe an efficient implementation of an algorithm that decides the implication problem for an expressive fragment of XML keys and thoroughly evaluate its performance. Our performance tests give first empirical evidence that reasoning about expressive notions of XML keys is practically efficient and scales well. Our work indicates that XML keys have great potential for database management tasks similar to their counterparts for relational data.

Exploiting our algorithm we compute non-redundant covers for sets of XML keys. A set Σ of keys is non-redundant if there is no key σ in Σ such that σ is implied by $\Sigma - \{\sigma\}$. Thus, considering such covers has the potential to reduce significantly the number of keys against which an XML document must be validated. This can result in enormous time savings. Our experiments show that the time to compute a cover for a given set of keys is just a small fraction of the average time needed to validate an XML document against a single key. Surprisingly, even though several algorithms that validate XML documents against sets of certain XML keys have been proposed and tested with promising results (see e.g. [5,12]), none of them makes use of the reasoning capabilities of XML keys as proposed in our work.

The paper is organized as follows. We recall basic notions in Section 2, including the central notion of XML keys which is used through this work. In Section 3, we present the algorithm for deciding XML key implication, and describe an implementation thereof in Section 4. In Section 5, we discuss how this implementation can be reused to speed up the validation of XML documents against sets of XML keys. Section 6 summarizes experimental results obtained from applying our implementations to publicly available XML data, including DBLP, the SIGMOD Record, and the Mondial database. We conclude the paper in Section 7 with final remarks.

2   Keys for XML

We use the common representation of XML data as ordered, node-labelled trees. Thus, an *XML tree* is a 6-tuple $T = (V, lab, ele, att, val, r)$ where V is a set of nodes, and lab is a mapping $V \to \mathcal{L} = \mathbf{E} \cup \mathbf{A} \cup \{S\}$ assigning a label to every node in V. A node $v \in V$ is an *element node* if $lab(v) \in \mathbf{E}$, an *attribute node* if $lab(v) \in \mathbf{A}$, and a *text node* if $lab(v) = S$. Here $\mathbf{E} \cup \mathbf{A} \cup \{S\}$ form a partition of $\mathcal{L}$. Moreover, ele and att are partial mappings defining the edge relation of T: for any node $v \in V$, if v is an element node, then $ele(v)$ is a list of element and text nodes in V and $att(v)$ is a set of attribute nodes in V. The partial mapping val assigns a string to each attribute and text node. Finally, r is the unique and distinguished root node.

The XML keys studied in this work are defined using the path language PL consisting of expressions given by the following grammar: $Q \to \ell \mid \varepsilon \mid Q.Q \mid \_^*$. Here $\ell \in \mathcal{L}$ is any label, ε denotes the empty path expression, "." denotes the concatenation of two path expressions, and "$\_^*$" denotes the *variable length "don't care"* wildcard. Let Q be a word from PL. A path $v_1, \ldots, v_n$ in an XML tree T is called a Q-path if $lab(v_1). \cdots .lab(v_n)$ can be obtained from Q by replacing variable length wildcards in Q by words from PL. For a node $v \in V$, $v[\![Q]\!]$ denotes the set of nodes in T that are reachable from v following any Q-path. We use $[\![Q]\!]$ as an abbreviation for $r[\![Q]\!]$ where r is the root node of T. We denote as PL_s the subset of PL expressions containing all words over the alphabet $\mathcal{L}$, i.e., we do not allow wildcards in PL_s expressions. $Q \in PL$ is *valid* if it does not have labels $\ell \in \mathbf{A}$ or $\ell = S$ in a position other than the last one.

We define formally the concept of XML key following [4]. For that, we need the concept value equality. Two nodes $u, v \in V$ are *value equal*, denoted by $u =_v v$, iff the subtrees rooted at u and v are isomorphic by an isomorphism that is the identity on string values. As an example, the third and fifth *employee*-nodes are not value equal while their respective child nodes labeled as *name* are.

Definition 1. *An XML key φ in the class $\mathcal{K}$ is an expression of the form $(Q_\varphi, (Q'_\varphi, \{P_1^\varphi, \ldots, P_{k_\varphi}^\varphi\}))$ where $k_\varphi \geq 1$, Q_φ and Q'_φ are PL expressions, and for all $i = 1, \ldots, k_\varphi$, P_i^φ is a PL_s expressions such that $Q_\varphi.Q'_\varphi.P_i^\varphi$ is a valid PL expression. An XML tree T satisfies the key $(Q, (Q', \{P_1, \ldots, P_k\}))$ if and only if for every node $q \in [\![Q]\!]$ and all nodes $q'_1, q'_2 \in q[\![Q']\!]$ such that there are nodes $x_i \in q'_1[\![Q_i]\!], y_i \in q'_2[\![P_i]\!]$ with $x_i =_v y_i$ for all $i = 1, \ldots, k$, then $q'_1 = q'_2$. Therefore, Q_φ is called the* context path, *Q'_φ is called the* target path, *and $P_1^\varphi, \ldots, P_{k_\varphi}^\varphi$ are called the* key paths *of φ.*

In particular, the four keys described informally in the introduction, belong to this class and can be expressed formally as follows: (a) $(\varepsilon, (project, \{pname\}))$; (b) $(project, (team\{tname\}))$; (c) $(\_^*.team, (employee, \{name\}))$; (d) $(\varepsilon, (project, \{pname, team\}))$.

3   Deciding XML Key Implication

Let $\Sigma \cup \{\varphi\}$ be a finite set of XML keys in a class $\mathcal{C}$. We say that Σ *implies* φ, denoted by $\Sigma \models \varphi$, if and only if every finite XML tree T that satisfies all $\sigma \in \Sigma$ also satisfies φ. The *implication problem* for $\mathcal{C}$ is to decide, given any finite set $\Sigma \cup \{\varphi\}$ of keys in $\mathcal{C}$, whether $\Sigma \models \varphi$.

A finite axiomatization for the implication of keys in the class of XML keys with nonempty sets of simple key paths $\mathcal{K}$, was established in [9]. The completeness proof of this axiomatization is based on a characterization of key implication in terms of the reachability problem for fixed nodes in a suitable digraph. This characterization, together with the efficient evaluation of Core XPath [8], resulted in a compact algorithm to decide XML key implication in time quadratic in the size of the input key. This algorithm, which is described next, forms the basis for our implementation. We need the following technical concepts.

Mini-trees and Witness Graphs. Let $\Sigma \cup \{\varphi\}$ be a finite set of keys in $\mathcal{K}$. Let $\mathcal{L}_{\Sigma,\varphi}$ denote the set of all labels $\ell \in \mathcal{L}$ that occur in path expressions of keys in $\Sigma \cup \{\varphi\}$, and fix a label $\ell_0 \in \mathbf{E} - \mathcal{L}_{\Sigma,\varphi}$. Let O_φ and O'_φ be the PL_s expressions obtained from the PL expressions Q_φ and Q'_φ, respectively, by replacing each wildcard "$\_^*$" by ℓ_0. Let p be an O_φ-path from a node r_φ to a node q_φ, let p' be an O'_φ-path from a node r'_φ to a node q'_φ and, for each $i = 1, \ldots, k_\varphi$, let p_i be a P_i^φ-path from a node r_i^φ to a node x_i^φ, such that the paths $p, p', p_1, \ldots, p_{k_\varphi}$ are mutually node-disjoint. From the paths $p, p', p_1, \ldots, p_{k_\varphi}$ we obtain the *mini-tree* $T_{\Sigma,\varphi}$ by identifying the node r'_φ with q_φ, and by identifying each of the nodes r_i^φ with q'_φ. The *marking* of the mini-tree $T_{\Sigma,\varphi}$ is a subset $\mathcal{M}$ of the node set of $T_{\Sigma,\varphi}$: if for all $i = 1, \ldots, k_\varphi$ we have $P_i^\varphi \neq \varepsilon$, then $\mathcal{M}$ consists of the leaves of $T_{\Sigma,\varphi}$, and otherwise $\mathcal{M}$ consists of all descendant nodes of q'_φ in $T_{\Sigma,\varphi}$.

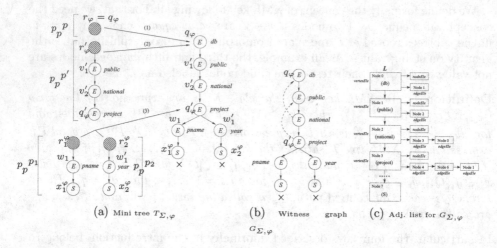

(a) Mini tree $T_{\Sigma,\varphi}$ (b) Witness   graph   (c) Adj. list for $G_{\Sigma,\varphi}$

$G_{\Sigma,\varphi}$

Fig. 2. Mini-tree, Witness-graph and Adjacency list

Example 4. Let $\Sigma = \{\sigma_1, \sigma_2\}$ where σ_1 and σ_2 are the XML keys $(\varepsilon, (public._\text{-}^*,$ $\{project.pname.S, project.year.S\}))$ and $(public, (\_^*.project, \{pname.S, year.S\}))$, respectively. Let $\varphi = (\varepsilon, (public._\text{-}^*.project, \{pname.S, year.S\}))$. The construction of the mini-tree $T_{\Sigma,\varphi}$ is schematized in Figure 2 (a).

The mini-trees are used in the algorithm as a base to calculate the impact of a key in Σ on a possible counter-example tree for the implication of φ by Σ. To distinguish keys that have an impact from those that do not, the following notion of *applicability* is needed. Let $T_{\Sigma,\varphi}$ be the mini-tree of the key φ with respect to Σ, and let $\mathcal{M}$ be its marking. A key σ is said to be *applicable* to φ if and only if there are nodes $w_\sigma \in [\![Q_\sigma]\!]$ and $w'_\sigma \in w_\sigma[\![Q'_\sigma]\!]$ in $T_{\Sigma,\varphi}$ such that $w'_\sigma[\![P_i^\sigma]\!] \cap \mathcal{M} \neq \emptyset$ for all $i = 1, \ldots, k_\sigma$. We say that w_σ and w'_σ *witness* the applicability of σ to φ.

We define the *witness graph* $G_{\Sigma,\varphi}$ as the node-labeled digraph obtained from $T_{\Sigma,\varphi}$ by inserting additional edges: for each key $\sigma \in \Sigma$ that is applicable to φ and for each pair of nodes $w_\sigma \in [\![Q_\sigma]\!]$ and $w'_\sigma \in w_\sigma[\![Q'_\sigma]\!]$ that witness the applicability of σ to φ, $G_{\Sigma,\varphi}$ contains the directed edge (w'_σ, w_σ) from w'_σ to w_σ.

Example 5. Let Σ and φ be as in Example 4. Both keys in Σ are applicable to φ. The witness graph $G_{\Sigma,\varphi}$ is shown in Figure 2 (b). It contains a witness edge from *national* to *db* that arises from σ_1 and a witness edge from *project* to *public* that arises from σ_2.

The algorithm. Algorithm 1 decides XML key implication. Its correctness is an immediate consequence of Theorem 1.

Theorem 1. ([9]) *Let $\Sigma \cup \{\varphi\}$ be a finite set of keys in the class $\mathcal{K}$. We have $\Sigma \models \varphi$ if and only if q_φ is reachable from q'_φ in $G_{\Sigma,\varphi}$.*

Algorithm 1. (XML key implication in $\mathcal{K}$)

Input: finite set of XML keys $\Sigma \cup \{\varphi\}$ in $\mathcal{K}$
Output: yes, if $\Sigma \models \varphi$; no, otherwise
1: Construct $G_{\Sigma,\varphi}$ for Σ and φ;
2: if q_φ is reachable from q'_φ in G **then return** yes;
   else return no; **end if**

---

4 An Efficient Implementation

In this section we discuss our implementation of Algorithm 1 and analyze its theoretical complexity. The implementation was developed in C++ using *gcc* version 4.4.3 from the GNU compiler collection.

Data Structures. We need data structures suitable to represent mini-trees and witness-graphs. The obvious candidates are adjacency matrices and adjacency lists. Since the algorithm does not require frequent determination of edge existence, we choose the latter in order to minimize the memory requirements. In our implementation, a mini-tree $T_{\Sigma,\varphi}$ is represented by using a list L of length $n = |V|$ where V is the vertex set of $T_{\Sigma,\varphi}$. Each element $e_i \in L$ is represented by an object of type *vertexEle* that has a pointer to the adjacency list of the i-th vertex v_i in some fixed enumeration of the vertices in V, a pointer to the data component of the vertex v_i, and a pointer to the next element e_{i+1} in the list. In turn, the data component of a vertex v_i is represented by an object of type *nodeEle*, and an element in the adjacency list of a vertex v_i is represented by an object of type *edgeEle*. An object of type *nodeEle* has an *id* component that uniquely identifies v_i, a *label* component with the label of v_i, a flag *visited*, and a *type* component with the type E (element), A (attribute) or S (PCDATA) of v_i. An object of type *edgeEle* has a pointer to an object of type *vertexEle* and a pointer to the next object of type *edgeEle* in the adjacency list. Witness graphs are represented likewise. Figure 2(b) shows a witness graph and Figure 2(c) a corresponding representation using adjacency lists.

The Implementation. We implemented Step 1 of Algorithm 1, using the following strategy:

  i. Construct $T_{\Sigma,\varphi}$;
 ii. Determine the marking of $T_{\Sigma,\varphi}$;
iii. For each $\sigma \in \Sigma$, add the edge (w'_σ, w_σ) to $T_{\Sigma,\varphi}$ whenever w_σ and w'_σ witness the applicability of σ to φ.

Substep (i) involves constructing the mini-tree $T_{\Sigma,\varphi}$ using the data structures defined at the beginning of this section. Note that we can find a label ℓ_0 that is not among the labels used in the XML keys in $\Sigma \cup \{\varphi\}$ in time $\sum_{\sigma_i \in \Sigma} |\sigma_i| + |\varphi|$, where $|\sigma_i|$ and $|\varphi|$ denote the sum of the lengths of all path expressions in σ_i and φ, respectively. Once we have got a suitable label, ℓ_0, $T_{\Sigma,\varphi}$ can be built in time $\mathcal{O}(|\varphi|)$, since the mini-tree $T_{\Sigma,\varphi}$ has only $|\varphi| + 1$ nodes.

Regarding Substep (ii), if $P_i^\varphi \neq \varepsilon$ we can determine the marking of the mini-tree $T_{\Sigma,\varphi}$ by simply traversing the list L marking the nodes whose adjacency list is empty. Note that those nodes correspond to leaves in $T_{\Sigma,\varphi}$. Otherwise, we mark all nodes in the adjacency list of the element e_i in L that represents q'_φ, and recursively mark all descendants of those nodes. This step takes $\mathcal{O}(|\varphi|)$ time.

In principle, Substep (iii) requires, for each $\sigma \in \Sigma$, to evaluate $w'_\sigma[P_i^\sigma]$ for $i = 1, \ldots, k_\sigma$, for all $w'_\sigma \in w_\sigma[Q'_\sigma]$ and all $w_\sigma \in [Q_\sigma]$. However, we do not need to determine all witness edges (w', w) to decide whether q_φ is reachable from q'_φ in the witness graph $G_{\Sigma,\varphi}$. Let W'_σ be the set of all nodes w' in $T_{\Sigma,\varphi}$ for which there exists some node w in $T_{\Sigma,\varphi}$ such that w and w' witness the applicability of σ to φ. Further, for each $w' \in W'_\sigma$, let $W_\sigma(w')$ be the set of all nodes w in $T_{\Sigma,\varphi}$ such that w and w' witness the applicability of σ to φ. The witness edges are just the pairs (w', w) with $w' \in W'_\sigma$ and $w \in W_\sigma(w')$. As shown in [9], it is not necessary to determine the entire set $W_\sigma(w')$ for each $w' \in W_\sigma$. We can actually restrict ourselves to the top-most ancestor of q'_φ in $T_{\Sigma,\varphi}$ that belongs to $W_\sigma(w')$, which we denote by $w_\sigma^{top}(w')$ (if it exists).

So, we need first to determine W'_σ, and then, for each $w' \in W'_\sigma$, we need to determine $w_\sigma^{top}(w')$ (if it exists). By definition, W'_σ consists of all nodes $w' \in [Q_\sigma.Q'_\sigma]$ in $T_{\Sigma,\varphi}$ such that, for each $i = 1, \ldots, k_\sigma$, there is a marked node in $w'[P_i^\sigma]$. Since a query of the form $v[Q]$ is a Core XPath query and can be evaluated on a node-labelled tree T in $\mathcal{O}(|T| \times |Q|)$ time, it follows that $[Q_\sigma.Q'_\sigma]$ can be evaluated in $T_{\Sigma,\varphi}$ in $\mathcal{O}(|\varphi| \times |Q_\sigma.Q'_\sigma|)$ time. Next, fix some $i \in \{1, \ldots, k_\sigma\}$. Let v be a marked node, and let u denote the ancestor of v that resides $|P_i^\sigma|$ levels atop of v in $T_{\Sigma,\varphi}$ (if it exists). We can then check whether $v \in u[P_i^\sigma]$, that is, whether the unique path from u to v is a P_i^σ-path. This can be done in $\mathcal{O}(\min\{|P_i^\sigma|, |\varphi|\})$ time, since P_i^σ is a PL_s expression. By inspecting all nodes $v \in \mathcal{M}$, we obtain the set U_i^σ of all nodes u in $T_{\Sigma,\varphi}$ for which $u[P_i^\sigma] \cap \mathcal{M} \neq \emptyset$. Overall, this takes $\mathcal{O}(|\mathcal{M}| \times |P_i^\sigma|\})$ time. Since W'_σ is the intersection of $[Q_\sigma.Q'_\sigma]$ with the sets U_i^σ, $i = 1, \ldots, k_\sigma$, we get that W'_σ can be determined in $\mathcal{O}(|\varphi| \times |\sigma|)$ time. Regarding $w_\sigma^{top}(w')$, note that if Q'_σ is a PL_s expression, then $w_\sigma^{top}(w')$ is the node $|Q'_\sigma|$ levels atop of w' in $T_{\Sigma,\varphi}$. Otherwise Q'_σ contains a _*, and thus has the form $A.\_^*.B$ where A is a PL_s expression and B is a PL expression. In this case, as shown in [9], $w_\sigma^{top}(w')$ is the top-most ancestor w of q'_φ in $T_{\Sigma,\varphi}$ that belongs to $[Q_\sigma]$ and for which $w[A]$ is non-empty. In particular, $w_\sigma^{top}(w')$ is independent from the choice of w' in W'_σ. Thus, we propose Algorithm 2 to determine $w_\sigma^{top}(w')$ for a given node w'.

Since $[Q_\sigma.A]$ can be evaluated in $\mathcal{O}(|\varphi| \times |Q_\sigma.A|)$ time, we can conclude from the previous algorithm that $w_\sigma^{top}(w')$ for a given w' can be determined in $\mathcal{O}(|\varphi| \times |Q_\sigma.A|)$ time. Thus, it takes us $\mathcal{O}(|\varphi| \times |\sigma|)$ time to determine all the witness edges arising from σ that are needed for deciding the reachability of q_φ from q'_φ in $G_{\Sigma,\varphi}$. Finally, Step 2 of Algorithm 1 can be implemented by applying a depth-first search algorithm to $G_{\Sigma,\varphi}$ with root q'_φ. This algorithm works in time linear in the number of edges of $G_{\Sigma,\varphi}$ [11]. Over all, our implementation can decide the implication problem $\Sigma \models \varphi$ in $\mathcal{O}(|\varphi| \times (\sum_{\sigma_i \in \Sigma} |\sigma_i| + |\varphi|))$ time.

Algorithm 2. (Determine $w_\sigma^{top}(w')$)

Input: a mini-tree $T_{\Sigma,\varphi}$, a set W'_σ, and a node $w' \in W'_\sigma$.

Output: $w_\sigma^{top}(w')$

1: **if** Q'_σ is a PL_s expression **then**
2:    **return** The node $|Q'_\sigma|$ levels atop of w' in $T_{\Sigma,\varphi}$
3: **else**
4:    Determine the set $[\![Q_\sigma.A]\!]$ of nodes in $T_{\Sigma,\varphi}$
5:    **if** $[\![Q_\sigma.A]\!] \neq \emptyset$ **then**
6:       Choose a topmost node v
7:       Select the node w that is $|A|$ levels atop of v in $T_{\Sigma,\varphi}$
8:       **if** w is an ancestor of q'_φ **then**
9:          **return** w
10:      **else**
11:         **return** $\perp$ //$w_\sigma^{top}(w')$ does not exist.
12:      **end if**
13:    **end if**
14: **end if**

---

5 Applying XML Key Reasoning to Document Validation

Fast algorithms for the validation of XML documents against keys are crucial to ensure the consistency and semantic correctness of data stored in databases or exchanged between applications. In this section we explain how our implementation of the implication algorithm for XML keys can be used to compute non-redundant cover sets of XML keys, which in turn can be used to significantly speed up the process of XML document validation against sets of XML keys. This is, up to our knowledge, the first time that the reasoning capabilities of XML keys are used in this context.

Cover Sets for XML Keys. We define the concept of *cover* set of XML keys following the notion given in [13] for functional dependencies in the relational model.

Definition 2. *Let* Σ^* *denote the set of all XML keys implied by a given set* Σ. *Two sets* Σ_1 *and* Σ_2 *of XML keys are* equivalent, *denoted by* $\Sigma_1 \equiv \Sigma_2$, *if* $\Sigma_1^* = \Sigma_2^*$. *If* Σ_1 *and* Σ_2 *are equivalent we call them a* cover *of one another. This means that* Σ_1 *and* Σ_2 *imply exactly the same XML keys.*

For all cover Σ_2 of Σ_1, if an XML tree T_D satisfies Σ_2 ($T_D \models \Sigma_2$), then $T_D \models \Sigma_1$ too. If $\Sigma_1 \equiv \Sigma_2$, then for each XML key ψ in Σ_1^*, $\Sigma_2 \models \psi$, because $\Sigma_2^* = \Sigma_1^*$. In particular, $\Sigma_2 \models \psi$ for each key ψ in Σ_1.

Definition 3. *A set* Σ_2 *of XML keys is* non-redundant *if it is not equivalent to any of its proper subsets.* Σ_2 *is a* non-redundant cover *for a set* Σ_1 *of XML keys if* Σ_2 *is non-redundant and a cover for* Σ_1.

An important property is that a non-redundant cover set has in most cases fewer keys that the original one (in the extreme case both sets are equal). This can result in enormous time saving when validating an XML document against a set of XML keys, as we will show in the experimental results.

A characterization of non-redundancy is that Σ is non-redundant if there is no key ψ in Σ such that $\Sigma - \{\psi\} \models \psi$. A key $\psi \in \Sigma$ is called redundant if $\Sigma - \{\psi\} \models \psi$. Thus, we propose Algorithm 3 to compute, given a set Σ of XML keys, a non-redundant cover Θ of Σ.

Algorithm 3. (Non-redundant Cover for XML keys)

Input: finite set Σ of XML keys
Output: a non-redundant cover for Σ
1: $\Theta = \Sigma$;
2: **for** each key $\psi \in \Sigma$ **do**
3:      **if** $\Theta - \{\psi\} \models \psi$ **then**
4:          $\Theta = \Theta - \{\psi\}$;
5:      **end if**
6: **end for**
7: **return** Θ;

It is important to note that a set Σ can have more than one non-redundant cover set and there can exist non-redundant cover sets that are not included in Σ.

The complexity of Algorithm 3 is determined by the complexity of the implication algorithm which is executed once for every key in Σ. Thus a non-redundant cover set for a set Σ of XML keys in $\mathcal{K}$ can be computed in $\mathcal{O}(|\Sigma| \times (max\{|\psi| : \psi \in \Sigma\})^2)$ time.

6 Experimental Results

In the following we present a performance analysis of the algorithms proposed in this work. Up to our knowledge, this is the first time that the theory on automated reasoning about XML keys is tested in practice. The running time results were obtained in an Intel Core 2 Duo 2.0 GHz machine, 3GB RAM, and Linux kernel 2.6.32.

The Data Set. We used a collection of large XML documents from [14]. The collection consists of the following XML documents. A characterization of the documents is shown in Table 1.

- *321gone.xml and yahoo.xml.* Auction data converted to XML.
- *dblp.xml.* Bibliographic information on computer science.
- *nasa.xml* Astronomical Data converted from legacy flat-file format into XML.
- *SigmodRecord.xml.* Index of articles from SIGMOD Record.
- *mondial-3.0.xml.* World geographic database from several sources.

We defined, for each document in the collection, a corresponding set of 5 to 10 appropriate (in the context of the document) XML keys.

Table 1. XML Documents

Doc ID	Document	No. of Elements	No. of Attributes	Size	Max. Depth	Average Depth
Doc1	321gone.xml	311	0	23 KB	5	3.76527
Doc2	yahoo.xml	342	0	24 KB	5	3.76608
Doc3	dblp.xml	29,494	3,247	1.6 MB	6	2.90228
Doc4	nasa.xml	476,646	56,317	23 MB	8	5.58314
Doc5	SigmodRecord.xml	11,526	3,737	476 KB	6	5.14107
Doc6	mondial-3.0.xml	22,423	47,423	1 MB	5	3.59274

Then, in order to test the scalability of the implication algorithm, we generated large sets of XML keys in the following two systematic ways. Firstly, using the manually defined sets of XML keys as seeds, we computed new implied keys by successively applying the inference rules from the axiomatization of XML keys presented in [9]. For instance, by applying the interaction rule to $(listing, (auction\_info, \{high\_bidder. bidder\_name.S, high\_bidder.bidder\_rating. S\}))$ and $(listing.auction\_info, (high\_bidder, \{bidder\_name.S, bidder\_rating.S\}))$, we derived the implied key $(listing, (auction\_info.high\_bidder, \{bidder\_name. S, bidder\_rating.S\}))$. Each key generated by this method was added to the original set. We applied the interaction, context-target, subnodes, context-path containment, target-path containment, subnodes-epsilon and prefix-epsilon rules whenever possible, since those are the rules which can produce implied keys with corresponding non trivial witness graphs (see the proof of Lemma 3.6 in [9]). Secondly, we defined some non-implied (by the keys defined previously) XML keys. We did that by taking non-implied XML keys φ, building their corresponding mini-trees $T_{\Sigma,\varphi}$, adding several witness edges to it while keeping q_φ not reachable from q'_φ, and finally defining new non-implied XML keys corresponding to those witness edges. As an example, let us take the mini-tree in Figure 3 which corresponds to the key $\varphi = (conference, (issue.\_^*.articles.article.author, \{first.S, last.S\}))$. From the witness edges (a), (b) and (c), we obtained the keys, $(conference.issue, (\_^*, \{ articles.article.author.first.S\}))$, $(conference.issue.\_^*, (articles.article, \{author.first.S\}))$ and $(conference.issue.\_^*.articles, (article. au-thor\{first.S\}))$, respectively.

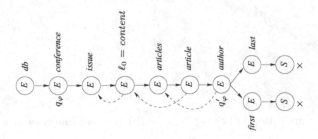

Fig. 3. Mini-tree corresponding to a non-implied key

This process gave us a robust collection of XML keys to thoroughly test the performance of the implication algorithm.

Deciding Implication of XML Keys: Tests Results. The results regarding running times for deciding the implication of XML keys are shown in Figures 4(a) and 4(b). In both figures, the x-axis corresponds to the number of keys in Σ, and the y-axis corresponds to the *average* running time required to decide whether Σ implies a given key φ. More precisely, let $time(\Sigma, \varphi)$ be the running time required to decide $\Sigma \models \varphi$ and let Φ be a set of XML keys such that $\Sigma \cap \Phi = \emptyset$, the running time shown in Figures 4(a) and 4(b), corresponds to $\left(\sum_{\varphi_i \in \Phi} time(\Sigma, \varphi_i)\right)/|\Phi|$. In our experiments the sets Φ were composed of 20 fixed XML keys each. We tested the scalability of the algorithm by adding, in each iteration, 5 new XML key to the corresponding Σ sets. The actual XML keys included in all these sets were created using the strategy explained above.

We consider Σ sets composed by (i) only absolute keys ("abs"), (ii) only relative keys ("rel") or (iii) both types of keys ("mix"). Given that an input key φ can be either absolute or relative, we have a total of six test cases. The results obtained in these experiments are summarized in Figure 4(a). For a small set Σ with about 5 XML keys, the execution takes 0.2ms in average, whereas for a large set of about 100 XML keys, the execution takes 1.7ms in average. This indicates that our implementation of the implication algorithm is practically efficient and scales well regardless of the type of XML keys considered.

Note that the resulting running time is slightly lower when φ is an absolute key and Σ is composed by either absolute or relative keys. This is mainly due to the fact that $Q_\varphi = \varepsilon$, which means that the construction of the mini-tree involves less steps and that the q_φ node corresponds to the root node, making it unnecessary to perform a search for such node. On the other hand, the performance shown by the "abs-rel" curve in Figure 4(a) is slightly degraded due to the fact that, in general, the algorithm needs to traverse more nodes to determine whether q_φ is reachable from q'_φ. This is consistent with the way in which the witness graphs are defined.

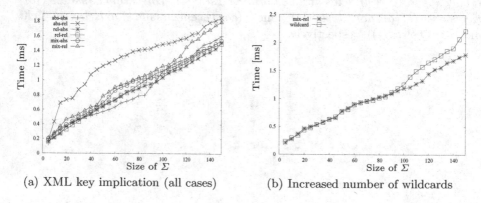

(a) XML key implication (all cases)   (b) Increased number of wildcards

Fig. 4. Performance of the Algorithm for the Implication of XML Keys

To isolate the effect of wildcards in the performance of the algorithm, we duplicated the number of wildcards in the keys of the test case "mix-rel", replacing some of the labels in the context and target paths of those keys by the variable length wildcard. Figure 4(b) shows the running times for the original set of keys (curve "mix-rel") and the set with increased number of wildcards (curve "wildcard"). The results show that the presence of wildcards in the path expressions increases the running time for the test sets with large number of keys, but such increase is not significant in practice.

Document Validation: Tests Results. We use the same data set as before. The aim is to determine the viability of computing non-redundant cover sets to speed up the validation of XML documents against XML keys. By validating an XML document against a set of XML key, we refer to the task of checking, for every XML key in the set, whether the document satisfies such key.

To validates XML keys, we use a naive algorithm that parses the XML document into a DOM tree and then evaluates the XML keys on the resulting tree, by using XPath queries to express their context, target and key paths. We do need to use sophisticated validation algorithms such as [5,12], since the proposed optimization based in cover sets is independent of the particular algorithm used for XML key validation.

The results (in milliseconds (ms)) obtained from the computation of non-redundant cover sets is summarized in Figure 5 (a). We emphasize that the behavior of the Algorithm 3 is linear in practice. For example, for a set of 146 keys, calculating a non-redundant cover set takes around 155ms. A total of 102 keys are discarded reducing the set to 44 keys.

Figure 5 (b) shows the optimization achieved by pre-calculating non-redundant cover sets during the validation process of the documents in Table 1. The results

XML doc.	Key Set	Time[ms]
321gone & yahoo (Doc1)	Processed Keys: 23 Discarded keys: 15 Cover set: 8 keys	3.458
DBLP (Doc2)	Processed Keys: 36 Discarded keys: 24 Cover set: 12 keys	12.757
nasa (Doc3)	Processed Keys: 35 Discarded keys: 28 Cover set size: 7 keys	9.23
Sigmod Record (Doc4)	Processed Keys: 24 Discarded Keys: 19 Cover set: 5 keys	5.294
mondial (Doc5)	Processed Keys: 26 Discarded Keys: 16 Cover set: 10	4.342

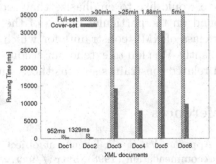

(a) Non-redundant Cover Sets.   (b) Validation Against Cover Sets.

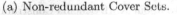

Fig. 5. Non-redundant Cover Sets of XML keys and Validation of XML Documents

indicate that the running time required to compute a non-redundant cover set is just a tiny fraction of the overall running time required to validate a single XML document against a key. Note that in most cases the validation time can be significantly reduced by pre-computing the non-redundant covers. This can be clearly observed in the case of the DBLP document ('Doc3') and Nasa document ('Doc4'). In these cases the running time of the validation against the original set of XML keys is approximately 63 times greater than the running time of the validation against its non-redundant cover-set.

7   Conclusion

Our research was motivated by two objectives. Firstly, we wanted to demonstrate that there are expressive classes of XML keys that are not only tractable in theory but can be reasoned about efficiently in practice. For that we studied a fragment of XML keys as originally introduced by Buneman et al. [3,4], namely the class $\mathcal{K}$ of XML keys with nonempty sets of simple key paths. For these keys it was known that their implication problem can be decided in quadratic time in theory [9]. Here we have presented an efficient implementation thereof, and our experiments show that it also runs fast in practice and scales well.

Secondly, we wanted to show that our observations on the problem of deciding implication is not only of interest for the problem itself but has immediate consequences for other perennial tasks in XML database management. As an example we study the problem of validating an XML document against a set of XML keys. We have presented an optimization method for this validation that computes a non-redundant cover for the set of XML keys given as input so that satisfaction only needs to be checked for the keys in this cover. This can reduce the number of keys significantly, and our experiments show that enormous time savings can be achieved in practice. This holds true even though the validation procedure is able to decide value equality among element nodes with complex content as this is required for the XML keys studied here (and distinguishes them from the keys defined in XML Schema). This illustrates the advantage of having efficient reasoning capabilities at hand for integrity constraints.

We would like to emphasize that the use of non-redundant covers does not depend on the particular choice of the XML fragment but can be tailored to any class of XML constraints for which the implication problem can be solved efficiently. We plan to extend our studies to other expressive classes of XML keys and related constraints such as those studied in [10,7].

References

1. Apparao, V., et al.: Document object model (DOM) level 1 specification, W3C recommendation (1998), http://www.w3.org/TR/REC-DOM-Level-1/
2. Arenas, M., Fan, W., Libkin, L.: What's Hard about XML Schema Constraints? In: Hameurlain, A., Cicchetti, R., Traunmüller, R. (eds.) DEXA 2002. LNCS, vol. 2453, pp. 269–278. Springer, Heidelberg (2002)

3. Buneman, P., Davidson, S., Fan, W., Hara, C., Tan, W.: Keys for XML. Computer Networks 39(5), 473–487 (2002)
4. Buneman, P., Davidson, S., Fan, W., Hara, C., Tan, W.: Reasoning about keys for XML. Inf. Syst. 28(8), 1037–1063 (2003)
5. Chen, Y., Davidson, S., Zheng, Y.: Xkvalidator: a constraint validator for XML. In: CIKM 2002: Proceedings of the 2002 ACM CIKM International Conference on Information and Knowledge Management, pp. 446–452. ACM (2002)
6. Clark, J., DeRose, S.: XML path language (XPath) version 1.0, W3C recommendation (1999), http://www.w3.org/TR/xpath
7. Ferrarotti, F., Hartmann, S., Link, S.: A Precious Class of Cardinality Constraints for Flexible XML Data Processing. In: Jeusfeld, M., Delcambre, L., Ling, T.-W. (eds.) ER 2011. LNCS, vol. 6998, pp. 175–188. Springer, Heidelberg (2011)
8. Gottlob, G., Koch, C., Pichler, R.: Efficient algorithms for processing XPath queries. Trans. Database Syst. 30(2), 444–491 (2005)
9. Hartmann, S., Link, S.: Efficient reasoning about a robust XML key fragment. ACM Trans. Database Syst. 34(2) (2009)
10. Hartmann, S., Link, S.: Numerical constraints on XML data. Inf. Comput. 208(5), 521–544 (2010)
11. Jungnickel, D.: Graphs, Networks and Algorithms. Springer (1999)
12. Liu, Y., Yang, D., Tang, S., Wang, T., Gao, J.: Validating key constraints over XML document using XPath and structure checking. Future Generation Comp. Syst. 21(4), 583–595 (2005)
13. Maier, D.: Minimum Covers in the Relational Database Model. J. ACM 27, 664–674 (1980)
14. Suciu, D.: XML Data Repository, University of Washington (2002), http://www.cs.washington.edu/research/xmldatasets/www/repository.html
15. Thompson, H., Beech, D., Maloney, M., Mendelsohn, N.: XML Schema Part 1: Structures Second Edition, W3C Recommendation (2004), http://www.w3.org/TR/xmlschema-1/

Finding Top-K Correct XPath Queries
of User's Incorrect XPath Query

Kosetsu Ikeda and Nobutaka Suzuki

University of Tsukuba
1-2, Kasuga, Tsukuba Ibaraki 305-8550, Japan
{lumely,nsuzuki}@slis.tsukuba.ac.jp

Abstract. Suppose that we have a DTD D and XML documents valid against D, and consider writing an XPath query to the documents. Unfortunately, a user often does not understand the entire structure of D exactly, especially in the case where D is very large and/or complex or D has been updated but the user misses it. In such cases, the user tends to write an incorrect XPath query q. However, it is difficult for the user to correct q by hand due to his/her lack of exact knowledge about the entire structure of D. In this paper, we propose an algorithm that finds, for an XPath query q, a DTD D, and a positive integer K, "top-K" XPath queries "most similar" to q among the XPath queries conforming to D so that a user select an appropriate query among the K queries. We also present some experimental studies.

1 Introduction

Suppose that we have a DTD D and XML documents valid against D, and let us consider writing an XPath query to the documents. Unfortunately, a user often does not understand the entire structure of D exactly, especially in the case where D is a very large and/or complex DTD or D has been updated but the user misses the update. In such cases, the user tends to write an incorrect XPath query q in the sense that q does not conform to D or the answer of q is disappointing due to his/her structural misunderstanding of D. However, it is difficult for the user to correct q by hand due to his/her lack of exact knowledge about the entire structure of D. On the other hand, a query q written by a user is at least an important "hint" in order to find a correct query, even if q is incorrect. Therefore, in this paper we propose an algorithm that finds, for an (possibly incorrect) XPath query q, a DTD D, and a positive integer K, top-K XPath queries "similar"(syntactically close) to q among the XPath queries conforming to D, in order that a user may select a desirable query from the top-K queries.

As a brief example of our algorithm, let us consider the following DTD D.

```
<!ELEMENT html (div)*>
<!ELEMENT div  (div|p)+>
<!ELEMENT p    (#PCDATA|span)*>
<!ELEMENT span (#PCDATA)>
<!ATTLIST p color CDATA "blue">
```

S.W. Liddle et al. (Eds.): DEXA 2012, Part I, LNCS 7446, pp. 116–130, 2012.

Suppose that a user wants every **span** element in a paragraph whose color is "red" and that he/she tries to use an XPath query $q =$ /p[@color = "red"]/spen, which does not conform to D. Our algorithm finds XPath queries similar to q based on the *edit distance* between XPath queries, introduced in this paper. In this example, our algorithm lists the following top-K XPath queries similar to q (assuming that $K = 3$). Each XPath query q' is followed by the edit distance between q and q', assuming that the cost of relabeling l with l' is the normalized string edit distance between l and l' [13].

1. //p[@color = "red"]/span   (0.75)
2. //div/p[@color = "red"]/span   (1.75)
3. /html/div/p[@color = "red"]/span   (2.25)

As above, by our algorithm the user can obtain top-K correct XPath queries similar to q without modifying q by hand. Although the above DTD D is very small, DTDs used in practice are larger and more complex [3]. In such a situation, a user tends not to understand the entire structure of a DTD exactly, and thus our algorithm is helpful to write correct XPath queries on such DTDs.

In this paper, we focus on an XPath fragment using child, descendant-or-self, following-sibling, preceding-sibling, and attribute axes. Although our XPath fragment supports no upward axes, this gives usually no problem since the majority of XPath queries uses only downward axes[8]. Thus, we believe that our algorithm is useful to correct a large number of XPath queries.

There have been a number of eminent studies related to this paper. Ref. [4] proposes an algorithm that finds valid tree pattern queries most similar to an input query. Their algorithm and ours are incomparable due to the underlying data models; in their data model a tree is unordered and a schema is represented by a DAG supporting multiple type for element name (as in XML Schema), while we use DTD (recursion is supported) and a tree is ordered. Note that Choi investigated 60 DTDs and 35 of the DTDs are recursive [3], which suggests that it is meaningful to support recursive schemas. Besides query correction, several related but different approaches have been studied for XML. Ref. [16] proposes the node insertion operation that is also proposed in this paper. Ref. [15] takes a query expansion approach instead of correcting queries. Refs. [1,2,7,6] deal with a top-K query evaluation for XML documents to derive inexact answers, i.e., evaluating a "relaxed" version of the input query, if it is unsatisfiable. Inexact querying is also studied in Refs. [10,11], in which a user can write an XQuery query without specifying exact connections between elements. Ref. [14] proposes an interactive system for generating XQuery queries. There has been a number of studies on XML keyword search (e.g., [19,9,18]), which are especially suitable for users that are not familiar with XML query languages.

2   Preliminaries

Let Σ_e be a set of labels (element names) and Σ_a be a set of attribute names with $\Sigma_e \cap \Sigma_a = \emptyset$. A *DTD* is a triple $D = (d, \alpha, s)$, where d is a mapping

Table 1. Syntax of XP

```
        XP ::= "/" RelativePath | "/" RelativePath "@" Attribute
RelativePath ::= LocationStep | LocationStep "/" RelativePath
LocationStep ::= Axis "::" Nodetest | Axis "::" Nodetest Predicate
        Axis ::= "↓" | "↓*" | "→⁺" | "←⁺"
    Nodetest ::= Label | "*"
       Label ::= (any label in Σₑ)
   Attribute ::= (any label in Σₐ)
   Predicate ::= "[" Exp "]"
         Exp ::= PredPath | PredPath Op Value
    PredPath ::= RelativePath | "@" Attribute | RelativePath "@" Attribute
          Op ::= "=" | "<" | ">" | "=<" "=>"
       Value ::= "'" (any string other than "'") "'"
```

from Σ_e to the set of regular expressions over Σ_e, α is a mapping from Σ_e to 2^{Σ_a}, and $s \in \Sigma_e$ is the *start label*. For example, the DTD in Section 1 is a triple (d, α, html), where $d(\text{html}) = \text{div}^*$, $d(\text{div}) = (\text{div}|\text{p})^+$, $d(\text{p}) = (\epsilon|\text{span})^*$, $d(\text{span}) = \epsilon$, $\alpha(\text{p}) = \{\text{color}\}$, and $\alpha(e) = \emptyset$ for any element $e \neq \text{p}$. By $L(d(a))$ we mean the language of $d(a)$. For labels b, c, if there is a string $str \in L(d(a))$ such that $str[i] = c$ and $str[j] = b$ with $i < j$ $(i > j)$, then we say that b can be *right* (resp., *left*) to c in $d(a)$, where $str[i]$ denotes the ith character of str. For example, e can be right to c in $d(a) = c(f|e)^*$. For a DTD $D = (d, \alpha, s)$ and labels $a, b \in \Sigma_e$, b is *reachable* from a in D if (i) $a = b$ or b appears in $d(a)$, or (ii) for some label a', a' is reachable from a and b appears in $d(a')$. In the following, we assume that any label in a DTD is reachable from the start label of the DTD.

The XPath fragment used in this paper, denoted XP, is a set of location paths using child ($\downarrow$), descendant-or-self ($\downarrow^*$), following-sibling ($\rightarrow^+$), preceding-sibling ($\leftarrow^+$), and attribute (@) axes. Formally, XP is defined in Table 1. Thus an *XPath query* (*query* for short) q in XP can be denoted

$$/ax[1] :: l[1][exp[1]]/ \cdots /ax[m] :: l[m][exp[m]], \tag{1}$$

where $ax[i] \in$ Axis and $l[i] \in \Sigma_e$ for $1 \leq i \leq m - 1$, $exp[i] \in$ Exp for $1 \leq i \leq m$, $ax[m] \in$ Axis $\cup \{@\}$, and $l[m] \in \Sigma_a$ if $ax[m] = @$, $l[m] \in \Sigma_e$ otherwise. If the ith location step has no predicate, then we write $exp[i] = \epsilon$.

Let q be a query in (1) containing no '$*$' as node test. For indexes i, j such that $ax[i] \in \{\downarrow, \downarrow^*\}$ and that $ax[i + 1], \cdots, ax[j] \in \{\rightarrow^+, \leftarrow^+\}$, we say that l is the *parent label* of $l[j]$ in q if (i) $ax[i] = \downarrow$ and $l = l[i - 1]$, or (ii) $ax[i] = \downarrow^*$, l is reachable from $l[i - 1]$, and $l[i]$ appears in $d(l)$. For example, if $q = / \downarrow :: a/ \downarrow :: b/ \rightarrow^+ :: c/ \leftarrow^+ :: d$, then a is the parent label of b, c, d in q. Let $D = (d, \alpha, s)$ be a DTD. Then q *conforms* to D if the following conditions hold.

– $ax[1] = \downarrow$ and $l[1] = s$, or, $ax[1] = \downarrow^*$ and $l[1] \in \Sigma_e$
– The following condition holds for every $2 \leq i \leq m$
  • $ax[i] = \downarrow$ and $l[i]$ appears in $d(l[i - 1])$,
  • $ax[i] = \downarrow^*$ and $l[i]$ is reachable from $l[i - 1]$ in D,

- $ax[i] = \rightarrow^+$ and $l[i]$ can be right to $l[i-1]$ in $d(l)$, where l is the parent label of $l[i]$ (the case where $ax[i] = \leftarrow^+$ is defined similarly), or
- $ax[i] = @$, $i = m$, and $l[i] \in \alpha(l[i-1])$.
- For every $1 \leq i \leq m$ with $exp[i] \neq \epsilon$, query $/\downarrow::l[i]/exp[i]$ conforms to DTD $(d, \alpha, l[i])$.

Let q be a query in (1) containing '*'s as node tests. Then q *conforms* to D if for some $l_1 \in L(l[1]), \cdots, l_m \in L(l[m])$, $/ax[1]::l_1[exp[1]]/\cdots/ax[m]::l_m[exp[m]]$ conforms to D.[1] By $|q|$ we mean the number of location steps in q, e.g., if $q = /\downarrow::a/\downarrow::*[\leftarrow^+::d]$, then $|q| = 3$. If a query q has neither predicate nor attribute axis, then we say that q is *simple*.

3 Edit Operations to XPath Query

In this section, we define edit operations to queries. We use the following four kinds of *edit operations*.

- *Axis substitution:* substitutes axis ax with ax', denoted $ax \rightarrow ax'$. For example, by applying $\downarrow \rightarrow \downarrow^*$ to $/\downarrow::a$ we obtain $/\downarrow^*::a$.
- *Label substitution:* substitutes label (possibly '*') l with l', denoted $l \rightarrow l'$. For example, by applying $a \rightarrow b$ to $/\downarrow::a$ we obtain $/\downarrow::b$.
- *Location step insertion:* inserts location step $ax::l$, denoted $\epsilon \rightarrow ax::l$. For example, by applying $\epsilon \rightarrow \downarrow::b$ to the tail of $/\downarrow::a$ we obtain $/\downarrow::a/\downarrow::b$.
- *Location step deletion:* deletes location step $ax::l$, denoted $ax::l \rightarrow \epsilon$. For example, by applying $\downarrow::a \rightarrow \epsilon$ to the first location step of $/\downarrow::a/\downarrow::b$ we obtain $/\downarrow::b$.

We next define the *position* of a location step ls, denoted $pos(ls)$. Let $q = /ax[1]::l[1][exp[1]]/\cdots/ax[m]::l[m][exp[m]] \in \mathrm{XP}$. We define that $pos(ax[i]::l[i]) = i$ for $1 \leq i \leq m$. As for location steps in predicates, let $exp[i] = ax'[1]::l'[1][exp'[1]]/\cdots/ax'[n]::l'[n][exp'[n]]$. Then we define that $pos(ax'[j]::l'[j]) = i.j$ for $1 \leq j \leq n$. The position of a location step in $exp'[j]$ can be defined similarly. For example, let $q = /\downarrow::a/\downarrow::b[\downarrow::*[\downarrow::g]]/\rightarrow^+::c$. Then $pos(\downarrow::b) = 2$, $pos(\downarrow::*) = 2.1$, and $pos(\downarrow::g) = 2.1.1$. By $[op]_{pos}$, we mean an edit operation op applied to the location step at position pos. If op is an edit operation inserting a location step ls, then $[op]_{pos}$ inserts ls just after the location step at pos.

Let $q \in \mathrm{XP}$. An *edit script* for q is a sequence of edit operations having a position in q. For an edit script s for q, by $s(q)$ we mean the query obtained by applying s to q. For example, let $s = [\epsilon \rightarrow \downarrow::b]_1 [c \rightarrow f]_3$ and $q = /\downarrow^*::a/\downarrow::d/\downarrow::c$. Then we have $s(q) = /\downarrow^*::a/\downarrow::b/\downarrow::d/\downarrow::f$. Throughout this paper, we assume the following. Let $U = \{\downarrow, \downarrow^*\}$, $S = \{\rightarrow^+, \leftarrow^+\}$, and $A = \{@\}$.

- An axis can be substituted with an axis of "same kind" only, that is, $ax \in U$ (resp., S, A) can be substituted with an axis in U (resp., S, A) only.
- A location step $ax::l$ can be inserted to a query only if $ax \in U$ and $l \in \Sigma_e$.

---

[1] $L(l[i]) = \{l[i]\}$ if $l[i]$ is a label, $L(l[i]) = \Sigma_e$ if $l[i] = $ '*'.

A *cost function* assigns a cost to an edit operation. By $\gamma(op)$ we mean the *cost* of an edit operation op, where γ is a cost function. In the following, we assume that $\gamma(op) \geq 0$. A cost function can be a general function as well as a constant. For example, $\gamma(op)$ can be a string edit distance between l and l' if $op = l \to l'$. For an edit script $s = op_1op_2 \cdots op_n$, by $\gamma(s)$ we mean the *cost* of s, that is, $\gamma(s) = \sum_{1 \geq i \geq n} \gamma(op_i)$. For a DTD D, a query q, and a positive integer K, the K *optimum edit script* for q under D is a sequence of edit operations $s_1, \cdots, s_K$ if (i) each of $s_1(q), \cdots, s_K(q)$ conforms to D, (ii) $\gamma(s_1) \leq \cdots \leq \gamma(s_K)$, and (iii) $s_1, \cdots, s_K$ are optimum, that is, for any edit script s for q such that $s(q)$ conforms to D, $s(q) \in \{s_1(q), \cdots, s_K(q)\}$ or $\gamma(s) \geq \gamma(s_K)$. We say that $s_1(q), \cdots, s_K(q)$ are *top-K* queries *similar* to q under D.

4   Xd-Graph Representing Queries Conforming to DTD

In this section, we introduce a graph called *xd-graph*, which forms the basis of our algorithm. Throughout this section, we assume that a query is simple.

To find top-K queries similar to a query q under a DTD D, we take the following approach.

1. We first construct an xd-graph for q and D. The graph is designed so that each path in the graph represents a simple query q' such that (a) q' is obtained by applying some edit script to q and that (b) q' conforms to D.
2. Then we solve the K shortest paths problem on the xd-graph. The result corresponds to top-K queries similar to q under D. The details of this step are presented in Section 5.

4.1   Xd-Graph Examples

To construct an xd-graph, we need a graph representation of DTD. The *DTD graph* $G(D)$ of a DTD $D = (d, \alpha, s)$ is a directed graph (V, E), where $V = \Sigma_e$ and $E = \{l \to l' \mid l'$ is a label appearing in $d(l)\}$. For example, Fig. 1(a) is the DTD graph of $D = (d, \alpha, s)$, where $d(s) = ba^*$, $d(a) = c|d$, $d(b) = d$, $d(c) = \epsilon, d(d) = b|\epsilon$.

Now let us illustrate xd-graph. We first present the following three cases by examples (assuming that no '$*$' can be used), then define xd-graph formally.

Case A) Only child ($\downarrow$) can be used as an axis.
Case B) Descendant-or-self ($\downarrow^*$) can be used as well as $\downarrow$.
Case C) Sibling axes ($\to^+$, $\leftarrow^+$) can be used as well as $\downarrow$ and $\downarrow^*$.

Case A). Let us first illustrate the xd-graph constructed from a simple query $q = / \downarrow$:: $a/ \downarrow$:: d and the DTD graph $G(D)$ in Fig. 1(a). Since only $\downarrow$ axis is allowed, it suffices to consider location step insertion, location step deletion, and label substitution. Fig. 1(b) shows xd-graph $G(q, G(D))$. The xd-graph is constructed from 3 copies of $G(D)$ with their nodes connected by several edges. Here, n_0, n_1, n_2 are newly added nodes, which correspond to the "root node" in

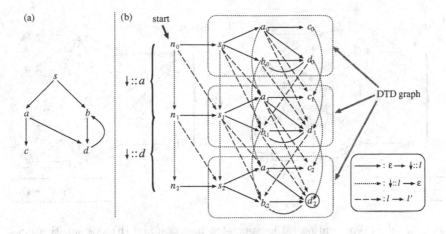

Fig. 1. (a) a DTD graph $G(D)$ and (b) an xd-graph $G(q, G(D))$

the XPath data model. Each node is subscripted, e.g., the node s in $G(D)$ is denoted s_0 on the topmost DTD graph of $G(p, G(D))$, s_1 on the second topmost DTD graph, and so on, as shown in Fig. 1(b).

We have the following three kinds of edges in an xd-graph.

- A "horizontal" edge $l \to l'$ corresponds to a location step insertion.
- A "slant" edge $l \dashrightarrow l'$ corresponds to a label substitution.
- A "vertical" edge $l \cdots\!\!\!\rightarrow l'$ corresponds to a location step deletion.

More concretely, let us first consider horizontal edge $n_0 \to s_0$ in Fig. 1(b). This edge means "moving from the root node to child node s, using no location step of q". In other words, the edge $n_0 \to s_0$ represents adding a location step $\downarrow:: s$, that is, the edge represents an edit operation $[\epsilon \to \downarrow:: s]_0$. Let us next consider slant edge $s_0 \dashrightarrow b_1$ in Fig. 1(b). This edge means "moving from node s to child node b using the first location step $\downarrow:: a$ of q". Since the target node is b rather than a, we have to substitute the label of $\downarrow:: a$ with b, that is, the edge $s_0 \dashrightarrow b_1$ represents $[a \to b]_1$. Finally, consider vertical edge $b_1 \cdots\!\!\!\rightarrow b_2$ in Fig. 1(b). This edge means "staying the same node b by ignoring (deleting) the second location step $\downarrow:: d$ of q". Thus the edge $b_1 \dashrightarrow b_2$ represents $[\downarrow:: d \to \epsilon]_2$.

In Fig. 1(b), n_0 is called *start node* and d_2 is called *accepting node*. Each path from the start node to the accepting node represents a simple query conforming to D obtained by correcting q. For example, let us consider a path $p = n_0 \to s_0 \dashrightarrow a_1 \dashrightarrow d_2$ in Fig. 1(b). Recall that $q = /\downarrow:: a/\downarrow:: d$. The first edge $n_0 \to s_0$ represents a location step insertion $[\epsilon \to \downarrow:: s]_0$. The second edge $s_0 \dashrightarrow a_1$ represents a label substitution $[a \to a]_1$, i.e., the first location step "$\downarrow:: a$" of q is unchanged. Similarly, the location step "$\downarrow:: d$" of q is unchanged. Thus, p represents a query $q' = /\downarrow:: s/\downarrow:: a/\downarrow:: d$, which is obtained by applying $[\epsilon \to \downarrow:: s]_0[a \to a]_1[d \to d]_2$ to q. Note that q' conforms to D.

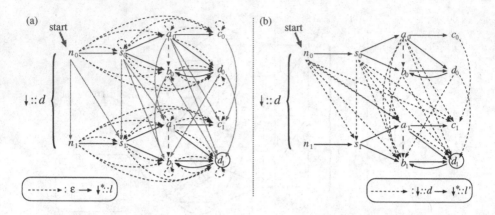

Fig. 2. Edges representing (a) location step insertion and (b) axis substitution

Case B). In this case, we can use $\downarrow^*$ axes as well as $\downarrow$ axes. Let us first consider an edit operation inserting location step $\downarrow^*:: l$ to a query. For this insertion, we add edges representing the edit operation to an xd-graph. Fig. 2(a) shows the xd-graph constructed from the DTD graph in Fig. 1(a) and a query $q = /\downarrow:: d$. Each dashed edge in Fig. 2(a) represents a location step insertion. For example, $s_0 \dashrightarrow d_0$ means "moving from node s to node d via $\downarrow^*$ axis, using no location step of q", that is, inserting a location step $\downarrow^*:: d$ at position 0 of q, i.e., $[\epsilon \rightarrow \downarrow^*:: d]_0$. As stated before, every path from the start node to the accepting node represents a simple query conforming to D, which is obtained by correcting q. For example, $n_0 \rightarrow s_1 \dashrightarrow d_1$ represents a simple query $/\downarrow:: s/\downarrow^*:: d$ obtained by applying $[d \rightarrow s]_1[\epsilon \rightarrow \downarrow^*:: d]_1$ to $q = /\downarrow:: d$.

Let us next consider axis substitution between $\downarrow$ and $\downarrow^*$. Fig. 2(b) shows the xd-graph constructed from the same DTD graph as above and the same query $q = /\downarrow:: d$. In the figure, for simplicity we omit some of the edges representing location step insertion, location step deletion, and label substitution. In Fig. 2(b), a dashed edge represents substituting $\downarrow:: a$ with $\downarrow^*:: l$. For example, $n_0 \dashrightarrow a_1$ means "moving from the root node to a with $\downarrow^*$ axis", i.e., substituting $\downarrow:: d$ with $\downarrow^*:: a$. Here, consider path $p = n_0 \rightarrow s_0 \dashrightarrow d_1$ in Fig. 2(b). p represents a query $/\downarrow:: s/\downarrow^*:: d$, which is obtained by applying $[\epsilon \rightarrow \downarrow:: s]_0[\downarrow \rightarrow \downarrow^*]_1$ to $q = /\downarrow:: d$.

Finally, substituting $\downarrow^*$ with $\downarrow$ can be represented by a slant edge similar to label substitution ($l \rightarrow l'$), and the deletion of a location step using $\downarrow^*$ axis can be handled similarly to the location step deletion in Case A.

Case C). Let us consider handling $\rightarrow^+$ and $\leftarrow^+$ axes. Fig. 3 shows the xd-graph constructed from the same DTD graph as above and a query $q = /\rightarrow^+:: d$. First, let us consider edges connecting the same labels having distinct subscripts, e.g., $s_0 \rightarrow s_1$ and $a_0 \rightarrow a_1$. Such an edge means that the position does not change (ignoring $\rightarrow^+:: d$ of q) and $\rightarrow^+:: d$ is deleted from q. Let us next consider dashed edges connecting "sibling labels". For example, we have four edges between a_0, b_0

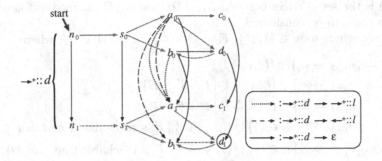

Fig. 3. Edges dealing with $\rightarrow^+$ and $\leftarrow^+$ axes

and a_1, b_1 (e.g., $a_0 \dashrightarrow b_1$, $b_0 \cdots\rightarrow a_1$) since a and b are siblings in $d(s) = ba^*$. A dashed edge $\cdots\rightarrow$ represents substituting a sibling axis ($\rightarrow^+$ or $\leftarrow^+$) with $\rightarrow^+$, and another dashed edge $\dashrightarrow$ represents substituting a sibling axis with $\leftarrow^+$. For example, $a_0 \dashrightarrow b_1$ means "moving from node a to b via $\leftarrow^+$ axis", that is, substituting the location step $\rightarrow^+ :: d$ of q with $\leftarrow^+ :: b$. An xd-graph has no edge violating a DTD, e.g., Fig. 3 does not have edge $b_0 \dashrightarrow a_1$ since $d(s) = ba^*$ and a cannot be left to b.

Wildcard Node Test). To handle wild card node test '$*$', we duplicate each "slant" edge in Fig. 1(b). For example, between s_0 and a_1 we use two edges $s_0 \dashrightarrow a_1$ and $s_0 \overset{*}{\dashrightarrow} a_1$ instead of a single edge $s_0 \dashrightarrow a_1$. The former of the two represents substituting a label with a as in Case A, and the latter represents substituting a label with '$*$' rather than a. Similarly, each dashed edge in Fig. 2(b) is duplicated. The details are skipped because of space limitation.

4.2 Formal Definition of Xd-Graph

Let $D = (d, \alpha, s)$ be a DTD, $G(D) = (V, E)$ be the DTD graph of D, and $q = /ax[1] :: l[1]/ \cdots /ax[m] :: l[m]$ be a simple query. Let $G_i(D) = (V_i, E_i)$ be a graph obtained by adding a subscript i to each node of $G(D)$, that is, $V_i = \{l_i \mid l \in V\}$ and $E_i = \{l_i \rightarrow l'_i \mid l \rightarrow l' \in E\}$ for $0 \leq i \leq m$. The xd-graph for q and $G(D)$, denoted $G(q, G(D))$, is a directed graph (V', E'), where

$$V' = \{n_0, \cdots, n_m\} \cup V_0 \cup \cdots \cup V_m,$$
$$E' = E_{insc} \cup (E'_0 \cup \cdots \cup E'_m) \cup (F_1 \cup \cdots \cup F_m). \qquad (2)$$

Here, E_{insc} in (2) is the set of edges inserting $\downarrow :: l$ (correspond to "$\epsilon \rightarrow\downarrow :: l$" in Fig. 1(b)), that is, $E_{insc} = \{n_0 \rightarrow s_0, \cdots, n_m \rightarrow s_m\} \cup (E_0 \cup \cdots \cup E_m)$, where E_i is the set of edges of $G_i(D)$. E'_i in (2) is the set of edges inserting $\downarrow^* :: l$ (corresponding to "$\epsilon \rightarrow\downarrow^* :: l$" in Fig. 2(a)) and define as follows.

$$E'_i = \{n_i \rightarrow l_i \mid l_i \in V_i\} \cup \{l_i \rightarrow l'_i \mid l' \text{ is reachable from } l \text{ in } D\}.$$

F_i in (2) is the set of edges between $G_{i-1}(D)$ and $G_i(D)$ defined as follows. We have two cases to be considered.

1) The case where $ax[i] \in \{\downarrow, \downarrow^*\}$: $F_i = D_i \cup C_i \cup C_i^* \cup A_i \cup A_i^*$, where

$$
\begin{aligned}
D_i &= \{n_{i-1} \to n_i\} \cup \{l_{i-1} \to l_i \mid l \in V\}, \qquad\qquad (3)\\
C_i &= \{n_{i-1} \to s_i\} \cup \{l_{i-1} \to l_i' \mid l \to l' \in E\},\\
C_i^* &= \{n_{i-1} \xrightarrow{*} s_i\} \cup \{l_{i-1} \xrightarrow{*} l_i' \mid l \to l' \in E\},\\
A_i &= \{n_{i-1} \to l_i \mid l_i \in V_i\} \cup \{l_{i-1} \to l_i' \mid l' \text{ is reachable from } l \text{ in } D\},\\
A_i^* &= \{n_{i-1} \xrightarrow{*} l_i \mid l_i \in V_i\} \cup \{l_{i-1} \xrightarrow{*} l_i' \mid l' \text{ is reachable from } l \text{ in } D\}.
\end{aligned}
$$

Here, D_i is the set of edges corresponding to "$\downarrow\!:: l \to \epsilon$" in Fig. 1(b), C_i is the set of edges corresponding to "$l \to l'$" in Fig. 1(b), and A_i is the set of edges corresponding to "$\downarrow\!:: d \to \downarrow^*\!:: l$" in Fig 2(b). C_i^* (A_i^*) is the set of "duplicated" edges of C_i (resp., A_i) to handle '$*$'.

2) The case where $ax[i] \in \{\leftarrow^+, \to^+\}$: $F_i = D_i \cup L_i \cup R_i$, where

$$
\begin{aligned}
L_i &= \{l_{i-1} \to l_i' \mid l' \text{ can be left to } l, \, l'' \text{ is the parent label of } l, l' \text{ in } d(l'')\},\\
R_i &= \{l_{i-1} \to l_i' \mid l' \text{ can be right to } l, \, l'' \text{ is the parent label of } l, l' \text{ in } d(l'')\},
\end{aligned}
$$

and D_i is the same as the previous case. L_i (resp., R_i) is the set of edges corresponding to "$\to^+\!:: d \to \leftarrow^+\!:: l$" (resp., "$\to^+\!:: d \to \to^+\!:: l$") in Fig. 3.

Finally, we define the cost of an edge in $G(q, G(D)) = (V', E')$. Suppose that $\gamma(l \to l')$, $\gamma(ax \to ax')$, $\gamma(\epsilon \to ax :: l)$, and $\gamma(ax :: l \to \epsilon)$ are defined for any $l, l' \in \Sigma_e$ and any axes ax, ax'. Then the cost of an edge $e \in E'$, denoted $\gamma(e)$, is defined as follows.

- The case where $e \in E_{insc}$: We can denote $e = l_i \to l_i'$. Since this edge represents inserting a location step $\downarrow\!:: l'$, $\gamma(e) = \gamma(\epsilon \to \downarrow\!:: l')$.
- The case where $e \in E_i'$: We can denote $e = l_i \to l_i'$. Since this edge represents inserting a location step $\downarrow^*\!:: l'$, $\gamma(e) = \gamma(\epsilon \to \downarrow^*\!:: l')$.
- The case where $e \in D_i$: We can denote $e = l_{i-1} \to l_i$. Since this edge represents deleting a location step $ax[i] :: l[i]$, $\gamma(e) = \gamma(ax[i] :: l[i] \to \epsilon)$.
- The case where $e \in C_i$: We can denote $e = l_{i-1} \to l_i'$. Since this edge represents substituting $ax[i]$ with $\downarrow$ and substituting $l[i]$ with l', $\gamma(e) = \gamma(ax[i] \to \downarrow) + \gamma(l[i] \to l')$. The case where $e \in C_i^*$ can be defined similarly.
- The case where $e \in A_i$: We can denote $e = l_{i-1} \to l_i'$. Since this edge represents substituting $ax[i]$ with $\downarrow^*$ and substituting $l[i]$ with l', $\gamma(e) = \gamma(ax[i] \to \downarrow^*) + \gamma(l[i] \to l')$. The case where $e \in A_i^*$ can be defined similarly.
- The case where $e \in L_i$: We can denote $e = l_{i-1} \to l_i'$. Since this edge represents substituting $ax[i]$ with $\leftarrow^+$ and substituting $l[i]$ with l', $\gamma(e) = \gamma(ax[i] \to \leftarrow^+) + \gamma(l[i] \to l')$. The case where $e \in R_i$ can be defined similarly.

5   Algorithm for Finding Top-K Queries

In this section, we present an algorithm for finding top-K queries similar to an input query under a DTD. We first consider the case where a query is simple, then present an algorithm for queries in XP. Because of space limitation, the proofs of the correctness and the running time estimations of the algorithm are skipped.

5.1 Method for Simple Query

Let D be a DTD, Σ_e be the set of labels in D, $q = /ax[1]::l[1]/\cdots/ax[m]::l[m]$ be a simple query, and $G(q, G(D)) = (V', E')$ be the xd-graph for q and $G(D)$. Moreover, let $n_0 \in V'$ be the start node and $(l[m])_m \in V'$ be the accepting node of $G(q, G(D))$. If $l[m] \notin \Sigma_e$ (due to user's typo), then the label $l \in \Sigma_e$ "most similar" to $l[m]$ is selected and $l_m \in V'$ is used as the accepting node.[2] Currently, we select $l \in \Sigma_e$ such that the edit distance between l and $l[m]$ is the smallest.

By the definition of xd-graph, in order to find top-K queries similar to q under D, it suffices to solve the K shortest paths problem over the xd-graph $G(q, G(D))$ between the start node and the accepting node. The resulting K shortest paths represent the top-K queries similar to q under D. Thus we have the following.

Theorem 1. *Let D be a DTD, q be a simple query, and K be a positive integer. Then the above method outputs top-K queries similar to q under D.*   □

Let us consider the time/space complexity of the method. First, the size of $G(q, G(D))$ is in $O(|q| \cdot |\Sigma_e|^2)$. Then the time complexity of the method depends on the algorithm to solve the K shortest paths problem. Among a number of algorithms for solving this problem (e.g., [12,5]), we currently use the extended Dijkstra's algorithm to implement our algorithm. In this case, the time complexity of the method is in $O(K \cdot |q| \cdot |\Sigma_e|^2 \cdot \log(|q| \cdot |\Sigma_e|))$.

5.2 Algorithm for General Query

We present an algorithm that finds, for a query $q \in$ XP and a DTD D, top-K queries similar to q under D. We first give some definitions. Let $q = /ax[1]::l[1][exp[1]]/\cdots/ax[m]::l[m][exp[m]] \in$ XP. By $sp(q)$ we mean the *selection path* of q obtained by dropping every predicate in q and the last location step of q if $ax[m] = @$; that is,

$$sp(q) = \begin{cases} /ax[1]::l[1]/\cdots/ax[m-1]::l[m-1] & \text{if } ax[m] = @, \\ /ax[1]::l[1]/\cdots/ax[m]::l[m] & \text{otherwise.} \end{cases}$$

Suppose that $ax[m] = @$. By definition the set of edit operations applicable to $ax[m]::l[m]$ is $S = \{ax[m]::l[m] \to \epsilon\} \cup \{l[m] \to l \mid l \in \alpha(l[m-1])\}$. We say that $op_1, \cdots, op_K$ are K optimum edit operations for $ax[m]::l[m]$ if $op_1, \cdots, op_K \in S$, $op_i \neq op_j$ for any $i \neq j$, $\gamma(op_1) \leq \cdots \leq \gamma(op_K)$, and $\gamma(op_K) \leq op$ for any $op \in S \setminus \{op_1, \cdots, op_K\}$ (we assume that $op_{|S|+1} = \cdots = op_K = nil$ with $\gamma(nil) = \infty$ if $|S| < K$).

---

[2] $G(q, G(D))$ can also have multiple accepting nodes. But since this approach tends to output "too diverse" answers, we currently use a single accepting node.

We now present the algorithm. To find top-K queries similar to a query q under a DTD D, we again construct an xd-graph $G(sp(q), G(D))$ and solve the K shortest paths problem on the xd-graph. But since q may not be simple, before solving the K shortest paths problem we modify $G(sp(q), G(D))$ as follows.[3]

- Suppose $exp[i] \neq \epsilon$. The cost of deleting location step $ax[i] :: l[i][exp[i]]$ should be $\gamma(ax[i] :: l[i] \to \epsilon) + \gamma(exp[i] \to \epsilon)$, where "$exp[i] \to \epsilon$" stands for the delete operations that delete every location step in $exp[i]$ (line (3-a) below).
  We also have to consider correcting $exp[i]$. To do this, we call the algorithm for query $/l[i]/exp[i]$ and DTD $(d, \alpha, l[i])$ recursively. The obtained result is incorporated into $G(sp(q), G(D))$ by using the gadget in Fig. 4 (node l_i corresponds to $l[m]$); the obtained K optimum edit scripts are assigned to the K edges $e_1, \cdots, e_K$ in the gadget (line (3-b)).
- If $ax[m] = @$, we have to modify $G(sp(q), G(D))$ in order to incorporate the K optimum edit operations for $ax[m] :: l[m]$ (line 4).

FINDKPATHS(D, q, K)

Input: A DTD $D = (d, \alpha, s)$, a query $q = /ax[1] :: l[1][exp[1]]/ \cdots /ax[m] :: l[m][exp[m]]$, and a positive integer K.
Output: K optimum edit scripts $s_1, \cdots, s_K$ for q under D.

1. Construct the DTD graph $G(D)$ of D.
2. Construct the xd-graph $G(sp(q), G(D))$ for q and $G(D)$.
3. For each $1 \leq i \leq m$ with $exp[i] \neq \epsilon$, modify $G(sp(q), G(D))$ as follows.
   (a) For each edge $e \in D_i$ (defined in Eq. (3)), let $\gamma(e) \leftarrow \gamma(e) + \gamma(exp[i] \to \epsilon)$.
   (b) For each node $l_i \in V_i$, do the following (i) – (iii).
      i. Replace l_i with its corresponding gadget (Fig. 4).
      ii. Call FINDKPATHS(D', q', K), where $D' = (d, \alpha, l_i)$ and $q' = /l_i/exp[i]$.[4]
         Let $s'_1, \cdots, s'_K$ be the result.
      iii. $\gamma(e_j) \leftarrow \gamma(s'_j)$ for every $1 \leq j \leq K$.
4. If $ax[m] = @$, modify $G(sp(q), G(D))$ as follows.
   (a) Replace the accepting node l_{m-1} of $G(sp(q), G(D))$ with its corresponding gadget (Fig. 4).
   (b) Let $op_1, \cdots, op_K$ be the K optimum edit operations for $ax[m] :: l[m]$.
   (c) $\gamma(e_j) \leftarrow \gamma(op_j)$ for every $1 \leq j \leq K$.
5. Delete the nodes unreachable from the accepting node in $G(sp(q), G(D))$.
6. Solve the K shortest paths problem on $G(sp(q), G(D))$ modified as above.
7. Let $s_1, \cdots, s_K$ be the result of line 6. Return $s_1, \cdots, s_K$.

The above algorithm runs in $O(K \cdot |q| \cdot |\Sigma_e|^2 \cdot \log(|q| \cdot |\Sigma_e|))$ time. We also have the following.

Theorem 2. *Let D be a DTD, $q \in$ XP a query, and K be a positive integer. Then the algorithm outputs K optimum edit scripts for q under D.*   □

---

[3] Since it is fairly difficult to correct the right hand side and the comparison operator of $exp[i]$ exactly, we focus on correcting the left hand side of $exp[i]$.

[4] Since l_i is added as the first location step of q', for each recursive call we assume that $\gamma(n_0 \to l) = 0$ if $l = (l_i)_0$ and $\gamma(n_0 \to l) = \infty$ otherwise, where n_0 is the start node of the constructed xd-graph in the recursive call.

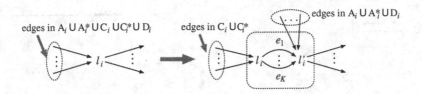

Fig. 4. Node l_i and its gadget, where l_i' is a new node and $e_1, \cdots, e_K$ are new edges

Pruning Xd-Graph. An xd-graph may contain unnecessary nodes, e.g., in Fig. 1(b) the accepting node d_2 is unreachable from c_0, c_1, and c_2, and thus these three nodes are unnecessary. By pruning such nodes (as in line 5 above) we can save space and time. Such a pruning is effective especially if a DTD has a tree-like structure. For example, suppose that the DTD graph $D(G)$ is a complete k-ary tree and that query q contains no sibling axis and no predicate. For a leaf node n in $D(G)$, the number of nodes from which n is reachable is in $O(\log |\Sigma_e|)$. Thus the size of the xd-graph can be reduced from $O(|q| \cdot |\Sigma_e|^2)$ to $O(|q| \cdot \log^2 |\Sigma_e|)$, and the time complexity of the algorithm in this subsection can be reduced to $O(K \cdot |q| \cdot \log^2 |\Sigma_e| \cdot \log(|q| \cdot \log |\Sigma_e|))$.

We also make an experiment to evaluate the effect of this pruning. This is shown in Section 6.1.

6   Experimental Results

In this section, we present two experimental results. The first experiment evaluates the execution time, and the second experiment evaluates the "quality" of the output of the algorithm. The algorithm is implemented in Ruby, and the experiments are performed on Apple Xserve with Mac OS X Server 10.6.8, Xeon 2.26GHz CPU, 6GB Mem, 2.16GB HDD, and Ruby-1.9.3.

6.1   Running Time of the Algorithm

Since the size of an xd-graph may become very large, pruning of xd-graph is important to obtain top-K queries efficiently. We evaluate the execution time of the algorithm, as follows.

1. We create a set Q of 10 queries (not shown because of space limitation). These queries are generated by XQGen [20], which is an XPath expression generator, under auction.dtd of XMark [17]. The average size of the queries in Q is 4.1. Two of the queries contains predicates and the others are simple.
2. For each query obtained above and for each $K = 1, \cdots, 10$, we execute the algorithm and measure its execution time. In this experiment, we use the following simple cost function.

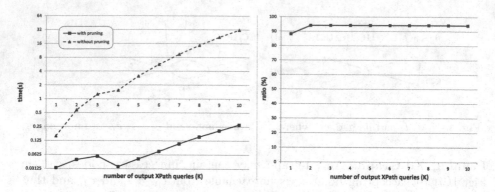

Fig. 5. Execution time and ratios at which the outputs contain correct answers

$$\gamma(l \to l') = \text{the normalized string edit distance between } l \text{ and } l'$$

$$\gamma(ax \to ax') = \begin{cases} 0 & \text{if } ax = ax' \\ 0.5 & \text{otherwise} \end{cases}$$

$$\gamma(\epsilon \to ax :: l) = \gamma(ax :: l \to \epsilon) = 1$$

Fig. 5(left) plots the average execution times for Q, with/without pruning. With pruning the average execution time for Q is about 30 to 250 milliseconds, but without pruning the average execution time is increased by a factor of 4 to 100. Thus, with pruning the algorithm runs efficiently and the pruning brings a huge reduction of the execution time of the algorithm.

6.2 Quality of the Output of the Algorithm

For a DTD D and an incorrect query q written by a user, there are a number of queries similar to q under D, and thus our algorithm need to output a result containing the "correct query" that the user requires. We evaluate the ratio at which the results of the algorithm contain the correct queries.

The outline of this experiment is as follows. We first prepare a set of pairs (q_c, q_i), where q_c is a correct query (a query a user should write) and q_i is an incorrect query (a query a user actually writes). Then for each pair (q_c, q_i), we execute the algorithm to obtain top-K queries similar to q_i and calculate the ratio at which the top-K queries contain q_c.

Let us give the details of the experiment. The experiment is achieved by the following five steps.

1. We generate 30 queries (not shown because of space limitation) by using XQGen under auction.dtd of XMark. The average size of these queries is about 5.4. There queries are treated as "correct queries".
2. For each query q_c obtained above, we make a "question", which describes the meaning of q_c in words. Fig. 6 shows an example of a simple question for `//interval/start`. Each question is carefully described so that it does not permit more than one correct queries. We obtain 30 questions.

Write a minimum XPath query q satisfying the following two conditions.

1. The target element of q is "start".
2. q must use an "interval" element.

Fig. 6. An example of a question

3. We request six people to solve the 30 questions obtained in step 2. That is, for each question they are asked to write a query whose semantics coincides with what the question means. In this step they can see auction.dtd at any time. We obtain 180 answers (i.e., queries written by users) in total.
4. We checked the 180 queries by hand and find 17 incorrect ones. Now we obtain 17 pairs (q_c, q_i) of correct queries and incorrect queries such that q_c and q_i share the same question.
5. For each query q_i of the 17 incorrect queries and each $K = 1, \cdots, 10$, we execute the algorithm for q_i and check whether the corresponding correct query q_c is contained in the output of the algorithm. We use the same cost function as the previous experiment. Fig. 5(right) illustrates the result.

As shown in the figure, the algorithm fairly succeeds in generating top-K queries containing correct queries. The reason why the ratio does not reach 100% is as follows. Auction.dtd contains a cycle and a correct query traverses the cycle, but a user write an incorrect query that "skips" the intermediate elements on the cycle and the algorithm cannot predict the correct query since too much elements are skipped. More concretely, the query written by a user is the following,

```
//closed_auctions/closed_auction/annotation/description/text
```

and the corresponding correct query is as follows. The algorithm does not predict it since four elements are skipped.

```
//closed_autcions/closed_auction/annotation
        /description/parlist/listitem/parlist/listitem/text
```

Thus, we may have to improve the algorithm to deal with cycles more appropriately.

7   Conclusion

In this paper, we proposed an algorithm that finds, for a query q, a DTD D, and a positive integer K, top-K queries similar to q under D. Experimental results suggests that the algorithm outputs "correct" answers efficiently in many cases.

This is an ongoing work and we still have a lot of things to do. One of the important future work is to devise a method for determining reasonable costs of edit operations automatically, since it may be difficult for users to specify the cost of each edit operation exactly. Another future work is to improve the efficiency of the algorithm.

References

1. Amer-Yahia, S., Cho, S., Srivastava, D.: Tree Pattern Relaxation. In: Jensen, C.S., Jeffery, K., Pokorný, J., Šaltenis, S., Bertino, E., Böhm, K., Jarke, M. (eds.) EDBT 2002. LNCS, vol. 2287, pp. 89–102. Springer, Heidelberg (2002)
2. Amer-Yahia, S., Lakshmanan, L.V., Pandit, S.: Flexpath: Flexible structure and full-text querying for xml. In: Proc. SIGMOD, pp. 83–94 (2004)
3. Choi, B.: What are real dtds like? In: Proc. WebDB, pp. 43–48 (2002)
4. Cohen, S., Brodianskiy, T.: Correcting queries for xml. Information Systems 34(8), 690–710 (2009)
5. Eppstein, D.: Finding the k shortest paths. SIAM J. Computing 28(2), 652–673 (1998)
6. Fazzinga, B., Flesca, S., Furfaro, F.: Xpath query relaxation through rewriting rules. IEEE Transactions on Knowledge and Data Engineering 23, 1583–1600 (2011)
7. Fazzinga, B., Flesca, S., Pugliese, A.: Retrieving xml data from heterogeneous sources through vague querying. ACM Trans. Internet Technol. 9(2), 7:1–7:35 (2009), http://doi.acm.org/10.1145/1516539.1516542
8. Ives, Z.G., Halevy, A.Y., Weld, D.S.: An xml query engine for network-bound data. The VLDB Journal 11(4), 380–402 (2002)
9. Li, G., Feng, J., Wang, J., Zhou, L.: Effective keyword search for valuable lcas over xml documents. In: Proc. ACM CIKM, CIKM 2007, pp. 31–40. ACM (2007)
10. Li, Y., Yu, C., Jagadish, H.V.: Schema-free xquery. In: Proc. VLDB, pp. 72–83 (2004)
11. Li, Y., Yu, C., Jagadish, H.V.: Enabling schema-free xquery with meaningful query focus. The VLDB Journal 17, 355–377 (2008)
12. Martins, E.: K-th shortest paths problem, http://www.mat.uc.pt/~eqvm/OPP/KSPP/KSPP.html
13. Marzal, A., Vidal, E.: Computation of normalized edit distance and applications. IEEE Transactions on Pattern Analysis and Machine Intelligence 15, 926–932 (1993)
14. Morishima, A., Kitagawa, H., Matsumoto, A.: A machine learning approach to rapid development of xml mapping queries. In: Proc. ICDE, pp. 276–287 (2004)
15. Schenkel, R., Theobald, M.: Feedback-Driven Structural Query Expansion for Ranked Retrieval of XML Data. In: Ioannidis, Y., Scholl, M.H., Schmidt, J.W., Matthes, F., Hatzopoulos, M., Böhm, K., Kemper, A., Grust, T., Böhm, C. (eds.) EDBT 2006. LNCS, vol. 3896, pp. 331–348. Springer, Heidelberg (2006)
16. Schlieder, T.: Schema-Driven Evaluation of Approximate Tree-Pattern Queries. In: Jensen, C.S., Jeffery, K., Pokorný, J., Šaltenis, S., Bertino, E., Böhm, K., Jarke, M. (eds.) EDBT 2002. LNCS, vol. 2287, pp. 514–532. Springer, Heidelberg (2002), http://dl.acm.org/citation.cfm?id=645340.650204
17. Schmidt, A., Waas, F., Kersten, M., Carey, M., Manolescu, I., Busse, R.: Xmark: A benchmark for xml data managemet. In: Proc. VLDB, pp. 974–985 (2002)
18. Termehchy, A., Winslett, M.: Using structural information in xml keyword search effectively. ACM Trans. Database Syst. 36(1), 4 (2011)
19. Xu, Y., Papakonstantinou, Y.: Efficient keyword search for smallest lcas in xml databases. In: Proc. ACM SIGMOD Conf., pp. 527–538. ACM (2005)
20. Wu, Y., Lele, N., Aroskar, R., Chinnusamy, S., Brenes, S.: Xqgen: an algebra-based xpath query generator for micro-benchmarking. In: Proc. CIKM, pp. 2109–2110 (2009)

Analyzing Plan Diagrams of XQuery Optimizers

H.S. Bruhathi and Jayant R. Haritsa

Database Systems Lab., SERC/CSA
Indian Institute of Science, Bangalore 560012, India

Abstract. The automated optimization of declarative user queries is a classical hallmark of database technology. XML, with its support for deep data hierarchies and powerful query operators, including regular expressions and sibling axes, renders the query optimization challenge significantly more complex. In this paper, we analyze the behavior of industrial-strength XQuery optimizers using the notion of "plan diagrams", which had hitherto been applied solely to relational engines. Plan diagrams visually characterize the optimizer's query plan choices on a parametrized query space, and extending them to the XML environment requires redesigned definitions of the parameters and the space. Through a comprehensive set of experiments on a variety of popular benchmarks, we demonstrate that XQuery plan diagrams can be significantly more dense and complex as compared to their relational counterparts. Further, they are more resistant to "anorexic reduction", requiring substantially larger cost-increase thresholds to achieve this objective. These results suggest that important research challenges remain to be addressed in the development of effective XQuery optimizers.

1 Introduction

Over the last decade, the flexibly structured XML language has become the de-facto standard for data representation and information exchange between applications. XML data was initially stored in traditional DBMS formats by shredding into relational tuples (e.g. [10]). However, in recent times most database vendors have augmented their SQL engines to provide *native* support for XML storage and XQuery interfaces, resulting in the so-called "hybrid" processors – examples include IBM DB2 [8], Oracle [14] and Microsoft SQL Server [15].

The automated optimization of declarative SQL queries is a classical hallmark of database technology. XML with its support for deep data hierarchies and powerful query operators, including regular expressions and sibling axes, has far more expressive power than SQL. Therefore, the optimization challenge becomes significantly more complex, motivating us to investigate, in this paper, the behavior of industrial-strength XQuery optimizers. For our analysis, we use the notion of "plan diagrams" developed in [9], which had hitherto been applied solely to relational engines, to drive the evaluation. Plan diagrams visually characterize the optimizer's query plan choices over an input parameter space, whose dimensions may include database, query and system-related features. In a nutshell, plan diagrams pictorially capture the geometries of the optimality regions of the *parametric optimal set of plans* (POSP) [5].

For a given database and system setup, the plan choices made by query optimizers are primarily a function of the *selectivities* of the predicates appearing in the query.

S.W. Liddle et al. (Eds.): DEXA 2012, Part I, LNCS 7446, pp. 131–146, 2012.
© Springer-Verlag Berlin Heidelberg 2012

Accordingly, our focus here is on plan diagrams obtained through selectivity variations on parametrized XQuery templates. In this process, we have to tackle a variety
of questions, including: (1) At what data granularity should the selectivity be varied –
specifically, *document* level and/or *node* level?; (2) What is the mechanism for varying
selectivities – specifically, through *structural* predicates and/or *value* predicates?; (3)
How are selectivities to be reliably estimated from the metadata available in these engines?; and (4) What constraints need to be imposed on the construction of XQuery
templates such that the resulting diagrams are semantically meaningful?

We begin this paper by describing our attempts to address the above issues. Subsequently, using the developed framework, we carry out a detailed analysis of a popular
commercial hybrid XQuery/SQL optimizer, which we refer to as **XOpt**, through an
extensive set of plan diagrams generated on three benchmark environments – XBench
[13], TPoX [7] and TPCH_X, the XML equivalent of the classical TPC-H [16] relational
benchmark. Our experimental results suggest that even two-dimensional plan diagrams
are often extremely dense, featuring hundreds of different plans, and further, exhibiting
intricate geometric patterns. This was especially the case when the XQuery template
featured *order by* clauses, *wild cards*, and *navigational axes*.

We also observed that when an XQuery template is rewritten in an equivalent XML/
SQL format [4], the plan diagrams produced are markedly different, clearly demonstrating that the choice of interface has a sizeable impact on the optimal plans' cost
behavior.

Finally, it had been empirically found in the relational world that even complex plan
diagrams could be simplified to retain just a few plans with only a marginal impact on
the query processing quality – this property was termed "anorexic reduction" in [3], and
has several useful applications, including providing robustness to selectivity estimation
errors. In the XML environment, however, we find instances wherein achieving anorexic
reduction incurs a very substantial deterioration of query processing quality.

Taken in toto, these results suggest that important challenges remain to be addressed
in the development of effective XQuery optimizers.

2   Background on Plan Diagrams

To set the stage, we first overview the notion of plan diagrams, developed in [9].
Consider a parametrized SQL query template that defines a relational selectivity
space – for example, QT8 shown in Fig. 1, which is based on Query 8 of the TPC-
H benchmark. Here, selectivity variations on the SUPPLIER and LINEITEM relations are
specified through the s_acctbal :varies and l_ extendedprice :varies predicates, respectively.[1]

The corresponding plan diagram for QT8, produced on a commercial database engine, is shown in Fig. 2(a). In this picture[2], each colored region represents a specific
plan, and a set of 42 different optimal plans (from the optimizer's perspective), P1
through P42, cover the selectivity space. The value associated with each plan in the

---

[1] We implement *:varies* using one-sided range predicates of the form *Relation.attribute≤const.*
[2] The figures in this paper should ideally be viewed from a color copy, as the grayscale version
may not clearly register the features.

```
select o_year, sum(case when nation = 'BRAZIL' then volume else 0 end) / sum(volume)
from (select YEAR(o_orderdate) as o_year, l_extendedprice * (1 - l_discount) as volume, n2.n_name as nation
      from part, supplier, lineitem, orders, customer, nation n1, nation n2, region
      where p_partkey = l_partkey and s_suppkey = l_suppkey and l_orderkey = o_orderkey and o_custkey =
          c_custkey and c_nationkey = n1.n_nationkey and n1.n_regionkey = r_regionkey and s_nationkey =
          n2.n_nationkey and r_name = 'AMERICA' and p_type = 'ECONOMY ANODIZED STEEL' and
          s_acctbal :varies and l_extendedprice :varies
      ) as all_nations
group by o_year
order by o_year
```

Fig. 1. Example SQL query template (QT8)

legend indicates the percentage area covered by that plan in the diagram – the biggest, P1, for example, covers about a quarter (26.86%) of the region.

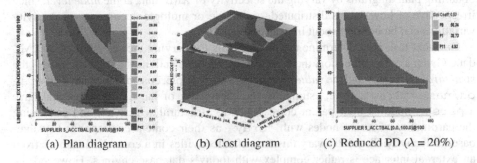

(a) Plan diagram   (b) Cost diagram   (c) Reduced PD ($\lambda = 20\%$)

Fig. 2. Plan, cost and reduced plan diagrams (QT8)

Related and complementary to the plan diagram is the "cost diagram", which quantitatively depicts the optimizer's (estimated) query processing costs of the plans featuring in the plan diagram. The cost diagram for the QT8 example is shown in Fig. 2(b).

Plan diagrams are often found to be complex and dense, featuring high plan cardinalities and intricate geometries, as can be observed in Fig. 2(a). However, these dense diagrams can typically be reduced to much simpler pictures retaining only a few plans from the POSP set, while ensuring that the processing quality of *any* individual query is not increased by more than a user-defined threshold λ. For example, the plan diagram in Fig. 2(a) can be reduced to that shown in Fig. 2(c), where only *three* of the original 42 plans are retained, while ensuring that no query suffering a plan replacement has had its cost increased by more than $\lambda = 20\%$. It has been empirically observed in [3] that, for templates based on the TPC-H and TPC-DS relational benchmarks, $\lambda = 20\%$ is typically sufficient to bring the plan cardinality in the final reduced picture down to around *ten plans* or fewer, referred to as "anorexic" (small absolute number) plan diagrams.

3   Generation of XML Plan Diagrams

As mentioned in the Introduction, a variety of issues crop up when we attempt to extend the concept of plan diagrams to the XML world. We discuss our approach to handling these issues in this section.

3.1 Varying XML Selectivity

In XML, information is organized in the form of *nodes* and *documents* containing these nodes. Therefore, selectivities can be computed at the granularity of nodes and/or of documents. For example, consider the scenario where 100 XML nodes are organized in a single document, and the other extreme where there are 100 documents, each containing one of these nodes. In the former case, the document selectivity will always be 0 (no node in the document satisfies the predicate) or 1 (at least one node in the document satisfies the predicate), whereas in the latter, the document selectivity will represent the fractional number of nodes satisfying the predicate. So, variations in the selectivity of XML data can potentially be achieved at the level of documents, or of nodes. In fact, it would even be possible to obtain plan diagrams among which selectivities of different dimensions are varied at different levels. However, in this paper, we only focus on obtaining plan diagrams by varying the selectivity of XML data at the *node level*, since irrespective of how data is distributed – in single or multiple documents – selectivity variations can be brought about in a desired range.

Our next task is to determine the mechanisms for varying the selectivities of XML data. Given a path expression, there are two kinds of selectivities to be estimated – the *structural selectivity* and the *value selectivity*. For example, consider the path `//person/name="Gray"`. Here, estimating the cardinality of the `name` nodes that have a `person` node as their parent corresponds to structural selectivity, and estimating the cardinality of `name` nodes with `"Gray"` as their content corresponds to value-based selectivity. To jointly vary this pair of selectivities in a controlled manner from an external interface is rather complex with today's database engines. However, the operator algorithms that are employed in XOpt are designed to enable the concurrent application of both predicates. We leverage this fact and vary the selectivity of chosen paths in the XML documents, through the application of parametrized value-based predicates to these paths. To ensure that both types of predicates are applied together, we create *value-based indexes* on the required paths. As a case in point, consider the path `/person/name` – here, we vary the path selectivity by applying suitable value-based predicates to `name` nodes that are direct descendants of `person` nodes emanating from the document root, after ensuring that a value-based index is present on the `/person/name` path.

It is to be noted that the above selectivity variations are brought about *externally*, through mechanisms that operate totally outside the database engine. Also, in our current work, we focus solely on varying the selectivities of predicates that are applied to document collections as the first step in the query execution plan. Selectivity variations of join predicates applied between document collections are not considered.

3.2 XQuery Template Construction

An XQuery template is used to specify the paths whose selectivities are to be varied through the application of value-based predicates. We will hereafter use the term **VSX** (Variable Selectivity XML element path)[3] to denote these varying dimensions.

---

[3] Our usage of the term *element* here denotes both XML elements as well as XML attributes.

```
1.  for $cust in xmlcol(CUSTOMER.CUSTOMER) /customers/customer[address_id :varies]
2.  for $order in xmlcol(ORDER.ORDER)
3.  /order[customer_id=$cust/@id]/order_lines/order_line[quantity_of_item :varies]
4.   return <Res> {$order} {$cust/first_name} {$cust/last_name} {$cust/phone_number} </Res>
```

Fig. 3. Example XQuery template

An example template is shown in Fig. 3, where there are two VSXs corresponding to customer addresses and customer orders, respectively. This template returns the names and phone numbers of customers located within a (parametrized) zip code and whose orders feature item quantities within a (parametrized) value.

The element path in a VSX predicate typically consists of a logical segment, denoted L, that defines the semantic object whose selectivity is desired to be varied, and a physical segment, denoted P, which is downstream of L and whose value is actually varied. To make this concrete, consider the VSX /order/order_lines/order_line [quantity_of_item :varies] in Fig. 3. Here L is the segment /order/order_lines/order_line, while P is the downstream segment order_line/quantity_of_item – the parametrized variation across order_lines is achieved through varying the values of quantify_of_item. As a final point, note that it is also acceptable to have templates where the logical segment itself terminates with the variable element, and there is no distinct P segment.

The value constants that would result in the desired selectivities of the VSXs are estimated by essentially carrying out an "inverse-transform" of the statistical summaries (histograms) corresponding to the VSXs. These path-specific value histograms are automatically built by the database engine whenever an index is declared on the associated paths, which as mentioned earlier, is important for our selectivity variation strategy. We employ linear interpolation mechanisms on these summaries to estimate the constants corresponding to the query locations in the selectivity space.

XQuery statements can often be complex, involving powerful constructs from second order logic such as regular expressions and navigational paths. Therefore, it is extremely important that the associated templates be constructed carefully, otherwise we run the risk of producing plan diagrams that are semantically meaningless since the intended selectivity variations were not actually realized in practice. Accordingly, we list below the conditions to be satisfied by a valid XQuery template, arising out of *conceptual* reasons.

1. *Many-to-one* relationships are not permitted to occur in the P segment (if present) of a VSX predicate. This translates to requiring, in the graphical representation of the XML schema [10], that there should be no '*' (wild card) node appearing in the path corresponding to P.
2. A collection C of XML documents adhering to a common schema can participate in at most one VSX. Further, they can also participate in *join-predicates* with other XML document collections, but are not permitted to be subject to additional selection predicates.
3. The VSXs should appear sufficiently frequently in the documents and their value predicates should be over dense domains.

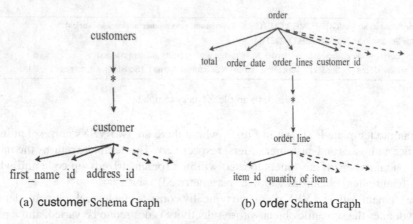

(a) **customer** Schema Graph   (b) **order** Schema Graph

Fig. 4. XQuery template constraints

To illustrate the constraints on XQuery templates, consider an XML database with **Customer** and **Order** document collections adhering to the schema graphs shown in Figs. 4(a) and 4(b), respectively. On this database, consider the XQuery template shown in Fig. 3 – this template is compatible with respect to all of the above conditions. However, if the query template were to be slightly altered to

`/order[customer_id=$cust/@id and order_lines/order_line/quantity_of_item :varies]`

in line 3, the template becomes invalid due to violating Requirement 1 above – a '*' node is present in the `order_lines/('*')order_line/quantity_of_item` physical segment.

4   Experimental Results

In this section, we describe the experimental framework on which the XOpt optimizer was evaluated, and discuss the results obtained on a variety of benchmark environments.

4.1   Experimental Setup

Our experiments were carried out on a vanilla hardware platform – specifically, a SUN ULTRA 20 system provisioned with 4GB RAM running Ubuntu Linux 64 bit edition 9.10. The XOpt engine was was run at its default optimization level, after enabling all options that enhanced the scope for optimization.

Databases. We operated with three different XML databases: **XBench**, **TPoX** and **TPCH_X**. TPoX and XBench are native XML benchmarks, while TPCH_X is an XML equivalent of the classical TPC-H relational benchmark. XBench and TPCH_X represent decision-support environments, whereas TPoX models a transaction processing domain – their construction details are given below.

XBench. This benchmark [13] supports four different kinds of databases, of which we chose the **DC/MD** (Data Centric, Multiple Documents) flavor, since it symbolises business data spread over a multitude of documents, and appears the most challenging from the optimizer's perspective. The DC/MD generator was invoked with the "large" option, resulting in a database size of around 1 GB, with all data conforming to a uniform distribution. For the physical schema, indexes were created for all paths involved in join predicates, and additionally for all paths appearing in VSXs.

TPoX. The data generator of the TPoX benchmark [7] was invoked at the **XXS scale**, resulting in 50000 CUSTACC, 500000 ORDER and 20833 SECURITY documents, collectively taking about 1GB of space. The data in these documents follow a variety of distributions, ranging from uniform to highly skewed. For the physical schema, all the 24 recommended indexes (10 on CUSTACC, 5 on ORDER and 9 on SECURITY) were created. These indexes covered all the paths involved in the join predicates and VSXs featured in our templates.

TPCH_X. Here, the relations of the TPCH [16] benchmark: NATION, REGION, SUPPLIER, CUSTOMER, PART, PARTSUPP, ORDERS, LINEITEM were first converted to their equivalent XML schemas, with ORDER and LINEITEM combined into the Order XML schema, and PART and PARTSUPP merged into Part XML schema. This merging was carried out since individual orders and parts are associated with multiple lineitems and suppliers, respectively, and these nested relationships are directly expressible in XML through its organic support for hierarchies.

Then, the TPC-H relational data at scale 1 was translated to these XML schemas using the Toxgene [1] tool, resulting in a database of around 1 GB size. For the physical schema, indexes were created on all paths that featured in join predicates and in VSXs.

XQuery Templates. For simplicity and computational tractability, we have restricted our attention to *two-dimensional* XQuery templates in this study. These templates were created from representative queries appearing in the above-mentioned benchmarks (X-Bench has a suite of 15 XQueries, TPoX has 11 XQueries and 11 SQL/XML queries, while TPCH_X has 22 XQueries). The templates were verified to be compatible with the constraints specified in Sect. 3, and the VSX value predicates are on floating-point element values (explicit indexes are created on the VSXs, resulting in value based comparisons).

The plan diagrams are produced at a resolution of 300 points in each dimension, unless specified otherwise – this means that close to a hundred thousand queries are optimized in each 2D diagram. Since optimizing an individual query takes between 100 to 200 milliseconds, generating the complete plan diagram typically requires a few hours (3-5 hours). Finally, for plan diagram reduction, which falls into the NP-hard complexity class, we employed the the Cost-Greedy heuristic algorithm described in [3], with the default cost-increase threshold λ set to 20%.

In the remainder of this section, we present results for XQuery plan diagrams produced on the XBench, TPoX and TPCH_X environments.

4.2 Plan Diagrams with XBench

We present here the results for two XQuery templates that cover the spectrum of query complexity: the first, referred to as QTXB1, features the basic constructs, whereas the second, referred to as QTXB2, includes a rich variety of advanced operators.

```
for $order in xmlcol(ORDER.ORDER)/order,
  $cust in xmlcol (CUSTOMER.CUSTOMER)/customers/customer
  where $order/customer_id = $cust/@id and $order/total :varies and $cust/address_id :varies
  order by $order/order_date
  return <Output> {$order/@id} {$order/order_status} {$cust/address_id}
    {$cust/first_name} {$cust/last_name} {$cust/phone_number} </Output>
```

Fig. 5. XQuery template for XBench (QTXB1)

Basic Template. The basic template, QTXB1, is based on Query 19 of the benchmark and is shown in Fig. 5. Its objective is to retrieve all purchases within a (parametrized) total value for which the associated customers are located within a (parametrized) address value, the result being sorted by the purchase date.

The plan diagram for QTXB1 is shown in Fig. 6(a), and we observe that, even for this basic template, as many as 42 plans are present with intricate spatial layouts. Further, the area distribution of the plans is highly skewed, with the largest plan occupying a little over 20% of the space and the smallest taking less than 0.001%, the overall Gini (skew) co-efficient being close to 0.9.

Most of the differences between the plans are due to operator *parameter switches*, rather than the plan tree structures themselves. For example, the difference between plan pairs is often solely attributable to the presence or absence of the TMPCMPRS switch, associated with the TEMP and SORT operators. When such switch differences are ignored, the number of plans comes down sharply to just 10! It is interesting to note that the switch by itself contributes very little to the overall plan cost.

In Fig. 6(a), the plans P1 (red) and P2 (dark blue) blend together in a wave-like pattern. The operator trees for these plans are shown in Figs. 7(a) and 7(b), respectively. We see here that the plans primarily differ on their join orders with respect to the document collections – P1 computes Order ⋈ Customer whereas P2 evaluates Customer ⋈ Order – that is, they differ on which document collection is outer, and which is inner, in the join.

Near and parallel to the Y-axis, observe the yellow vertical strip corresponding to plan P4, sprinkled with light orange spots of plan P16. The P4 and P16 plans differ only in their positioning of an NLJOIN-XSCAN operator pair, where the expressions $ord/order_status, $ord/@id and $ord/order_date of the return clause are evaluated. In XOpt, the XSCAN operator parses the input documents with a SAX parser and evaluates complex XPath expressions in a single pass over these documents.

The associated cost diagram is shown in Fig. 6(b), and we observe a steep and affine relationship for the expected execution cost with regard to the selectivity values.

When plan diagram reduction with $\lambda = 20\%$ is applied, we obtain Fig. 6(c), where the cardinality is brought down to 21 from the original 42. Although there is an appreciable degree of reduction, it does not go to the extent of being "anorexic" (around 10 plans or less) – this is in marked contrast to the relational world, where anorexic reduction was invariably achieved with this λ setting [3]. More interestingly, anorexia could not be achieved even after substantial increases in the λ setting – in fact, only at the impractically large value of $\lambda = 150\%$ was this objective reached!

Complex Template. We now turn our attention to QTXB2, the complex XQuery template shown in Fig. 8. This template attempts to retrieve the names, number of items bought, and average discount provided for customers who live within a (parametrized)

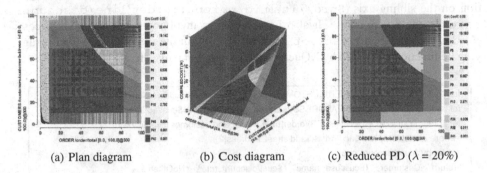

(a) Plan diagram   (b) Cost diagram   (c) Reduced PD ($\lambda = 20\%$)

Fig. 6. Plan, cost and reduced plan diagrams for XBench QTXB1
(X-Axis: ORDER /order/total, Y-Axis: /customers/customer/address_id)

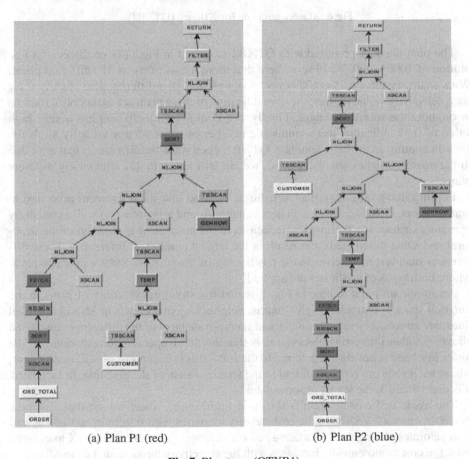

(a) Plan P1 (red)   (b) Plan P2 (blue)

Fig. 7. Plan trees (QTXB1)

address value and whose total purchases are within a (parametrized) value, with the output sorted on the discount rates. QTXB2 is based on XQueries 9, 14 and 19 of XBench and incorporates most features provided by the XQuery language. Specifically, it has all the FLWOR clauses, expressions involving wild cards ($ord//item_id), and navigation on the sibling axis ($ord//item_id/parent::order_line/discount_rate) in the *return* clause. It also has predicates involving positional node access ($add/exists($add/street_address[2])) in the *where* clause. Finally, aggregate functions provided by XQuery, such as *count* and *avg* are also employed.

```
for $cust in xmlcol (CUSTOMER.CUSTOMER)/customers/customer[address_id :varies]
let $add := xmlcol(ADDRESS.ADDRESS)/addresses/address[@id=$cust/address_id]
let $order in xmlcol(ORDER.ORDER)/order[total :varies and customer_id=$cust/@id]
  where exists($ord) and $add/exists($add/street_address[2])
  order by $cust/discount_rate
  return <Customer> {$cust/user_name} {$cust/discount_rate} <NoOfItems>
    {count($ord//item_id)} </NoOfItems> <AvgDiscount>
    {avg($ord//item_id/parent::order_line/discount_rate)} </AvgDiscount> </Customer>
```

Fig. 8. XQuery template for XBench (QTXB2)

The plan diagram produced with QTXB2 is shown in Fig. 9(a), produced at a resolution of 1000*1000. We observe here that there are as many as 310 different plans! With a cursory glance, it would almost seem as though a different plan is chosen for each selectivity value. Further, the spatial layouts of these plans are extremely complex, throughout the selectivity range. Finally, the Gini skew co-efficient has a very high value of 0.95 indicating that a miniscule number of plans occupy virtually all of the area. In contrast to the basic template, QTXB1, even when the differences that arise due to parameter switches are disregarded, we are still left with 222 structurally different plans.

Drilling down into these plan structures, we find that the differences arise due to many factors, including changes in access methods and join orders, as well as ancillary operators such as SORT. However, variations in the position of application of the structural and value-based predicates result in the largest number of differences. These differences manifest themselves in the positioning of the NLJOIN-XSCAN operator pairs, where both types of predicates are applied.

An important point to note in Fig. 8 is that the structural volatility of plans in the diagram space is extremely high – that is, neighboring plans, even in adjacent parallel lines, are structurally very dissimilar and incorporate most of the differences discussed above. Another interesting observation is that the SORT operator corresponding to the *order by* clause is not always deferred to the end – this is in contrast to SQL query plans, where such sorts are typically found in the terminal stem of the plan tree. In fact, here, it is even found at the leaves of some plans!

The associated cost diagram is shown in Figure 9(b), where we see that along the /order/total VSX axis, the cost steadily increases with selectivity until 50%, and then saturates. Along the /customers/customer/address_id VSX axis, however, the cost monotonically increases with the selectivity throughout the range.

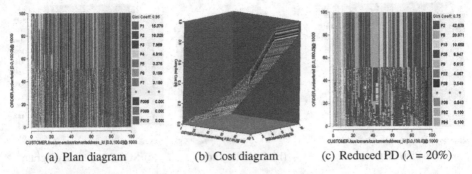

(a) Plan diagram   (b) Cost diagram   (c) Reduced PD (λ = 20%)

Fig. 9. Plan, cost and reduced plan diagrams for XBench QTXB2
(X-Axis: /customers/customer/address_id, Y-Axis: /order/total)

Finally, when reduction at $\lambda = 20\%$ is applied to the plan diagram, Fig. 9(c) is obtained wherein 13 plans are retained. The geometries of these surviving plans continue to be intricate even after reduction, whereas in the relational world, cardinality reduction was usually accompanied by a corresponding simplification in the geometric layout as well.

4.3 Plan Diagrams with TPoX

Here, we present the results obtained with an XQuery template on the TPoX XML transaction processing benchmark. The query template, which is shown in Fig. 10, is based on the cust_sold_security.xqr XQuery given as part of the benchmark. This template, which we will hereafter refer to as QTX_SEC, returns details of customers with account(s) whose working balance(s) are within a (parametrized) value, and have traded one or more securities, with the trading amounts of the associated orders being within a (parametrized) value – the final result is alphabetically sorted on customer last names.

The plan diagram (at a resolution of 1000*1000) for QTX_SEC is shown in Fig. 11(a). It consists of 23 plans with plan P1 occupying about three-quarters of the space and plan P23 present in less than 0.001% of the diagram, resulting in an overall Gini coefficient of 0.34. Further, the plan cardinality decreases to 10, when differences between plan trees due to parameter switches, such as PREFETCH and SORT, are not considered. We also see that the plan diagram predominantly consists of vertical blue bands (plan P2) on the red region (plan P1). Only close to the X and Y-axes do we find other plans,

```
declare default element namespace "http://www.fixprotocol.org/FIXML-4-4";
declare namespace c="http://tpox-benchmark.com/custacc";
for $cust in xmlcol(CUSTACC.CADOC)/c:Customer/c:Accounts/c:Account[c:Balance/c:WorkingBalance :varies]
for $ord in xmlcol(ORDER.ODOC)/FIXML/Order[@Acct=$cust/@id/fn:string(.) and OrdQty/@Cash :varies]
 order by $cust/../../c:Name/c:LastName/text()
 return <Customer> {$cust/../../c:Name/c:LastName} {$cust/../../c:Nationality} </Customer>
```

Fig. 10. XQuery template for TPoX (QTX_SEC)

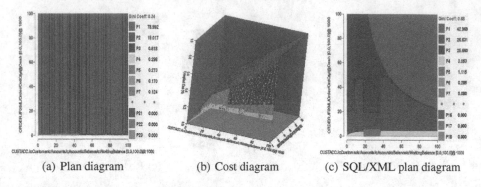

(a) Plan diagram   (b) Cost diagram   (c) SQL/XML plan diagram

Fig. 11. Plan and cost diagrams for TPoX QTX_SEC; Equivalent SQL/XML plan diagram
(X-Axis: /c:Customer/c:Accounts/c:Account/c:Balance/c:WorkingBalance, Y-Axis: /FIXML/Order/OrdQty/@Cash)

such as the yellow vertical stripe of plan P4, and the brown and purple horizontal bands of plans P3 and P5, respectively.

The associated cost diagram is shown in Fig. 11(b), where we observe a strongly non-linear behavior with respect to the VSX selectivities. When the plan diagram is reduced with $\lambda = 20\%$, the number of plans comes down to 11 plans, with the blue bands (P2) being swallowed by the red plan (P1).

The TPoX benchmark also features a semantically equivalent SQL/XML version of the XQuery example used above. We converted this version into an equivalent template, for which we obtained the plan diagram shown in Fig. 11(c). The operators found in the plans of the SQL/XML plan diagram are the same as those found in the XQuery plan diagram. However, the choice of optimal plans vastly differs across the diagrams – note that the SQL/XML plan diagram throws up only 18 plans, and that too, without any striking patterns.

Further, although the end query results are identical, there are substantial cost variations between the two query template versions. Specifically, the SQL/XML template has minimum and maximum costs of 2.77E3 and 1.14E7, respectively, whereas the XQuery template has minimum and maximum costs of 2.76E3 and 1.65E6, respectively. These differences in costs are due to the application of the SORT operator at different positions in the plan trees. Sorting is always carried out at the stem of the plan tree, after applying all predicates, in the case of SQL/XML, whereas the application of sorting is sometimes done earlier in the case of XQuery. For the cases where sorting is applied at the stem level for both types of queries, there is also a difference in the estimated cardinalities, which is reflected in the cost estimates. This indicates that, even when the underlying data and the query semantics are the same, the specific choice of query interface may have a material impact on the runtime performance.

4.4   Plan Diagrams with TPCH_X

An XQuery template based on Query 10 of TPCH_X is shown in Fig. 12(a). This template retrieves the names, home nations, and marketing segments of customers, and the identifiers of all their purchases, with the results ordered on the marketing segments. The plan diagram for this template is shown in Fig. 13(a), where we again observe

for $c in xmlcol(CUSTOMERS.CUSTOMERS)
/Customers/Customer/[**AcctBal :varies**]
let $n:=xmlcol(NATIONS.NATIONS)/Nations/Nation[@key=$c/NationKey]
let $o := xmlcol(ORDERS.ORDERS)
/Orders/Order[$c/@key=CustKey and **TotalPrice :varies**]
where fn:exists($o)
order by $c/MktSegment
return <Customer> {$c/Name} {$n/Name} {$c/MktSegment}
{$o/OrderKey} </Customer>

(a) XQuery template

select c_name, n_name,
c_mktsegment, o_orderkey
from customer, nation, orders
where **c_acctbal :varies** and
c_natkey=n_natkey and
o_custkey=c_custkey
and **o_totalprice :varies**
order by c_mktsegment

(b) SQL query template

Fig. 12. XQuery and SQL templates for TPCH_X/TPCH

an extremely complex diagram populated with 61 different plans, appearing mostly as rapidly alternating bands of colors. When this diagram is subject to reduction with $\lambda = 20\%$, we obtain Fig. 13(b), which retains 19 plans and is therefore not anorexic in nature. In fact, λ had to be increased to as much as 50% to obtain an anorexic diagram.

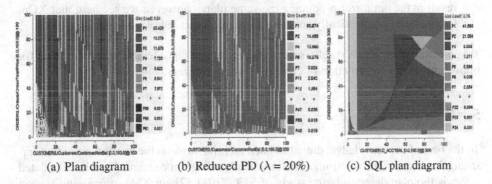

(a) Plan diagram   (b) Reduced PD ($\lambda = 20\%$)   (c) SQL plan diagram

Fig. 13. Plan and reduced plan diagrams for TPCH_X and plan diagram for TPCH
(X-Axis: /Customers/Customer/AcctBal, Y-Axis: /Orders/Order/TotalPrice)

As a matter of curiosity, we also investigated the behavior of the equivalent (in terms of the result set) SQL query template, shown in Fig. 12(b). The associated plan diagram in Fig. 13(c) throws up a much simpler picture, both in terms of cardinality (34 plans) and in the spatial layouts of the optimality regions. Further, the estimated execution costs for the XQuery template are *orders of magnitude* higher in comparison to those obtained with the SQL template! Assuming that the optimizer's modeling quality is similar in both environments, these results indicate that database administrators of hybrid systems must make a careful choice of data representation to provide the best performance for their users.

4.5 General Observations

During the course of our experimentation on XQuery plan diagrams, a few general observations emerged, which are highlighted below:

- The presence of an *order by* clause in the XQuery templates results in a dramatic increase in the richness of plan diagrams, with respect to both the density and geometric complexity. The reason is as follows: XML is inherently ordered, and results are always produced in document order (without the presence of *order by*). With the presence of *order by* on a path, a low-cost sort can potentially be accomplished at several steps in the optimization process, and hence there is a large set of similarly-costed alternative plans to choose from, many of which surface as the locally-optimal plan at one or the other location in the selectivity space.
- The complexity of the plan diagrams increases with the complexity of the predicates involved in the XQuery template, with advanced features such as navigation on different axes (sibling and parent), wild cards and positional node access triggering this behavior.
- The position of predicates – whether appearing in the XPath expression or in the *where* clause of the XQuery templates – has a significant impact on the complexity of plan diagrams in terms of both plan cardinality and spatial distribution. Further, and very importantly, this shift of position also results in plans with substantially changed costs. Ideally, in a truly declarative world, all equivalent queries should result in the optimizer producing the same plan – however, we see here that XOpt is not able to automatically sniff out these important rewriting opportunities.

Taken in toto, the above results seem to suggest that considerable scope remains for improving on the construction of current hybrid optimizers.

5   Related Work

To the best of our knowledge, there has been no prior work on the analysis of *industrial-strength* native XQuery optimizers using the plan diagram concept. The closest related effort is the plan-diagram-based study of SUCXENT [2], an XML processing system that uses Microsoft SQL Server as the backend relational storage engine. They studied the behavior of this optimizer in the context of XPath processing, by first converting all XPath queries to their equivalent SQL versions.

An XML plan diagram *instance* was also shown in [11], using IBM DB2, to motivate the need for accurate cardinality estimations of XQuery expressions – in their experimental setup, the optimal plan choice is highly volatile, varying with small changes in selectivity, and inaccurate estimations of cardinalities result in choosing plans that are worse than the optimal by orders of magnitude.

A flexible optimization framework, incorporating both rules and cost estimates, was presented in [12] for visualizing the XQuery optimization process in a native XML DBMS environment. The framework supports implementing, evaluating, and reconfiguring optimization techniques, even during run-time, through a visual tool. However, all these features are applicable to individual queries, and are not intended for visualization over a parameter space.

Finally, an evaluation of open-source XQuery processors was carried out in [6], but their focus was on characterizing the response time performance for specific benchmark queries.

6   Conclusions and Future Work

In this paper, we have attempted to analyze the behavior of XOpt, an industrial-strength XQuery/SQL hybrid optimizer, using the concept of plan diagrams proposed some years ago in [9]. We first addressed the issue of what comprises a XML selectivity space, and the mechanisms to be used to bring about the desired selectivity variations. Then, we enumerated the constraints that need to be satisfied in formulating XQuery templates so as to obtain meaningful plan diagrams. Subsequently, we provided a detailed report on XOpt's plan diagram behavior over a representative set of complex XQuery templates constructed over popular XML benchmarks.

Our experimental results indicate that XML plan diagrams are significantly more complex than their relational counterparts in terms of plan cardinalities, densities and spatial layouts. In particular, we observe a pronounced "banding" effect resulting in wave-like patterns. Further, the presence of even syntactic expressions, such as *order by*, visibly increase the complexity of the resulting diagrams. We also find that these diagrams are not always amenable to anorexic reduction at the 20% cost-increase threshold found sufficient in the relational literature, often requiring substantially higher thresholds to achieve the same goal. Another interesting facet is that equivalent XML and SQL queries typically produce substantially different cost estimations from the optimizer. Overall, these results suggest that important challenges remain to be addressed in the development of effective hybrid optimizers.

While our analysis was restricted to XOpt in this study, it would be interesting to profile the plan diagram behavior of other industrial-strength XML query processing engines as well. Further, going beyond the two-dimensional query templates evaluated here to higher dimensions will provide greater insight into addressing the design issues underlying these systems.

References

1. Barbosa, D., Mendelzon, A., Keenleyside, J., Lyons, K.: ToXgene: An extensible template-based data generator for XML. In: Proc. of 5th Intl. Workshop on the Web and Databases (WebDB 2002) (June 2002)
2. Bhowmick, S., Leonardi, E., Sun, H.: Efficient Evaluation of High-Selective XML Twig Patterns with Parent Child Edges in Tree-Unaware RDBMS. In: Proc. of 16th ACM Conf. on Information and Knowledge Management (CIKM) (November 2007)
3. Harish, D., Darera, P., Haritsa, J.: On the Production of Anorexic Plan Diagrams. In: Proc. of 33rd Intl. Conf. on Very Large Data Bases (VLDB) (September 2007)
4. Eisenberg, A., Melton, J.: SQL/XML and the SQLX Informal Group of Companies. SIGMOD Record (30) (September 2001)
5. Hulgeri, A., Sudarshan, S.: Parametric Query Optimization for Linear and Piecewise Linear Cost Functions. In: Proc. of 28th Intl. Conf. on Very Large Data Bases (VLDB) (August 2002)
6. Manegold, S.: An empirical evaluation of XQuery processors. Information Systems Journal (April 2008)
7. Nicola, M., Kogan, I., Schiefer, B.: An XML Transaction Processing Benchmark (TPoX). In: Proc. of 2007 ACM SIGMOD Intl. Conf. on Management of Data (June 2007)

8. Nicola, M., Linden, B.: Native XML support in DB2 universal database. In: Proc. of 31th Intl. Conf. on Very Large Data Bases (VLDB) (August 2005)
9. Reddy, N., Haritsa, J.: Analyzing Plan Diagrams of Database Query Optimizers. In: Proc. of 31st Intl. Conf. on Very Large Data Bases (VLDB) (August 2005)
10. Shanmugasundaram, J., Tufte, K., Zhang, C., He, G., DeWitt, D., Naughton, J.: Relational Databases for Querying XML Documents: Limitations and Opportunities. In: Proc. of 25th Intl. Conf. on Very Large Data Bases (VLDB) (September 1999)
11. Teubner, J., Grust, T., Maneth, S., Sakr, S.: Dependable Cardinality Forecasts for XQuery. In: Proc. of 34th Intl. Conf. on Very Large Data Bases (VLDB) (August 2008)
12. Weiner, A., Härder, T., Silva, R.O.: Visualizing Cost-Based XQuery Optimization. In: Proc. of 26th Intl. Conf. on Data Engineering (ICDE) (March 2010)
13. Yao, B., Ozsu, M., Khandelwal, N.: XBench Benchmark and Performance Testing of XML DBMSs. In: Proc. of 20th Intl. Conf. on Data Engineering (ICDE) (March 2004)
14. http://docs.oracle.com/cd/B14117_01/appdev.101/b10790/xdb01int.htm
15. http://msdn.microsoft.com/en-us/library/ms345117(v=sql.90).aspx
16. http://www.tpc.org/tpch

Spreadsheet Metadata Extraction:
A Layout-Based Approach

Somchai Chatvichienchai

Dept. of Information and Media Studies, University of Nagasaki,
`somchaic@sun.ac.jp`

Abstract. Metadata is an essential part of modern information system since it helps people to find relevant documents from disparate repositories. This paper proposes an innovative metadata extraction method for spreadsheets. Unlike traditional methods which concern only content information, this paper considers both layout and content. The proposed method extracts metadata from the spreadsheets whose metadata is stored under certain conditions. Data types (such as date, number, etc.) of metadata are taken into account in order to realize document search based on metadata of various data types. Furthermore, the extracted metadata is semantically classified and hierarchically grouped in order to allow end-users to define complex search queries and the meanings of search keywords. System implementation of the proposed method is also discussed.

Keywords: Metadata, schema, spreadsheet, data model, XML.

1   Introduction

Spreadsheet programs, such as Microsoft Excel[11], Lotus 1-2-3[8] and Calc[13], are used by millions of users as a routine all-purpose data management tool. As more and more spreadsheets become electronically available, finding the spreadsheets that fit users' needs from disparate repositories is becoming increasingly important. Current search tools, such as Google Desktop Search[5], X1 Professional Client[20], create an index of all the words in a document. However, just indexing words that appears in the documents is not enough to provide an effective search because they will return the document results including those that do not match with user's search objective. Consider a search of the purchase orders, which have been ordered by Tanaka Book Store after 2011/8/10. The first problem is that current search tools did not know which documents contain purchase order data and which fields are customer name and order date. They basically find the files which include inputted keywords. As the result, the search tools return many irrelevant document files. The second problem is that current search tools cannot find the spreadsheets when data formats of search keywords (such as date, number etc.) do not match with those of metadata of the documents.

In order to solve the above problems, my previous work[2] proposed a metadata extraction method that (1) binds the cells storing metadata of Excel template with XML schema[17] which defines classes of metadata of a document type, and (2)

S.W. Liddle et al. (Eds.): DEXA 2012, Part I, LNCS 7446, pp. 147–160, 2012.

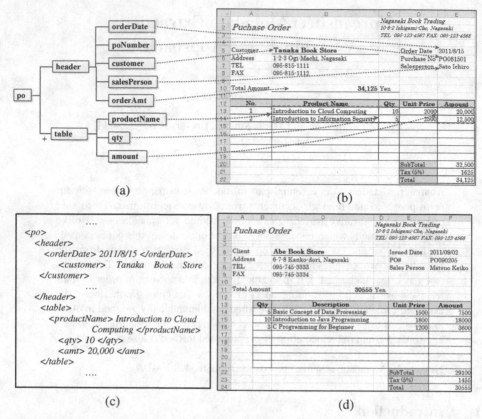

Fig. 1. (a) An example of a v-metadata schema, (b) an example of a spreadsheet whose cells are bound with the v-metadata schema elements, (c) metadata of the spreadsheet of (b), and (d) another spreadsheet whose layout is similar to that of (b).

embeds a VBA macro[19] into the template to output metadata into XML[18] files when users save spreadsheets created from the template. Figure 1(c) illustrates an example of metadata which is generated from a spreadsheet of Fig.1(b). The metadata is stored as XML data so that XML tags present classes of the metadata and the nested structure of XML presents hierarchical structure of metadata classes. For example, *<orderDate>2011/8/15</orderDate>* is a sub-element of *header* element. It states that "2011/8/15" is classified to the order date which is a sub-class of the header. The metadata files, which are outputted by the VBA macro, will be read by indexer for search index creation. However, the previous method has two drawbacks. The first is that it cannot extract metadata from the spreadsheets created before implementing it. The last is that VBA macro of these spreadsheets may be disabled by users who are not sure of security of the macro. Therefore, metadata of these spreadsheets cannot be outputted by the macro.

The objective of this paper is to propose an innovative metadata extraction method that can solve the above problems of the previous work. Many organizations usually use templates which are pre-designed documents formatted for common purposes such as a purchase order, invoice or sales report. A new template may be created from the existing template by changing some layout and data fields. In case of minor layout modification, the spreadsheets created from the new template tend to have layouts which are similar to those created from the old template. The proposed method will be effectively applied for the spreadsheets having similar layout. Figure 1(d) depicts a spreadsheet having a worksheet whose layout is similar to that of Fig.1(b). By considering the synonym of the word "customer" of spreadsheet of Fig.1(b) and the word "client" of spreadsheet of Fig.1(d), it can be deduced that "Abe Book Store" be metadata that should be classified as the customer name. The proposed method discovers the spreadsheets whose metadata is organized in the patterns defined by users. This paper focuses on the spreadsheets that are edited by Microsoft Office Excel 2003/2007/2010 (Excel, for short) since Excel is used in many public and private organizations. However, the proposed method can be extended to handle the spreadsheets of other software. Here, the term "spreadsheets" denotes the spreadsheets edited by Excel.

The rest of the paper is organized as follows. Section 2 introduces basic concepts of spreadsheet data model, metadata and XML. Section 3 presents a method of locating metadata of a spreadsheet by schema binding. In section 4, issues in metadata extraction of spreadsheets are discussed. Metadata extraction algorithm is presented in section 5. Section 6 discusses implementation of the proposed algorithm. In section 7, the related work is discussed. Finally, the last section concludes this paper and future work.

2   Basic Concept

2.1   Spreadsheet Data Model

A spreadsheet is a set of worksheets. A worksheet consists of cells. A cell of a worksheet is defined as follows.

Definition 2.1 (Worksheet Cells). A worksheet cell c is represented by the coordinate $<w, x, y>$ where w is the name of a worksheet, x is the column number, and y is the row number.   □

Following spreadsheet conventions, cell coordinates x and y are numbered starting from 1 and can also be denoted using capital letters for column numbering followed by numbers for rows numbering. For example, $<sheet1, 1, 3>$ can also be denoted as cell A3 of worksheet $sheet1$. Cell A1 is the upper-leftmost cell of a worksheet. A rectangular subset of cells is called a cell range (or a range, for short) and is conventionally denoted by its upper-leftmost and lower-rightmost cells separated by a colon. For example, a range A13:E14 of worksheet w is represented by $\{<w, x, y> \mid 1 \leq x \leq 5; 13 \leq y \leq 14\}$. A range $<w, x_1, y_1>:<w, x_2, y_2>$ always verifies $1 \leq x_1 \leq x_2$ and $1 \leq y_1 \leq y_2$. Note that if $x_1 = x_2$ and $y_1 = y_2$, then $<w, x_1, y_1>:<w, x_2, y_2> = <w, x_1, y_1>$ is derived. This means that a cell can be seen as a range of one row and one column. Merge cell is a function of Excel that allows multiple adjacent cells to be combined into a single larger cell which is called a *merged cell*. In this paper, the location of a merged cell is defined by the upper-leftmost of the cells that are combined.

Let $r_1 = <w, x_1, y_1>:<w, x_2, y_2>$ and $r_2 = <w, x_3, y_3>:<w, x_4, y_4>$ be cell ranges of worksheet w. Distance from r_1 to r_2 is defined as follows.

- Column distance from r_1 to r_2 represented by $c\_dist(r_1, r_2)$ is equal to $x_1 - x_3$. Therefore, the column distance between r_1 and r_2 is equal to $|c\_dist(r_1, r_2)|$.
- Row distance from r_1 to r_2 represented by $r\_dist(r_1, r_2)$ is equal to $y_1 - y_3$. Therefore, the row distance between r_1 and r_2 is equal to $|r\_dist(r_1, r_2)|$.

2.2   Metadata

Metadata is data about data, more specifically a collection of key information about a particular content, which can be used to facilitate the understanding, use and management of data. In this paper, the metadata defined for searching spreadsheets is classified into the following two categories: *K-metadata* is the data used to compare with inputted search keywords, and *F-metadata* is the data used to locate the corresponding K-metadata. Location relationship between K-metadata and F-metadata is defined as following conditions. For example, "Product Name" of cell B12 of the spreadsheet of Fig.1(b) is F-metadata for K-metadata stored at the range B13:B14. "Customer" of cell A5 is F-metadata for K-metadata stored at cell B5.

2.3   XML

XML provides a way to describe metadata. XML tags are used to define the structure and types of the data itself. Users can define an unlimited set of XML tags. XML uses a set of tags to delineate elements of data. Each element encapsulates a piece of data that may be very simple or very complex. XML schema is a document used to define and validate the content and structure of XML data, just as a database schema defines and validates the tables, columns, and data types that make up a database. XML Schema defines and describes certain types of XML data by using the XML Schema definition language (XSD). In this paper, an XML schema is logically viewed as a tree which is defined as follows.

Definition 2.2 (XML Schema Trees) An XML schema tree is a finite node labelled and edge labelled tree $T = (V, E, Nm, r, lbl, C, f_C)$, where V is a set of vertices (nodes) representing elements and attributes of the schema, $E \subseteq V \times V$ is a set of edges, Nm is a set of element and attribute names, r is the root of the schema, lbl is a labelling function assigning a name from Nm to a node in V, C is the set $\{+, *, ?\}$ called cardinality, and f_C is a partial function assigning a label from C to an edge in Ed. It should be clear that edge labels indicate the occurrence of sub-element to be expected: $+$, $*$, $?$

meaning "one or more", "zero or more", and "zero or one", respectively. An edge with no label indicates that the occurrence of sub-element is "exactly one".   □

3   Locating Metadata of a Spreadsheet by Schema Binding

3.1   K-Metadata Schemas

A K-metadata schema defines K-metadata for searching spreadsheets of the same type (such as purchase order, sales report, etc.). In this paper, it is assumed that each organization assigns a person who is responsible for managing document search system of the organization. In this paper, this person is called system manager. K-metadata schema is defined by the system manager under by the following criteria.

- Criteria 1: The name of a leaf node of K-metadata schema tree denotes the class of corresponding K-metadata.
- Criteria 2: In order to enforce that a set of K-metadata stored in the same worksheet, the leaf nodes presenting these metadata should be defined as child nodes of the same parent node. The cardinality of the parent node under the root node is one.
- Criteria 3: In order to define a set of K-metadata stored in the same table, the leaf nodes presenting these metadata should be defined as child nodes of the same parent node. The cardinality of the parent node under the root node is '+' or '*'.

Figure 1(a) depicts the schema of K-metadata for the spreadsheet of Fig.1(b). The root node is *po* which states that the document type is purchase order. The root node has two child nodes: *header* and *table* nodes. According to criteria 2, *Header* node contains child nodes denoting K-metadata stored in the same worksheet. For example, *orderDate* node is bound to cell E5 of spreadsheet of Fig.1(b). Therefore, the string "2011/8/15" of cell E5 is classified to the order date. The cardinality of *table* node is '+' which denotes that number of *table* nodes under *po* node is greater than or equal to one. Therefore, each *table* node has child nodes which define K-metadata stored in the same table row. For example, *productName* is bound to cells B13:B14. Then, the strings "Introduction to Cloud Computing" and "Introduction to Information Security" are classified to the product name.

Let *limit_dist* be the limited distance between K-metadata and the corresponding F-metadata. This value is defined by the system manager to match with layout of spreadsheets of the organization.

Definition 3.1 (Well-formed Spreadsheets). A spreadsheet S is a *well-formed spreadsheet* if S holds the following properties.

- For each range r of S that does not store K-metadata as table data, (1) r's F-metadata stored in the same row as r, (2) r is the nearest range for its F-metadata, and (3) the column distance from r to its F-metadata is less than *limit_dist*.
- For each range r of S that stores K-metadata as table data, its F-metadata is the column header of the table data.   □

3.2   Schema Binding

Excel 2003, Excel 2007 and Excel 2010 have the ability to attach an XML Schema file to any spreadsheet by using the XML Source task pane to map ranges of the spreadsheet to elements of the schema. Once users have mapped the XML elements to a spreadsheet, users can seamlessly import and export XML data into and out of the mapped ranges. Location of K-metadata of a spreadsheet can be defined by binding each schema element of K-metadata schema to a range of the spreadsheet.

Definition 3.2 (Well-formed Spreadsheets bound with K-metadata Schema). Let S be a spreadsheet, $T = (V, E, Nm, r, lbl, C, f_C)$ be a K-metadata schema tree, $P \subset V$ be the set of the parent nodes of leaf nodes whose cardinality is one, and $P' \subset V$ be the set of the parent nodes of leaf nodes whose cardinality is '+' or '*'. S is a *well-formed spreadsheet bound by T* if and only if

- (a) S is a well-formed spreadsheet;
- (b) $\forall p_i \in P$, $\forall v_j \in child(p_i)$, v_j is bound to a range of the same worksheet of S; and
- (c) $\forall p'_i \in P'$, $\forall v'_j \in child(p'_i)$, v'_j is bound to column data of the same table of S.   □

Condition (b) and (c) enforce binding schema elements to ranges of spreadsheet according to criteria (2) and (3), respectively of K-metadata schema definition. Note that Excel provides a function that outputs mapped ranges of spreadsheet S as an instance of T.

4   Issues on Metadata Extraction of Spreadsheets

4.1   How to Handle the Difference of Presenting Format of the Same Data Type

The difference of presentation format of the same data type makes search engines unable to compare metadata with search keywords precisely. For example, the format of order date of the PO shown in Fig.1(b) is *yyyy/m/d* while that of the PO shown in Fig.1(d) is *yyyy/mm/dd*. If users input a query searching the purchase orders which were ordered before 2011/8/30. Based on string comparison of general search engines, string 2011/09/02 is decided to be smaller than the string 2011/8/30. This makes the name of this irrelevant document to be included into the query result. The same problem also occurs in case comparing number having comma separator with those having no comma separator. In order to solve the above problem, standard presentation format for each data type is defined (see TABLE I). For example, the string 2010/8/15 is converted to 2010/08/15. The string 34,125 is converted to 00034125. The string "Sato Ichiro" is converted to "sato ichiro". Metadata extraction can perform data format conversion properly by checking data types of K-metadata from the K-metadata schema.

Table 1. Presentation Format for Each Data Type

Data type	Standard Presentation Format
Date	*yyyy/mm/dd*
Number	*n*-digit number (which leading zeros are added to and comma separators are removed) Note that *n* is the number of digits defined by the system manager.
String	All-lowercase string

4.2   How to Handle Synonyms

In order to extract metadata precisely from other spreadsheets, metadata extraction program has to identify the ranges which store F-metadata before identifying the ranges that store the corresponding K-metadata. However, users may use synonyms to present F-metadata. For example, As "Order Date" of *PO* of Fig.1(b) is synonymous to "Issued Date" of *PO* of Fig.1(d). In order to identify F-metadata of a spreadsheet properly, the system manager needs to define the synonyms of F-metadata of the same class. An example of synonym definition file that is stored in XML format is shown in Fig.2. Based on this example, both "Order Date" and "Issued Date" are defined as synonyms of the *orderDate* class.

```
<?xml version="1.0" encoding="UTF-8"?>
<Def>
    <synonym class="orderDate">
        <data>Order Date</data>
        <data>Issued Date</data>
    </synonym>
    <synonym class="poNumber">
        <data>Purchase No.</data>
        <data>PO#</data>
    </synonym>
    ...
    <synonym class="productName">
        <data>Product Name</data>
        <data>Description</data>
    </synonym>
    ...

</Def>
```

Fig. 2. An example of synonym definition file

4.3 How to Identify the Spreadsheets Whose Metadata Can Be Extracted According to a Given K-Metadata Schema

In this paper, a spreadsheet whose metadata can be extracted according to a K-metadata schema is called a *candidate spreadsheet*. The definition of a candidate spreadsheet is given as follows.

Definition 4.1 (Candidate Spreadsheets). Let $S = \{w_1, w_2, .. , w_q\}$ be a spreadsheet where w_k is a worksheet and $1 \leq k \leq q$, *limit_disc* be the limited distance between K-metadata and corresponding F-metadata. Let $T = (V, E, Nm, r, lbl, C, f_C)$ be a K-metadata schema treee, $P \subset V$ be the set of the parent nodes of leaf nodes of T whose cardinality is one, and $P' \subset V$ be the set of the parent nodes of leaf nodes of T whose cardinality is '+' or '*'. Let *syn_def* be the synonym definition file. S is a *candidate spreadsheet according to T, syn_def and limit_disc* if and only if

- Condition 1: $\forall p_i \in P$, $\forall v_j \in child(p_i)$ such that there is a range r of w_k storing a text string and a range r' of w_k storing F-metadata which is defined as a synonym classified by $lbl(v_j)$ of *syn_def*   and $c\_dist(r, r') < limit\_dist$.
- Condition 2: $\forall p'_i \in P'$, $\forall v_j \in child(p'_i)$, there exists a table t of w_k such that t's column data part stores text strings and the corresponding column header storing F-metadata which is defined as a synonym classified by $lbl(v_j)$ of *syn_def*.   □

Assume that *limit_dist* is two. According to the above definition, the spreadsheet of Fig.1(b) is a candidate spreadsheet according to K-metadata schema of Fig.1(a) and the synonym definition file of Fig.2.

5 Metadata Extraction Algorithm

This section presents the *MetadataExtract* algorithm that extracts K-metadata from a given spreadsheet.

MetadataExtract (S, T, syn_def, limit_disc, flag, result).
Input:
- $S = \{w_1, w_2, .. , w_q\}$ be a spreadsheet whose metadata will be extracted where w_k is a worksheet and $1 \leq k \leq q$,
- $T = (V, E, Nm, r, lbl, C, f_C)$ be a K-metadata schema tree,
- *syn_def*   is the synonym definition file,
- *limit_disc* is the limited distance between K-metadata and corresponding F-metadata.

Output:
- *flag* = 'yes' if S is justified to be a candidate spreadsheet according to T, *syn_def* and *limit_disc*. Otherwise, *flag* = 'no'.
- *result* stores K-metadata of S.

Process:
1.   `flag = 'no'`
2.   Let $P \subset V$ be the set of the parent nodes of leaf nodes of T, and `child(`p_i`)` be the set of child nodes of p_i where $p_i \in P$.
3.   For each $p_i \in P$ do the following
4.   If the cardinality of p_j is "+" or '*' then
5.   /* Process table data */
            If S has table t_k (where $k \geq 1$) satisfied by the following condition:
                $\forall v_j \in$ `child(`p_i`)`, there exists a column header of t_k such that the column header stores F-metadata of v_j class defined by `syn_def` and its column data stores text strings.
            Then
6.   For each $v_j \in$ `child(`p_i`)` do the following
7.   Bind column data of t_k with v_j where its column header stores F-metadata of v_j class defined by `syn_def`.
8.   Else return.
9.   Else /* Process non-table data */
10.   If there exists worksheet w_m (where $1 \leq m \leq q$) of S satisfied by the following condition:
                $\forall v_j \in$ `child(`p_i`)`, there exist a range r_k of w_m such that r_k stores F-metadata of v_j class defined by `syn_def`.
            Then
11.   For each $v_j \in$ `child(`p_i`)` do the following
12.   Let r_k be a range of w_m storing F-metadata of v_j class defined by `syn_def`.
13.   If there exists a range $r'_{k'}$ such that
                    (1) $r'_{k'}$ stores a text string and
                    (2) $r'_{k'}$ is in the same row as r_k and
                    (3) $r'_{k'}$ is the nearest range of r_k and
                    (4) `c_dist(`$r'_{k'}$, r_k`) < limit_disc` then
14.   Bind $r'_{k'}$ with v_j.
15.   Else return.
16.   Else return.
17.   `flag = 'yes'`
18.   Output the data of S which is bound to T as XML data into `result`.
19.   Return.

Theorem 1.   The result of *MetadataExtract* is complete and correct.
Proof: Based on definition 4.1, *MetadataExtract* justifies whether S is a candidate spreadsheet according to T, *syn_def* and *limit_disc*. If S is not the candidate spreadsheet, then *flag* is set to 'no' and the algorithm terminates. Otherwise, *MetadataExtract* binds S with T. Based on the properties of a candidate spreadsheet, S is a well-formed spreadsheet bound with T. This makes *result* outputted by *MetadataExtract* is an instance of T. Therefore, the result of *MetadataExtract* is complete and correct.   □

Time Complexity of MetadataExtract
Let r be the number of ranges storing strings of spreadsheet S, n be the number of nodes of K-metadata schema tree T, and d be the size of synonym definition file *syn_def*. Gottlob et al.[6] has proposed an XPath query processing algorithm that works in quadratic time and linear space with respect to the size of the XML document. Therefore, the query time on the synonym definition file is bound to $O(d^2)$.

Given S, T, and *syn_def*, the process time for *MetadataExtract* to output the result is bound to $O(d^2 nr)$.

6   Implementation of the Proposed Algorithm

MetadataExtract is implemented as a part of crawler of office document search system[2]. The K-metadata outputted by the algorithm are inputted by indexer to generate XML-based search index which is completely different from conventional search index. The main reason of storing metadata as XML data is that tag names of XML are used to present the classes of metadata. For sake of scalability, the search index is implemented into PostgreSQL v8.3[14] which was enhanced with integrated XML data type. PostgreSQL v8.3 includes implementation of standard SQL/XML functions and support of XPath[16] expressions. Each file type (e.g. spreadsheet file, word file etc.) has the data structure that is different from that of other type. Based on this fact and program maintainability, a crawler is designed to extract metadata from documents of the same file type (such as .doc, .xls, etc.). Since this paper focuses on spreadsheets, the specification of prototype of spreadsheet crawler will be explained.

As shown in Fig.3, the inputs of the program are (1) a sample spreadsheet bound with K-metadata schema, (2) the file path of the folder containing spreadsheets whose K-metadata is extracted, and (3) the synonym definition file. Note that the sample spreadsheet is necessary to generate a query form for searching spreadsheet of that type. The F-metadata of the sample spreadsheet will appear in the query form. The outputs of this program are (a) a list processed files, (b) XML-based K-metadata files, and (c) a log file that records the process result. The system manager can investigate the files that the crawler cannot generate K-metadata. Information of the log file is used to improve *MetadataExtract*. The program is developed by using Visual Basic .NET programing language[15] of Microsoft Visual Studio 2010[12].

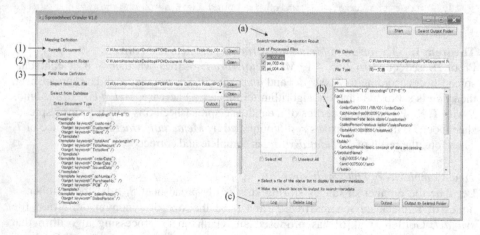

Fig. 3. A sample screenshot of metadata extraction program for spreadsheets

Metadata-based Search Index

The first design goal of search index is to enable users specifying the meaning of inputted keywords to attain high precision search result. This leads to design of special search index containing K-metadata and their semantic definitions. The second goal is that the search index should be able to address files of various application types. XML data is viewed as a tree where a sub-tree under the root presents K-metadata and file property of an indexed document file. The outline of the search index is shown in Fig.4. As described in the beginning of this section, the search index is stored in PostgreSQL V8.3 database in order to return search result as fast as possible. The search index is separated into the following two tables.

- *t_doc_kind* table stores document types and index sizes and number of files to be indexed.
- *t_xml_data* table stores the name of indexed file and K-metadata of each document type. The attribute *xml_data* stores the K-metadata set which is coded in XML format. PostgreSQL V8.3 provides XML functions that allow programmers to query XML data stored in the table attributes.

Query User Interface

In order to eliminate the need to remember how metadata is classified, this paper also proposes a search user interface that facilitates non-expert users in posing query. This interface shows a list of document types recorded in the search index. After a user selects a document type that she wants to search, the interface generates a query form for the selected document type. The query form shows user familiar field names

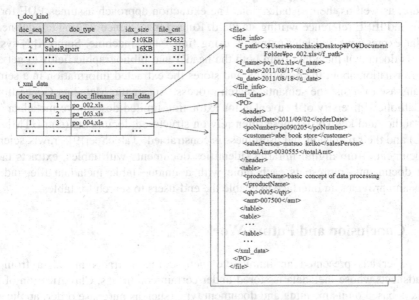

Fig. 4. The outline of the search index

(e.g. *Ordered date*, *Product Amount*) by looking up metadata definition file of the selected document type. This file also contains data type definitions of the input fields in order to transform them to match to standard presentation format (see TABLE I). The query form allows users defining options to sort search result by file creation date or file modification date in ascending or descending order. The reader can find more detail of the query interface at the previous work[2]. The query response time of this work for searching 10 target files from 100,000 files is about 0.03 seconds which is about 3,600 times faster than that of the previous work[2].

7   Related Work

Recently, metadata-based search has received attention as an effective approach to retrieve pertinent digital documents. Much work has been done in metadata extraction which is one of the most important components of metadata-based search. The harvesting of metatags (from generic HTML and <meta> tags) is the simplest form of metadata generation. DC-dot[4] is the most well-known of a group of tools that extracts metatags from web pages in order to generate Dublin Core metadata. The effectiveness of these tools is obviously constrained by the number and quality of metatags found in the source document. Clearly, such tools are not useful for auto-generating metadata values for properties that are not already described. Metadata Miner Pro[10] is a commercial application developed by Soft Experience. Its primary purpose is to extract descriptive information, such as title, author, subject, keyword, from common document formats (Microsoft Office applications, OpenOffice, HTML, Adobe PDF, and Apple Mac file comments).

The approach proposed by Cvetković et al.[3] extracted metadata from scientific literature, as well as their visualization. The extraction approach assumes PDF format of file and IEEE reference writing standard. Rules for reference writing are defined as regular expressions and later extracted using finite state machine. SemreX[7] system extracts document metadata from both the header and bibliographic fields to get deeper information about the document, and stores the extracted information in a semantic database to assist the semantic search process. Automatic Metadata Generation at the Catholic University of Leuven[1] for extracting limited descriptive metadata (e.g. title, author and keywords) these often rely on structured documents (e.g. HTML and XML) and their precision and usefulness is constrained. TableSeer[9] crawls scientific documents from digital libraries, identifies documents with tables, extracts tables from documents, represents each table with a unique table metadata file, indexes tables and provides an interface to enable the end-users to search for tables.

8   Conclusion and Future Work

This paper has presented an innovative method that extracts metadata from the spreadsheets whose metadata is stored under certain conditions. Classification of metadata is based on its meaning and document type (such as purchase order, application forms, etc.). Metadata of spreadsheets of the same type is defined by XML schema.

By this way the names of schema elements define classes of the metadata. Further-more, hierarchical structure of metadata classes can also be defined. Given a spread-sheet, metadata schema tree, synonym definition file, the metadata extraction algo-rithm of the proposed method justifies whether metadata conformed to the metadata schema tree can be extracted from the given spreadsheet. Data types (such as date, number, etc.) of metadata are taken into account in order to realize document search based on metadata of various data types. The metadata outputted from the algorithm will be used to create an XML-based search index. By implementing the XML-based search index in PostgreSQL V8.3 that allows programmers to query XML data stored in the table attributes, the query response time is much faster than that of my previous work. Based on metadata classification of this work, end-users can define the mean-ings of search keywords. Therefore, they can make queries that meet to their search requirements better than those of conventional keyword search.

As the future work, I plan to make an experiment that compares the proposed me-thod with other work in terms of search precision and recall. Furthermore, I plan to extend the proposed metadata extraction method to extract metadata stored in pic-tures, text boxes and shapes of spreadsheets.

References

1. Automatic Metadata Generation (2011), http://ariadne.cs.kuleuven.be/SAmgI/design/SimpleAmgInterface_1_0_prealpha.pdf
2. Chatvichienchai, S., Tanaka, K.: Office Document Search by Semantic Relationship Ap-proach. International Journal of Advances on Information Sciences and Service Sciences 3(1), 30–40 (2011)
3. Cvetković, S., Stojanović, M., Stanković, M.: An Approach for Extraction and Visualiza-tion of Scientific Metadata. In: ICT Innovations 2010 Web Proceedings, pp.161–170 (2010)
4. DC-dot (2011), http://www.ukoln.ac.uK-metadata/dcdot/
5. Google, Google Desktop Search (2011), http://desktop.google.com/
6. Gottlob, G., Koch, C., Pichler, R.: Efficient algorithms for processing XPath queries. ACM Trans. Database Syst. 30(2), 444–491 (2005)
7. Guo, Z., Jin, H.: A Rule-Based Framework of metadata Extraction from Scientific Papers. In: 10th International Symposium on Distributed Computing and Applications to Business, Engineering and Science, JiangSu, China, pp. 400–404 (2011)
8. IBM. Lotus 1-2-3 (2011), http://www-01.ibm.com/software/lotus/products/123/
9. Liu, Y., Bai, K., Mitra, P., Giles, C.: Searching for tables in digital documents. In: 9th Int'l Conf. on Document Analysis and Recognition (ICDAR 2007), pp. 934–938 (2007)
10. Metadata Miner Pro (2011), http://peccatte.karefil.com/software/Catalogue/MetadataMiner.htm
11. Microsoft Excel (2010), http://office.microsoft.com/en-us/excel/excel-2010-features-and-benefits-HA101806958.aspx
12. Microsoft Visual Studio (2010), http://www.microsoft.com/visualstudio/en-us
13. OpenOffice, Calc: The all-purpose spreadsheet (2011), http://www.openoffice.org/product/calc.html

14. PostgreSQL (2012),
    http://www.postgresql.org/docs/8.3/static/functions-xml.html
15. Vick, P.: The Visual Basic.NET Programming Language. Addison-Wesley Professional (March 2004)
16. W3C, XML Path Language (XPath) Version 1.0, REC-xpath-19991116 (1999),
    http://www.w3.org/TR/1999/
17. W3C, XML Schema (2001), http://www.w3c.org/XML/Schema
18. W3C, Extensible Markup Language (XML) 1.0, 4th edn. (2006),
    http://www.w3.org/TR/2006/REC-xml-20060816/
19. Walkenbach, Excel 2007 Power Programming with VBA. Wiley (2007)
20. X1 Technologies, X1 Professional Client (2011),
    http://www.x1.com/products/professional-client

Automated Extraction of Semantic Concepts from Semi-structured Data: Supporting Computer-Based Education through the Analysis of Lecture Notes

Thushari Atapattu, Katrina Falkner, and Nickolas Falkner

School of Computer Science,
University of Adelaide, Adelaide, Australia
{thushari,katrina,jnick}@cs.adelaide.edu.au

Abstract. Computer-based educational approaches provide valuable supplementary support to traditional classrooms. Among these approaches, intelligent learning systems provide automated questions, answers, feedback, and the recommendation of further resources. The most difficult task in intelligent system formation is the modelling of domain knowledge, which is traditionally undertaken manually or semi-automatically by knowledge engineers and domain experts. However, this error-prone process is time-consuming and the benefits are confined to an individual discipline. In this paper, we propose an automated solution using lecture notes as our knowledge source to utilise across disciplines. We combine ontology learning and natural language processing techniques to extract concepts and relationships to produce the knowledge representation. We evaluate this approach by comparing the machine-generated vocabularies to terms rated by domain experts, and show a measurable improvement over existing techniques.

Keywords: ontology, POS tagging, lecture notes, concept extraction.

1 Introduction

Computer-based intelligent education systems have been an area of research for the past decade. Early systems, such as SCHOLAR [1], provided intelligent assistance without human intervention by presenting a single set of digital materials to all students. Subsequently, research efforts have focused on creating student-centered learning environments, supporting learner's diversity and individual needs. Intelligent Tutoring systems, question answering systems and student-centered authoring systems are examples of 'one-to-one teaching', which provide individual attention for each student [2-5]. These systems are capable of generating customised questions for students, responding to unanticipated questions, identifying incorrect answers, providing immediate feedback and guiding students towards further knowledge acquisition.

The foremost effort in intelligent system development is allocated for knowledge-base modeling. Traditionally, knowledge engineers and domain experts formulate the domain knowledge manually [13] and the success of any intelligent system is heavily

S.W. Liddle et al. (Eds.): DEXA 2012, Part I, LNCS 7446, pp. 161–175, 2012.

dependent on the quality of the underlying knowledge representation. Manual efforts are constrained in their usefulness due to their error-prone nature and time-delays in the ability to incorporate new domain knowledge.

Accordingly, research has focused on overcoming the knowledge acquisition bottleneck, including the use of authoring shells [4-6]. This semi-automatic approach allows teachers to define pedagogical annotations of teaching materials, which includes pedagogical purpose, difficulty level, evaluation criteria and performance level [6]. These authoring environments provide natural language interfaces for teachers, and knowledge engineering tools are required to transform them into a machine comprehensible form. A significant problem with both manual and semi-automated processes is that any extracted knowledge is not reusable due to its domain-specific nature.

The goal of our research is to further automate the knowledge acquisition process from domain independent data, using digital lecture notes as our knowledge source. In this paper, we discuss terminology extraction from PowerPoint slides written in natural language. The purpose is to migrate from a full-text representation of the document to a higher-level representation, which can then be used to produce activity generation, automated feedbacks, etc. According to Issa and Arciszewski [7], an ontology is "a knowledge representation in which the terminologies have been structured to capture the concepts being represented precisely enough to be processed and interpreted by people and machines without any ambiguity". Therefore, our system extracts domain-specific vocabularies from lecture notes and stores them in an ontology which complies with an underlying knowledge model comprising concepts, instances, relations and attributes.

Generally, lecturers (domain experts) dedicate significant effort to produce semantically-rich, semi-structured lecture slides based on extensive knowledge and experience. Our research is based upon reusing this considerable effort and expertise, enabling computers to automatically generate activities for students.

We combine natural language processing techniques such as part-of-speech tagging (POS tagging) [8], lemmatisation [9] with pre and post-processing techniques to extract domain-specific terms. Information-retrieval based weighting models are utilised to arrange the precedence of concepts, placing a higher weight upon those that are more important [10]. In order to evaluate our approach, a comparison has been carried out between machine generated vocabularies and concepts identified by independent human evaluators.

The domain knowledge extracted from the lecture notes can be used to model the knowledge base of intelligent education systems (e.g. intelligent tutoring systems, question answering systems). Moreover, the ontologies generated from this research can be used as a knowledge source for semantic web search [12] and support for knowledge transfer for tutors and teaching assistants [13]. According to Rezgui [15], ontologies provide a perspective to migrate from a document to a content-oriented view, where knowledge items are interlinked. This structure provides the primitives needed to formulate queries and necessary resource descriptions [15]. Therefore, we use ontology reasoning techniques to generate questions for students and answer student questions in the forums of Learning Management System. Simultaneously, students can ask discipline-oriented questions from the system, with algorithms defined to traverse within the particular ontology to find suitable answers. If we can make use of lecture slides as a source, then this will be a valuable technique, as lecture slides are a commonly used and widely available teaching tool.

This paper includes a background study in section 2. A detailed description of methodology including concept and hierarchy extraction can be found in section 3, 4 and 5 respectively. In section 6, we evaluate the concept extraction algorithm using human judgment as a reference model. We discuss the overall work and provide a conclusion in section 7.

2   Background

At present, people share an enormous amount of digital academic materials (e.g. lecture notes, books and scientific journals) using the web. Educators and students utilise these knowledge resources for teaching and learning, however this digital data is not machine processable unless manipulated by humans. For instance, a computer cannot construct activities from a book available on the web without the involvement of a teacher; there is no simple transition pathway from semi-structured content (book) to teaching activities. Artificial intelligence researchers have attempted to automate knowledge acquisition, with the purpose of improving computer-based education, semantic web search and other intelligent applications.

Hsieh et al. [11] suggest the creation of a base ontology from engineering handbooks. They utilise semi-structured data, such as table of contents, definitions and the index, from earthquake engineering handbooks for ontology generation. Although the glossary development is automatic, the authors cite the importance of the participation of domain experts and ontology engineers in defining upper level concepts and complex relationships. The evaluation and refinement of this research is a manual process, which they believe to be an essential part to improve the quality of the base ontology. Similar to our approach, Gantayat and Iyer [12] indicate the automated generation of dependency graphs from lecture notes in courseware repositories. Although our concern is to extract contents of particular lecture notes, this research focuses on extracting only the *lecture topics* to build dependency graphs. During the online courseware search, this system recommends 'prerequisites' and 'follow-up' modules. They use the 'tf-idf' measure for terminology extraction, and their relationship building only corresponds to determine whether the other concept is prerequisite or follow-up (hierarchical order). Similarly, Rezgui [15] uses 'tf-idf' and metric cluster techniques to extract concepts and relations from the documents in the construction engineering domain. Although this automated approach reduces the knowledge acquisition bottleneck, this knowledge is not reusable over a diverse range of domains.

The authors of [13] propose the construction of a semantic network model from PowerPoint lecture notes. Teaching assistants gain knowledge of the sequence of teaching contents from the proposed semantic network model, creating a graph structure from each slide and combining all the graphs to form a semantic network reporting an agreement rate of 28%, which is the degree of agreement between human experts and the generated semantic network, measured by correlation of concepts. Ono et al. [13] argue that some teachers disagree with this model as the semantic network does not contain the original ideas of lecturers. Instead, it contains a machine generated overview of the lecture. Furthermore it also does not contain the knowledge of graphs and figures as in the lecture, resulting in a further negative attitude towards the new model. Concept map generation in [16] indicates the automatic construction of concept maps from an e-learning domain. The e-learning domain is an emergent and expanding field of research. The manual construction of the e-learning domain over months or years can produce obsolete knowledge. Therefore, automating the domain construction from e-learning journal and conference articles is significant not only for novice users but also for experts [16]. The extracted key word list from scientific articles is used to build the concept map and indicate the relations to guide learners to other areas which related to current topic.

Kerner et al. [14] suggest base line extraction methods and machine learning for key phrases acquisition from scientific articles. The results found Maximal Section Headline Importance (MSHI) to be the best base line extraction method over Term frequency, Term Length, First N Terms, Resemblance to Title, Accumulative Section Headline Importance, etc. Besides, in machine learning approach, optimal learning results are achieved by C4.5 algorithm over multilayer perceptron and naïve Bayes [14].

The majority of related works utilise the popular 'tf-idf' measure for concept extraction, filtering the most frequently occurred concepts in the domain over linguistically important nouns. Despite contemporary efforts to combine natural language processing and earlier techniques, the field is still open to considerable development.

3   Our Model

The proposed methodology, illustrated in Figure 1, consists of concept extraction techniques to construct the domain-specific vocabulary and concept hierarchy extraction algorithm to arrange the extracted vocabularies.

We have implemented a PowerPoint reader to process text and multimedia contents of Microsoft PowerPoint documents using Apache POI API [17]. This reader is capable of acquiring rich text features such as title, bullet offset, font color, font size and underlined text. We make use of these features in our research with the purpose of identifying emphasised key terms. Later sections of this paper discuss the importance of emphasised key terms in selecting concepts in the domain. In this paper, we address text-based extraction; multimedia processing will be addressed in a later publication.

We assume all the PowerPoint slides presented to the system are well structured (e.g. include titles, text/multimedia content) and contain no grammatical errors in natural text. Our system automatically corrects spelling errors using the in-built Microsoft spell checker.

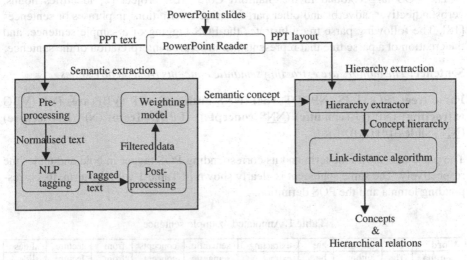

Fig. 1. An overview of our methodology

4   Concept Extraction

This section consists of four stages of concept extraction: pre-processing; natural language processing (NLP) tagging; post-processing and weighting model.

4.1   Pre-processing

In order to improve the acquisition, we normalise the PowerPoint text contents before preparing them for linguistic annotation. This normalisation includes;

1. splitting statements at the occurrence of periods, comma or semi colon,
2. replacing non-alphanumeric symbols,
3. processing punctuation marks (e.g. hyphen, &),
4. expanding abbreviations, and
5. removing white spaces.

4.2   Natural Language Processing Tagging

NLP tagging is categorised into two parts: lemmatisation and Part-of-speech (POS) tagging. In this stage, we first annotate the normalised statements using lemmatisation [9], where words are mapped to their base form (e.g. *activities => activity, normalisation =>normalise*). We have selected lemma annotation over the most popular stemming or morphological analysis techniques as the latter two techniques will entirely remove suffixes of term. This often gives different meaning of the term [9] (e.g. *Computation, computing, computer => compute*).

The POS tagger found in the Stanford Core NLP project [8] identifies nouns, verbs, adjectives, adverbs and other part-of-speech definitions in phrases or sentences [18]. The following parse tree illustrates the POS tagging of a sample sentence, and the creation of a parse tree that represents the structured interpretation of the sentence.

Sentence: *The authors are extracting semantic concepts from lecture slides.*

Parse tree: (ROOT (S (NP (<u>DT</u> **The**) (<u>NNS</u> **authors**)) (VP (<u>VBP</u> **are**) (VP (<u>VBG</u> **extracting**) (NP (<u>JJ</u> **semantic**) (<u>NNS</u> **concepts**)) (PP (<u>IN</u> **from**) (NP (<u>NN</u> **lecture**) (<u>NNS</u> **slides**))))) (. **.**)))

The parse tree shows the term and its corresponding POS tagger in bold and underline respectively. The same statement is clearly shown in Table 1 with the term, its corresponding lemma and the POS definition.

Table 1. Annotated example sentence

Word	The	authors	are	extracting	semantic	concepts	from	lecture	slides
Lemma	The	author	be	extract	semantic	concept	from	lecture	slide
POS	DT	NNS	VBP	VBG	JJ	NNS	IN	NN	NNS

The definitions of POS tags can be found in the Brown Corpus [18]. The Brown Corpus includes a list of POS tags and their definitions (e.g. VB => verb, IN => preposition, CC => coordinating conjunction like and, or). Our algorithm extracts adjectives (JJ), comparative adjectives (JJR), singular or mass noun (NN), possessive singular noun (NN$), plural noun (NNS), proper noun (NP), possessive proper noun (NP$) and plural proper noun (NPS) for annotations. There are some verbs (VB) which rarely indicate as concepts in particular domains. We plan to integrate the extractions of such verbs as a future work.

Our retrieval results have indicated that domain-specific concepts are combinations of numerous POS tag patterns (e.g. nouns followed by adjectives, compound nouns). A careful analysis confirms the requirement for a priority-based processing mechanism for these retrieved term or phrases. We define a list of regular expressions to arrange the n-grams (i.e. contiguous sequence of *n* items from given text or speech) on the basis of matching patterns. This list is processed in order and when the first one that matches is applied, the algorithm will eliminate that phrase from the sentence and apply the regular expressions recursively until the sentence has no more singular nouns (because singular nouns are the least prioritised pattern).

For example, let us assume a particular sentence includes a noun followed by an adjective, four consecutive nouns and a singular noun. According to the defined regular expressions list, we extract four consecutive nouns (four-grams) at the beginning. Then we eliminate that pattern from the sentence. Then the algorithm will return the noun followed by an adjective (bi-grams) and a singular noun (uni-gram) respectively.

4.3   Post-processing

We notice that the NLP tagger returns the terms such as *example, everything, somebody* as nouns. However, in the computing education context, we can identify that these terms are not of high importance. In order to improve our results, we have implemented our own stop-words filter which includes 468 stop-words (e.g. a, also, the, been, either, unlike, would), enabling the elimination of common words. Further, stop-words are not permitted in bi-grams and can only be used as a conjunction word in between tri-grams (e.g. point-to-point communication) or any other n-grams (n > 2). Thus, stop-words are not allowed at the beginning or end of a phrase.

4.4   Weighting Model

Although our algorithm returns nouns and compound nouns, it may include phrases which do not belong to domain-specific categories (e.g. 'authors' in table 1). In order to refine the results, a weighting model has been employed which places highest weighted terms as the most important concepts. We have developed a new model by building on term frequency-based approaches and incorporating this with n-gram count and typography analysis. We then present four weighting models that are based on these weighting factors as discuss below.

Term Frequency
The system counts the occurrence of each concept in the lecture notes (t_f). We assign the 'log frequency weighting' [10] for each term to normalise the occurrences within a controlled range. For instance, if the term frequency is 1, weight will be 0.3 and for 100, weight will be 2.0. This prevents a bias towards high frequency terms in determining the threshold value and important concepts. This will result in the term frequency being an important and influential factor in choosing important concepts rather than the only factor.

$$\text{Term}_{weight} = \text{Log} (1 + t_f) \qquad (1)$$

N-gram Count and Location
In our analysis, we identify that the majority of noun phrases that are confirmed as concepts are brief PowerPoint statements rather than a fragment of long sentences. Therefore, we count the number of tokens (n-grams) included in each item of bullet text and assigns a weight to the bullet text based on its n-gram count (n=1 to 5). If the count is more than five, which implies the bullet text is a moderate to long sentence. According to the evaluation results, shorter statements are more likely to be chosen as concepts over long statements and our experiments verified that text locations (e.g. title, topic, bullet or sub-bullet) are not effective in the judgment of concept selection.

Typography Analysis
In PowerPoint slides, lecturers often emphasise terms or phrases to illustrate their importance in the given domain, frequently employing larger fonts, different colors or different font faces. Underlined, bold or italic terms are also considered as emphasised. In our algorithm, we introduce a probability model (illustrated in Table 2) to assign unique weights for these terms.

Table 2. Weight assignment of emphasised text based on probability

Probability of emphasised text (%)	Weight (out of 1)
50 < P <= 100	0
20 < P <= 50	0.05
10 < P <= 20	0.15
5 < P <= 10	0.3
0 < P <= 5	0.5

Model Selection

Finally, we have defined weighting models based on the weighting factors discussed above. The indication of letters - N: No weight, L: Logarithm term frequency weight, T: weight of n-gram count, E: emphasized probabilistic weight.

- NNN: We do not apply any weighting models at this instance and measure the performance from the concepts extracts from linguistic annotation (i.e. all noun phrases and adjectives).
- LNN: This model considers log frequency weight as the factor for refining the extracted concept list. The terms with high frequency occurrence will be selected as important concepts in this setup.
- LTN: This model considers the accumulated weights of log frequency and n-gram count as the judgment factor for filtering the most important concepts
- LTE: This model calculates the accumulated weight of log frequency, n-gram count and emphasised probability as the selection factor.

We compare the performance of each of these weighting models in the evaluation section to select the best model for concept extraction algorithm.

5   Concept Hierarchy Extraction

The success of automated ontology construction depends not only on the accuracy of concept extraction but also the capability of building concept hierarchies and lexical relations effectively. In this paper, we propose an approach to extract the concept hierarchies based on PowerPoint document layout [20].

The PowerPoint document layout has been used as a key feature in extracting hierarchies based on a top-down approach [19]. In concept extraction, we store the text features (e.g. topic, title, bullet, sub bullet) of each term and arrange the terms based on the sequence of topic, title, bullet and sub bullet respectively.

Figure 2 shows the relationship defined between the different indentation levels of bullets. The bold text shows the returned terminology from the concept extraction algorithm (Figure 2a) and the hierarchy extraction algorithm arranges sub points as children (is-a) of bulleted-points according to subsumption relationship (Figure 2b). This relation indicates that the *IP, TCP and HTTP* are specialization of *Protocol*. Thus, *Protocol* is a generalization of *IP* and can be specified as 'IP is a Protocol'.

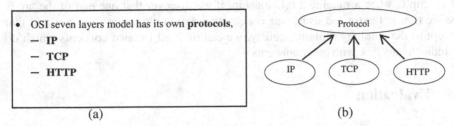

Fig. 2. (a) Example slide (b) hierarchy of concepts

When considering the structure of PowerPoint slides, a concept can occur in multiple levels in the hierarchy (e.g. title, sub-bullet). In order to avoid the conflicts of multiple occurrences in tree-levels, we have introduced and implemented the following algorithm (link-distance algorithm), which determines the correct occurrence of the concept by its number of corresponding links and distance from the root node. This defines the concept as a more specific one rather than general concept with the most number of links.

```
Document (d) has more than one occurrences of similar
term (s)
For s_i (i=1 to n; n ∈ R)
   Calculate number of links term s_i has - L_i
   Consider s_i corresponds to L_i where L_i= max (L_1,L_2,..,Ln)
If (number of max(m)) > 1
   For s_j (j=1 to m; m ∈ R)
      Calculate number of nodes from root to term s_j - t_j
      Consider s_j corresponds to t_j where t_j= max(t_1,..,t_m)
Else
   Get s_i
```

Due to the semi-structured nature of lecture slides, it is possible for there to exist multiple concepts within a single level, for instance, where the lecturer has joined two or more concepts in a title or descriptive sub-point. When considering the case of multiple concepts within a title, sub-points will then inherit features from multiple parents, introduced multiple inheritance. Although allowable in ontology specification, the additional complexity introduced by multiple inheritance makes it undesirable. When considering ontology development, we must also consider the need to specify a single root node, or most general concept, and restrictions on the expansion of our ontology, as it is constrained to have only four levels (in this case: topic, title, bullet, sub-bullet). Because of these additional constraints, and to preserve the simplicity and understandability of our ontology, we eliminate the concepts with more than four parents of multiple inheritance when applying our relationship extraction process.

Although the relationships are usually of a hierarchical nature, our expectation is to improve lexical relationship extractions based on computational linguistics and syntactic textual patterns [19] in the next stage of our work.

According to the ontology layer cake, identifying concepts, hierarchies and relations is an iterative process, which modifies each stage recursively [19]. As a result, we can improve the concept acquisition at the relationship extraction stage.

For example, when arranging a relationship, if we discover that one end of the arc is missing (i.e. not extracted using our concept extraction algorithm), we can add that concept to the ontology. Simultaneously, we can omit all isolated concepts which do not indicate any link with other concepts.

6   Evaluation

This section discusses the details of our data set, performance matrices, evaluation and analysis of the results.

6.1   Data Set

In this stage, we analysed 40 randomly selected lecture slide sets as training data from various sources such as web, open course ware, university lectures and book slides. As discussed in section 4.4, we obtained the performance of each weighting models (e.g. NNN, LNN, LTN, LTE) and increased their performance further by adjusting weights and threshold value (an experimental value used to filter least important concepts).

After those improvements, we tested the algorithm for thirty lecture slide sets chosen from two core computer science courses (computer networking and software engineering), comprising a combination of text, figures, examples, case studies, scenarios and source code. Evaluation was carried out by six independent evaluators. The evaluators are teaching assistants in computer science, and are experienced in tutoring, paper marking or completing the particular subject with highest grades. The task of the evaluator is to judge the terminology of the given lectures having the concept map in their mind. Each evaluator is given five lecture slide sets (each slide set contains approximately fifty slides).

6.2   Evaluation Matrices

We utilise common evaluation techniques from the information retrieval domain, including the confusion matrix used in predictive analysis for comparing two models [10] (see Table 3).

Table 3. Confusion matrix

	True	False
Positive	tp	fp
Negative	tn	fn

Based upon this matrix, we use precision, recall and F-measure to find the performance of our algorithm.

$$\text{Precision} = tp / (tp + fp) \qquad (2)$$

$$\text{Recall} = tp / (tp + fn) \qquad (3)$$

$$\text{F-measure} = 2 \text{ x (precision * recall)/ (precision + recall)} \qquad (4)$$

For inter-human agreement, we have selected positive specific agreement (P_{pos}) [21] over the more popular kappa statistic [10], as the latter technique considers the number of valid concepts that neither evaluator selected. Although it is possible to identify this in document relevancy measures in information retrieval applications, the number of potential concepts that neither evaluator marked cannot be defined objectively in our approach.

Table 4. Agreement between two independent evaluators

Evaluator 1's Judgement	Evaluator 2's judgement	
	Positive	Negative
Positive	a	b
Negative	c	d

Table 4 demonstrates an example of applying the positive specific agreement technique, where two evaluators are asked to select potential concepts from the same data source, and agree as follows:

a: the number of concepts that both evaluators agree as potential concepts
d: the number of concepts that both evaluators agree as not potential concepts
b and c: the number of concepts that the evaluators disagree on.

We calculate the positive specific agreement as follows,

$$\text{Positive specific agreement} = 2a / (2a + b + c) \qquad (5)$$

In this research, we used three independent evaluators per each lecture. Accordingly, we calculate average of pairwise between all three evaluators in Section 6.3 and included in Table 6.

6.3   Analysis

This section includes a comparison between machine generated terms and human rated terms in order to measure the machine performance over traditional manual extraction. Additionally, it summarises the comparison of different weighting models and the inter-human agreement between independent evaluators.

Table 5. Evaluation results of randomly selected lecture slides; L: Log term frequency, N: No weight, T: N-gram count, E: emphasised weight, P: Precision, R: Recall, F: F-measure

Slide set	NNN			LNN			LTN			LTE		
	P	R	F	P	R	F	P	R	F	P	R	F
1	0.238	0.876	0.375	0.636	0.157	0.252	0.535	0.426	0.475	0.534	0.528	0.531
2	0.242	0.893	0.381	0.35	0.148	0.208	0.355	0.340	0.347	0.278	0.723	0.402
3	0.367	0.925	0.526	0.678	0.158	0.256	0.711	0.658	0.683	0.611	0.733	0.666
4	0.219	0.847	0.348	0.305	0.239	0.268	0.349	0.630	0.449	0.329	0.630	0.432
5	0.244	0.801	0.374	0.583	0.143	0.230	0.477	0.438	0.457	0.471	0.458	0.465

Table 5 shows the precision, recall and F-measure results for five randomly selected lecture slide sets, using each of the weighting models identified in Section 4.4.

According to Table 5, it is evident that relatively low precision and very high recall
is discovered from the model which does not apply any weights (NNN). This model
returns almost all the terms that match with the NLP tagger without a restriction. As a
result, it includes the highest number of true positive values. However, it measures the
precision based on the ratio between true positive to number of computer generated
terms, resulting in a low precision value. Consequently, we get the highest amount of
matches between true positive to number of relevant terms indicated by domain experts. This confirms the high recall values from experiments.

The second model (LNN) has given priority to the highest frequency terms. But the
analysis of recall identified that the highest frequency occurrence did not produce any
impact for the selection of concepts.

Based on the summary of the results, we obtain best precision, recall and F-
measure from the LTN and LTE models. Both of these models provide equivalent
performance for all three matrices, however, both the NNN and LNN models indicate
a diverse range of values for each measure. For instance, the NNN model provides
very low precision values and very high recall values, whereas, the LNN model
presents average values for precision and very low values for recall and F-measure
respectively.

From our evaluation, lectures with plain text (which are not emphasised) produce
the best results from the LTN model. However, our algorithm occasionally underper-
forms when considering lecture slides with long statements or very short statements
(token count of statement). Overall, the LTE model performs best when considering
lecture slides with varied content.

We achieved 35%, 60% and 42% of overall performance for precision, recall and
F-measure respectively for a collection of 70 lecture slides from the LTE model. This
is a considerable improvement over other existing researches completed to date, as
shown in [13].

Inter-Human Agreement
Table 6 presents indication for the agreement on same task between all allocated independent evaluators based on LTE weighting model.

Table 6. Comparison of independent evaluators

Lecture notes	1	2	3	4	5	6	7	8	9	10
Agreement (average)	0.518	0.598	0.540	0.54	0.516	0.428	0.454	0.465	0.353	0.388

In Table 6, the first five lecture notes refer to software engineering lectures which
comparatively include meaningful sentences, satisfactory hierarchy and good summa-
rization. The three evaluators who judged these lecture notes have fair agreement
around 0.55 since the potential concepts have a clear separation from the redundant
data. The last five lecture notes refer to networking lectures which contain brief
phrases where some evaluators consider the entire slide is important without

considering any of the individual slide concepts, etc. This results in a lower rate of agreement between independent evaluators. This occurred repeatedly throughout their evaluations, indicating coarseness in granularity when using human involvement. In addition, each evaluator had to deal with more than 50 slides in one lecture, causing them to miss some important concepts in the review. These issues emphasise the error-prone and variable nature of manual knowledge extraction and ontology creation, stressing the importance of machine extraction to generate explicit knowledge.

6.4   Discussion

The analysis results indicate pros and cons of utilising the PowerPoint lecture notes as our knowledge source. The PowerPoint lecture notes regularly contain brief phrases (the number of tokens per PowerPoint statement/bullet is generally between 2 to 5). As discussed in Section 4.4 and evaluation, we have provided evidence that this PowerPoint pattern improves the automatic extraction to some extent.

The machine interpretation of natural text frequently deals with the issue of anaphora expressions, which occur when one sentence refers to another preceding sentence (e.g. John and Arthur went to city. They saw the Aquarium). In the example, the term 'they' refers to John and Arthur. In order to identify this, we need an improved anaphora resolution algorithm. However, the natural hierarchy of PowerPoint documents includes contextual information which is related to each slide title. As a result, we do not need to apply anaphora resolution techniques to concept extraction and instead can utilise concept hierarchies to arrange contextual information within a slide.

Although there are certain benefits in using PowerPoint lecture notes as our knowledge source, PowerPoint often contains grammatically incomplete sentences. As discussed above, when the contextual content of a slide has a relation with its title, lecturers tend to include such incomplete sentences since humans can interpret the natural language and its conceptual relations easily. However, machines need improved natural language processing algorithms to interpret these ambiguous sentences.

7   Conclusion

Computer-based education approaches provide valuable support for traditional teaching methods, but are constrained by our current abilities to automatically extract knowledge from educational domains. In this paper, we have described our approach to the automatic extraction of concepts and concept hierarchies from digital lecture notes, enabling the reduction of semantic gap between natural language and formalised knowledge. According to our analysis, we have achieved 42% of overall machine performance (F-measure) in our initial knowledge extraction. This represents a measurable increase over existing techniques.

There is considerable further work that can be achieved in this area. In addition to enhancements to the core algorithm, in order to achieve higher levels of performance, we intend to consider additional multimedia contents that are included within digital lecture notes, such as figures and graphs. We expect to integrate image feature extraction from PowerPoint slides as a future expansion.

Further, human judgement on concept extraction is a subjective process, and to confirm the accuracy and utility of our approach, we must consider improvements to the effectiveness of computer-based learning and teaching. Evaluating the real success of our approach depends on the capability of the generated ontology to be used in this context, through assisting lecturers in organising teaching materials to best facilitate human concept extraction, supporting teaching assistants to understand key course objectives and materials, and semantic web searching. We intend to utilise our results to further research into the automated generation of activities for students, including the generation of question banks, automated forum participation, and the generation and identification of resources based on personalised learning needs.

References

1. Carbonell, J.R.: AI in CAI: An Artificial-Intelligence Approach to Computer-Assisted Instruction. IEEE Transactions on Man-machine Systems 11(4), 190–202 (1970)
2. McArthur, D., Stasz, C., Hotta, J., Peter, O., Burdorf, C.: Skill-oriented task sequencing in an intelligent tutor for basic algebra. RAND Note 17(4), 281–307 (1988)
3. Butz, C.J., Hua, S., Maguire, R.B.: A Web-Based Intelligent Tutoring System for Computer Programming. In: IEEE/WIC/ACM International Conference on Web Intelligence, pp. 159–165. IEEE Computer Society, USA (2004)
4. Stankov, S., Rosic, M., Itko, B., Grubisic, A.: TEx-Sys model for building intelligent tutoring systems. Computer and Education 51(3), 1017–1036 (2008)
5. Zitko, B., Stankov, S., Rosic, M., Grubisic, A.: Dynamic test generation over ontology-based knowledge representation in authoring shell. Expert Systems with Applications 36(4), 8185–8196 (2009)
6. Zhuge, H., Li, Y.: KGTutor: A Knowledge Grid Based Intelligent Tutoring System. In: Yu, J.X., Lin, X., Lu, H., Zhang, Y. (eds.) APWeb 2004. LNCS, vol. 3007, pp. 473–478. Springer, Heidelberg (2004)
7. Issa, R., Arciszewski, T.: Ontology: An Introduction, Teaching Modules (PowerPoint presentation). In: ASCE Global Center of Excellence in Computing (2011)
8. Toutanova, K., Klein, D., Manning, C., Singer, Y.: Feature-Rich Part-of-Speech Tagging with a Cyclic Dependency Network. In: North American Chapter of the Association for Computational Linguistics on Human Language Technology, pp. 252–259. Association for Computational Linguistics, Canada (2003)
9. The Stanford NLP (Natural Language Processing) Group, http://nlp.stanford.edu/software/corenlp.shtml
10. Manning, C.D., Raghavan, P., Schutze, H.: Introduction to Information Retrieval. Cambridge University Press, New York (2008)
11. Hsieh, S., Lin, H., Chi, N., Chou, K., Lin, K.: Enabling the development of base domain ontology through extraction of knowledge from engineering domain handbooks. Advanced Engineering Informatics 25, 288–296 (2011)
12. Gantayat, N., Iyer, S.: Automated building of domain ontologies from lecture notes in courseware. In: IEEE International Conference on Technology for Education, pp. 89–95. IIT Madras, India (2011)
13. Ono, M., Harada, F., Shimakawa, H.: Semantic Network to Formalize Learning Items from Lecture Notes. International Journal of Advanced Computer Science 1(1), 10–15 (2011)

14. HaCohen-Kerner, Y., Gross, Z., Masa, A.: Automatic Extraction and Learning of Keyphrases from Scientific Articles. In: Gelbukh, A. (ed.) CICLing 2005. LNCS, vol. 3406, pp. 657–669. Springer, Heidelberg (2005)
15. Rezgui, Y.: Text-based domain ontology building using tf-idf and metric clusters techniques. The Knowledge Engineering Review 22(4), 379–403 (2007)
16. Chen, N., Kinsuk, Wei, C., Chen, H.: Mining e-Learning Domain Concept Map from Academic Articles. In: Sixth International Conference on Advanced Learning Technologies, pp. 694–698. IEEE Computer Society, Netherlands (2006)
17. Apache POI- the Java API for Microsoft Documents, http://poi.apache.org/
18. Brown Corpus, http://en.wikipedia.org/wiki/Brown_Corpus
19. Cimiano, P.: Ontology Learning and Population from Text: Algorithms, Evaluation and Applications. Springer, New York (2006)
20. Understanding the PowerPoint MS-PPT Binary File Format, http://msdn.microsoft.com/en-us/library/gg615594.aspx#UnderstandMS_PPT_Overview
21. Hripcsak, G., Rothschild, A.S.: Agreement, the F-measure, and reliability in information retrieval. J. Am. Med. Inform. Assoc. 12(3), 296–298 (2005)

A Confidence–Weighted Metric for Unsupervised Ontology Population from Web Texts

Hilário Oliveira[1], Rinaldo Lima[1], João Gomes[1], Rafael Ferreira[1],
Fred Freitas[1], and Evandro Costa[2]

[1] Informatics Center, Federal University of Pernambuco, Recife, Brazil
{htao,rjl4,jeag,rflm,fred}@cin.ufpe.br
[2] Computing Institute, Federal University of Alagoas, Maceió, Brazil
evandro@ic.ufal.br

Abstract. Knowledge engineers have had difficulty in automatically constructing and populating domain ontologies, mainly due to the well-known knowledge acquisition bottleneck. In this paper, we attempt to alleviate this problem
by proposing an unsupervised approach for extracting class instances using the
web as a big corpus and exploring linguistic patterns to identify and extract ontological class instances. The prototype implementation uses shallow syntactic
parsing for disambiguation issues. In addition, we propose a confidence-
weighted metric based on different versions of the classical PMI metric, Word-
Net similarity measures, and heuristics to calculate the final confidence score
that can altogether improve the ranking of candidate instances retrieved by the
system. We conducted preliminary experiments comparing the proposed confidence metric against some versions of the PMI metric. We obtained promising
results for the final ranking of the candidate instances, achieving a gain in precision up to 24%.

Keywords: Ontology Population, Ontology-Based Information Extraction,
Similarity Measure, Heuristics.

1   Introduction

In recent years, there has been an increasing interest in ontologies, mainly because
they have become very popular as a means for representing and sharing machine-
readable semantic knowledge. From the computer science point of view, ontologies
can also be defined as logical theories that are able to encode knowledge about a certain domain in a declarative way. In addition, ontologies can provide conceptual and
terminological agreement among the members of a group (humans or computational
agents) that need to share electronic documents and information, independently of the
way this knowledge can be used. Currently ontologies are extensively used in applications for information integration, knowledge management, information retrieval, and
specially, the Semantic Web.

The Semantic Web [1] is a global initiative aiming to achieve a semantically annotated Web, in which search engines are able to process resources from a semantic
point of view. In this scenario, the Semantic Web can intensely increase the quality of

S.W. Liddle et al. (Eds.): DEXA 2012, Part I, LNCS 7446, pp. 176–190, 2012.

the information to users by providing a uniform data access and integration layer for elaborated services. However, the first step to this end equally requires a global consensus in defining the appropriate semantic structures for representing any possible domain of knowledge, which implies in the development of domain or task-specific ontologies. Thus, once the ontology for a specific domain is available, the next step is to semantically annotate related web resources.

On the other hand, although domain or task-based ontologies are recognized as essential resources for the Semantic Web, the development of such ontologies relies on domain experts or knowledge engineers that typically adopt a manual construction process. It turns out that this manual construction process is very time-consuming and error-prone [2]. Thus, an automated or semi-automated mechanism to convert the information contained in existing web pages into ontologies is highly desired. Ontology-Based Information Extraction (OBIE) [3], a subfield of Information Extraction, is a promising candidate of such a mechanism. According to [3] an OBIE system can process unstructured or semi-structured natural language text through a mechanism guided by ontologies to extract certain types of information, and present the output using ontologies. As had been pointed out by [4, 5], OBIE systems can be used for semantic annotation of web pages as well.

This paper describes our approach that is instantiated by an OBIE system for extracting class instances from web pages. The proposed system integrates linguistic patterns to identify text realizations of ontological classes, and shallow syntactic information for disambiguation purposes, as well as for increasing coverage of our pattern matching mechanism. Additionally, the system employs a confidence-weighted metric based on different versions of the classical Pointwise Mutual Information (PMI) [4] metric, WordNet similarity measures, and simple heuristics to calculate the final confidence score which can improve the ranking of candidate instances. The prototype operates in a fully automated, unsupervised manner, which means that it does not need any training corpus with annotated examples. Moreover, it considers the web as an enormous corpus to overcome data sparseness problems.

The rest of this paper is organized as follows: Section 2 presents related work. Section 3 presents the assumptions and a detailed description of our approach. We report our experimental setup and results in Section 4. Finally, Section 5 concludes this paper.

2   Related Work

Much research work has already followed the idea of using the web as a big corpus [4, 5, 6, 7, 11] applying domain-independent linguistics patterns on the web, such as Hearst´s patterns [8] through the use of search engines. In the following, we present some systems that adopt a similar approach to ours.

KnowItAll [4] learns instances and other relations from the web. This system is domain-independent and extracts information in an automated manner relying on the redundancy of the web to bootstrap its information extraction process. To assess the candidate instances, KnowItAll uses the PMI metric on the web to compute features, and a Naive Bayes classifier to combine those features for achieving a rough estimate of the probability that each candidate instance is correct.

Cimiano et al. [5] have implemented an OBIE system, named Pattern-based Annotation through Knowledge on the Web (PANKOW). This system semantically annotates web pages using web-based search engines. It extracts all proper nouns from the text and conducts searches for every combination of identified proper nouns with all concepts of the input ontology by using a set of linguistic patterns. Later an improved version of PANKOW, the C-PANKOW system, operated on the same principles, but improving its classification results by considering the context of the extracted sentences. The computational efficiency was also improved by reducing the number of queries to the search engine. Both systems are concerned with semantic annotation tasks and assess the candidate instances only using basic PMI information.

On the other hand, the OntoSyphon system [7] focuses on particular parts of an input ontology and tries to learn instances about those ontological classes from two representative corpora: a local one, containing 60 million web pages; and the whole web. In [7], the authors have performed many experiments for evaluating several normalized confidence metrics based on simple hit counts for instance classification and instance ranking, in an analogous way as we did in the present work. Besides the exact precision, they also used another evaluation metric called Learning accuracy [6] that takes into account the ontology hierarchy in the evaluation process.

The current work focuses on the evaluation of a confidence-weighted metric that, besides PMI information as it has been done in all related work above, also considers semantic features and some simple heuristics for improving the classification performance of candidate instances. That constitutes the main difference between the method presented in this paper and previous work.

3   The Chop System

The CHOP (Combined Heuristics for Ontology Population) system implements our unsupervised approach to automatically acquire ontological class instances from the web. This approach shows that it can yield promising results by using the web as a big corpus. The main idea is to take profit of the high redundancy present in the web content. Indeed, several authors pointed it out as an important feature because of the amount of redundant information can represent a measure of its relevance [9, 10, 11]. Moreover, we take into account the portability issue, i.e., the approach has to be able to perform independently of the domain ontology. Taking a domain ontology as input, the knowledge engineer or the ontology expert selects the most relevant concepts that will be populated with instances provided by the extraction component from our functional architecture shown in Fig. 1.

The CHOP System relies on a set of domain-independent linguistic patterns from which it creates extraction rules for each class in an input ontology. Actually, the input ontology guides both the selection and the extraction process of candidate instances. Furthermore, the same ontology may be used as a repository for the extracted instances, which characterizes a typical task of Ontology Population [12].

In addition, the system uses a shallow syntactic parser for English (OpenNLP parser[1]). This parser performs the preprocessing tasks we need for identifying Noun Phrases (NP) as candidate instances of ontological classes. In order to assess such candidate instances, i.e., to decide which one to choose as an actual instance for the output ontology, we propose a combination of different measures and heuristics that explore different levels of evidence. Fig. 1 illustrates the main components of the proposed solution. The whole process described in Fig.1 follows an iterative cycle that can be executed for each class of the input ontology.

The rest of this section describes the four main components of our prototype: (i) *Corpus Retrieval*, (ii) *Extraction and Filtering of Candidate Instances*, (iii) *Classification of Candidate Instances*, and (iv) *Ontology Population*.

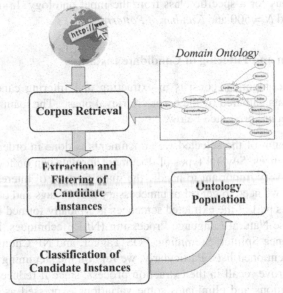

Fig. 1. Functional architecture of the proposed approach

3.1   Corpus Retrieval

The corpus retrieval process starts retrieving relevant documents from web in order to automatically establish a working corpus. We use a set of domain-independent linguistic patterns for collecting documents from the web. Then, a web search engine assists this component with the searches. Table 1 presents the queries needed to gather some relevant documents for the class labeled as "SPORT" in our domain ontology.

The left column in Table 1 represents the linguistic patterns originally proposed in [8]. These hand-crafted patterns denote a certain relation to find candidate instances of ontological classes using the web as a corpus. On the other hand, the right column specifies the instantiated queries derived from Hearst´s patterns.

---

[1] http://opennlp.apache.org

Table 1. Patterns for retrieving relevant documents

Extraction Pattern	Search Query
class such as candidates	"*sports* such as"
such class as candidates	"such *sports* as"
candidates and other class	"and other *sports*"
candidates or other class	"or other *sports*"
class especially candidates	"*sports* especially"
class including candidates	"*sports* including"

Due to performance reasons, for each pattern given in Table 1, we had to limit the number of web documents that the system retrieved to the first N documents. At the end of this phase, we will have ($N * Numbers\ of\ Patterns$) documents which makes up the working corpus for a specific class from the input ontology. In our experimental setup we specified $N = 500$ and *Number of Patterns* = 6.

3.2 Extraction and Filtering of Candidate Instances

The aim of this component consists in extracting and filtering candidate instances from the documents gathered by the corpus retrieval phase. The main steps performed by this component are presented below.

1. The preprocessing of the selected web documents is done in order to extract only clean text sentences. Several types of document formats can be found on the web; therefore, it is very important to handle the main formats of interest. For instance, for web pages, we need to get rid of unnecessary HTML tags and other elements.
2. The **candidates** part of the extracted sentences is typically formed by a list of noun phrases. We use Natural Language Processing (NLP) techniques, including Tokenization, Sentence Splitter, Stemming, POS Tagger, and NP Chunking for analyzing the aforementioned lists. Particularly, we rely on the stemming technique as an attempt to improve recall in the extraction process, since it reduces words to their root representations and eliminates some variations expressed as plural and verb tenses, for example.

   The list of noun phrases are extracted as candidates instances. In addition, each candidate instance keeps a list containing the extraction patterns that were responsible to extract it. This step is iterated over all extraction patterns until all documents have been processed. Table 2 shows two examples of candidate instances for the Sport class. In each row, we have the name of the candidate instance followed by the patterns that extracted it.

Table 2. Candidate instances produced for the Sport class

Class	Candidate Instance	List of Extraction Patterns
Sport	Basketball	[class such as candidates] [class including candidates]
Sport	Soccer	[such class as candidates] [candidates or other class]

3. In order to avoid invalid and repeated candidate instances, the following filters are applied in this preprocessing step:
   - *Stop word filtering.* Some words in the retrieved sentences are useless. In this way, this filter removes words that appear in a Stop List. In addition, candidate instances with invalid characters, such as ">", ")", etc. are removed as well.
   - *Redundant candidates*: The system picks up a candidate instance and verifies if it was previously found; or if the candidate instance actually represents a class on the input ontology. In either case, the candidate instance is removed.
   - *Syntactic filtering*: By using a stemming algorithm, the system identifies syntactic variations of candidate instances. In this case, if two candidate instances are syntactically equivalent, just one of the syntactic forms is maintained (e.g., the singular form takes precedence over plural), and the lists of extraction patterns of both forms are merged.
   - *Semantic filtering*: two candidate instances can be syntactically different, but semantically equivalent. For instance, the candidate instances "USA" and "The United States of America" both refer to the same instance of the Country class. In order to identify these cases among the set of candidate instances, the system retrieves a list of synonyms of the two candidates using the WordNet repository. Then, in case of one candidate is a synonym of the other, they are considered semantically equivalent. This semantic filtering is restricted only to candidates having an entry in the WordNet[2]. The list of extraction patterns are merged in analogous way as in Syntactic Filtering.

3.3   Classification of Candidates Instances

The previous component produces a set of candidate instances. Next, the system has to decide which candidate instance to choose as an actual instance for a given class of the input ontology. The CHOP system uses several web-scale statistics and semantic metrics for evaluating confidence scores to be applied to the specific task of ontology population. Such a confidence metric should estimate the likelihood that a candidate instance is an actual instance of the related class.

Instead of using just one alternative, i.e., PMI scoring for ranking candidate instances as it has been done in related work presented in Section 2, we looked for an approach that would combine several alternatives aiming at yielding better results than simply using one of them separately. Hence, we propose a confidence-weighted metric based on variations of the PMI metric found in [7], WordNet similarity measures, and simple heuristics to calculate a final confidence score that can improve altogether the ranking of candidate instances retrieved by the system. In the following, we explain the constituents of our proposed confidence score and how they are put together.

Pointwise Mutual Information. In this work, we use web-scale statistics to assess the likelihood between words and phrases that is estimated from the hit counts returned by web search engines. For instance, for estimating the likelihood that "Pox" is

---

[2] http://wordnet.princeton.edu

an instance of the *Disease* class, we have to consider the number of hits between a phrase such as "Diseases such as Pox" and "Pox". In our experiments, we have chosen three variations of the PMI measure presented in [7].

In the following formulas, *hits(ci, c, p)* is the number of hits returned by the search engine for the pair *(ci, c)* corresponding to a candidate instance *ci*, and a class *c*, using the pattern *p*; whereas *hits(ci)*   is the hit count for the candidate instance *ci* alone.

1. *Strength.* This metric gives the number of times a candidate pair *(ci, c)* was observed for each pattern *p* belonging to the set of extraction patterns *P* listed in Section 3.1. Thus, the higher the Strength score, the higher is the likelihood that the pair *(ci, c)* is correct.

$$\text{Strength(ci,c)} = \sum_{p \in P} hits(ci, c, p) \qquad (1)$$

2. *Str-INorm-Thresh.* As pointed out in [7], the Strength metric is biased towards very frequent instances. In order to compensate this, we normalize the above metric by the number of hits of the candidate instance *ci, hits(ci)*. However, even using this normalized version, ones can be mislead when the candidate instance is very rare or misspelled. Thus, the normalization factor *hits(ci)* is modified to be constrained to have at least a minimum value. This value is determined by sorting the candidate instances by *hits(ci)* and then selecting *Count_{25}*, the hit count that appears at the 25th percentile (see Formula 2.). As reported in [7], other percentiles could also work well for addressing this problem.

$$\text{Str-INorm-Thresh(ci,c)} = \frac{\sum_{p \in P} hits(ci, c, p)}{\max(hits(ci), Count_{25})} \qquad (2)$$

3. *Str-ICNorm-Thresh.* Continuing with the normalization idea, we also adopted the normalized version of the Strength metric that combines normalization factors for both the candidate instance *ci* and the class *c* at the same time. The Formula 3 shows how to calculate this metric.

$$\text{Str-ICNorm-Thresh(ci,c)} = \frac{\sum_{p \in P} hits(ci, c, p)}{\max(hits(ci), Count_{25}) \cdot hits(c)} \qquad (3)$$

WordNet Similarity. Semantic similarity measures based on WordNet have been widely used in NLP applications [13], and they typically take into account the Word-Net structure to produce a numerical value for assessing the degree of the semantic similarity between two concepts. In what follows, we briefly describe two distinct similarity measures based on WordNet adopted in this research work:

1. *Lin* [14]. Lin defines the similarity between two concepts as the ratio of the shared information content to the information content that separately describe each concept. This measure estimates the specificity of a concept, deriving empirical information from corpora, and only exploiting concepts in the WordNet *is-a* hierarchy.

2. *Wu and Palmer* [15]. This similarity measure relies on finding the most specific concept that subsumes both the concepts under measurement. The path length from the shared concept to the root is scaled by the sum of the distances of the concepts to the subsuming concept. This measure has the advantage of both being easy for implementation and having competitive performance against other similarity measures.

We refer the reader to [14, 15] for more information about the above similarity measures.

In the CHOP System, these similarity measures provide the degree of similarity between the class c and the candidate instance ci. The two measures above-mentioned range from 0 to 1. We use the sum of them (*WNS*) as our semantic similarity score. Therefore, the maximum similarity score (*MaxWNS*) assigned for each candidate instance is 2.

Number of Extra Patterns. This heuristic is based on the main idea that if a candidate instance is extracted by many extraction patterns, this gives a strong evidence that this candidate instance is a valid instance for the related class. Based on this assumption, we defined the *Extra Pattern Score* (*EPS*) as the number of extraction patterns that extracted a particular candidate instance. In this manner, the maximum score (*MaxEPS*) assigned to a candidate instance is equal to the size of the initial list of extractions patterns, i.e., MaxEPS = 6.

Direct Matching. This last heuristic is based on the idea of finding the label of the class within the instance candidate [16]. To this end, we employ a classical stemming algorithm on both labels of the class and the candidate instance. If they match, then the system assigns 1 as its *Direct Matching Score* (*DMS*), or 0 otherwise. For example, given the University class and the candidate instance 'University of London', then the candidate is flagged as an admissible positive instance.

At this point, we are finally able to define the proposed confidence-weighted score function. For simplicity, we define the *Confidence Score* (*ConfScore*) of a candidate instance as the weighted sum of all constituent scores shown above, i.e., PMI variant (*PMI_Var*), combined with WordNet Similarity measures, Number of Extra Patterns utilized, and Direct Matching of candidate instances. The Formula 4 shows how to calculate the final score *ConfScore* for a given candidate instance ci of a class c.

$$ConfScore(ci,c) = \frac{EPS + WNS + DMS + 1}{MaxEPS + MaxWNS + MaxDMS + 1} \cdot \text{PMI\_Var}(ci,c) \qquad (4)$$

The Formula 4 reflects our central idea of combining different measures in order to obtain better results than using only one of them isolatedly. We have distributed the weights based on our initial hypotheses according to what we believed that were the most reliable heuristic. On the other hand, since we are in an ongoing project, we have not tried other alternative scoring formula yet. As future work, after analyzing the results obtained in the experiments, we expect to assess the impact of each above-mentioned measure on the final confidence score value, and then empirically evaluate different alternatives to the confidence score function.

By the end of this phase, the system generates a list of candidate instances sorted in decreasing order based of their confidence values determined by the Formula 4.

3.4 Ontology Population

Some incorrect candidate instances are resulted from typical noisy information found on the web. Spurious candidate instances may also take place for other reasons, such as incorrect parsing of noun phrases, misspelled instance names, etc. We argue that it is imperative to provide reliable estimation of the quality of the extracted instances. Accordingly, the CHOP system removes candidate instances having a confidence value below a given threshold.

It is worth mentioning that the CHOP system does not take into account the class hierarchy when assigning class instances to the classes in the input ontology. It restricts itself to only populating the class chosen at the beginning of the population process.

4   Experimental Evaluation

This section describes the experimental setup conducted in this research work in order to evaluate the CHOP system. We wanted to experimentally evaluate the effect that our confidence-weighted metric had on the final ranking of the candidate instances generated by the system.

4.1 Experimental Setup

Dataset Description. The dataset used for the experiments reported in this section was constructed using the six extraction patterns listed in Table 1 (Section 3.1). We used an ontology containing 10 classes[3] for our experiments and used each pattern to query up to 500 web pages, totalizing 3000 documents for each class in the input ontology. The documents were gathered using the Bing Search Engine Application Programming Interface (API)[4].

The primary reason why we used 500 web pages was due to the computational effort required to perform both the web searches and the preprocessing tasks. When we compare this number of pages against 1.800 ones used in the experiments in [7], it may seem insufficient. However, the researchers in [7] were allowed to use the output of the BE Engine[5], which returned all pages already preprocessed at no cost. Contrarily, we had to perform all the preprocessing work from scratch. Nevertheless, we argue that 500 web pages is a reliable and significant basis for validating our approach.

After running the CHOP system on this dataset, we selected the first 300 candidate instances sorted by confidence score in descending order for each pair (*class, PMI score*).

---

[3] Mammals, Birds, Cities, Diseases, Foods, Fruits, Movies, Sports, TV Series, and Universities
[4] http://www.bing.com/developers/s/APIBasics.html
[5] The Bindings Engine (BE) is an efficient search engine for natural language applications that enables information extraction, by using a generalized query language based on POS tags.

Finally, a considerable annotation effort was necessary in which three humans evaluators were in charge of manually confirming the system predictions.

Evaluation Measures and Preliminary Results. With the aim of validating the effectiveness of our approach, we compared the system results using the proposed confidence-weighted score (*ConfScore*) against the three PMI variations (*Strength, Str-INorm-Thresh, Str-ICNorm Thresh*) presented in Section 3.3. The classification precision of our system can be evaluated at a given cut-off rank, considering only the *n* topmost candidate instances returned by the system. In our experimental setup, we determined four cut-off points corresponding to the top 25, 50, 100, and 300 in the list of candidate results. Thus, Precision at the top *n* candidate instances, *P(n)*, is defined as follows:

$$P(n) = \frac{number\ of\ correct\ system\ predictions}{n} \tag{5}$$

The grouped bar graphs depicted in Figures 2-5 report our experimental results for 10 classes. In all graphs, we compare the precision values of each PMI metric variant (*Strength, Str-INorm-Thresh, and Str-ICNorm-Thresh*) with its weighted version, i.e., *CS(Str), CS(Str-INorm)*, and *CS(Str-ICNorm)*, respectively. For instance, considering the second group in the Figure 2 which corresponds to the class *Bird*, the first black bar represents the precision value obtained by the *Strength* PMI metric (indicated by *Str)*, whereas the second black bar represents the precision value of the weighted version of it, i.e., *ConfScore(Strength)*, indicated by *CS(Str)*. As shown in the legend, the other compared pairs, Str-INorm-Thresh vs. CS(INorm) and Str-ICNorm-Thresh vs. CS(ICNorm), are depicted in different grayscale colors.

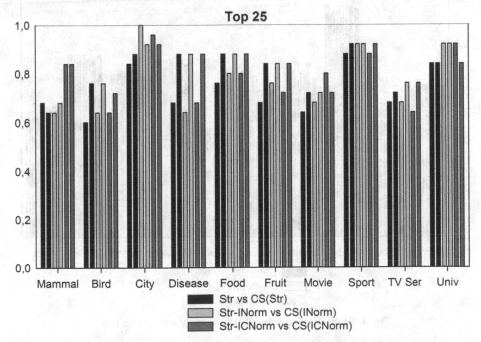

Fig. 2. Precision results for all classes evaluated in Top 25

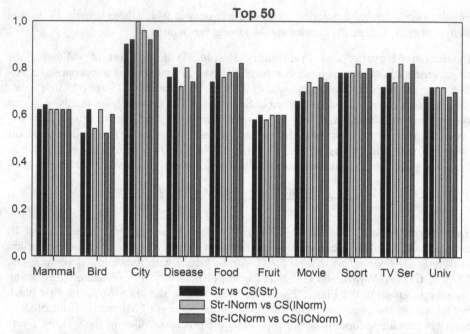

Fig. 3. Precision results for all classes evaluated in Top 50

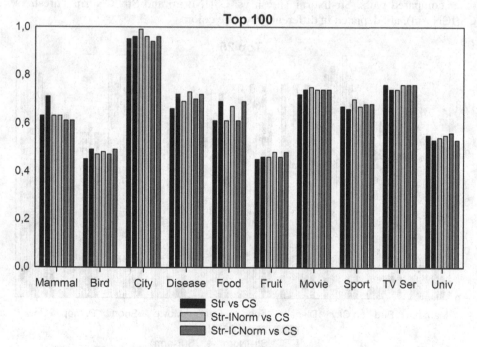

Fig. 4. Precision results for all classes evaluated in Top 100

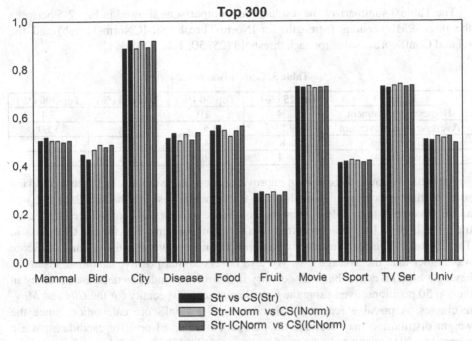

Fig. 5. Precision results for all classes evaluated in Top 300

4.2   Discussion

Analyzing the results shown in Fig. 2-5, one can notice that the system was able to successfully extract a considerable amount of positive instances for most classes. Due to the use of domain-independent linguistic patterns, some classes yielded better results than others. In our experiments, the best score results were obtained by the *City*, *Movie*, and *TV Series* classes in the top 300 candidate instances. One possible reason is that it is quite common to find text-fragments that match the patterns "cities such as" and "such movies as" followed by positive instances of the respective classes. Other classes like Fruit, Food, and Bird did not present the same behavior.

Another reason for the decreasing of precision for some classes was due to the presence of noisy information as a hampering factor during the extraction process. The main difficulty lies in extracting noun phrases matching the pattern "*NP and/or NP*", because the sentences extracted by this pattern usually represent a single noun phrase, which, in most cases, generates erroneous candidate instances. However, during the evaluation phase, some of the candidates extracted by the previous pattern showed a high PMI value, resulting in misclassification of such candidates usually putting them between the top 50 and 300. Yet, the simple separation of the two noun phrases using the delimiter "*and/or*" would result in an erroneous fragmentation of some candidate expressed by compound nouns. This happens, for instance, with the candidate instance for the *TV Series* class "Two and Half Men", in which it would result in two false candidates like "Two" and "Half Men". This problem is a limitation of the proposed approach and we intend to investigate how to solve it in future work.

The Table 3 summarizes the results of all comparisons showed in Fig. 2-5 between the three PMI versions (Strength, Str-INorm-Thresh, Str-ICNorm-Thresh) and the related ConfScore results, for each threshold (25, 50, 100, 300).

Table 3. Comparison results

Results	Top 25 (%)	Top 50 (%)	Top 100 (%)	Top 300 (%)
Highest improvement	24	10	8	3,4
Average of improvement	9,71	5	3,44	1,60
Highest loss	8	4	3	2,7
Average loss	6,4	2,66	2,14	1,21

As Table 3 shows, there was an improvement on precision results in most comparisons, with an encouraging gain observed for some classes. For the Top 25, there was an improvement up to 24% comparing ConfScore against Str-INorm-Thresh. By contrast, the highest loss (8%) occurred in a few comparisons, e.g. for the *City* class in Top 25. In all comparisons, the average precision improvement using our ConfScore was higher than the average loss. We also observed that positive candidate instances having a high hit count, but with no entry in WordNet, tended to decrease accuracy in the top 50 positions, even using the ConfScore metric, specially for the *City* and *Movie* classes. A possible reason for that lies in the ConfScore calculation, since the weight distribution may decrease the assessment value of positive candidates, while promoting false positives in the top 50 positions. On the other hand, as we expand the sample up to the top 300, the *City* and *Movie* classes had a considerable gain in precision for the ConfScore metric. In general, classes containing an entry in WordNet had large variation in precision values when contrasting ConfScore against all PMI variations separately, which was observed for the *Disease* and *Bird* classes. Thus, the results suggest that the WordNet similarity measure, when applied, has a strong impact on the final classification.

On the one hand, the Extra Pattern heuristic caused a significant impact on the ConfScore metric, as we observed that candidate instances extracted by more than four patterns confirmed to be positive instances. On the other hand, the DM heuristic did not have a notable influence on ConfScore, unless for the *University* class.

By analyzing the Fig. 2-5, the influence of the ConfScore metric is clearly diminishing as N increases. The bottom line is that the ConfScore metric actually promotes positive instances to the top positions, which was confirmed by the highest precision values achieved by the system for the Top 25 and 50. Indeed, statistical significance tests were performed for each possible combination of PMI variants and the ConfScore. They have demonstrated a significance difference at $\alpha = 0,05$ for the Top 25 and 50, whereas this also occurred for the Top 100 (CS vs Str) and Top 300 (CS vs ICNorm) at $\alpha = 0,1$. Moreover, this assessment showed that the ConfScore (Str-INorm-Thresh) incremented precision for most classes analyzed.

It is important to mention that we did not compare the CHOP system with other systems listed in related work because that would be only possible if the compared systems were under the same experimental setup (same corpus, same domain ontology, etc). Since we use the web as a big corpus, and considering the dynamic nature of it, we cannot provide a fair and direct comparison against other OBIE systems.

5   Conclusion and Future Work

We have proposed an unsupervised approach for ontology population based on a con-fidence-weighted metric that assess candidate instances extracted from web texts. Additionally this paper described the full implementation of our approach - the CHOP system - using classical linguistic patterns and a combination of PMI metrics, Word-Net similarity measures, and simple heuristics for tackling the specific problem of finding candidate instances of ontological classes. We have conducted some experi-ments in order to evaluate the impact that the proposed confidence-weighted metric would have on three PMI variants.

Although we have achieved encouraging results in this research work, there are still many opportunities for improvement. Thus, we intend to provide a detailed quantitative analysis of the actual contribution of each weight factor used in our confidence-weighted metric, aiming to select the more reliable factors in the instance ranking process. More work is needed for improving the extraction of candidate instances by using more efficient filters, enabling the system to eliminate false candidate derived from typical noisy data found on the web. Another expected improvement concerns the integration of a new component in the system architecture for suggesting specific lin-guistic patterns in order to expand the initial set of extraction patterns.

Acknowledgment. The authors would like to thank the National Council for Scientif-ic and Technological Development (CNPq/Brazil) for financial support (Grant No. 140791/2010-8).

References

1. Berners-Lee, T., Hendler, J., Lassila, O.: The Semantic Web. Scientific American 284(5), 34–43 (2001)
2. Cimiano, P.: Ontology Learning and Population from Text: Algorithms, Evaluation and Applications. Springer, New York (2006)
3. Wimalasuriya, D.C., Dou, D.: Ontology-based information extraction: An introduction and a Survey of Current Approaches. J. Information Science 36(3), 306–323 (2010)
4. Etzioni, O., Cafarella, M., Downey, D., Kok, S., Popescu, A., Shaked, T., Soderland, S., Weld, D., Yates, A.: Web-Scale Information Extraction in KnowItAll. In: Proc. of the 13th Inter. WWW Conference (WWW 2004), New York City, New York, pp. 100–110 (2004)
5. Cimiano, P., Handschuh, S., Staab, S.: Towards the self-annotating web. In: Proceedings of the 13th International Conf. on World Wide Web, pp. 462–471. ACM, New York (2004)
6. Cimiano, P., Ladwig, G., Staab, S.: Gimme The Context: Context driven Automatic Se-mantic Annotation with CPANKOW. In: Proc. of the 14th Inter. Conf. on WWW, Japan, pp. 332–341 (2005)
7. McDowell, L.K., Cafarella, M.: Ontology-Driven, Unsupervised Instance Population. Web Semantics: Science, Services and Agents on the World Wide Web 6(3), 218–236 (2008)
8. Hearst, M.A.: Automatic Acquisition of Hyponyms from Large Text Corpora. In: 14th Conference on Computational Linguistics, COLING 1992, Nantes, France, vol. 2, pp. 539–545. Morgan Kaufmann (1992)

9. Wu, F., Weld, D.S.: Autonomously Semantifying Wikipedia. In: CIKM, pp. 41-50. ACM (2007)
10. Brill, E.: Processing Natural Language without Natural Language Processing. In: Gelbukh, A. (ed.) CICLing 2003. LNCS, vol. 2588, pp. 360–369. Springer, Heidelberg (2003)
11. Ciravegna, F., Dingli, A., Guthrie, D., Wilks, Y.: Integrating Information to Bootstrap Information Extraction from Web Sites. In: IJCAI 2003 Workshop on Intelligent Information Integration, pp. 9–14 (2003)
12. Petasis, G., Karkaletsis, V., Paliouras, G., Krithara, A., Zavitsanos, E.: Ontology Population and Enrichment: State of the Art. In: Paliouras, G., Spyropoulos, C.D., Tsatsaronis, G. (eds.) Multimedia Information Extraction. LNCS, vol. 6050, pp. 134–166. Springer, Heidelberg (2011)
13. Pedersen, T.: Information Content Measures of Semantic Similarity Perform Better Without Sense-Tagged Text. In: Proc. of the 11th Annual Conf. of the North American Chapter of the Association for Computational Linguistics, Los Angeles, pp. 329–332 (2010)
14. Lin, D.: An Information-Theoretic Definition of Similarity. In: Proceedings of International Conference on Machine Learning, Madison, Wisconsin (1998)
15. Wu, Z., Palmer, M.: Verb Semantics and Lexical Selection. In: 32nd Annual Meeting of the Association for Computational Linguistics, Las Cruces, New Mexico, pp. 133–138 (1994)
16. Monllaó, C.V.: Ontology-based Information Extraction. Dissertation Thesis, Polytechnic University of Catalunya (2011)

Situation-Aware User's Interests Prediction for Query Enrichment

Imen Ben Sassi[1], Chiraz Trabelsi[1], Amel Bouzeghoub[2], and Sadok Ben Yahia[1,2]

[1] Faculty of Sciences of Tunis, University Tunis El-Manar, 2092 Tunis, Tunisia
[2] Department of Computer Science, Télécom SudParis,
UMR CNRS Samovar, 91011 Evry Cedex, France
imen.bsassi@gmail.com, {chiraz.trabelsi,sadok.benyahia}@fst.rnu.tn,
Amel.Bouzeghoub@it-sudparis.eu

Abstract. Situation-Aware User's Interest Prediction aims at enhancing the information retrieval (IR) capabilities by expanding explicit user requests with implicit user interests, to better meet individual user needs. However, not all user interests are the same in all situations, especially for the case of a mobile environment. Thus, user interests are complex, dynamic, changing, and even contradictory. Consequently, they should be adapted to the user's specific search context. In this paper, we introduce a new approach that aims at building a dynamic representation of the semantic situation of ongoing mobile environment retrieval tasks. The semantic situation is then used to activate different classification rules of user's past interests at run time. Doing so, the best interest class's is proposed to expand the user's request. Our approach makes use of a semantic enrichment using DBPEDIA, providing enriched descriptions of the semantic situations involved for discovering user interests, and enabling the definition of effective means to related contexts. Carried out experiments, undertaken versus Google search, emphasize the relevance of our proposal and open many promising issues.

Keywords: Mobile information retrieval, Classification rules, Situation-Aware, DBPEDIA.

1 Introduction

The emergence of smartphones - mobile phones capable of functions typically associated with personal digital assistants (PDAs) or even personal computers - has made mobile information access an everyday reality. Thus, mobile computing emerged as a new paradigm of personal computing and communications. Facing the large volume of information available in the internet and the constraints of mobile devices such as difficulties of query input and limited display zone, user queries are more likely to be shorter and more ambiguous. Indeed, studies of the query of mobile users [8] showed that the average length of mobile applications is 2.56 words and 16.8 characters. As a primary means of addressing many of the issues in mobile computing, situation-awareness offers a way to adapt the search result to each usage situation, location, environment, user goal, etc. More generally, determining the user's situation serves as a convenient means of limiting the

S.W. Liddle et al. (Eds.): DEXA 2012, Part I, LNCS 7446, pp. 191–205, 2012.

search space. Indeed, the mobile user may merely "have questions" and will need very specific (and thus potentially terse) answers. Hence, using situation information to limit the scope will make it easier to provide more user-friendly answers.

The aim of contextual retrieval is to "combine search technologies and knowledge about query and user context into a single framework in order to provide the most appropriate answer for a user's information needs". In a typical retrieval environment, we are given a query and a large collection of documents. The basic IR problem is to retrieve relevant documents to the query. A query gathers all the information that we have to guess about user's needed information and to determine its relevance. Typically, a query contains only a few keywords, which are not always a good descriptor of content. Given this absence of adequate query information, it is important to consider how other information sources could be exploited to grasp the information need, such as user's situation. Situational retrieval is based on the hypothesis that situation information be of help to describe a user's needs and consequently improve retrieval performance. Especially, in situation-aware systems, the additional information of mobile users' locations creates vast opportunities for personalization of information retrieval results.

Example 1. Suppose that a user submits a query *Mona Lisa*. It is not clear whether the user is interested in the famous Leonardo Da vinci's painting or the Julia Robert's movie. Without understanding the user's search intent, many existing methods may classify the query into both categories "Arts" and "Movies". However, if we find that the user's location is museum, it is likely that he is interested in the category of "Arts". Conversely, if the user's location is cinema, it may suggest that the user is interested in the topics related to "Movies".

In fact, any modern information retrieval system is based on an autocompletion engine and can hence obtain the most popular query terms or phrases from indexed pages. Such elements can be used to interactively complete user's queries. Indeed, an autocompletion tool can be helpful both by saving typing time and by finding new, serendipitous terms. However, a pure syntactic approach can present difficulties, because it will be based on the matching of string representations (words), but not in concepts and their relationships. The corollary is that, the autocompletion mechanism will be helpful if the user intends to use the same word that the system is expecting. But if, for instance, a synonym is in user's mind, the system will be unable to recognize and autocomplete it. The previous facts have led us to introduce a concept-driven semantic and situation-aware user's interest prediction approach.

Hence, in this paper, we investigate a new approach of Information retrieval based on the **prediction of the user's interests**. This approach aims at building a dynamic representation of the semantic situation of ongoing mobile environment retrieval tasks. The semantic situation, is then used to activate different classification rules of user's past interests at run time, in such a way that the better interest class's is proposed to expand the user's request.

The rest of the paper is organized as follows. Section 2 scrutinizes the related work dedicated to IR personalization. In Section 3, we thoroughly describe, SA-IRI, our approach for situation-aware personalized IR system based on interests prediction. Section 4 presents the experimental results of the introduced approach. Finally, Section 5 concludes and points out avenues of future work.

2 Background and Related Work

In this section, we start by introducing the key concepts that will be of use in the remainder.

2.1 Key Concepts

Situation Awareness. The situation awareness may be defined as: *The perception of the elements in the environment within a volume of time and space, the comprehension of their meaning and the projection of their status in the near future* [5]. In fact, mobile devices have more features than their computer counterparts, including location information, time and social networks. We can use information from these additional features to create new research areas. IR and data mining of the new information will require novel technologies specifically developed to process information such as time, location and semantics [15]. Specifically, in situation-aware systems, the additional information of mobile users' locations creates vast opportunities for personalization of information retrieval results.

Query Enrichment. The query enrichment process consists in reformulating the initial user query by applying a pre-treatment and adding concepts describing his/her interests. In our work, the query enrichment process is defined as a mapping between a user's initial query Q and an enriched one $Q \cup I$, where I represents user's extracted interest.

In the following, we present a scrutiny of the related work.

2.2 Scrutiny of the Related Work

The expansion of mobile devices has promoted the research on the benefit of contextual information, introducing new requirements and expectations. In this respect, the main challenge introduced by context-awareness is to come up with a flexible and unambiguous representation of the Context. Indeed, Dey in [4], suggests the following general definition of Context and context-awareness: *Context is any information that can be used to characterize the situation of an entity. An entity is a person, place, or object that is considered relevant to the interaction between a user and an application, including the user and application themselves. A system is context-aware if it uses the Context to provide relevant information and/or services to the user, where relevancy depends on the users task.* According to this definition, the Context could be considered the current data displayed on the screen, the surrounding environment or even the whole application. For that reason, every context-aware application must explicitly specify what information is part of the Context. Although, contextual information has been previously exploited for improving web search, there is no agreement on context interpretation and scope.

Thus, in [9,10], the authors consider the history of visited pages as web search context. Following that criteria, they try to analyze and extract the most important keywords found in those pages, and use them to ensure that further queries will keep on subject. While these solutions can improve web search on the desktop, mobile search should go beyond, considering additional variables like location, weather situation or guessed user intentions [6]. In fact, the Context is inherently dynamic and constantly changing. A few properties will be barely modified over time, like user name or age, but other ones may frequently vary, like time or location. Context properties must be thus, kept up to date [13]. In addition, White *et al.*, in [17] uses the current page and 5 distinct sources of information for modeling user interests during Web interaction: (i) *social*: the combined interests of other users that also visited the current page; (ii) *collection*: pages with hyperlinks to the current page; (iii) *task*: pages related to the current page by sharing the same search engine queries; (iv) *historic*: the longterm interests for the current user, and; (v) *interaction*: recent interaction behavior preceding the current page.

Other research studies have attempted to include context information during the search process by exploiting query expansion techniques in order to complete the submitted query with additional terms [6,14]. Such terms can be synonyms, disambiguating subject-related words or special keywords that aim at clarifying the intention, like "how to", "ways to", "what is", etc. Another way to contextualize the search is by re-ranking the obtained result set according to given properties: subject, proximity to user's location or intentions. These aforementioned approaches ensure that the most contextually-relevant results will appear on the top of the result set, which reduces the number of interactions needed to find restricted results [8]. A context-aware approach is developed within the UP-CASE project in [12], it uses different type of sensors (sound, GPS, temperature, etc.) explored to define users contexts (walking, hot temperature, etc.). In this work, an inference process is used to generate relations between environment and contexts. However, this approach does not provide a solution to reduce the size of the decision tree containing different contexts, which makes its maintenance a difficult process. While, Sohn *et al.*, in [16] conduct a two-week diary study to better understand mobile information needs and how they can be explored in information retrieval personalization. In fact, user contexts can be described by four categories: Activity, Location, Time, and Conversation. (i) *Activity* reflects what the user was doing at that time; (ii) *location* is the place where the person was at and includes any additional artifacts at that specific location; (iii) *time* is the time when the need arose, and; (iv) *conversation* is any phone or in-person conversation the participant was involved in at that time.

The context concept can be seen as particulary, based on a limited number of dimensions, as *user situation*. In fact, in [18], the authors use this term to describe their technique, called SAP-IR, combining situation-based adaptation and profile-based personalization in a mobile environment. In this work, user's actions are generated to learn his behaviors, interests and intentions, then to generate his situation using its context (time, location, light intensity and noise level). The main limit of this study is the lack of semantics in the generation of the user's situation. More recently, in [11], the authors have addressed this problem by introducing a situation-aware mobile search to personalize user's search results. This latter exploits a case-based reasoning (CBR) method to modelize the past research. Indeed, when a new situation S is constructed, the CBR nodes are explored to check for the existence of a similar situation to S and to extract its related interest. Otherwise, this situation is added to the knowledge base as a new entry without checking its accuracy level.

To sum up, mobile Web Search introduces new thriving challenges in traditional Information retrieval field. Indeed, users normally own modern smart phones which allow them to be permanently online anywhere, anytime. A typical mobile web search scenario consists of a user outdoors with an information need. At this point he/she takes his/her phone and uses a web search engine to find an answer to a query. The main moan that can be addressed to the above surveyed approaches, stands on the fact they do not actually take advantage of semantic information and the situation of the user to generate useful rules that can be reused in future research sessions, in order to predict the user's information needs. To palliate such drawbacks, we introduce in the following a new approach, called SA-IRI, towards a situation-aware user's predicted interests for IR personalization. The main thrust of this approach stands in the conjunction of the associative classification as well as the relying on the semantic web through its sweeping of DBPEDIA[1].

3   SA-IRI: Situation Aware Information Retrieval Based Interests

The idea of IR personalization, proposed and developed in the remainder, responds to the fact that human interests are multiple, heterogeneous, changing, and even contradictory, and should be understood in context with the user search intention. In this context, the problem to be addressed includes how to represent the user situation? how to determine it at runtime? and how to use it to influence the activation of user interests and to personalize the search results? To answer these questions, we propose our approach for IR personalization based situation-aware (*c.f.*, figure 1). In order to select the most adequate user profile to be used for personalization, we compare the similarity between a new search situation and the past ones.

In the following, we introduce a new situation-aware approach used to personalize search results in mobile environment. This situation is generated by a mapping between the user situation, and the semantic concepts extracted from the DBPEDIA ontology contained in Linked Data [2]. For example, if the user is at the location " 36.851111, 10.226944 " and the time is " Mond February 27 11:00:00 2012 ", then we can guess that he is " at museum, winter, morning ". Thus, SA-IRI approach operates through three steps: (i) *User's semantic situation construction*: Semantic information is extracted from the user's physical

---

[1] http://dbpedia.org/About

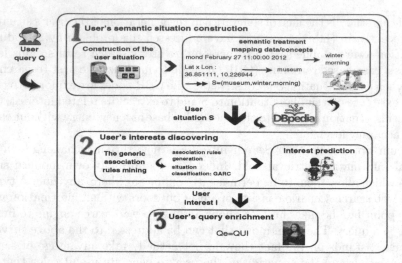

Fig. 1. SA-IRI approach

situation (location, time and season), in order to guess its interest behind such situation; (ii) *User's interests discovering*: The interest related to the user's semantic situation is predicted, using the DBPEDIA dataset, and; (iii) *User's query enrichment*: The initial user's query is extended with the interest previously predicted. These steps are detailed in the following:

3.1   Step1: User's Semantic Situation Construction

To model the user situation, we propose to concentrate on a semantic representation the user's current location and time. Our hypothesis is that users interests are related to the semantic signification of physical entities. For example, it does not matter where the user is actually located in "Petit Palace Museum" or "Louvre museum" but it is of importance to guess that he/she is in a "museum". Thus, we propose to model the user's situation by a vector of three dimensions:

- **Location:** refers to the type of user's position (beach, museum, university, etc.) extracted from a Linked data ontology;
- **Season:** refers to a year's season (autumn, winter, spring or summer). This dimension can influence the preferences of the user;
- **Time of the day:** refers to a day's part (morning, noon or night). This data determines the type of activities that can be performed by the user (work activities, entertainment, etc.).

Thus, the user's situation is presented by the 3-dimensional vector S= (S_l, S_s, S_t) where : S_l (resp. S_s and S_t) refers to location type (resp. season and time of the day). This situation is used, in the step2, to extract the user's interest.

Example 2. If the user is standing in the location defined by the following GPS coordinates " 35.1877778, 8.655 " and the time is " Sun January 29 13:00:00 2012 ", then we can describe its situation by the following 3-dimensional vector:

$$S = (S_l, S_s, S_t) = (mountain, winter, noon)$$

Assuming that this user is running a web search "sport". Then, the name of the mountain from which he launched its query is irrelevant compared to the type of the location (beach, mountain, etc.).

In general, two types of sporting activities can be distinguished in this situation:

- If the season is winter, then we can restrict the search to "winter sports" such as: ski, snowboard, etc.
- Otherwise, we can consider that the user is interested in other types of mountain sports like rock climbing.

3.2   Step 2: User's Interests Discovering

Based on the nature of mobile environment, user interests may change anytime due to changes in their situation (location, time, etc.). Static approaches for building the user profile are poorly useful, so we rather focus on more dynamic techniques, that are capable of adjusting the user interests to the current search situation. In this context, based on multiple users' situations and the correlation situation/interest, this step fuses a classification technique with a semantic treatment.

Firstly, in order to minimize the number of user's situations and interests we apply the informative generic association rules extracted by $\mathcal{IGB}$ [1]. In particular, the $\mathcal{IGB}$ basis generates rules with reduced-size premise parts presenting the user's situation (S_l, S_s and S_t) and large conclusion parts presenting its actual interest. Secondly, we apply a generic association classification technique [3], with an aim of obtaining a concentration of only useful prediction rules, that associate an interest to each user's situation. Thus, a set of intuitive generic classification prediction rules, whose conclusion's part present a user's interest is generated, e.g., $S_l \wedge, \ldots, \wedge S_t \Rightarrow Interest_i$.

The choice of the association rules is based on the intuitive relation between users' situations and interests, and a comparison with the Case Based Reasoning that shows many limitations of this learning technique, i.e., (1) *Handling noisy data*: Users might rely on previous experience without validating it in the new situation; and (2) *Handling large case bases*: The maintenance of rule bases becomes a difficult process as the size of the rule base increases and the problem of redundant rules. The results of this comparison are shown in section 4.

Example 3. By applying the generic association classification technique on the context given by Table 1 describing a set of interests, each one associated to a situation, we can derive the following generic classification rules:

- R_1: mountain, summer $\Rightarrow$ camping
- R_2: beach, summer $\Rightarrow$ surf
- R_3: night, summer $\Rightarrow$ shopping
- R_4: spring, night $\Rightarrow$ art
- R_5: noon $\Rightarrow$ art

In the following, we briefly remind the definition of the formal context.

Definition 1. FORMAL CONTEXT: *A formal context is a triplet* $\mathcal{K} = (\mathcal{O}, \mathcal{I}, \mathcal{R})$, *where* $\mathcal{O}$ *represents a finite set of transactions stored in a learning base,* $\mathcal{I}$ *is a finite set of items and R is a binary (incidence) relation (i.e.,* $\mathcal{R} \subseteq \mathcal{O} \times \mathcal{I}$). *Each couple* $(o, i) \in \mathcal{R}$ *expresses that the transaction* $\{o\} \in \mathcal{O}$ *contains the item* $\{i\} \in \mathcal{I}$.

Example 4. Let us consider the formal context sketched by table 1, where:

- $\mathcal{O}$ is the set of transactions such that $\mathcal{O}=\{1,\dots, 5\}$;
- $\mathcal{I}$ is the set of items {attribute1, attribute2, attribute3, Class}, where:
  - attribute1 $\in$ { "autumn", "winter", "spring", "summer"} defined by S_s the 2nd dimension in the situation vector;
  - attribute2 $\in$ { "morning", "noon", "night"} defining S_t the 3rd dimension in the situation vector;
  - attribute3 is the user's current type of location defining S_l the 1st dimension in the situation vector;
  - Class is the user's interest extracted from DBPEDIA.
- $\mathcal{R}$ is the relation that links the user's interest to its actual situation. This binary relation aims to associate each transaction (new situation) to an item (single interest).

Table 1. An example of an extraction context $\mathcal{K}$

	Attribute1				Attribute2			Attribute3				Class
	autumn	winter	spring	summer	morning	noon	night	museum	theatre	beach	mountain	mall	Interest
1	×					×		×					art
2			×				×		×				art
3				×	×					×			surf
4		×			×						×		camping
5				×	×							×	shopping

Based on the classification rules previously generated, the new user's situation S is compared to each rule premise to select the most similar past situation. In the case where a similar situation is found, the rule conclusion is extracted to explain the user's interest related to his/her actual situation. Note that two situations are similar if and only if they share at least two dimensions.

Example 5. Considering the context given by Table 1, and based on rules generated in example 3. Assuming that $S = (beach, summer, morning)$ is the new user's situation, the most similar premise to our new situation is obtained thanks to R_2. In this case, the surf conclusion is extracted to explain the interest related to the user's current situation.

In the case of no rule with similar situation to the current one is extracted, a semantic treatment is carried out to discover the new user's interest. Thus, the user's query is mapped into concepts using DBPEDIA. Such concepts are identified by the property **rdfs:label**, *i.e.*, to those concepts for which there is an RDF [2] triple in DBPEDIA. Thus, the concept stands for the subject, the data property is **rdfs:label** and the object stands for a string exactly matching the user's query. Two cases have to be distinguished:

- The user's query matches a category in DBPEDIA such as sport, art, etc. In this case, we extract all subcategories that have a semantic relationship with the user's type of location previously defined. This relation is defined by the **skos:broader** property. For example, if the user submits the query "sport" at beach location, then the subcategories returned by our system are the beach-sports (beach soccer, beach polo, rowing, beach volleyball, etc.) presenting the interests concepts.
- The user's query defines a subcategory in DBPEDIA such as a film title, a painting name, etc. In this case, we have to disambiguate its query. To do so, we look for super-classes in connection with the user's type of location that will be defined as user's interest using the **dcterms:subject** property. For example, suppose that the user submits a query "Mona Lisa". It is not clear whether the user is interested in the famous Leonardo DA VINCI painting or the Julia ROBERTS movie "Mona Lisa Smile". Interestingly, if we find that the user's type of location is cinema, then it is likely that he is interested in "films". Conversely, if the user's type of location is museum, then we may guess that the user is interested in topics related to "art".

If many concepts have been found, then the most frequent concept is retained as the better candidate to mimic the user's new interest. Finally, a new transaction is added to the learned base in order to extract other classification rules in the next execution of the system (launch of a new query).

Example 6. Let us consider the context given by Table 1, and the rules set generated with GARC given in example 3. Assume that $S = (cinema, winter, night)$ (resp. Mona Lisa) is the new user's situation (resp. query). In this case, no rule with similar situation to the current one is extracted, then using DBPEDIA, a set of interests is returned to extract the most frequent one having as label "film". Then the vector, **(winter,night,cinema,film)** is added to the learning base, as a new transaction, in order to enrich the knowledge base with the next execution of our system.

3.3   Step 3: User's Query Enrichment

User queries are usually short and ambiguous in order to accurately describe an information need. One way to tackle this problem is to enrich the original query by adding terms. So, having as goal the improvement of the user's results, we enrich his/her query with the interest related to his/her actual situation and extracted during step 2.

Example 7. With respect to the interest predicted in Example 6, the user's query "Mona Lisa" is enriched by its interest. Thus, the new expanded query "Mona Lisa **film**" is executed to provide a list of personalized results to the user.

---

[2] http://www.w3.org/RDF

In the following, we present the *QueryEnrichment* algorithm, whose pseudo-code is given by algorithm 1. It shows the different steps of user's query enrichment.

Algorithm 1. *QueryEnrichment*

    Input: Q: A new query.
    Output: Q_e: An enriched query
1 **begin**
2   S= SITUATIONCONSTRUCTION();
3   {r}= SITUATIONCLASSIFICATION(S);
4   **if** $\{r\} \neq \emptyset$ **then**
5   **if** $\{r\}.size=1$ **then**
6   I= r.conclusion;
7   **else**
8   I=MOSTFREQUENTCLASSEXTRACTION(R);
9   **else**
10   I=DBPEDIAINTERESTEXTRACTION(Q,S_l);
11      $LB = LB \bigcup \{(S,I)\}$;
12   return $Q_e = Q \bigcup I$;

QueryEnrichment algorithm takes as input the query Q and outputs an enriched query $Q_e = (Q \bigcup I)$. The algorithm operates as follows: firstly, it constructs the user's situation S described by the 3-dimensional vector (S_l, S_s, S_t) (*c.f.* line 2). Then, in line 3, this situation is used to generate a set of classification rules containing similar situations, by invoking the SITUATIONCLASSIFICATION() function, that generates a set of association rules of use during the classification process. Therefore, if this set of rules is not empty, then the most frequent rule conclusion is used to enrich user's interest (*c.f.* line 8). Otherwise, a new interest is predicted by invoking the DBPEDIAINTERESTEXTRACTION() function, and the learning base is extended with a new transaction T=(S,I) (*c.f.* line 11). Once identified, the user's interest I is used to enrich his/her query Q (*c.f.* line 12).

4   Experimental Results

4.1   Evaluation Framework

– **The 1st Evaluation Methodology: The Diary Study**
   The main limit we encountered to evaluate our approach, is the absence of an evaluation benchmark for evaluating situation-aware approaches for IR. Thus, we conducted a dairy study, where each user is asked to save the time, the date, the location and the query. 6 participants (3 men and 3 women) whose ages vary between 24 and 34 years, have participated in our study (mainly people from our laboratory) and all of them have already an experience in web research. This study lasted 2 months and has generated 60 entries, with an average of 10 requests per participant (minimum = 2 and maximum = 25).

Table 2. An example of entries retrieved from the dairy study

User	Time	Place	Query
1	Sat Dec 31 13:04:00 2011	Mall	Puma
2	Sun Feb 5 16:30:00 2012	The Louvre museum	MonaLisa
3	Mond Jan 2 11:00:00 2012	The Marsa beach	Sport

Table 2 shows some examples of queries collected, each one described by the user identifier, the time, the date and the place of request transmission. For the evaluation of our approach, we split the gathered data, *i.e.*, the set of user transaction, in two subsets, *i.e.*, a training set (of 50 entries) and a test set (of 10 entries).

– **The 2nd Evaluation Methodology: The Quaero Challenge**
In the following, we describe the dataset of Quaero Evaluation for Task 2.6 on Contextual Retrieval Version 3.1 [3] used in our second evaluation approach. This dataset comprises 25 topics representing information needs of real users, where such topic includes: (i) The title presenting the user's information need or its query; (ii) The geolocation of the user at the time he/she submitted its query; (iii) The search history presenting the user's past queries and the clicked resources, *i.e.*, resources jugged pertinent. Thus, for each topic, we split user's queries in a training set gathering user's search history, and a test set containing user's current query.

The training set is used to extract a set of classification rules that will be used during the interest discovering step, *i.e.*, when a new query is submitted (test query), the user's situation is classified to extract its interest describing its research intention. In the case where the result of the classification algorithm is negative (no similar situation is found by the classification rules), then information collected from DBPEDIA is of use to predict the user's new interest.

4.2   Results and Discussion

In our experiments, we aim at evaluating the effectiveness of situation-aware information retrieval that integrates the user interests.

– **Evaluating the SA-IRI precision based on the diary study**
First, the number of resources returned was compared to that of Google [4] results as shown in Table 3. Then, to perform an information retrieval test, we used 10 keyword queries and computed the precision values of both systems, *i.e.*, SA-IRI vs those of Google Search engine. To do so, each participant tried manually the results of its requests., with those shown in Table 4. The judgments relevance were established on a scale of two values: relevant and irrelevant.
   The results obtained by our approach were compared with the interests representing the actual research intentions of users, as shown in Table 5.

---

[3] http://quaero.profileo.com/modules/movie/scenes/home
[4] https://www.google.com

As detailed by the statistics of Table 3, we highlight that the total number of resources returned by our approach is much lower compared to those returned by Google, except for the case of the query 5 where the number of resources is increased due to the inaccuracy of the predicted interest. Thus, the SA-IRI approach ensures the reduction of the research space.

Table 3. Comparison of SA-IRI and Google Results

Query	# Resources retrieved by Google	#Resources retrieved by SA-IRI
Query1	920,000,000	21,900,000
Query2	2,250,000	2,250,000
Query3	13,200,000	8,480,000
Query4	1,860,000,000	822,000,000
Query5	3,640,000,000	3,800,000,000
Query6	5,750,000	571,000
Query7	73,900,000	73,100,000
Query8	177,000,000	8,830,000
Query9	48,400,000	1,490,000
Query10	48,400,000	634,000

Table 4. Comparison of the accuracy of SA-IRI and Google results

Query	Top10 retrieved by Google	Top10 retrieved by SA-IRI
Query1	8 of 10 relevant resources	10 of 10 relevant resources
Query2	1 of 10 relevant resources	1 of 10 relevant resources
Query3	0 of 10 relevant resources	10 of 10 relevant resources
Query4	1 of 10 relevant resources	4 of 10 relevant resources
Query5	2 of 10 relevant resources	8 of 10 relevant resources
Query6	3 of 10 relevant resources	0 of 10 relevant resources
Query7	1 of 10 relevant resources	1 of 10 relevant resources
Query8	0 of 10 relevant resources	1 of 10 relevant resources
Query9	2 of 10 relevant resources	2 of 10 relevant resources
Query10	2 of 10 relevant resources	10 of 10 relevant resources

Table 4 sketches a comparison, in terms of number, of the resources deemed pertinent returned using SA-IRI approach and that returned by Google. Interestingly enough, the accuracy of the results is highly improved thanks to the use of SA-IRI.

Considering that the Precision metric is defined as follows:

$$Precision = \frac{\{relevant\ resources\} \bigcap \{retrieved\ resources\}}{\{retrieved\ resources\}} \quad (1)$$

Figure 2 shows a comparison between the average precision values of precision based on the number of the test queries obtained by both SA-IRI and Google. Indeed, for $\#queries=1$, the average precision of our approach is 0.95, while, the average precision of Google is 0.8, $i.e.$, an improvement in the average precision of 15%.

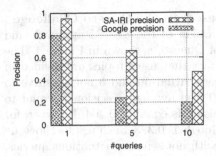
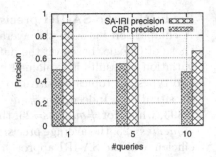

Fig. 2. Variation of the Average precision values vs those of Google

Fig. 3. Variation of the Average precision values vs those of CBR

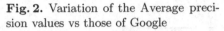

Table 5. Precision of interest prediction

Query	Relevant predicted interests	Total number of predicted interests
Query1	1	1
Query2	1	5
Query3	3	4
Query4	1	2
Query5	3	3
Query6	1	1
Query7	2	3
Query8	2	3
Query9	3	5
Query10	3	3

The results presented in Table 5 show the level of precision of our interest prediction. Indeed, the total number of the predicted interests sketches the interests with the maximum weight (the highest number of appearance) returned by SA-IRI, while relevant predicted interests are those considered by the user.

Figure 3 shows the average values of precision of our approach and a Case Based Reasoning approach [11], in interest prediction, based on the number of test queries. Indeed, for $\#queries=1$, the average precision is equal to 0.92, while, with the CBR technique the prediction precision is equal to 0.5. Whereas, for $\#queries=5$, we have an average precision of 0.73 with our approach, *i.e.*, a decrease in the average precision of 19%. These results show the quality of the SA-IRI precision compared to the CBR technique precision. They are justified by the compactness and the flexibility of the knowledge representation using the generic associative rules, opposing to the CBR model limitations. The case based technique satisfy the *"Good for me now so always good for all"* assumption. In fact, based on mobile environment specificities, the user can have two different interests in the same situation. In this case, using the CBR technique and without the update of the case base, only the 1st predicted interest is considered, and the 2nd one is ignored. Thus, user's interests are assumed to be invariant, witch contradicts the mobile context properties.

- **Evaluating the SA-IRI precision based on the Quaero Challenge**
  A comparison between the average precision of the SA-IRI approach and those of Google, based on the dataset of Quaero, is shown in Figure 4. The variation of results is explained by the fact that users topics are divided in two types: 20 topics sensitive to the user's situation and 5 not sensitive to its situations. Indeed, for $\#queries=20$, the average precision is equal to 0.393, while, for $\#queries=25$, the precision is equal to 0.404. However, for $\#queries=25$, the average precision of Google is 0.328. Such results show the efficiency of the SA-IRI approach even with not sensitive situations queries.

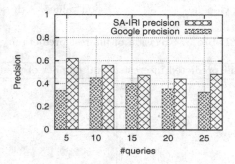

Fig. 4. The average precision of resources returned by our approach vs those of Google

5 Conclusion and Future Work

Situation is an increasingly common notion in Information Retrieval. This is not surprising since it has been long acknowledged that the whole notion of relevance, at the core of IR, is strongly dependent from context. In fact, several authors in the IR field have explored approaches that are similar to our in that they find indirect evidence of searcher interests by extracting implicit meanings in information objects manipulated by users in their retrieval tasks [7,10,14]. A first distinctive aspect in our approach is its snugness connection with semantic concepts, DBPEDIA, to describe the user's situation and interests. In particular, these concepts allow to identify the inherent relationship that links the situation of the user and his search intentions, and to predict its interests related to the current situation. In addition, our approach presents an originality since it combines the semantic IR and a data mining technique (*i.e.*, the classification rules), in order to reduce the size of the knowledge base and improve the prediction of user's interests. In the near future, we plan to take into consideration the social aspect in the user's interests prediction by using one of the social ontologies, *i.e.*, The FOAF ontology, to take advantage of the user's social circle, *e.g.*, activities, interests, relationships, etc.

Acknowledgements. This work was partially supported by the project Utique CMCU 11G1417.

References

1. Ben Yahia, S., Gasmi, G., Nguifo, E.M.: A new generic basis of factual and implicative association rules. Intelligent Data Analysis (IDA) 13(4) (2009)
2. Berners-Lee, T.: Design issues: Linked data (2009),
   http://www.w3.org/DesignIssues/LinkedData.html
3. Bouzouita, I., Elloumi, S., Ben Yahia, S.: GARC: A New Associative Classification Approach. In: Tjoa, A.M., Trujillo, J. (eds.) DaWaK 2006. LNCS, vol. 4081, pp. 554–565. Springer, Heidelberg (2006)
4. Dey, A.K.: Providing architectural support for building context-aware applications. Phd thesis, Georgia Institute of Technology, Atlanta, GA, USA (2000)
5. Endsley, M.R.: Design and evaluation for situation awareness enhancementl. In: Proceedings of the Human Factors Society 32nd Annual Meeting Santa Monica, pp. 97–101 (1988)
6. Hattori, S., Tezuka, T., Tanaka, K.: Context-aware query refinement for mobile web search. In: Proceedings of the 2007 International Symposium on Applications and the Internet Workshops. IEEE Computer Society, Washington, DC (2007)
7. Jones, G., Brown, P.: The role of context in information retrieval. In: Proc. of the SIGIR Information Retrieval in Context Workshop, Sheffield, UK (2004)
8. Kamvar, M., Baluja, S.: Deciphering trends in mobile search. Computer 40(5), 58–62 (2007)
9. Kraft, R., Maghoul, F., Chang, C.: Y!q: contextual search at the point of inspiration. In: Proceedings of the 14th ACM International Conference on Information and Knowledge Management, pp. 816–823. ACM, New York (2005)
10. Finkelstein, L., Gabrilovich, E., Matias, Y., Rivlin, E., Solan, Z., Wolfman, G., Ruppin, E.: Placing search in context: The concept revisited. ACM Transactions on Information Systems 20(1), 116–131 (2002)
11. Bouidghaghen, O., Tamine, L., Boughanem, M.: A Diary Study-Based Evaluation Framework for Mobile Information Retrieval. In: Cheng, P.-J., Kan, M.-Y., Lam, W., Nakov, P. (eds.) AIRS 2010. LNCS, vol. 6458, pp. 389–398. Springer, Heidelberg (2010)
12. Santos, A.C., Cardoso, J.A.M.P., Ferreira, D.R., Diniz, P.C., Chaínho, P.: Providing user context for mobile and social networking applications. Pervasive Mob. Comput. 6(3), 324–341 (2010)
13. Schmidt, A.: A Layered Model for User Context Management with Controlled Aging and Imperfection Handling. In: Roth-Berghofer, T.R., Schulz, S., Leake, D.B. (eds.) MRC 2005. LNCS (LNAI), vol. 3946, pp. 86–100. Springer, Heidelberg (2006)
14. Lawrence, S.: Context in web search. IEEE Data Engineering Bulletin 23(3), 25–32 (2000)
15. Tsai, F.S., Etoh, M., Xie, X., Lee, W.C., Yang, Q.: Introduction to mobile information retrieval. IEEE Intelligent Systems 10, 1541–1672 (2010)
16. Sohn, T., Li, K., Griswold, W.: Diary study of mobile information needs. In: Proceedings of the Twenty-sixth Annual SIGCHI Conference on Human Factors in Computing Systems (CHI 2008), Florence, Italy, pp. 433–442 (2008)
17. White, R.W., Bailey, P., Chen, L.: Predicting user interests from contextual information. In: Proceedings of the 32nd International ACM SIGIR Conference on Research and Development in Information Retrieval, vol. (8), pp. 363–370 (2009)
18. Yau, S.S., Liu, H., Huang, D., Yao, Y.: Situation-aware personalized information retrieval for mobile internet. In: Proceedings of the 27th Annual International Conference on Computer Software and Applications, pp. 638–645 (2003)

The Effective Relevance Link
between a Document and a Query

Karam Abdulahhad[1], Jean-Pierre Chevallet[2], and Catherine Berrut[1]

[1] UJF-Grenoble 1
[2] UPMF-Grenoble 2,
LIG laboratory, MRIM group
{karam.abdulahhad,jean-pierre.chevallet,catherine.berrut}@imag.fr

Abstract. This paper proposes to understand the retrieval process of relevant documents against a query as a two-stage process: at first an identification of the reason why a document is relevant to a query that we called the Effective Relevance Link, and second the valuation of this link, known as the Relevance Status Value (RSV). We present a formal definition of this semantic link between d and q. In addition, we clarify how an existing IR model, like Vector Space model, could be used for realizing and integrating this formal notion to build new effective IR methods. Our proposal is validated against three corpuses and using three types of indexing terms. The experimental results showed that the effective link between d and q is very important and should be more taken into consideration when setting up an Information Retrieval (IR) Model or System. Finally, our work shows that taking into account this effective link in a more explicit and direct way into existing IR models does improve their retrieval performance.

1 Introduction

Information Retrieval Systems (IRSs) are supposed to classify documents in two sets: the set of relevant documents to a query q, and the set of documents that are not relevant. An IRS computes a *machine* relevance that is supposed to be closed to a *human* relevance judgment, i.e. the judgment from the author of the query, also called *user relevance*. Moreover, IRSs usually compute a relevance score: a Relevance Status Value (RSV), against all documents, or against only those that are retrieved[1].

This distinction is very important because, that means there are two different notions in IR: the *relevance* notion and the *valuation* of this relevance computed by the RSV. Unfortunately, if the RSV is computed against all documents of a corpus, the first notion disappears. In this case, only a ranking of documents based on the RSV is computed.

---

[1] The set of retrieved documents is implicitly the set of relevant documents from the machine point of view.

S.W. Liddle et al. (Eds.): DEXA 2012, Part I, LNCS 7446, pp. 206–218, 2012.
© Springer-Verlag Berlin Heidelberg 2012

For practical reason, many IRSs compute RSV only against the set of documents that share terms with the query: we interpret this as the minimal constraint often chosen to build the set of retrieved document is a non empty term intersection between a retrieved document and a query. We feel that this poor constraint hides a more semantical constraint that a document should fulfill in order to be retrieved by a system: there should be a hidden *semantic link* and a reasoning process that could be followed to *demonstrate* this relevance. We call this link the *Effective Relevance Link* and we denoted it $d \leftrightarrow q$ in this paper. Hence the effective link between a document d and a query q is related to the reason or reasons that make d a candidate document to be retrieved for q. We also think that this Effective Semantic Link can be expressed in some logic. For example, we can state that for having $d \leftrightarrow q$, the system should explicitly has a *logical* reason like: if d is relevant to q, then there should exist a logical deduction chain that starts from d and ends at q, written as $d \rightarrow q$.

Information Retrieval (IR) models include more or less this effective link, each model in its own way. Vector Space Model (VSM) [18] assumes that both d and q are vectors in a specific term space and the link $d \leftrightarrow q$ is simply equivalent to a non empty terms intersection. Probabilistic Models (PM) [14] propose the probabilistic ranking principle for ranking documents by decreasing probability of their relevance to queries. Different estimations of the previous probability mean different variants of PM. Language Models (LM) [13] borrow their notion from the speech recognition community. LMs suppose that each document is a language and then they estimate the ability of document's language to reproduce the query.

Actually, the two notions are mixed in most of classical IR models: the effective link $d \leftrightarrow q$, or in other words, *why* d is a candidate answer for q, and *how much* $d \leftrightarrow q$ is strong, i.e. the relevance score $RSV(d, q)$. Therefore, we think one should study $d \leftrightarrow q$ and $RSV(d, q)$ that characterizes it, as separate notions.

Moreover, most IRS compute first an $RSV(d, q)$ and then deduce $d \leftrightarrow q$, for example by an implicit thresholding of the RSV value of the ordered documents list. We think the correct view should first be the study and computation of $d \leftrightarrow q$, and then computation of the strength of this link.

In this work, we separate the two notions through introducing a formal-logical definition for $d \leftrightarrow q$. In addition, we study the influence of this separation on the performance of some IR methods.

The paper is structured as follow: In section 2, we present a more formal definition of $d \leftrightarrow q$ using the two notions *Exhaustivity* and *Specificity*. We show the importance of $d \leftrightarrow q$ and the attempts of using it in IR literatures, in section 3. Then in section 4, we show a practical view of $d \leftrightarrow q$. In section 5, we describe our concrete framework for integrating $d \leftrightarrow q$ in IR methods. We show our experiments for validating our hypothesis, in section 6. We conclude the paper in section 7.

2   A Tentative Formal-Logical Definition of $d \leftrightarrow q$

We introduced $d \leftrightarrow q$ in a subjective way. In order to build a more formal and general definition, we model it in a logical framework.

Many researchers argued that the retrieval process could be formalized as a logical implication from a document d to a query q noted $d \rightarrow q$. One of the earliest studies in this direction is the one of *Rijsbergen* [21], who introduced the use of logic as a theoretical IR foundation. He proposes to see d and q as a set of logical sentences in a specific logic. Then d is a candidate answer for q *iff* it logically implies q noted $d \rightarrow q$. In other words, if q is deducible based on d. However, IR is an uncertain process [5], because:

- The query q is an imperfect representation of the user needs;
- The document model d is also an imperfect representation of the semantic content of the document:
- Relevance judgment depends on an external factor, which is the user knowledge.

Therefore, another component, beside the logic, should be added to estimate the certainty of an implication and to offer a ranking mechanism. In other words, a measure P should exist and be able to measure the certainty of the logical implication between d and q, noted $P(d \rightarrow q)$. This formulation already split the matching computation into two stages: first establish the truth of the effective link $d \rightarrow q$ and then compute a score P, on this link.

This proposal exhibits a non symmetrical Effective Link, and leads to this question: are the two implications $d \rightarrow q$ and $q \rightarrow d$ the same? *Nie* [12], distinguishes the *Exhaustivity* of a document to a query $d \rightarrow q$, which means that d satisfies all *themes* of q, from the *Specificity* of a document to a query $q \rightarrow d$, which means that d's themes are all related to q.

In other words, *Exhaustivity* means that all themes of q should be referred in d. In this manner, suppose that we have a document d where $d \rightarrow q$ is valid, if we build another document d' by adding more themes to d then $d' \rightarrow q$ should be still valid. However, d is more relevant to q because it is more specific. Therefore, we need the other notion of "*Specificity*" in order to retrieve the most specific or precise document that already covers q [6]. The ideal case is when d contains all and only all the themes of q, that means, we can prove both $d \rightarrow q$ and $q \rightarrow d$.

Formal logic has a syntax but can also have semantics[2]. This semantic translates the formal sentences of that logic into another mathematical world. For example, we get the semantic of a logical sentence in the Propositional Logic by assigning a truth value (T or F) to each proposition in that sentence. Hence, each logical sentence s can have several translations or interpretations I_s depending on the truth values assignments. The subset M_s of I_s that make s true is called the set of *models* of s.

For any two sentences s_1 and s_2, we say that s_1 logically entails s_2, written as $M_{s_1} \models s_2$, or simply $s_1 \models s_2$, if s_2 is true in all models of s_1. In other words, any interpretation that makes s_1 true should also make s_2 true. Obviously, $\models$ is not commutative. In this manner, the Exhaustivity $d \rightarrow q$ could be translated into $d \models q$ whereas the Specificity into $q \models d$.

---

[2] We would like to warn the reader unfamiliar with logic formalisms that this notion of semantics (called formal semantics) is not related to "human" meaning.

The more terms a document has, the less number of models validating that document exist. For example, with the indexing vocabulary $V = \{t_1, t_2, t_3, t_4, t_5\}$, one can have 2^5 different interpretations over V. If a document d is indexed by the terms $\{t_1, t_2, t_3\}$, then one can associate with d the set of 4 models M_d that make t_1, t_2, t_3 true. Another document d' indexed by $\{t_1, t_2, t_3, t_4\}$ is associated with the set $M_{d'}$ of only 2 models[3]. In this example any model of d' is also a model of d which means $M_{d'} \subseteq M_d$. In other words[4], if $d \models q$ then $d' \models q$.

By taking the uncertainty into account, the two notions Exhaustivity and Specificity could be rewritten as follow:

- **Exhaustivity** $P(d \rightarrow q)$: means to which limit M_d and $M_d \cap M_q$ are close, or in other words, $\overline{P(d \rightarrow q)}$ could be equivalent to evaluate an other function $P(M_d, M_d \cap M_q)$. The best case is when $M_d \subseteq M_q$, which means $M_d = M_d \cap M_q$.
- **Specificity** $P(q \rightarrow d)$: means to which limit M_q and $M_d \cap M_q$ are close, or in other words, $P(q \rightarrow d)$ could be equivalent to $P(M_q, M_d \cap M_q)$. The best case is when $M_q \subseteq M_d$, which means $M_q = M_d \cap M_q$.

After this detailed description of $d \leftrightarrow q$ and after clarifying the potential interaction between Exhaustivity and Specificity, instead of calculating the relevance score between d and q as a degree of certainty of the logical implication $P(d \rightarrow q)$, now the relevance score is a function of the two implications [12] (1):

$$RSV(d, q) = F[P(d \rightarrow q), P(q \rightarrow d)] \tag{1}$$

3 $d \leftrightarrow q$ in IR Models

Many IR models, one way or another, try to integrate $d \leftrightarrow q$ in the process of computing the Relevance Status Value (RSV) between d and q. Abdulahhad (et al.) [1] exploit the semantic relations between document's concepts and query's concepts and use the attached weights of those relations for computing the final matching value between d and q.

Other studies *Rocchio* [16], *Salton (et al.)* [19] and *Buckley (et al.)* [4], need a second round of evaluation for integrating $d \leftrightarrow q$, through query reformulation using several prejudged documents. In fact after using RSV for sorting retrieved documents, they make the hypothesis that only some of them are really relevant, i.e. satisfies the effective relevance link. This technique is known as relevance feedback.

In classical bag-of-terms based IR methods, $d \leftrightarrow q$ is implicitly integrated. For example, in BM25 [15] and Pivoted Normalization Method [20], $d \leftrightarrow q$ appears timidly through the sum over shared terms ($\sum_{t \in d \cap q}$). The same thing

---

[3] One that makes t_1, t_2, t_3, t_4 true and t_5 false, and the other that make t_1, t_2, t_3, t_4 and t_5 true.

[4] Note that this is not necessarily true for all logical IR models. For example, this does not hold for the classical IR boolean model because documents are associated with only one interpretation.

for Language Models [13], but instead of sum, it is the product ($\prod_{t \in d \cap q}$). It is also true for Information-based methods [2] [7].

Several studies *Wilkinson (et al.)* [22] and *Rose (et al.)* [17] show that users prefer documents sharing more distinct terms with queries. Moreover, Fang (et al.) [10] determine several retrieval constraints for building effective retrieval methods. The second constraint $TFC2$ implies another constraint, which encourages promoting documents with more distinct query terms.

Historically, one of the earliest methods of ranking was the number of shared terms between d and q ($|d \cap q|$). This method is added to the Boolean Model in order to rank retrieved documents. In addition, the ranking formula of VSM could be restricted to $|d \cap q|$ when using binary weights for document and query terms (1 if t occurs in d, 0 otherwise).

From the previous presentation, we can see that in spite of the importance of $d \leftrightarrow q$, represented by $d \cap q$, it is not sufficiently integrated in the classical IR methods. In this study, we try to explicitly integrate the $d \leftrightarrow q$ in the process of estimating the retrieval score between d and q: $RSV(d, q)$, in order to build a more precise and effective retrieval method.

4 $d \leftrightarrow q$ and Weighting

In all IR models, e.g. Language Models [13], Probabilistic Models [14], Vector Space Models (VSM) [18], etc. the weight of an indexing term t is usually estimated depending on three sources of information:

1. The document d is usually used for estimating the descriptive power or the local weight w_t^d of t in d. For example, the term frequency of t in d.
2. The query q: the weight w_t^q is whether manually assigned by users or estimated through the term frequency of t in q.
3. The corpus or document collection D is used for estimating the discriminative power w_t^D of t in D. For example, the Inverse Document Frequency (IDF), or the smoothing component of Language Models [23].

In general, at the time of computing the matching value between a specific document d and a specific query q:

1. The value of w_t^d is independently estimated of q. For certain d and t, w_t^d is constant whatever is q.
2. The value of w_t^q is independently estimated of d. For certain q and t, w_t^q is constant whatever is d.
3. The value of w_t^D is independently estimated of both d and q. For certain D and t, w_t^D is constant whatever is d and q.

We think that this is an insufficient modeling because each weight is independently computed from the effective link $d \leftrightarrow q$. Hence, we propose the matching score computation to take into account the $d \leftrightarrow q$ in an explicit manner, in addition to w_t^d, w_t^q and w_t^D.

We illustrate this problem using one of the classical IR methods, the Pivoted Normalization Method [20] (2).

$$RSV(d,q) = \sum_{t \in d \cap q} \frac{1 + ln\left(1 + ln\left(tf_{t,d}\right)\right)}{(1-s) + s\frac{|d|}{avdl}} \times tf_{t,q} \times ln\frac{N+1}{n_t} \qquad (2)$$

where $tf_{t,d}$ is the term frequency of t in d, $tf_{t,q}$ is the term frequency of t in q, s is a constant (normally $s = 0.2$), $|d|$ is the length of d, $avdl$ is the average document length in the corpus, N is the total number of documents in the corpus, and n_t is the number of documents that contain t.

$$RSV(d,q) = \sum_{t \in d \cap q} w_t^d \times w_t^q \times w_t^D$$

$$w_t^d = \frac{1 + ln(1 + ln(tf_{t,d}))}{(1-s) + s\frac{|d|}{avdl}} \qquad w_t^q = tf_{t,q} \qquad w_t^D = ln\frac{N+1}{n_t} \qquad (3)$$

(3) shows that w_t^d is independent of q, w_t^q is independent of d, and w_t^D is independent of both d and q.

As most of IR methods are based on the bag-of-terms paradigm, the most evident indication to $d \leftrightarrow q$ could be the shared terms between d and q: $d \cap q$, because what makes d a candidate answer for q is having shared terms with q: $d \cap q \neq \emptyset$. The shared terms compose the ground where both d and q interact with each other. Without shared terms ($d \cap q = \emptyset$), there is no explicit link between d and q, hence d is not potentially a relevant document. Actually this is not quite correct because of the *term-mismatch* problem [8], where two terms are used for expressing on the same meaning, e.g. flat vs. apartment. However, the term-mismatch problem is out of the scope of this study.

5   Revisiting the VSM with $d \leftrightarrow q$

The Vector Space Model (VSM) is a well known model that can benefit from an explicit integration of the Effective Relevance Link. Before revisiting the VSM model, let's analyze the relationship between the logical description of $d \leftrightarrow q$ and a term set representation.

In the previous section, we rewrote the two implications $d \to q$ and $q \to d$ using $M_d \cap M_q$. One can associate a set of terms d with a set of models M_d in the following way: M_d are the models where each term of d is true. Hence adding a term to the set d reduces the model set M_d. Moreover, if $q \subseteq d$ then $M_d \subseteq M_q$, and finally $M_d \cap M_q$ is equivalent to $d \cap q$: see Fig. 1.

For example, with the vocabulary set $\{t_1, t_2, t_3, t_4\}$, given the document $d = \{t_1, t_2, t_3\}$. Then using the VSM notation (with 1 for true):

$$M_d = \{\langle 1,1,1,0\rangle, \langle 1,1,1,1\rangle\}$$

If $q = \{t_1, t_2\}$ then:

$$M_q = \{\langle 1,1,0,0\rangle, \langle 1,1,0,1\rangle, \langle 1,1,1,0\rangle, \langle 1,1,1,1\rangle\}$$

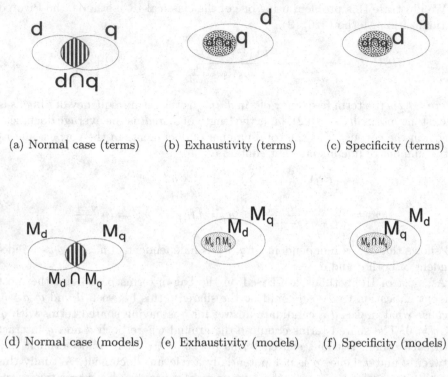

(a) Normal case (terms)   (b) Exhaustivity (terms)   (c) Specificity (terms)

(d) Normal case (models)   (e) Exhaustivity (models)   (f) Specificity (models)

Fig. 1. The different cases of interaction between d and q

In this example $q \subseteq d$ whereas $M_d \subseteq M_q$. In this case the document cover the query in an exhaustive manner. See Figs. 1(b) and 1(e).

According to *Nie* [12], the *RSV* value between d and q could be estimated as a function F of the degree of certainty of the two implications (1).

According our definition of Exhaustivity and Specificity, the (1) could be rewritten as follow (4):

$$RSV(d, q) = F[P(M_d, M_d \cap M_q), P(M_q, M_d \cap M_q)] \qquad (4)$$

By assuming that documents and queries are sets of terms instead of assuming that they are logical sentences with sets of models, (1) could be written as (5). There is always a possibility to go from a logical sentence to a set of terms and vice-versa [11], through: 1- using the Propositional Logic, 2- assuming that each term is a proposition, and 3- assuming that each document is a logical sentence of conjunctive propositions or it is a set of terms.

$$RSV(d, q) = F[P(q, d \cap q), P(d, q \cap d)] \qquad (5)$$

Actually, we need a concrete framework for computing the $RSV(d, q)$. Therefore, we need to realize the following abstract elements:

- The document d.
- The query q.
- The shared terms $d \cap q$.
- The function F.
- The uncertainty measure P.

In IR field, there are many frameworks for doing that, e.g. Vector Space Framework [18], Probabilistic Framework [14], Language Models [13], etc. In this study, we choose the Vector Space Framework. Therefore, the previous abstract elements become:

- The document $\vec{d}$ is a vector in term space T. For each term $t \in T$, there is a correspondent component in $\vec{d}$: w_t^d, where $w_t^d > 0$ is the weight of t in d if t occurs in d or 0 otherwise.
- The query $\vec{q}$ is a vector in term space T. For each term $t \in T$, there is a correspondent component in $\vec{q}$: w_t^q, where $w_t^q = 1$ if t occurs in q or 0 otherwise.
- The shared terms $\overrightarrow{d \cap q}$ is a vector in term space T. For each term $t \in T$, there is a correspondent component in $\overrightarrow{d \cap q}$: $w_t^{d \cap q}$, where $w_t^{d \cap q} = 1$ if t occurs in both d and q or 0 otherwise.
- The function F: there are many choices for F, e.g. sum, weighted sum for favoring Exhaustivity over Specificity or vice-versa, product, etc. In this study, we suppose that both Exhaustivity and Specificity are equally important and we choose the product ($\times$).
- The uncertainty measure P: in Vector Space Framework there are many choices for computing the distance between two vectors [9]. Here, we choose the inner-product measure.

Finally, $P(q, d \cap q)$ is the distance between $\vec{q}$ and $\overrightarrow{d \cap q}$, same for $P(d, d \cap q)$. The (5) could be rewritten as follow (6):

$$RSV(d, q) = \left(\vec{q} \cdot \overrightarrow{d \cap q} \right) \times \left(\vec{d} \cdot \overrightarrow{d \cap q} \right) \tag{6}$$

where ($\cdot$) is the inner-product (dot-product). Then the retrieval formula becomes (7):

$$RSV(d, q) = \left[\sum_{t \in T} w_t^q \times w_t^{d \cap q} \right] \times \left[\sum_{t \in T} w_t^d \times w_t^{d \cap q} \right] = |d \cap q| \times \sum_{t \in d \cap q} w_t^d \tag{7}$$

where $|d \cap q|$ is the number of shared terms between d and q.

The only remaining component that should be clarified is the weight of a term t in a document d or w_t^d. Several weighting formulas exist e.g. Pivoted Normalization [20], BM25 [15], DFR [2], TF-IDF, etc. Here we will use a version of the *TF-IDF* formula. Our final retrieval formula becomes (8):

$$RSV(d, q) = |d \cap q| \times \left[\sum_{t \in d \cap q} \frac{tf_{t,d}}{tf_{t,d} + |d|} \times \frac{N}{n_t} \right] \tag{8}$$

6 Experiments

To validate our hypothesis about the utility of integrating the $d \leftrightarrow q$ into IR models, we apply (8) on corpuses and compare the performance against the performance of some classical IR methods. We use for the comparison the Mean Average Precision (MAP) metric.

6.1 Experiments Setup

We use in our experiments three different corpuses and three types of indexing terms.

The types of indexing terms: each type of indexing terms represents a different facet of documents and queries.

- 5Grams ($5G$) / 4Grams ($4G$): we used five-characters-wide / four-characters-wide window for extracting 5grams / 4grams with shifting the window one character each time.
- Words (W): we eliminated the stop words and stemmed the remaining words using Porter algorithm to get finally the list of words that indexes documents and queries.
- Concepts (C): we mapped the text into UMLS's concepts using MetaMap, where UMLS[5] is a multi-source knowledge base in the medical domain. Whereas, MetaMap[6] [3] is a tool for mapping text into UMLS concepts.

Corpuses: we validate our hypothesis against three corpuses. One from Image-CLEF2010[7] and two from ImageCLEF2011 (Table 1):

Table 1. Corpuses statistics. *avdl* and *avql* are the average length of documents and queries.

Corpus	#d	#q	Type	avdl	avql
image2010	77495	16	5G	627.23	29.88
			W	62.12	3.81
			C	157.27	12.0
image2011	230088	30	5G	468.86	32.1
			W	44.83	4.0
			C	101.92	12.73
case2011	55634	10	4G	30380.17	192.4
			W	2594.5	19.7
			C	5752.38	57.5

---

[5] Unified Medical Language System.
  http://www.ncbi.nlm.nih.gov/bookshelf/br.fcgi?book=nlmumls
[6] http://metamap.nlm.nih.gov/
[7] http://www.imageclef.org/

- *image2010*: contains short medical documents and queries.
- *image2011*: also contains short medical documents and queries. However, it is larger than image2010.
- *case2011*: contains long medical case description documents and long queries.

IR models: from one side, the performance of TF-IDF$_{d\cap q}$ (8) is compared to the performance of the same formula but without $|d \cap q|$ component (TF-IDF). We did that for showing the positive effect of integrating $d \leftrightarrow q$ into weighting formulas. From another side, we compare the performance of TF-IDF$_{d\cap q}$ to the performance of Pivoted Normalization Method PIV (2), BM25 method (9), and Dirichlet language model DIR (10), where $p(t, D)$ is the probability of t given by the collection language model D. Through this comparison, we show the validity of our hypothesis.

s, k_1, b, k_3, and μ are all constants. They usually have the following values: $s = 0.2$ [20]. $k_1 = 1.2$, $b = 0.75$, and $k_3 = 1000$ [10]. $\mu = 2000$ [23].

$$RSV(d, q) = \sum_{t \in d \cap q} ln \frac{N - n_t + 0.5}{n_t + 0.5} \times \frac{(k_1 + 1) \times tf_{t,d}}{k_1 \times ((1-b) + b \times \frac{|d|}{avdl}) + tf_{t,d}}$$

$$\times \frac{(k_3 + 1) \times tf_{t,q}}{k_3 \times tf_{t,q}}$$

(9)

$$RSV(d, q) = \sum_{t \in d \cap q} tf_{t,q} \times ln \left(1 + \frac{tf_{t,d}}{\mu \times p(t, D)}\right) + |q| \times ln \frac{\mu}{|d| + \mu}$$

(10)

6.2 Results and Discussion

Table (2) shows the experimental results of (8), applying on the three corpuses and using the three types of terms. Table (2) shows that using $|d \cap q|$, or in other words, explicit $d \leftrightarrow q$ integration into IR methods, improves considerably the average precision. This conclusion is valid for all corpuses and all types of terms. That means, our hypothesis is valid for short and long documents and queries, in addition, it is also valid for different facets of documents and queries.

Table (3) show the experimental results of (8) and some classical IR methods (2, 9, and 10). They show that for all types of terms and for all corpuses, (8)

Table 2. The experimental results of applying (8) to the three corpuses and using three types of terms.

Type	Formula	image2010		image2011		case2011	
		MAP	gain	MAP	gain	MAP	gain
5G / 4G	TF-IDF$_{d\cap q}$	0.3165	+16%	0.1474	+26%	0.0755	+26%
	TF-IDF	0.2739		0.1169		0.0599	
W	TF-IDF$_{d\cap q}$	0.3332	+14%	0.2069	+51%	0.1044	+33%
	TF-IDF	0.2916		0.1368		0.0786	
C	TF-IDF$_{d\cap q}$	0.3248	+13%	0.1672	+13%	0.1605	+19%
	TF-IDF	0.2883		0.1484		0.1347	

Table 3. The experimental results of applying (8) and some classical IR methods (2, 9, and 10) to the three corpuses and using the three types of terms

Type	Formula	image2010	image2011	case2011
5G / 4G	TF-IDF$_{d \cap q}$	**0.3165**	**0.1474**	0.0755
	PIV	0.2872	0.1069	0.0759
	BM25	0.2733	0.1302	0.0062
	DIR	0.2947	0.1241	**0.0775**
W	TF-IDF$_{d \cap q}$	**0.3332**	**0.2069**	0.1044
	PIV	0.2992	0.1546	0.1023
	BM25	0.2745	0.1995	0.0964
	DIR	0.2960	0.1534	**0.1295**
C	TF-IDF$_{d \cap q}$	**0.3248**	**0.1672**	**0.1605**
	PIV	0.2530	0.1096	0.1037
	BM25	0.2123	0.1552	0.0956
	DIR	0.2455	0.1228	0.1036

performs better than the other formulas, except when using words with long documents and queries. In other words, even a simple non-parametric formula (8) performs better than classical IR methods, through simple integration of $|d \cap q|$ into the formula, where $|d \cap q|$ is an indication to $d \leftrightarrow q$.

In conclusion, the effective link between d and q ($d \leftrightarrow q$) is a very important component, and it should be correctly exploited for improving the performance of IR methods.

7   Conclusion

We study in this paper the explicit integration of the effective link $d \leftrightarrow q$ into an IR matching model. We have presented a formal definition of $d \leftrightarrow q$ based on logical framework through two notions: Exhaustivity and Specificity. Those notions describe an interesting relevance link between d and q. According to Exhaustivity and Specificity, the best answer for a query q is the most specific (smallest) document that fully contains q.

We revisit the Vector Space Model, and test the effect of integrating $d \leftrightarrow q$ into the matching formula. Experimental results on three test corpuses show that our hypothesis about the importance of integrating $d \leftrightarrow q$ into IR models is valid. We also validated our hypothesis against three types of indexing terms and we get similar positive results.

The next steps of this work concern the revisiting of other IR models like the probabilistic and language models, and some experimentation on other test collections, not specifically in the medical domain.

References

1. Abdulahhad, K., Chevallet, J.-P., Berrut, C.: Solving concept mismatch through bayesian framework by extending umls meta-thesaurus. In: la huitième édition de la COnférence en Recherche d'Information et Applications (CORIA 2011), Avignon, France, March 16–18 (2011)
2. Amati, G., Van Rijsbergen, C.J.: Probabilistic models of information retrieval based on measuring the divergence from randomness. ACM Trans. Inf. Syst. 20(4), 357–389 (2002)
3. Aronson, A.R.: Metamap: Mapping text to the UMLS metathesaurus (2006)
4. Buckley, C., Salton, G., Allan, J., Singhal, A.: Automatic Query Expansion Using SMART: TREC 3. In: TREC (1994)
5. Chiaramella, Y., Chevallet, J.P.: About retrieval models and logic. Comput. J. 35, 233–242 (1992)
6. Chiaramella, Y., Mulhem, P., Fourel, F.: A model for multimedia information retrieval. Technical report (1996)
7. Clinchant, S., Gaussier, E.: Information-based models for ad hoc ir. In: Proceedings of the 33rd International ACM SIGIR Conference on Research and Development in Information Retrieval, SIGIR 2010, pp. 234–241. ACM, New York (2010)
8. Crestani, F.: Exploiting the similarity of non-matching terms at retrievaltime. Inf. Retr. 2(1), 27–47 (2000)
9. Dominich, S.: Mathematical Foundations of Information Retrieval, 1st edn. Mathematical Modelling: Theory and Applications. Springer (March 2001)
10. Fang, H., Tao, T., Zhai, C.: A formal study of information retrieval heuristics. In: Proceedings of the 27th Annual International ACM SIGIR Conference on Research and Development in Information Retrieval, SIGIR 2004, pp. 49–56. ACM, New York (2004)
11. Losada, D.F., Barreiro, A.: A logical model for information retrieval based on propositional logic and belief revision. The Computer Journal 44, 410–424 (2001)
12. Nie, J.: An outline of a general model for information retrieval systems. In: Proceedings of the 11th Annual International ACM SIGIR Conference on Research and Development in Information Retrieval, SIGIR 1988, pp. 495–506. ACM, New York (1988)
13. Ponte, J.M., Bruce Croft, W.: A language modeling approach to information retrieval. In: Proceedings of the 21st Annual International ACM SIGIR Conference on Research and Development in Information Retrieval, SIGIR 1998, pp. 275–281. ACM, New York (1998)
14. Robertson, S.E.: The probability ranking principle in IR. In: Readings in Information Retrieval, pp. 281–286. Morgan Kaufmann Publishers Inc., San Francisco (1997)
15. Robertson, S.E., Walker, S.: Some simple effective approximations to the 2-poisson model for probabilistic weighted retrieval. In: Proceedings of the 17th Annual International ACM SIGIR Conference on Research and Development in Information Retrieval, SIGIR 1994, pp. 232–241. Springer-Verlag New York, Inc., New York (1994)
16. Rocchio, J.: Relevance Feedback in Information Retrieval, pp. 313–323 (1971)
17. Rose, D.E., Stevens, C.: V-twin: A lightweight engine for interactive use. In: TREC (1996)
18. Salton, G., Wong, A., Yang, C.S.: A vector space model for automatic indexing. Communications of the ACM (18), 613–620 (1975)

19. Salton, G., McGill, M.J.: Introduction to Modern Information Retrieval. McGraw-Hill, Inc., New York (1986)
20. Singhal, A., Buckley, C., Mitra, M.: Pivoted document length normalization. In: Proceedings of the 19th Annual International ACM SIGIR Conference on Research and Development in Information Retrieval, SIGIR 1996, pp. 21–29. ACM, New York (1996)
21. van Rijsbergen, C.J.: A non-classical logic for information retrieval. Comput. J. 29(6), 481–485 (1986)
22. Wilkinson, R., Zobel, J., Sacks-Davis, R.: Similarity measures for short queries. In: TREC (1995)
23. Zhai, C., Lafferty, J.: A study of smoothing methods for language models applied to ad hoc information retrieval. In: Proceedings of the 24th Annual International ACM SIGIR Conference on Research and Development in Information Retrieval, SIGIR 2001, pp. 334–342. ACM, New York (2001)

Incremental Computation of Skyline Queries with Dynamic Preferences

Tassadit Bouadi[1], Marie-Odile Cordier[1], and René Quiniou[2]

[1] IRISA - University of Rennes 1
[2] IRISA - INRIA Rennes
Campus de Beaulieu, 35042 RENNES, France
{tassadit.bouadi,marie-odile.cordier}@irisa.fr, rene.quiniou@inria.fr

Abstract. Skyline queries retrieve the most interesting objects from a database with respect to multi-dimensional preferences. Identifying and extracting the relevant data corresponding to multiple criteria provided by users remains a difficult task, especially when the data are large. In 2008-2009, Wong et al. showed how to avoid costly skyline query computations by deriving the skyline points associated with any preference from the skyline points associated with the most preferred values. They propose to materialize these points in a structure called *IPO-tree* (*Implicit Preference Order Tree*). However, its size is exponential with respect to the number of dimensions. We propose an incremental method for calculating the skyline points related to several dimensions associated with dynamic preferences. For this purpose, a materialization of linear size which allows a great flexibility for dimension preference updates is defined. This contribution improves notably the computation cost of queries. Experiments on synthetic data highlight the relevance of EC^2Sky compared to *IPO-Tree*.

1 Introduction

Skyline queries aim at retrieving the most interesting objects from a database with respect to given criteria. In a multidimensional space where the dimension domains are ordered, skyline queries return the points which are not dominated by any other point. A point p dominates a point q if p is strictly better than q on at least one dimension and p is better or equal than q on the remaining dimensions. Skyline queries can formulate multi-criteria queries [1], for example to find the cheapest hotels close to the beach. Identifying and extracting relevant data according to many criteria is often a difficult task especially when dealing with large volumes of data. Several studies [2,3,4,5,6,7,8,9,10] were carried out on skyline analysis as a retrieval tool in a decisional context.

However, most of the work mentioned above assume that there exists a predefined order on the domain of each dimension. When users are allowed to define or to change their own preferences online, the order may change dynamically on some dimensions and the skyline evolves accordingly. A naive solution is to recalculate the skyline from scratch for each dynamic preference that has changed. However, it is too expensive on large databases of high dimensionality. The challenge is thus the following: how to efficiently recalculate the least amount of skyline points while minimizing the required memory space.

S.W. Liddle et al. (Eds.): DEXA 2012, Part I, LNCS 7446, pp. 219–233, 2012.
© Springer-Verlag Berlin Heidelberg 2012

Wong et al. [3] propose a semi-materialization method based on a specific data structure called IPO-tree (*Implicit Preference Order Tree*). An IPO-tree stores partial useful results corresponding to every combination of first order preferences. A first order preference states that one value is most preferred in some dimension and that the other values are left unordered. An n order preference specifies an order over n values from some dimension, whereas the other values are less preferred and left unordered. Wong et al. also introduced the *merging property* which makes possible to derive skyline of any n-th order preference by simple operations on the skyline related to the first order preferences on the same dimension. However, this approach has a main drawback: the *merging property* is applicable to only one dimension at a time. The size of an IPO tree is thus in $O(c^m)$ (where m is the number of dimensions associated with dynamic preferences and c is the cardinality of a dimension). In the context of large databases of high dimensionality this structure becomes very complex. In [4], Wong et al. propose another structure, called CST (*Compressed Ordered skyline Tree*), to materialize all possible preference orders. However, this method turns to be incomplete and very complex and, so, it cannot be used.

While reusing the *merging property* proposed by Wong et al. to deal with the refinement of preferences on a single dimension, we propose an incremental method, called EC^2Sky, for calculating the skyline points related to several dimensions associated with dynamic preferences. This work improves and extends the short contribution presented in [11]. As a side effect, EC^2Sky can return the most relevant knowledge by emphasizing the compromises associated with the specified preferences. The benefits of this proposition are twofold. On the one hand, the complexity in space of the materialization of precomputed skylines reduces to $O(c*m)$. On the other hand, the number of dominance tests decreases significantly. We proved experimentally, that the total computation cost is much lower than in Wong et al.'s method.

The rest of the paper is organized as follows. In Section 2, we introduce the basic concepts related to skyline queries and dynamic preferences. We develop the formal aspects of EC^2Sky in Section 3 and its implementation in Section 4. In Section 5, we present the results of the experimental evaluation performed on synthetic datasets and highlight the relevance of the proposed solution. We conclude the paper in Section 6.

2   Basic Concepts

Let $D = \{d_1, ..., d_n\}$ be an n-dimensional space, E a data set defined in space D and $p, q \in E$. We denote by $p(d_i)$ the value of p on dimension d_i. The following definitions concern a subspace $D' \subseteq D$. Obviously, these definitions can be generalized to the full dimension space D. In a subspace $D' \subseteq D$, a point p is said to *dominate* another point q, denoted by $p \prec_{D'} q$, if $\forall d_i \in D'$, $p(d_i) \leq_{d_i} q(d_i)$ (i.e. p is preferred or equal to q on D') and $\exists d_i \in D'$, $p(d_i) <_{d_i} q(d_i)$ (i.e. p is strictly preferred to q on d_i). When $D' = D$, $p \prec_{D'} q$ is simply noted $p \prec q$.

A preference on the domain of a dimension d_i is defined by a partial order $\leq_{d_i}$. We distinguish two types of preferences: static preferences which correspond to a predefined order relation, and dynamic preferences which correspond to an order relation that can vary from one user to another or from one user session to another. By abuse of language we write *dynamic (resp. static) dimension* instead *dimension associated with dynamic (resp. static) preferences*. In the sequel, S denotes the subspace associated with static preferences and Z the subspace associated with dynamic preferences: $D = S \bigcup Z$ and $S \bigcap Z = \emptyset$.

Definition 1. *(Skyline) The skyline set of the dataset E on the subspace D' with Z being the subspace associated with the dynamic preferences $\wp$, is defined by $Sky(D', E)_{(Z,\wp)} = \{p \in E \mid \forall q \in E,\, q \not\prec_{D'} p\}$.*
If $\wp = \emptyset$, $Sky(D', E)_{(Z,\wp)}$ is simply written $Sky(D', E)$.

The set $Sky(D', E)$ contains points, denoted $MaxSky(D', E)$, that are the best along at least one dimension. It contains also points denoted $CompSky(D', E)$ that are not dominant on any dimension of D' while being better than any point of E on at least one dimension. These *compromise* points represent interesting solutions for the user from a decision making point of view.

Definition 2. *(MaxSky, CompSky)*
Let $D' \subseteq D$. $Sky(D', E) = MaxSky(D', E) \bigcup CompSky(D', E)$ with :
$MaxSky(D', E) = \{p \in Sky(D', E) \mid \exists D'' \subseteq D',\, \forall q \in E,\, p \preceq_{D''} q\}$
$CompSky(D', E) = \{p \in Sky(D', E) \mid \forall q \in E,\, \exists D'' \subseteq D',\, p \prec_{D''} q \vee p =_{D'} q\}$.
If D' is reduced to one dimension $(D' = \{d_i\})$, $Sky(D', E) = MaxSky(D', E)$.

Example 1. *(Running example) In this paper, we use as a running example the dataset E in Table 1 that contains 8 points described by 4 dimensions*
$S = \{Price, Distance\}$: the values of these two dimensions follow the order relation $\leq$ specifying that the lower the price (resp. the distance), the more preferable the hotel (e.g. $a(Price) <_{Price} d(Price)$). This order is accepted by any user, so dimensions Price and Distance are static.
$Z = \{Gr, Air\}$: no order relation is defined a priori on dimensions Gr, Air which are dynamic. The definition of an order is left to users and may vary from one user to another.
$Sky(S, E) = \{a, b, e\}$. $MaxSky(S, E) = \{a, b\}$ since $a(Price)$ (resp. $b(Price)$) is the most preferred value on dimension Price (resp. Distance).
$CompSky(S, E) = \{e\}$ since the value of e is not the most preferred either on Price or on Distance. However, $e(Price) <_{Price} b(Price)$ and $e(Distance) <_{Distance} a(Distance)$. So, e is better then any MaxSky point on at least one dimension.

Table 1. A set of hotels

Hotel ID	Price	Distance	Hotel group (Gr)	Airline (Air)
a	1600	4	T (Tulips)	G(Gonna)
b	2400	1	T(Tulips)	G(Gonna)
c	3000	5	H(Horizon)	G(Gonna)
d	3600	4	H(Horizon)	R(Redish)
e	2300	2	T(Tulips)	R(Redish)
f	3000	3	M(Mozilla)	W(Wings)
g	3600	4	M(Mozilla)	R(Redish)
h	3000	3	M(Mozilla)	R(Redish)

$Sky(S \bigcup Z, E) = \{a, b, e, c, d, f, h\}$ when no preferences are given on the dynamic dimensions Gr and Air.

Conventional skyline queries retrieve the most interesting objects of a multidimensional dataset. Our goal is to aid a user explore his dataset by letting him express various preferences on dynamic dimensions and assess the consequences of such choices by retrieving the skyline points. Different users may have different preferences on a dynamic dimension, e.g Hotel group. When a customer prefers Hotel group $Horizon$ to other values, hotels c, d, f and h are added in the skyline since they become the best along dimension Hotel group. However, for another customer preferring $Tulips$ to other values, c, d, f and h don't belong to the skyline since they are dominated by a, b and e. An interesting observation is that hotels a, b and e are always in the skyline no matter which preference order on the Hotel group is chosen (because a has the lowest price, b has the smallest distance from the beach and e represent a good compromise for the dimensions Price and Distance).

A user who formulates a query involving a dynamic dimension d_i can specify the preference order on the $|d_i|$ values of this dimension. The order is total if all these values are ordered. But this is not always possible and the user may order only n of the $|d_i|$ values. Implicitly, the user considers that they are more preferred than the $(|d_i| - n)$ remaining values which are left unordered. This corresponds to the notion of n-th order implicit preference introduced in [3].

Definition 3. *(n-th order preference)* Let $d_i \in Z$ and $|d_i| = m$. $\wp_i$ is a n-th order preference on d_i iff :

- $\wp_i = v_1 <_{d_i} \cdots <_{d_i} v_n <_{d_i} *$, with $v_j \in dom(d_i)$, $j \in \{1, .., n\}$ and $n \leq m$,
- $\forall k \in \{n+1, .., m\}, v_n <_{d_i} v_k$.
- When $n = 1$, $\wp_i = v_1 <_{d_i} *$ is called a first order preference.

$\wp_i = v_1 <_{d_i} \cdots <_{d_i} v_n <_{d_i} *$ denotes the set of binary preferences $\wp_i = \{v_1 <_{d_i} v_2, v_2 <_{d_i} v_3, \ldots, v_n <_{d_i} *\}$. In the sequel, we use both notations for $\wp_i$. The absence of preference on dimension d_i is denoted by $\wp_i = \emptyset$. Note the importance of first order preferences: they are sufficient to determinate the dominant points of a dimension.

Example 2. *For the dimension Hotel group in Table 1, a user prefers $T(Tulips)$ to $M(Mozilla)$, T to $H(Horizon)$ and M to H (i.e. $T <_{Group} M <_{Group} H$). This preference is a third order preference and defines a total order. Some other user could prefer Hotel group T to any other group (i.e., $T <_{Group} *$). In this case, the preference is a first order preference which defines a partial order.*

We give below some useful properties of the preference relationship. These properties will be used later to reduce the number of domination tests during skyline computation. In the following, $\wp = \bigcup_{i=1}^{|Z|} \wp_i$ denotes the set of dynamic preferences associated with $Z \subseteq D$ combined implicitly with the set of static preferences associated with $S \subseteq D$.

Property 1. *(Monotonicity of preference refinement)* Let $\wp'$ and $\wp''$ two preferences sets on Z. If $\wp''$ is a refinement of $\wp'$ (i.e. $\wp' \subseteq \wp''$) then, $Sky(D, E)_{(Z, \wp'')} \subseteq Sky(D, E)_{(Z, \wp')}$.

Property 1 indicates that when preferences are refined, the skyline may become smaller and so, some skyline points may be disqualified. Also, if a point is not in the skyline related to some preference it won't belong to the skyline related to a refined preference.

The following theorem formulates an important property called the *merging property* that was introduced by Wong et al. [3]. This property provides a means to derive the skyline related to any possible n-th order preference by operations on sets related to first order preferences on the same dimension.

Theorem 1. *(Merging property) Let $\wp'$ and $\wp''$ be two preferences differing only on dimension d_i, i.e. $\wp'_j = \wp''_j$ for all $j \neq i$. Let $\wp'_i = v_1 <_{d_i} \ldots < v_{k-1} <_{d_i} *$ and $\wp''_i = v_k <_{d_i} *$. Let $PSky(D, E)_{(Z, \wp')}$ be the set of points in $Sky(D, E)_{(Z, \wp')}$ with d_i values in $\{v_1 \ldots v_{k-1}\}$. Let $\wp'''$ be a preference differing from $\wp'$ and $\wp''$ only on dimension d_i and $\wp'''_i = v_1 < \ldots < v_{k-1} < v_k < *$. Then the skyline associated with $\wp'''$ is*

$$Sky(D, E)_{(Z, \wp''')} = (Sky(D, E)_{(Z, \wp')} \bigcap Sky(D, E)_{(Z, \wp'')}) \bigcup PSky(D, E)_{(Z, \wp')}.$$

Wong et al. have proposed successively two methods, IPO-tree [3] and CST [4], for skyline computation based on the properties and theorem 1 above. However, the implementation of these two proposals raises several problems. First, the size of an IPO-Tree is $O(c^m)$, where m is the number of dynamic dimensions and c the cardinality of a dimension. So it is intractable in the context of large databases with high dimensionality and does not allow scaling. It is worth noting that the *merging property* is applicable to only one dimension at a time. Second, the CST method does not solve the *IPO-Tree* problems: its algorithm is incomplete (it disqualifies points which should be in the skyline).

To cope with several dynamic dimensions, Wong et al. propose to store the skyline of every combination of the first order preferences on dynamic dimensions. So, the size of the proposed materialization structure is exponential, which is prohibitive when dealing with several dimensions. We propose in Section 3 an incremental method which makes possible to introduce dynamic dimensions one by one.

3   EC²Sky: An Incremental Skyline Computation

In the following, we assume that the subset of dimensions D^i is such that $D^i = D^{i-1} \cup d_i$, with $d_i \in Z$, $i \in \{1, .., |Z|\}$, $D^i \subseteq D$ and $D^0 = S$. This notation represents the incremental addition of dimensions in skyline computation. Consider the addition of a dynamic dimension d_i to a set D^{i-1}. The first task is to compute $Sky(d_i \cup S, E)_{(Z, \wp)}$, the skyline related to dimension d_i as if it were independent of the other dynamic dimensions. Note that every skyline of $d_i \cup S$ related to a first-order preference on d_i can be pre-computed. Wong et al.'s method can be used to achieve this task. However, this set may contain skyline points that are disqualified i.e. they are dominated on the dynamic dimensions of the subspace D^{i-1}. Precisely, let $p, q \in Sky(d_i \cup S, E)_{(Z, \wp)}$ be two skyline points with the same values on every dimension of $d_i \cup S$. If q is preferred on D^{i-1} it will dominate p and disqualify it from the skyline $Sky(D^i, E)_{(Z, \wp)}$. This set of points is denoted $CutSky(d_i \cup S, E)$.

Definition 4. *(Disqualified skyline points from $d_i \cup S$)* *The set of skyline points related to the subspace $d_i \cup S$ that are disqualified by the introduction of the subspace D^{i-1} is defined by $CutSky(d_i \cup S, E) =$*
$\{p \in Sky(d_i \cup S, E)_{(Z,\wp)} \mid \exists q \in Sky(d_i \cup S, E)_{(Z,\wp)}, p =_{d_i \bigcup S} q \wedge q \prec_{D^{i-1}} p\}$.

On the other hand, the old skyline $Sky(D^{i-1}, E)_{(Z,\wp)}$ may contain points that are disqualified by dominant points brought by the new dimension d_i. Precisely, let $p, q \in Sky(D^{i-1}, E)_{(Z,\wp)}$ and having the same values on every dimension of D^{i-1}. If q is preferred on the new dimension d_i it will dominate p and disqualify it from the skyline $Sky(D^i, E)_{(Z,\wp)}$. This set of points is denoted $CutSky(D^{i-1}, E)$.

Definition 5. *(Disqualified skyline points from D^{i-1})* *The set of skyline points related to the subspace D^{i-1} that are disqualified by the introduction of dimension d_i is defined by $CutSky(D^{i-1}, E) =$*
$\{p \in Sky(D^{i-1}, E)_{(Z,\wp)} \mid \exists q \in Sky(D^{i-1}, E)_{(Z,\wp)}, p =_{D^{i-1}} q \wedge q \prec_{d_i \bigcup S} p\}$.

Example 3. *Let $D^{i-1} = D^1 = \{Price, Distance, Gr\}$, $d_i = Air$, the new dimension, and the preferences: $\{M <_{Gr} H <_{Gr} T\}$ and $\{G <_{Air} R <_{Air} W\}$.*
$Sky(D^1, E)_{(Z,\wp)} = \{a, b, e, f, h\}$. $CutSky(\{Air\} \cup S, E) = \{\}$ and
$CutSky(D^1, E) = \{f\}$ as f should not be in the skyline $Sky(D^2, E)_{(Z,\wp)}$ since it is dominated by h on the new dimension $d_i = Air$.

Finally, some points should appear in the new skyline. Precisely, before taking into account the new dimension d_i, some points may be dominated on every dimension of $D^{i-1} \bigcup S$ and, so, are not in the skyline. But, when dimension d_i is introduced, being better on d_i than some skyline points they were dominated by, they may well be no longer dominated by any skyline point from $Sky(D^{i-1}, E)_{(Z,\wp)}$ on some dimensions from D^{i-1}: they are new compromise skyline points. This set of points is denoted $NewCompSky(D^i, E)$.

Definition 6. *(New compromise skyline)* *Let $C = (Sky(D^{i-1}, E)_{(Z,\wp)} \bigcup Sky(d_i \bigcup S, E)_{(Z,\wp)}) - (CutSky(D^{i-1}, E) \bigcup CutSky(d_i \bigcup S, E))$.*
The set of new compromise skyline points is defined by
$NewCompSky(D^i, E) = \{p \in E - C \mid \forall q \in C, \exists d_k \in D^{i-1}, p \prec_{d_k} q\}$.

Example 4. *Let $D^{i-1} = D^1 = \{Price, Distance, Gr\}$, $d_i = Air$, the new dimension, and the preferences: $\{M <_{Gr} H <_{Gr} T\}$ and $\{G <_{Air} R <_{Air} W\}$.*
$Sky(D^1, E)_{(Z,\wp)} = \{a, b, e, f, h\}$, $Sky(\{Air\} \bigcup S, E)_{(Z,\wp)} = \{a, b, e\}$,
$CutSky(\{Air\} \bigcup S, E) = \{\}$ and $CutSky(D^1, E) = \{f\}$. But, if we consider simultaneously the two dimensions Gr and Air then c is no longer dominated by f. As f was the only point c was dominated by, c becomes a new skyline point. Since c is the only such "promoted" point, $NewCompSky(D^2, E) = \{c\}$.

The following theorem states that the skyline of the extended subspace can be computed by removing disqualified skyline points from the old skyline and by adding the new skyline points brought by the preference on the new dimension. The new skyline points are either dominant points on the new dimension or new compromise skyline points introduced by the new preference.

Theorem 2. *(Incremental skyline)*[1]
Let E *be a* $|D|$-*dimensional dataset,* $Z \subseteq D$ *the subspace of size* $|Z| = m$ *with dynamic preferences* $\wp = \{\wp_j\}_{j=1,...,m}$ *on* D, $Sky(d_i \cup S, E)$ *the skyline of the subspace* $\{d_i\} \cup S$ *and* $D^i = D^{i-1} \bigcup \{d_i\}$, *with* $i = \{1, .., m\}$.

$$Sky(D^i, E)_{(Z,\wp)} = (Sky(D^{i-1}, E)_{(Z,\wp)} \bigcup Sky(d_i \cup S, E)_{(Z,\wp)})$$
$$- (CutSky(D^{i-1}) \bigcup CutSky(d_i \cup S)) \bigcup NewCompSky(D^i, E)$$

Example 5. *(Illustration of Theorem 2)* Let $D^1 = \{Price, Distance, Gr\}$, $D^2 = \{Price, Distance, Gr, Air\}$ *and the preferences of the previous example.*
$Sky(D^2, E)_{(Z,\wp)}$
$$= (Sky(D^1, E)_{(Gr, H <_{Gr} M <_{Gr} T)} \bigcup Sky(\{Air\} \cup S, E)_{(Air, G <_{Air} W <_{Air} R)})$$
$$- ((CutSky(D^1, E) \bigcup CutSky(\{Air\} \cup S, E)) \bigcup NewCompSky(D^2, E)$$
$$= (\{a, b, e, f, h\} \bigcup \{a, b, e\}) - (\{f\} \bigcup \{\}) \cup \{c\} = \{a, b, c, e, h\}.$$

Theorem 2 provides a scheme for an incremental computation of skyline queries associated with several dynamic dimensions. In the following, we describe more precisely the structure EC^2Sky that stores the precomputed sets and we give the algorithms.

4   EC²Sky Implementation

In this section, we present the implementation of incremental skyline computation and we introduce definitions to characterize the points that are involved in the incremental computation of skyline points and facilitate the specification of the algorithms and of the materialization structure. To ensure an efficient and online computation of skyline, we provide an effective materialization structure detailed in the sequel. We propose a trade-off between (i) *materialize* all the skyline points for all possible preferences and (ii) *calculate*, for each user query, the skyline points associated with the preferences formulated in the query. Our approach is based on three steps:

1. compute and store the skyline on static dimensions. One can adopt any existing algorithm (e.g. [12]) that computes the skyline for partially ordered domains;
2. for each dimension with dynamic preferences, compute and store the candidate skyline points according to any possible first order preference;
3. rely on the information stored in step 1 and 2 to compute the skyline points, for user query with any preferences on dynamic dimensions.

4.1   Skyline Associated with Static Dimensions

In step 1 we compute all the skyline points corresponding to the defined on static dimensions of D. Two concepts introduced by Wong et al. in [4] are helpful. They decompose the set $Sky(D, E)$, corresponding to the defined static preferences of D and denoted by $\wp_\emptyset$, into two subsets: the *global skyline set GSky(D,E)* and the *order-sensitive skyline set OsSky(D,E)*.

The points in the global skyline set $GSky(D, E)$ remain in the skyline whenever any preference on some dimension of Z is added.

---
[1] Due to limited space we skip the proof of the Theorem.

Definition 7. *(Global skyline points) The global skyline set of the space* $D = S \bigcup Z$ *on the dataset* E, *is defined by* $GSky(D,E) =$ $\{p \in Sky(D,E) \mid \forall\, q \in Sky(D,E),\, \nexists\, d_i \in Z,\, p =_S q \wedge p(d_i) \neq_{d_i} q(d_i)\}$

Some skyline points are qualified order-sensitive skyline points because, depending on the preferences associated with dynamic dimensions, these points may be skyline or not. Note first that no global skyline points is order sensitive. Second, *CutSky* points have to be searched among order sensitive skyline points.

Definition 8. *(Order-sensitive skyline points) The order-sensitive skyline set of the space* D *on the dataset* E, *is defined by* $OsSky(D,E) = \{p \in Sky(D,E) \mid p \notin GSky(D,E)\}$ *or equivalently* $OsSky(D,E) = Sky(D,E) - GSky(D,E)$.

Example 6. *Let* $S = \{Price, Distance\}$ *and* $Z = \{Gr, Air\}$. *Then* $GSky(D,E) = \{a,b,e\}$ *and* $OsSky(D,E) = \{c,d,f,h\}$.

4.2   Skyline Associated with Dynamic Dimensions

This section details step 2 of our approach. In this step, we pre-compute the useful information that does not depend on the dynamic preferences provided by users. For each dimension d_i with dynamic preferences, we introduce *the candidate skyline point* (CP_{d_i}), *the new skyline point set* $(NewSky_{(d_i, \wp_i^j)})$ and *the compromise candidate point set* $(CandComp_{(d_i, \wp_i^j)})$.

   The set CP_{d_i} represents the points that may become skyline points over the dimension d_i. It is the set of points from $OsSky(D,E)$:

(1) having on $d_i \in Z$ a value different from any point of $GSky(D,E)$ that dominates them,
(2) having the same value on the static dimensions but different values on $d_i \in Z$.

In the following, $p \prec_{d_i \bigcup S}^j q$ indicates that p dominates q on the subspace $d_i \bigcup S$ according to the first order preference $\wp_i^j$ of the dimension d_i.

Definition 9. *(Candidate skyline points) The candidate skyline point set of the dynamic dimension* d_i, *is defined by* $CP_{d_i} =$ $\{p \in OsSky(D,E) \mid \exists\, q \in GSky(D,E),\, q \prec_S p,\, p(d_i) \neq_{d_i} q(d_i)\} \bigcup$ $\{p \in OsSky(D,E) \mid \exists\, q \in OsSky(D,E),\, q =_S p,\, p(d_i) \neq_{d_i} q(d_i)\}$

Example 7. *Let* $S = \{Price, Distance\}$ *and* $d_i = \{Gr\}$. *Then* $CP_{Gr} = \{c,d,f,h\}$.

To find the new skyline after the introduction of the new dimension d_i, it is sufficient to test the points in CP_{d_i} instead of all non-skyline points. This can reduce significantly the number of domination tests.

   $NewSky_{(d_i, \wp_i^j)}$ is the set of points in CP_{d_i} that are preferred to the points in $GSky(D,E)$ according to the first order preference $\wp_i^j = v_j <_{d_i} *$. Intuitively, NewSky points are equivalent to MaxSky points on d_i according to $\wp_i^j$.

Definition 10. *(New skyline points)*
The new skyline point set of the dynamic dimension d_i, *is defined by* $NewSky_{(d_i, \wp_i^j)} = \{p \in CP_{d_i} \mid \forall\, q \in GSky(D,E) \bigcup \{CP_{d_i} - p\},\, q \not\prec_{d_i \bigcup S}^j p\}$

Algorithm 1. Calculate $CandComp_{(d_i, \wp_i^j)}$

---

input : d_i: a dimension, $\wp_i^j$: a first order preference on d_i, $NewSky_{(d_i, \wp_i^j)}$: skyline points
       added by d_i for $\wp_i^j$, $GSky(D, E)$: global skyline set, CP_{d_i}: candidate skyline set
output: $CandComp_{(d_i, \wp_i^j)}$

1 $CandComp_{(d_i, \wp_i^j)} \leftarrow \emptyset$
2 **foreach** $p \in \{CP_{d_i} - NewSky_{(d_i, \wp_i^j)}\}$ **do**
3 $\quad$ $Set_p \leftarrow \emptyset$
4 $\quad$ **foreach** $q \in \{GSky(D, E) \bigcup NewSky_{(d_i, \wp_i^j)}\}$ **do**
5 $\quad\quad$ **foreach** $d_k \in \{d_i\} \bigcup S$ **do**
6 $\quad\quad\quad$ **if** $p \prec_{d_k}^j q$ **then**
7 $\quad\quad\quad\quad$ $Set_p \leftarrow Set_p \bigcup q$
8 $\quad$ $CandComp_{(d_i, \wp_i^j)} \leftarrow CandComp_{(d_i, \wp_i^j)} \bigcup \{(p, Set_p)\}$

Example 8. $NewSky_{(Gr, H < _{Gr} *)} = \{c, d, f, h\}$.

$CandComp_{(d_i, \wp_i^j)}$ represents the set of points that may become skyline compromises (i.e. compromise candidates) when considering a new dimension (cf. Algorithm 1). They are computed for each first order preference $\wp_i^j$ on d_i.

Definition 11. *(Compromise candidate points)*
Let $E' = (CP_{d_i} - NewSky_{(d_i, \wp_i^j)})$ and $E'' = (GSky(D, E) \bigcup NewSky_{(d_i, \wp_i^j)})$.
The compromise candidate points associated with the preference $\wp_i^j$ is a set of pairs (p, Set_p) defined by $CandComp_{(d_i, \wp_i^j)} =$

$$\{(p, Set_p) \in E' \times \mathcal{P}(E'') \mid \forall q \in \mathcal{P}(E''), \exists d_k \in \{d_i\} \bigcup S, p \prec_{d_k}^j q\}.$$

Example 9. $CandComp_{(Air, R < _{Air} *)} = \{(f, \{a\})\}$ *where the notation* $\{(f, \{a\})\}$ *means that* f *belongs to* $CandComp_{(Air, R < _{Air} *)}$ *because* f *dominates* a *on at least one dimension from* $\{Air\} \bigcup S$ *(here Distance).*

4.3 EC²Sky Structure

Now, let us consider how to construct an EC^2Sky data structure to store efficiently all the precomputed information. Our aim is to avoid building a data structure containing all the combinations of the dynamic preferences on all dimensions as proposed in [3] . In section 5.1 and 5.2 and thanks to theorem 2, we have shown that the skyline of an extended dimensional subspace can be computed by taking into account first order preferences only. We propose to store in the EC^2Sky structure all the sets $NewSky$ and $CandComp$ associated to each first order preference in each dimension.

For each dimension d_i of Z, we compute and store CP_{d_i} and for each first order preference on d_i, we compute and store the two sets: $NewSky_{(d_i, \wp_i^j)}$ and $CandComp_{(d_i, \wp_i^j)}$ related to first order preference j on dimension d_i. The sets $NewSky_{(d_i, \wp_i^j)}$ and $CandComp_{(d_i, \wp_i^j)}$ associated with any possible first order preference on dimension *Hotel group* or *Airline* are presented in Table 2.

Now, we evaluate the space complexity of the EC^2Sky structure. Let m be the number of dynamic dimensions and c be the maximal cardinality of a dynamic dimension. The space complexity of the EC^2Sky structure is given by:

$$\sum_{i=0}^{m}(c) = O(c.m)$$

We can note that the size of the EC^2Sky structure is significantly smaller than the number of possible $n-th$ order preferences given by:

$$(\sum_{i=0}^{c-1}(P_i(c)))^m = O((c.c!)^m)$$

Where $P_i(c)$ is the number of permutations of ordering i elements from c elements. The space complexity of the EC^2Sky structure is also significantly smaller than the space complexity of IPO-tree structure given by:

$$\sum_{i=0}^{m}(c+1)^i = O(c^m)$$

For example, when $m = 3$ and $c = 40$, the number of stored preferences in the EC^2Sky structure is 123 only, while in IPO-tree structure is 70,644, and the number of all possible $n-th$ order preferences ($n \in 1, .., c$), is $4.1 * 10^9$. This is 574.35 times smaller than the IPO-tree and 714,502,572 times smaller than the number of all possible $n-th$ order preferences. The difference is more obvious when the number of dimensions m is high.

4.4 Query Evaluation

In this section, we describe step 3 of our proposal. The information precomputed and stored in step 1 and 2 is used in step 3 to calculate, interactively, the skyline set according to the specified preferences in the user query.

One dimension with dynamic preferences. First, we consider only one dimension with dynamic preferences in the dimensional space D. According to the user query, we are faced with two cases:

(i) *Query with first order preferences*: the skyline associated with a first-order preference $\wp_i^j$ is : $Sky(\{d_i \bigcup S, E)_{(Z,\wp_i^j)} = GSky(D, E) \bigcup NewSky_{(d_i,\wp_i^j)}$. The two latter sets are stored in step 2. Recall that, when dealing with one dimension only, there is no compromise points ($CompSky = \emptyset$).

Example 10. *We use the EC^2Sky structure in Table 2 to illustrate the different steps of a query evaluation. The skyline (stored in Table 2) associated with the*

Table 2. The EC^2Sky structure of the running example

$GSky = \{a, b, e\}$					
$CP_{Gr} = \{c, d, f, h\}$			$CP_{Air} = \{f\}$		
$\wp = M <_{Gr} *$	$\wp = T <_{Gr} *$	$\wp = H <_{Gr} *$	$\wp = R <_{Air} *$	$\wp = G <_{Air} *$	$\wp = W <_{Air} *$
$NewSky_{\{Gr,\wp\}}$			$NewSky_{\{Air,\wp\}}$		
$\{f, h\}$	$\{\}$	$\{c, d, f, h\}$	$\{\}$	$\{\}$	$\{f\}$
$CandComp_{\{Gr,\wp\}}$			$CandComp_{\{Air,\wp\}}$		
$\{\}$	$\{(f, \{a\}), (h, \{a\})\}$	$\{\}$	$\{(f, \{a\})\}$	$\{(f, \{a\})\}$	$\{\}$

Algorithm 2. Calculate $NewCompSky(D^i, E)$

---

input : EC^2Sky structure, $GSky(D, E)$: global skyline points
output: $NewCompSky(D^i, E)$

1 $CandCompSet = \bigcup^i \bigcup^j CandComp_{(d_i, \wp_i^j)}$

2 $NewCompSky(D^i, E) \leftarrow \emptyset$

3 **foreach** $p \in CandCompSet$ **do**
    // Compute the set of points dominated by p on at least one dimension of Z

4      $Dominated(p) \leftarrow \{q | \exists d_i \in Z, p \prec_{d_i} q\}$

5   **if**
    $Dominated(p) = Sky(D^{i-1}, E) \bigcup Sky(d_i \bigcup S, E) - (CutSky(D^{i-1}) \bigcup CutSky(d_i \bigcup S))$
    then

6         $NewCompSky(D^i, E) \leftarrow NewCompSky(D^i, E) \bigcup \{p\}$

7 Dominance test over all the elements of the set $NewCompSky(D, E)$

---

preference $\wp^1 = \{M <_{Gr} *\}$ is computed as follow:
$$Sky(\{Gr\} \bigcup S, E)_{(Z, M <_{Gr}*)} = GSky(D, E) \bigcup NewSky_{\{Gr, M <_{Gr}*\}}$$
$$= \{a, b, e\} \bigcup \{f, h\} = \{a, b, e, f, h\}.$$

(ii) *Query with $n - th$ order preferences*: in this case we use the *merging property* of Wong et al. [3] (see Theorem 1). This is illustrated by the following example.

Example 11. *Suppose now that preference $\wp^1$ is refined to $\wp' = \{M <_{Gr} H <_{Gr} *\}$. The resulting skyline can be computed from the skyline related to the preferences $\wp^1 = \{M <_{Gr} *\}$ and $\wp^2 = \{H <_{Gr} *\}$ (stored in Table 2), as follow:*
$Sky(\{Gr\} \bigcup S, E)_{(Z, M <_{Gr}*)} = \{a, b, e, f, h\}$ *(cf. example 10), which is the skyline for $\wp^1$. In the same way, $Sky(\{Gr\} \bigcup S, E)_{(Z, H <_{Gr}*)} = \{a, b, e, c, d, f, h\}$, which is the skyline for $\wp^2$. Finally, to compute $\wp' = \{M <_{Gr} H <_{Gr} *\}$, we use the merging property (Theorem 1):* $Sky(D, E)_{(Z, M <_{Gr} H <_{Gr}*)} =$
$(Sky(\{Gr\} \bigcup S, E)_{(Z, M <_{Gr}*)} \bigcap Sky(\{Gr\} \bigcup S, E)_{(Z, H <_{Gr}*)}) \bigcup$
$PSky(D, E)_{(Z, M <_{Gr}*)} = (\{a, b, e, f, h\} \bigcap \{a, b, e, c, d, f, h\}) \bigcup \{f, h\} = \{a, b, e, f, h\}$

Several dimensions with dynamic preferences. Second, we consider the case of several dynamic dimensions which is more complex. According to definition 4 and 6, some skyline points ($CutSky$ points) may be disqualified when a new dimension is introduced, while new skyline points ($CompSky$ points) may appear (the computation of $CompSky$ points is described by Algorithm 2).

We are now in position to detail the EC^2Sky method. Algorithm 3 describes the general process of EC^2Sky dedicated to step 3 (the computation of changing elements of the skyline). The sets $CompSky$ (Algorithm 3, line 11) and the union of $CutSky$ (Algorithm 3, line 10) are computed. As stated by the *incremental skyline theorem* (Theorem 2), the final skyline is obtained by eliminating all the $CutSky$ points and by adding all the $CompSky$ points to the union of skylines related to queries involving one dynamic preference (Algorithm 3, line 12).

Example 12. *The skyline associated with the preferences $\wp = \{M <_{Gr} H <_{Gr} *, G <_{Air} *\}$ is computed from the skyline associated with the preferences $\wp_1 = \{M <_{Gr} H <_{Gr} *\}$ and $\wp_2 = \{G <_{Air} *\}$. Let $D^1 = \{Price, Distance, Gr\}$ and $D^2 = \{Price, Distance, Gr, Air\}$. The skyline associated with $\wp_1$ and $\wp_2$ is*

Algorithm 3. $EC^2Sky(Sky(D, E)_{(Z, \wp)})$

input : $Sky(D, E)_{(Z, \wp)}$:skyline query, EC^2Sky structure
output: $Sky(D, E)_{(Z, \wp)}$

1 $Sky(D^0, E)_{(Z, \wp)} \leftarrow GSky(D, E)$
2 **for** $i \leftarrow 1$ **to** $m = |Z|$ **do**
3   **if** $\wp_i = \emptyset$ **then**
4          $Sky(d_i \bigcup S, E)_{(Z, \wp)} \leftarrow GSky(D, E) \bigcup CP_{d_i}$
5   **else**
6   **if** $\wp_i = \wp_i^j$ **then**
7              $Sky(d_i \bigcup S, E)_{(Z, \wp)} \leftarrow GSky(D, E) \bigcup NewSky_{(d_i, \wp_i^j)}$
8   **else**
9   Use merging property

10   Computation of $CutSky(D^{i-1})$ (resp. $CutSky(d_i \bigcup S)$)
11   Computation of $NewCompSky(D^i, E)$;   // Algorithm 2
12     $Sky(D^i, E)_{(Z, \wp)} \leftarrow Sky(D^{i-1}, E)_{(Z, \wp)} \bigcup Sky(d_i \bigcup S, E)_{(Z, \wp)} \bigcup$
      $NewCompSky(D^i, E) - (CutSky(D^{i-1}) \bigcup CutSky(d_i \bigcup S))$
13 $Sky(D, E)_{(Z, \wp)} \leftarrow Sky(D^m, E)_{(Z, \wp)}$

---

computed as in example 11. $Sky(D^1, E)_{(Z, \wp_1)} = \{a, b, e, f, h\}$ and $Sky(\{Air\} \bigcup S, E)_{(Z, \wp_2)} = \{a, b, e\}$.

Since, we have two dimensions with dynamic preferences we compute the sets $CutSky(D^1)$, $CutSky(\{Air\} \bigcup S)$ and $NewCompSky(D^2, E)$. For this example, $CutSky(D^1) = \emptyset$, $CutSky(\{Air\} \bigcup S) = \emptyset$ and $NewCompSky(D^2, E) = \emptyset$. Finally, $Sky(D^2, E)_{(Z, \wp)} = Sky(D^1, E)_{(Z, \wp_1)} \bigcup Sky(\{Air\} \bigcup S, E)_{(Z, \wp_2)} = \{a, b, e, f, h\}$.

Our proposal provides the user with a way to express preferences and with the ability to change them without being penalized by long response times. A good performance is achieved by storing only the minimal amount of information required to enable quick and easy updates. The experimental evaluation presented in the following highlights the relevance of the proposed solution.

5 Experiments

In this section, we report an experimental evaluation of our algorithm EC^2Sky on synthetic data sets. EC^2Sky is implemented in $C/C++$ and the experiments were performed on a 3GHz CPU with 16-Gbyte memory on a Linux platform. For the static dimensions, the data were produced by the generator released by the authors of [8]. Three kinds of data sets were generated: independent data, correlated data and uncorrelated data. The description of these data sets can be found in [8]. Like in [3], we only show the experimental results for the uncorrelated data sets. The results for independent data sets and correlated data sets are similar, but the execution times are much shorter for correlated data sets. The dynamic dimensions were generated according to a Zipfian distribution [13]. By default, we set the Zipfian θ parameter to 1. We obtained $200,000$ tuples for 4 dimensions with static order. The number of dynamic dimensions varied from 5 to 8 and the cardinality of these dimensions from 2 to 5. We chose a query template such that the most frequent value of some dynamic dimension has the highest priority over all other values. This represents a parameter that becomes more difficult to manage as the skyline tends to be larger.

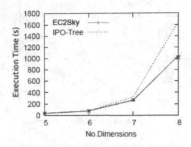
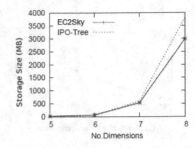

Fig. 1. Scalability with respect to dimensionality

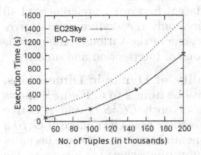
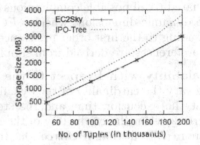

Fig. 2. Scalability with respect to database size

In the following experiments, we compare the performance of our algorithm EC^2Sky with algorithm IPO-tree [3], in terms of the execution time and the storage size.

Scalability with Respect to Dimensionality. We fix the number of static dimensions to 4 and we vary the number of dynamic dimensions from 1 to 4. Figure 1 shows that the execution time and the storage size of both EC^2Sky and IPO-tree increase with the number of dynamic dimensions. However, the increase rate of IPO-tree is greater than the increase rate of EC^2Sky. This is because of the complexity of the preferences tree built by IPO-tree. In contrast with the size of the IPO-tree which is in $O(c*m)$. The table built by EC^2Sky has a size in $O(c*m)$. This induces a substantial increase of the storage size but which evolves more slowly than the IPO-tree. We have also studied the variation of the number of dimensions with static preferences when the number of dynamic dimensions is fixed to be 2. The results are similar to those in Figure 1.

Scalability with Respect to the Database Size. In this experiment, the number of tuples of the dataset varies from $50,000$ to $200,000$. Figure 2 shows that the execution time and the storage size of both EC^2Sky and IPO-tree increase with the size of the dataset. This is because the size of the information stored and analyzed increases with the size of the dataset. However, our method

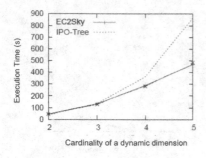
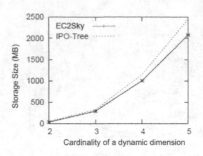

Fig. 3. Scalability with respect to the cardinality of the dynamic dimensions

is more efficient than the IPO-tree method. IPO-tree stores all the skylines associated to all possible combinations of the different first order preferences of all the dynamic dimensions whereas EC^2Sky stores only the skyline points corresponding to the first order preferences. The skyline of the various combinations of preferences are derived from simple operations of intersection and union.

Scalability with Respect to the Cardinality of Dynamic Dimensions.
We vary the cardinality of the dynamic dimensions from 2 to 5. Figure 3 shows that the execution time and the storage size of both EC^2Sky and IPO-tree increase when the cardinality of the dimensions increases. Once more, EC^2Sky is more efficient than IPO-tree. Due to its exponential size, the IPO-tree becomes more complex and larger when the cardinality of dimensions (c) increases. We can also observe a significant increase of the related execution time.

6   Conclusion

In this paper, we have proposed a new efficient method to compute skyline queries in the presence of dimensions associated with dynamic user preferences. We have investigated preferences on dimension values that can be expressed by any partial or complete order, with a particular focus on the compromise points which are important in decision making. Our approach, is based on a materialization of the first order preferences, that can respond efficiently to skyline queries related to user preferences even in the context of large volumes of data. The experimentations presented in this paper highlight the performance improvements of EC^2Sky compared to IPO-tree [3].

The consideration of dimensions with dynamic preferences opens several promising future research directions. First, to demonstrate the usefulness of our method, we want to experiment our algorithm on a real data set. We are particularly interested in the analysis of simulation results from a biophysical model to extract the most polluting plots in a watershed with respect to different analysis criteria. Another possible future direction is to investigate how to compute skyline queries in the context of hierarchical and aggregated data. The adopted approach would search the best compromises along the set of axes. However, this approach raises several problems. One is to define a computation adapted to the level of the explored hierarchy. Another is to define the semantics of skyline points at different levels of granularity.

Acknowledgments. This work is part of the ACASSYA project supported by the French National Agency for Research (ANR SYSTERRA).

References

1. Sawaragi, Y., Nakayama, H., Tanino, T.: Theory of Multiobjective Optimization. Academic Press, Orlando (1985)
2. Raïssi, C., Pei, J., Kister, T.: Computing closed skycubes. PVLDB 3(1), 838–847 (2010)
3. Wong, R.C.W., Fu, A.W.C., Pei, J., Ho, Y.S., Wong, T., Liu, Y.: Efficient skyline querying with variable user preferences on nominal attributes. PVLDB 1(1), 1032–1043 (2008)
4. Wong, R.C.W., Pei, J., Fu, A.W.C., Wang, K.: Online skyline analysis with dynamic preferences on nominal attributes. IEEE Trans. Knowl. Data Eng. 21(1), 35–49 (2009)
5. Yuan, Y., Lin, X., Liu, Q., Wang, W., Yu, J.X., Zhang, Q.: Efficient computation of the skyline cube. In: VLDB, pp. 241–252 (2005)
6. Tao, Y., Xiao, X., Pei, J.: Efficient skyline and top-k retrieval in subspaces. IEEE Trans. on Knowl. and Data Eng., 1072–1088 (2008)
7. Huang, Z., Guo, J., Sun, S., Wang, W.: Efficient optimization of multiple subspace skyline queries. J. Comput. Sci. Technol, 103–111 (2008)
8. Borzsonyi, S., Kossmann, D., Stocker, K.: The skyline operator. In: Proc. of the 17th ICDE, pp. 421–430. IEEE Computer Society (2001)
9. Mindolin, D., Chomicki, J.: Preference elicitation in prioritized skyline queries VLDB J 20(2), 157–182 (2011)
10. Brando, C., Goncalves, M., González, V,: Evaluating Top k Skyline Queries over Relational Databases. In: Wagner, R., Revell, N., Pernul, G. (eds.) DEXA 2007. LNCS, vol. 4653, pp. 254–263. Springer, Heidelberg (2007)
11. Bouadi, T., Bringay, S., Poncelet, P., Teisseire, M.: Requêtes skyline avec prise en compte des préférences utilisateurs pour des données volumineuses. In: EGC 2010, pp. 399–404 (2010)
12. Balke, W., Guntzer, U., Siberski, W.: Exploiting indifference for customization of partial order skylines. In: Proc. of the 10th IDEAS, pp. 80–88. IEEE Computer Society (2006)
13. Trenkler, G.: Univariate discrete distributions : N.L. Johnson, S. Kotz and A.W. Kemp: 2nd edn. John Wiley, New York (1992) ISBN 0-471-54897-9; Computational Statistics & Data Analysis, 240–241 (1994)

Efficient Discovery of Correlated Patterns in Transactional Databases Using Items' Support Intervals

R. Uday Kiran and Masaru Kitsuregawa

Institute of Industrial Science, The University of Tokyo, Komaba-ku, Tokyo, Japan
{uday_rage,kitsure}@tkl.iis.u-tokyo.ac.jp
http://www.researchweb.iiit.ac.in/~uday_rage
http://www.tkl.iis.u-tokyo.ac.jp/Kilab/Members/memo/kitsure_e.html

Abstract. Correlated patterns are an important class of regularities that exist in a transactional database. CoMine uses pattern-growth technique to discover the complete set of correlated patterns that satisfy the user-defined minimum support and minimum all-confidence constraints. The technique involves compacting the database into FP-tree, and mining it recursively by building conditional pattern bases (CPB) for each item (or suffix pattern) in FP-tree. The CPB of the suffix pattern in CoMine represents the set of complete prefix paths in FP-tree co-occurring with itself. Thus, CoMine implicitly assumes that the suffix pattern can concatenate with all items in its prefix paths to generate correlated patterns of higher-order. It has been observed that such an assumption can cause performance problems in CoMine. This paper makes an effort to improve the performance of CoMine by introducing a novel concept known as items' support intervals. The concept says that an item in FP-tree can generate correlated patterns of higher-order by concatenating with only those items in its prefix-paths that have supports within a specific interval. We call the proposed algorithm as CoMine++. Experimental results on various datasets show that CoMine++ can discover high correlated patterns effectively.

Keywords: Data mining, Knowledge Discovery in Databases, Correlated patterns and Pattern-growth technique.

1 Introduction

Mining frequent patterns [2] from transactional databases has been actively and widely studied in data mining [3,5]. A major obstacle for the popular adoption of frequent pattern mining in real-world applications is its failure to capture the true correlation relationship among data objects [4]. To confront the obstacle, researchers have made efforts to discover correlated patterns using alternative measures [4,9,11,12]. Although there exists no universally accepted best measure to judge the interestingness of a pattern, *all-confidence* [9] is emerging as a measure that can disclose true correlation relationships among data objects [13].

S.W. Liddle et al. (Eds.): DEXA 2012, Part I, LNCS 7446, pp. 234–248, 2012.
© Springer-Verlag Berlin Heidelberg 2012

An interesting application of correlated patterns is the activity recognition for people with dementia [10]. The model of frequent and correlated patterns is as follows [8].

Let $I = \{i_1, i_2, \cdots, i_n\}$ be a set of items, and DB be a database that consists of a set of transactions. Each transaction T contains a set of items such that $T \subseteq I$. Each transaction is associated with an identifier, called TID. Let $X \subseteq I$ be a set of items, referred as an itemset or a *pattern*. A pattern that contains k items is a k-pattern. A transaction T is said to contain X if and only if $X \subseteq T$. The support of a pattern X in DB, denoted as $S(X)$, is the number of transactions in DB containing X. The pattern X is **frequent** if it occurs no less frequent than the user-defined minimum support ($minSup$) threshold, i.e., $S(X) \geq minSup$. The *all-confidence* of a pattern X, denoted as *all-conf(X)*, can be expressed as the ratio of its support to the maximum support of an item within it. That is, $all\text{-}conf(X) = \dfrac{S(X)}{max\{S(i_j)| \forall\ i_j \in X\}}$. A pattern X is said to be **all-confident** or **associated** or **correlated** if $S(X) \geq minSup$ and $all\text{-}conf(X) \geq minAllConf$, where $minAllConf$ is the user-defined minimum all-confidence threshold.

Example 1. Consider the transactional database of 20 transactions shown in Table 1. The set of items $I = \{a,\ b,\ c,\ d,\ e,\ f,\ g,\ h\}$. The set of items a and b, i.e., $\{a,\ b\}$ is a pattern. It is a 2-pattern. For simplicity, we write this pattern as "ab". It occurs in 8 transactions ($tids$ of $1, 2, 7, 10, 11, 13, 16$ and 19). Therefore, the support of "ab," i.e., $S(ab) = 8$. If the user-specified $minSup = 3$, then "ab" is a frequent pattern because $S(ab) \geq minSup$. The *all-confidence* of "ab", i.e., $all\text{-}conf(ab) = \frac{8}{max(11,9)} = 0.72$. If the user-specified $minAllConf = 0.63$, then "ab" is a correlated pattern because $S(ab) \geq minSup$ and $all\text{-}conf(ab) \geq minAllConf$.

Table 1. Transactional database

TID	ITEMS	TID	ITEMS	TID	ITEMS	TID	ITEMS
1	a, b	6	a, c	11	a, b	16	a, b
2	a, b, e	7	a, b	12	a, c, f	17	c, d
3	c, d	8	e, f	13	a, b, e	18	a, c
4	e, f	9	c, d, g	14	b, e, f, g	19	a, b, h
5	c, d	10	a, b	15	c, d	20	c, d, f

The problem statement of mining correlated patterns is as follows. Given a transactional database (DB), and user-defined *minimum support* ($minSup$) and *minimum all-confidence* ($minAllConf$) thresholds, discover the complete set of correlated patterns in DB that satisfy both $minSup$ and $minAllConf$ thresholds.

CoMine [8] extends FP-growth [6] to discover the complete set of correlated patterns in a database. In particular, CoMine uses pattern-growth technique to discover the patterns. The technique involves the following steps.

  i. The given database is compressed into a tree known as frequent pattern-tree
      (FP-tree). The FP-tree retains the pattern association information.
 ii. Using each item in the FP-tree as an initial suffix item (or pattern)[1], CoMine
      constructs its Conditional Pattern Base (CPB) consisting of the set of com-
      plete prefix paths in the FP-tree co-occurring with the suffix item, then
      construct its conditional FP-tree and perform mining recursively on such
      a tree. The pattern-growth is achieved by the concatenation of the suffix
      pattern with the frequent patterns generated from the conditional FP-tree.

The construction of initial CPB for the suffix item in FP-tree is a key step
because it reduces the search space effectively. Since the CPB of a suffix pattern
represents the set of complete prefix paths in the FP-tree co-occurring with
itself, the CoMine implicitly assumes that the suffix item can concatenate with
all items in its prefix paths to form correlated patterns of higher-order. This
assumption causes performance problems in CoMine. The reason is that *all-
confidence* facilitates the suffix item to generate correlated patterns of higher-
order by combining with only those items in its prefix-paths that have supports
within a specific interval range.

    This paper makes an effort to discover correlated patterns effectively. The key
contributions of this paper are as follows:

  i. This paper introduces a novel concept known as items' support intervals. It
      states that an item can combine with only those items having supports within
      a specific interval to form correlated (or interesting) patterns of higher-order.
      A methodology to find items' support intervals for the correlated pattern
      model has also been discussed in the paper.
 ii. An improved CoMine, called CoMine++, has also been presented in the
      paper using items' support intervals. Unlike CoMine, the CPB of the suffix
      item in CoMine++ represents the set of partial prefix-paths (i.e., involving
      only those items that have support within the support interval of suffix item)
      in FP-tree co-occurring with itself.
iii. Using the prior knowledge regarding the construction and mining of FP-tree,
      CoMine++ uses a novel pruning technique to construct the CPB of the suffix
      item effectively.
 iv. Experimental results on both synthetic and real-world datasets show that
      mining high correlated patterns with CoMine++ is time and memory effi-
      cient, and highly scalable as well.

The rest of this paper is organized as follows. Section 2 discusses previous works
on mining correlated patterns. Section 3 describes the working of CoMine. Section 4 discusses the performance problems of CoMine and introduces CoMine++
to mine correlated patterns. Experimental evaluations of CoMine and CoMine++
are reported in Section 5. Finally, Section 6 concludes with future research
directions.

---

[1] In this paper, the suffix pattern with one item has been referred as suffix item for
   simplicity purpose.

2   Related Work

Since the introduction of frequent patterns in [2], numerous algorithms have been discussed in the literature to mine frequent patterns from transactional databases [5]. FP-growth [6] is a popular algorithm to discover frequent patterns. It uses pattern-growth technique (discussed in Section 1) to discover frequent patterns. The initial CPB of a suffix item in FP-growth represents the set of complete prefix paths in the FP-tree co-occurring with itself. Such a CPB for the suffix item is valid for FP-growth. It is because every suffix item (or pattern) in FP-tree can concatenate with all other items in its prefix-paths to form frequent patterns of higher-order.

Brin et al. [4] have introduced correlated pattern mining using $lift$ and χ^2 as the interestingness measures. Lee et al. [8] have shown that correlated patterns can be effectively discovered with *all-confidence* measure as it satisfies both *null-invariance* and *downward closure properties*. The *null-invariance* property facilitates the measure to disclose genuine correlation relationships without being influenced by the object co-absence in a database. The *downward closure property* facilitates in the reduction of search space as all non-empty subsets of a correlated pattern must also be correlated. An FP-growth-like algorithm known as CoMine was proposed in [8] to discover the patterns using both *all-confidence* and *support* values. The performance issues of CoMine and the techniques to improve the same are discussed in later parts of this paper.

CCMine is a variant of CoMine to discover confidence-closed correlated patterns. Please note that CCMine also suffers from the same performance problems as the CoMine. However, this paper focuses only on CoMine algorithm.

Kim et al. [7] have made an effort to discover top-k correlated patterns using the measures that satisfy null-invariance property. Since some of those measures (e.g. *cosine*) do not satisfy anti-monotonic property, an apriori-like algorithm, NICOMINER, has been proposed by introducing new pruning techniques. Although their work is closely related to our work, their pruning techniques cannot be directly extendable to discover correlated patterns using support and all-confidence measures. The reason is that their work have not discussed any methodology to determine the support intervals for items if correlated patterns are defined using both *support* and *all-confidence* measures. Moreover, NICOMINER suffers from the same performance problems as the Apriori algorithm, which includes the generation of huge number of candidate patterns and multiple scans on the dataset.

CoMine performs better than NICOMINER as it is specifically designed only for *support* and *all-confidence* measures using downward closure property. In the next section, we describe the working of CoMine.

3   Working of CoMine

The CoMine uses pattern-growth technique to discover the complete set of correlated patterns in a database. The technique involves: (i) compressing the database into FP-tree and (ii) recursive mining of FP-tree to discover the complete set of correlated patterns.

The FP-tree is a compact data structure for storing all necessary information about frequent patterns in a database. Every branch of the FP-tree represents a pattern, and the nodes along the branch are ordered decreasingly by the frequency of the corresponding item, with leaves representing the least frequent items. Compression is achieved by building the tree in such a way that overlapping patterns are represented by sharing pre-fixes of the corresponding branches. It has a header table containing two fields: item name (I), and their support count (S). A row in the header table also contains the head of a list that links all the corresponding nodes of the FP-tree.

The FP-tree is constructed with only two scans on the database. In the first scan, all frequent items are found and sorted in support descending order. Using the sorted order, the second scan constructs the FP-tree which stores all frequency information of the original dataset. Mining the database then becomes mining the FP-tree. For the user-defined $minSup = 3$ and $minAllConf = 0.63$, the set of sorted frequent items found after first scan on the database of Table 1 are $\{\{a : 11\}, \{b : 9\}, \{c : 9\}, \{d : 6\}, \{e : 5\}, \{f : 5\}\}$. Figure 1 shows the FP-tree constructed after performing the second scan on the database.

Mining correlated patterns using FP-tree of Figure 1 is shown in Table 2 and detailed as follows. Consider the item f, which is the last item in FP-tree, as a suffix item. The item f occurs in four branches of the FP-tree of Figure 1. The paths formed by these branches are $\langle acf : 1 \rangle$, $\langle cdf : 1 \rangle$ $\langle ef : 2 \rangle$ and $\langle bef : 1 \rangle$. Therefore, considering f as a suffix, its corresponding four prefix paths

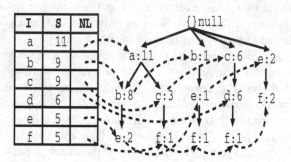

Fig. 1. FP-tree. The terms 'I', 'S' and 'NL' respectively denote item, support and node-link.

Table 2. Mining correlated patterns with CoMine

Suffix item	Conditional Pattern Base	Conditional FP-tree	Correlated Patterns
f	$\{\{ac : 1\}, \{cd : 1\}, \{e : 2\},$ $\{be : 1\}\}$	$\langle e : 3 \rangle$	$\{ef : 3\}$
e	$\{ab : 2\}$	-	-
d	$\{c : 6\}$	$\langle c : 6 \rangle$	$\{cd : 6\}$
c	$\{a : 3\}$	-	-
b	$\{ea : 8\}$	$\langle a : 8 \rangle$	$\{ab : 8\}$

are $\langle ac : 1 \rangle$, $\langle cd : 1 \rangle$, $\langle e : 2 \rangle$ and $\langle be : 1 \rangle$, which form its conditional pattern base. Its conditional FP-tree contains only a single path, $\langle e : 3 \rangle$. The items a, b, c and d are not included in Conditional FP-tree because their support of 1 is less than $minSup$ (also, the all-confidence values of af, bf, cf and df are less than $minAllConf$). The concatenation of suffix pattern with conditional FP-tree generates the correlated pattern $\{ef : 3\}$. Similar process is repeated for other items in the FP-tree to discover the complete set of correlated patterns.

In the next section, we discuss the performance issues of CoMine and propose techniques to improve its performance.

4   Proposed Algorithm

In this section, we first introduce the concept of items' support intervals. This term simplifies the understanding of performance issue in CoMine, which is discussed subsequently. Next, we propose the basic idea of incorporating the proposed concept in CoMine to improve its performance. We call the improved CoMine as CoMine++.

4.1   Items' Support Intervals

The concept of *items' support intervals* say that every item can generate interesting patterns by combining with only those items having supports within a specific interval. Finding the support interval for each item in the database is non-trivial because it depends upon the following two factors: (i) the support of an item and (ii) the user-defined threshold values' for the measures. In this paper, we make an effort to identify the items' support interval for correlated pattern model defined using *support* and *all-confidence* measures.

For the user-defined $minAllConf$ and $minSup$ thresholds, the support interval of an item $i_j \in I$ is $\left[max \left(\dfrac{S(i_j) \times minAllConf,}{minSup} \right), max \left(\dfrac{S(i_j)}{minAllConf}, \dfrac{}{minSup} \right) \right].$
The correctness is shown in Theorem 1, and is based on Properties 1 and 2 and Lemmas 1 and 2.

Property 1. If X and Y are two patterns such that $X \subset Y$, then $S(X) \geq S(Y)$.

Property 2. The maximum support a pattern X can have is $min(S(i_j)|\forall i_j \in X)$. Therefore, the maximum all-confidence a pattern X can have is
$$\dfrac{min(S(i_j)|\forall i_j \in X)}{max(S(i_j)|\forall i_j \in X)}.$$

Lemma 1. *Let $minAllConf$ be the user-defined minimum all-confidence threshold value. The lower limit of a support interval for an item $i_q \in I$ having support $S(i_q)$ is $S(i_q) \times minAllConf$.*

Proof. Let 'i_p' and 'i_q' be the two items (or 1-patterns) having supports such that $S(i_p) < S(i_q)$. The maximum all-confidence the pattern '$i_p i_q$' ($=\{i_p, i_q\}$) can have is $\frac{S(i_p)}{S(i_q)}$ (Property 2). If $S(i_q) \times minAllConf > S(i_p)$, then $minAllConf > \frac{S(i_p)}{S(i_q)}$ ($= all\text{-}conf(i_p i_q)$). If $S(i_q) \times minAllConf = S(i_p)$, then $minAllConf = \frac{S(i_p)}{S(i_q)}$ ($= all\text{-}conf(i_p i_q)$). Therefore, the lower limit of a support interval for an item $i_q \in I$ with support $S(i_q)$ is $S(i_q) \times minAllConf$.

Lemma 2. *Let $minAllConf$ be the user-defined minimum all-confidence threshold value. The upper limit of a support interval for an item $i_q \in I$ having support $S(i_q)$ is $\dfrac{S(i_q)}{minAllConf}$.*

Proof. Let 'i_q' and 'i_r' be the two items (or 1-patterns) having supports such that $S(i_q) < S(i_r)$. Using Property 2, the maximum all-confidence for the pattern '$i_q i_r$' ($=\{i_q, i_r\}$) can be $\frac{S(i_q)}{S(i_r)}$. If $S(i_r) > \dfrac{S(i_q)}{minAllConf}$, then $minAllConf > \dfrac{S(i_q)}{S(i_r)}$ ($= all\text{-}conf(i_q i_r)$). Therefore, the upper limit of a support interval for an item $i_q \in I$ is $\dfrac{S(i_q)}{minAllConf}$.

Theorem 1. *For the user-defined minSup and minAllConf thresholds, the support interval for an $i_q \in I$ with support $S(i_q)$ is*

$$\left[max \left(\frac{S(i_q) \times minAllConf}{minSup} \right), \ max \left(\frac{\frac{S(i_q)}{minAllConf}}{minSup} \right) \right].$$

Proof. The support interval for an item $i_q \in I$ for *support* measure is $[minSup, \infty)$. In other words, items having support no less than $minSup$ can combine with one another in all possible ways to generate frequent patterns of higher-order. Using Lemmas 1 and 2, the support interval of an item $i_q \in I$ for *all-confidence* measure is $\left[S(i_q) \times minAllConf, \ \dfrac{S(i_q)}{minAllConf} \right]$. Therefore, the support interval for an item $i_q \in I$ for both *support* and *all-confidence* measures is $\left[max \left(\dfrac{S(i_q) \times minAllConf}{minSup} \right), \ max \left(\dfrac{\frac{S(i_q)}{minAllConf}}{minSup} \right) \right].$

Example 2. The support of f in Table 1 is 5. If $minAllConf = 0.63$ and $minSup = 3$, then f can generate correlated patterns by combining with only those items having frequencies in the range of $[3, \ 8]$ ($= [max(5 \times 0.63, 3), max(\frac{5}{0.63}, 3)]$).

4.2   Performance Problems in CoMine

Since the initial CPB of a suffix item constitutes of the set of complete prefix paths in the FP-tree co-occurring with itself, CoMine implicitly assumes that the suffix item can generate correlated patterns of higher-order by concatenating with all items in its prefix path. However, this is not the seldom case because *all-confidence* facilitates the suffix item to concatenate with only those items in its prefix paths that have support within a specific interval range to generate the patterns. As a result, CoMine suffers from the performance problems pertaining to both memory and runtime requirements.

Example 3. Since the CPB of f contains a, b and c, CoMine implicitly assumes that f can concatenate with a, b and c items to generate correlated patterns. However, any correlated pattern containing f can never have the items, a, b and c. It is because their supports do not lie within the support interval of f.

4.3   Basic Idea: Pruning Technique

Using the concept of items' support intervals, the complete set of correlated patterns can be discovered from FP-tree by building the CPB of the initial suffix item with the set of partial prefix paths involving only those items that have supports within its support interval.

To construct such CPB for a suffix item, one can assume that the support of every item present in the prefix path of the suffix item needs to tested, which is same as constructing CPB with complete set of prefix paths (as in CoMine). This paper argues that such a process is not necessary. It is because of the following pruning technique. *If an item at level k in the prefix path of the suffix item fails to have its support within the support interval of corresponding suffix item, then all items lying in between the root node and the failed item (i.e., items lying in levels 1 to k−1) will also fail to have their supports within the support interval of corresponding suffix item.* The correctness is shown in Lemma 3 and illustrated in Example 4.

Lemma 3. *Let* $\langle i_1, i_2, \cdots, i_k, i_{k+1} \rangle$, $1 \le k < n$, *be a branch in FP-tree such that* $S(i_1) \ge S(i_2) \ge \cdots \ge S(i_k) \ge S(i_{k+1})$. *If all-conf*$(i_k i_{k+1}) < minAllConf$, *then all-conf*$(i_j i_{k+1}) < minAllConf$, *where* $1 \le j < k$.

Proof. The prefix path for the suffix item i_{k+1} is $\langle i_1, i_2, \cdots, i_k \rangle$. If *all-conf*$(i_k, i_{k+1}) < minAllConf$, then $S(i_k) > \dfrac{S(i_{k+1})}{minAllConf}$ (using Lemma 2). Since FP-tree is constructed in support descending order of items, it turns out that $S(i_j) \ge S(i_k) > \dfrac{S(i_{k+1})}{minAllConf}$, where $1 \le j < k$. In other words, *all-conf*$(i_j i_{k+1}) < minAllConf$, $1 \le j < k$.

Example 4. A prefix path for the suffix item f in FP-tree of Figure 1 is $\langle a, c : 1 \rangle$. The support of c does not lie in the support interval of f, i.e., $S(c) > \dfrac{S(f)}{minAllConf}$. As a result, it can be stated without testing that the support of a will not lie in the support interval of f, i.e., $S(a) > \dfrac{S(f)}{minAllConf}$.

4.4 CoMine++ Algorithm

The proposed CoMine++ involves two steps: (*i*) Construction of FP-tree and (*ii*) Mining FP-tree to discover the complete set of correlated patterns. The first step is same as in CoMine. However, the second step is different, and is as follows. Considering each frequent item in the FP-tree as a suffix item, CoMine++ constructs its CPB with the set of partial prefix paths containing only those items having supports within its support interval. Next, compress the CPB by performing tree-merging operations, which involves merging the common nodes in the CPB by summing up their respective supports. Next, conditional FP-tree is built using $minSup$ and $minAllConf$ constrains. The conditional FP-tree contains only those items that have generated correlated patterns by concatenating with the suffix item. Thus, subsequent recursive mining on conditional FP-tree involves constructing CPBs with the set of complete prefix paths co-occurring with the suffix pattern. The pattern-growth is achieved by the concatenation of the suffix pattern with the correlated patterns generated from a conditional FP-tree.

The working of CoMine++ is shown in Algorithm 1, Procedure 2 and Procedure 3. The input parameters to CoMine++ are transactional database and the user-defined $minSup$ and $minAllConf$ thresholds. The Algorithm 1 constructs FP-tree using $minSup$ threshold and calls Procedure 2 for mining correlated patterns from FP-tree. Procedure 2 builds CPB for the suffix item with the set of partial prefix paths involving only those items having supports within its support interval, performs *tree-merging* operation to compress the CPB, constructs conditional FP-tree for the suffix item and calls Procedure 3 for recursive mining of conditional FP-tree of the suffix item. The working of Procedure 3 resembles Procedure 2. However, the variation is that Procedure 3 builds the CPB of the suffix pattern with the complete set of prefix paths.

We now explain the working of CoMine++ using the database shown in Table 1. Let $minSup = 3$ and $minAllConf = 0.63$. First, FP-tree shown in Figure 1 is created using $minSup = 3$. Next, mining correlated patterns using FP-tree is shown in Table 3 and detailed as follows. Consider the item f as a **suffix item**. The support interval of f is [3, 7]. The item f occurs in four branches of the FP-tree of Figure 1. The paths formed by these branches are $\langle acf : 1 \rangle$, $\langle cdf : 1 \rangle$ $\langle ef : 2 \rangle$ and $\langle bef : 1 \rangle$ (see Figure 2(a)). Therefore, considering f as a suffix, its corresponding four prefix paths are $\langle ac : 1 \rangle$, $\langle cd : 1 \rangle$, $\langle e : 2 \rangle$ and $\langle be : 1 \rangle$ (see Figure 2(b)). In the prefix path $\langle ac : 1 \rangle$, the item c fails the test (i.e., its support does not lie in the support interval of the suffix item f). Using the pruning technique discussed in the basic idea, we neglect testing of a in the prefix path (line 6 in Procedure 2) and directly select another prefix path for testing. In the prefix path $\langle cd : 1 \rangle$, the item d satisfies the test. As a result, the item c is tested and is found that it fails the test. Similar process is repeated for the other prefix paths of f. Finally, we neglect the items a, b and c and construct CPB of f with only the items d and e, resulting in three branches $\langle e : 1 \rangle$, $\langle d : 1 \rangle$ and $\langle e : 2 \rangle$ (see Figure 2(c)). Since two branches in the CPB of f contain a common item e, tree-merging operation (line 8 in Procedure 2)

Algorithm 1. CoMine++

INPUT: Transactional database DB, minimum support ($minSup$) and minimum all-confidence ($minAllConf$)

OUTPUT: Complete set of correlated patterns.

1: The FP-tree is constructed in the following steps:

    i. Scan the transactional database D once. Collect F, the set of frequent items, and their support counts. Sort F in support descending order as L, the list of frequent items.

    ii. Create the root of an FP-tree, and label it as "null." For each transaction t in D do the following. Select and sort the frequent items in t according to the order of L. Let the sorted frequent item list in t be $[p|P]$, where p is the first element and P is the remaining list. Call **insert_tree**($[p|P], T$), which is performed as follows. If T has a child N such that $N.item\text{-}name = p.item\text{-}name$, then increment N's count by 1; else create a new node N, and let its count be 1, its parent link be linked to T, and its node-link to the nodes with the same item-name via the node-link structure. If P is non-empty, call **insert_tree**(P, N) recursively.

2: The FP-tree is mined by calling Co-mine_1($Tree, null$).

---

is performed on the CPB of f to create two branches $\langle d : 1 \rangle$ and $\langle e : 3 \rangle$ (see Figure 2(d)). The conditional FP-tree of f contains only a single path, $\langle e : 3 \rangle$, d is not included because its support of 1 in less than $minSup$. The concatenation of suffix pattern with conditional FP-tree generates the correlated pattern $\{ef : 3\}$. Similar process is repeated for other items in the FP-tree to discover the complete set of correlated patterns.

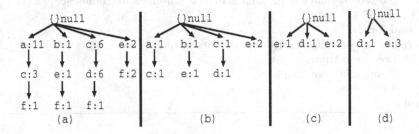

(a)   (b)   (c)   (d)

Fig. 2. Generating conditional pattern base of f. (a) Branches of FP-tree containing f, (b) prefix paths of f, (c) partial prefix paths of f and (d) Final conditional pattern base of f after merging its partial prefix paths.

5   Experimental Results

In this section, we evaluate the perform of CoMine and CoMine++ algorithms. We show that CoMine++ is a better algorithm to mine highly correlated patterns in different types of datasets. The algorithms are written in GNU C++ and run

Procedure 2. Co-mine_1($Tree, \alpha$); Constructing the conditional pattern base for frequent item or length-1 suffix pattern.

1: **for** each a_i in the header of $Tree$ **do**
2:     Generate pattern $\beta = \alpha \cup a_i$ with $all\text{-}conf = \dfrac{S(\beta)}{max\_item\_sup(\beta)} \cdot \{S(\beta) = S(a_i)$
       in α-projected database$\}$
3:     Construct β-projected database as follows.
4:     **for** each prefix path PP_k of a_i **do**
5:         Starting from the last item in PP_k and moving up the prefix path, test whether
           their support is no less than $max(\frac{S(a_i)}{minAllConf}, minSup)$.
6:         Skip testing other items in PP_k if an item at level l in PP_k fails the test. It
           is because other items lying in between the root node and the failed item in
           PP_k will fail the test.
7:         Create a temporary branch involving only those items in PP_k that satisfy the
           test. Preserve the node counts of the items and their order in the temporary
           branch.
8:         Add the temporary branch in β-projected database. This step is similar to
           insert_tree function in Procedure 1. Reduce the size of β-projected database
           by merging the common nodes and summing up their node counts.
9:     **end for**$\{$CoMine do not perform the above steps (i.e., lines from 4 to 9). It
       simply constructs β-projected database with every item in a prefix path of β.$\}$
10:     Let I_β be the set of items in β-projected database.
11:     For each item in I_β, compute its count in β-projected database;
12:     **for** each $b_j \in I_\beta$ **do**
13:         **if** $S(\beta \cup b_j) < minSup$ **then**
14:             delete b_j from I_β; $\{$pruning based on minimum support$\}$
15:         **end if**
16:         **if** $all\text{-}conf(\beta \cup b_j) < minAllConf$ **then**
17:             delete b_j from I_β; $\{$pruning based on minimum all-confidence$\}$
18:         **end if**
19:     **end for**
20:     construct β-conditional FP-tree with items in I_β $Tree_\beta$.
21:     **if** $Tree_\beta \neq \emptyset$ **then**
22:         Co-mine_k($Tree_\beta, \beta$);
23:     **end if**
24: **end for**

Table 3. Mining correlated patterns with CoMine++

Item	Conditional Pattern Base	Conditional FP-tree	Correlated Patterns
f	$\{\{d:1\}, \{e:3\}\}$	$\langle e:3 \rangle$	$\{ef:3\}$
e	-	-	-
d	$\{c:6\}$	$\langle c:6 \rangle$	$\{cd:6\}$
c	$\{a:3\}$	-	-
b	$\{a:8\}$	$\langle a:8 \rangle$	$\{ab:8\}$

Procedure 3. Co-mine_k$(Tree, \alpha)$; Constructing the conditional pattern base for length-k suffix pattern, where $k \geq 2$.

1: **for** each a_i in the header of $Tree$ **do**

2:     Generate pattern $\beta = \alpha \cup a_i$ with $all\text{-}conf = \dfrac{S(\beta)}{max\_item\_sup(\beta)}$. $\{S(\beta) = S(a_i)$

     in α-projected database$\}$

3:     Get a set I_β of items to be included in β-projected database. $\{$No need to check

     whether $S(i_j) \leq \dfrac{min\_item\_sup(\beta)}{minAllConf}$ $\}$

4:     for each item in I_β, compute its count in β-projected database;

5:     **for** each $b_j \in I_\beta$ **do**

6:        **if** $S(\beta \cup b_j) < minSup$ **then**

7:           delete b_j from I_β; $\{$pruning based on minimum support$\}$

8:        **end if**

9:        **if** $all\text{-}conf(\beta \cup b_j) < minAllConf$ **then**

10:        delete b_j from I_β; $\{$pruning based on minimum all-confidence$\}$

11:        **end if**

12:     **end for**

13:     construct β-conditional FP-tree with items in I_β $Tree_\beta$.

14:     **if** $Tree_\beta \neq \emptyset$ **then**

15:        Co-mine_k$(Tree_\beta, \beta)$;

16:     **end if**

17: **end for**

---

with the Ubuntu 10.04 operating system on a 2.66 GHz machine with 1GB memory. The runtime is expressed in seconds and specifies the total execution time, i.e., CPU and I/Os. We pursued experiments on synthetic (T10I4D100K) and real-world (BMS-WebView-1,BMS-WebView-2, Mushroom and Kosarak) datasets. The datasets are available at Frequent Itemset Mining repository [1]. The details of these datasets are shown in Table 4.

Please note that we are not comparing the proposed CoMine++ with the NICOMINER [7]. It is because of two reasons: (i) CoMine performs better than the NICOMINER as the latter suffers from the same performance problems as the Apriori and (ii) NICOMINER is meant to discover top-k correlated patterns and not the complete set of correlated patterns in a database.

Table 4. Dataset Characteristics. The terms "Max.", "Avg." and "Tran." are respectively used as the acronyms for "maximum", "average" and "transaction"

Dataset	Transactions	Distinct Items	Max. Trans. Size	Avg. Trans. Size	Type
$T10I4D100k$	100000	870	29	10.102	sparse
$BMS\text{-}WebView\text{-}1$	59602	4971	267	2.5	sparse
$BMS\text{-}WebView\text{-}2$	77512	33401	161	2	sparse
$Mushroom$	8124	119	23	23.0	dense
$Kosarak$	990002	41270	2498	8.1	sparse

5.1   Memory Tests on CoMine and CoMine++ Algorithms

Figure 3 (a), (b), (c) and (d) respectively show the number of nodes generated by CoMine and CoMine++ algorithms at different $minAllConf$ values in T10I4D100k, BMS-WebView-1, BMS-WebView-2 and Mushroom datasets, respectively. The $minSup$ used in these datasets are 0.1%, 0.1%, 0.1% and 25%, respectively. The following observations can be drawn from this graph: (i) Increase in $minAllConf$ (keeping $minSup$ constant) has decreased the number of nodes getting generated in both CoMine and CoMine++. It is because of the decrease in correlated patterns with the increase in $minAllConf$. (ii) CoMine++ generates relatively fewer numbers of nodes than CoMine with the increase in $minAllConf$ value. It is because of the decrease in the width of items' support intervals with the increase in $minAllConf$ value.

Since the memory requirement of an algorithm depends on the number of nodes being generated by an algorithm, it turns out that CoMine++ is more memory efficient than CoMine.

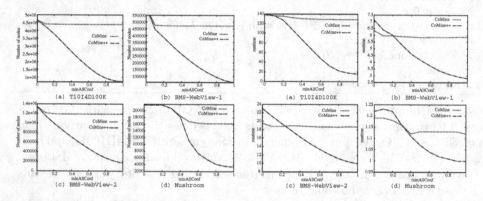

Fig. 3. Number of nodes generated in CoMine and CoMine++ algorithms

Fig. 4. Runtime (in seconds) comparison of CoMine and CoMine++ algorithms

5.2   Runtime Tests on CoMine and CoMine++ Algorithms

Figure 4 (a), (b), (c) and (d) respectively show the runtime taken by CoMine and CoMine++ algorithms at different $minAllConf$ values in T10I4D100k, BMS-WebView-1, BMS-WebView-2 and Mushroom datasets, respectively. The $minSup$ used in these datasets are 0.1%, 0.1%, 0.1% and 25%, respectively. The following three observations can be drawn from these graphs. (i) Increase in $minAllConf$ has decreased the runtime in both CoMine and CoMine++ algorithms. It is because of decrease in the number of correlated patterns with the increase in $minAllConf$ value. (ii) CoMine++ has outperformed CoMine at higher $minAllConf$ values. It is because of the decrease in items' support interval which resulted in the construction of the initial CPB of a suffix item timely. (iii) CoMine has performed better than CoMine++ at low $minAllConf$ values. It is because of increase in items' support interval which resulted in CoMine++ to construct the CPB of the suffix item with almost all items in its prefix paths.

5.3   Scalability Test on CoMine and CoMine++ Algorithms

In this experiment, we evaluate the scalability performance of CoMine and CoMine++ algorithms on memory and runtime requirements by varying the number of transactions in a database. We use real-world *kosarak* dataset for the scalability experiment, since it is a huge sparse dataset. We divided the dataset into ten portions of 0.1 million transactions in each part. Then we investigated the performance of CoMine and CoMine++ algorithms after accumulating each portion with previous parts while performing correlated pattern mining each time. We fixed $minSup = 0.1\%$ and $minAllConf = 0.5$ for each experiment. The experimental results are shown in Figure 5. The memory (represented in number of nodes) and time in y-axes of the left and right graphs in Figure 5 respectively specify the required memory and total runtime with the increase of database size. It is clear from the graphs that as the database size increases, overall tree construction and mining time, and required memory increase. However, CoMine++ requires relatively less memory (nearly half of the memory requirements of CoMine) and runtime with respect to the database size. Therefore, it can be observed from the scalability test that CoMine++ can efficiently mine correlated patterns over large datasets and distinct items with considerable amount of runtime and memory.

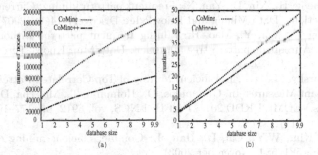

Fig. 5. Scalability test. (a) Memory requirement and (b) Runtime

In many real-world applications, users are generally interested in highly correlated patterns. Thus, the proposed CoMine++ algorithm is a better choice over the existing CoMine algorithm.

6   Conclusion

This paper argues that a pattern-growth algorithm that simply constructs the CPB of the suffix item with every item in its prefix path can suffer from performance problems. It is because some measures facilitate an item to combine with only those items having supports within a specific interval to generate interesting patterns of higher order. This paper introduced the concept of items' support intervals and proposed a methodology to determine it for the correlated pattern model defined using *support* and *all-confidence* measures. A pattern-growth algorithm, called CoMine++, has also been proposed to discover correlated patterns effectively. Unlike the traditional pattern-growth algorithms (such as CoMine), CoMine++ discovers the complete set of correlated patterns by constructing the initial CPB of the suffix item with only those items in its prefix paths that have support within its support interval. A novel pruning technique has also been discussed to construct CPB of the suffix item effectively. Experimental results have shown that proposed CoMine++ algorithm can efficiently mine highly correlated patterns over the existing CoMine algorithm.

As a part of future work, we would like to extend the notion of items' support intervals to improve the performance of CCMine algorithm to discover closed coverage patterns effectively. In addition, we would like to investigate the methods to determine the items' support intervals for other measures.

References

1. Frequent itemset mining repository, http://fimi.cs.helsinki.fi/data/
2. Agrawal, R., Imieliński, T., Swami, A.: Mining association rules between sets of items in large databases. In: SIGMOD 1993, pp. 207–216. ACM (1993)
3. Agrawal, R., Srikant, R.: Fast algorithms for mining association rules in large databases. In: VLDB 1994, pp. 487–499 (1994)
4. Brin, S., Motwani, R., Silverstein, C.: Beyond market baskets: generalizing association rules to correlations. SIGMOD Rec. 26(2), 265–276 (1997)
5. Han, J., Cheng, H., Xin, D., Yan, X.: Frequent pattern mining: Current status and future directions. Data Mining and Knowledge Discovery 14(1) (2007)
6. Han, J., Pei, J., Yin, Y., Mao, R.: Mining frequent patterns without candidate generation: A frequent-pattern tree approach. Data Min. Knowl. Discov. 8(1), 53–87 (2004)
7. Kim, S., Barsky, M., Han, J.: Efficient Mining of Top Correlated Patterns Based on Null-Invariant Measures. In: Gunopulos, D., Hofmann, T., Malerba, D., Vazirgiannis, M. (eds.) ECML PKDD 2011, Part II. LNCS, vol. 6912, pp. 177–192. Springer, Heidelberg (2011)
8. Lee, Y.K., Kim, W.Y., Cai, D., Han, J.: Comine: efficient mining of correlated patterns, pp. 581–584 (November 2003)
9. Omiecinski, E.R.: Alternative interest measures for mining associations in databases. IEEE Trans. on Knowl. and Data Eng. 15(1), 57–69 (2003)
10. Sim, K., Phua, C., Yap, G., Biswas, J., Mokhtari, M.: Activity recognition using correlated pattern mining for people with dementia. In: Conf. Proc. IEEE Eng. Med. Biol. Soc. (2011)
11. Tan, P.-N., Kumar, V., Srivastava, J.: Selecting the right interestingness measure for association patterns. In: KDD 2002, pp. 32–41. ACM, New York (2002)
12. Wu, T., Chen, Y., Han, J.: Re-examination of interestingness measures in pattern mining: a unified framework. Data Min. Knowl. Discov. 21(3), 371–397 (2010)
13. Kim, W.-Y., Lee, Y.-K., Han, J.: CCMine: Efficient Mining of Confidence-Closed Correlated Patterns. In: Dai, H., Srikant, R., Zhang, C. (eds.) PAKDD 2004. LNCS (LNAI), vol. 3056, pp. 569–579. Springer, Heidelberg (2004)

On Checking Executable Conceptual Schema Validity by Testing

Albert Tort, Antoni Olivé, and Maria-Ribera Sancho

Universitat Politècnica de Catalunya – Barcelona Tech
{atort,olive,ribera}@essi.upc.edu

Abstract. Ensuring the semantic quality of a conceptual schema is a fundamental goal in conceptual modeling. Conceptual schema testing is an emerging approach that helps to achieve this goal. In this paper, we focus on "what to test" and, more specifically, on the properties that test sets of conceptual schemas should have. We propose and formally define a set of four adequacy criteria which can be automatically checked in order to ensure, by testing, the necessary conditions for schema validity (correctness and relevance). The proposed criteria are independent from the languages of the schema and of the testing program. The criteria have been implemented in a prototype of a test processor able to execute test sets. The criteria have been applied to the test sets of large conceptual schemas.

Keywords: Conceptual modeling, Validation, Semantic quality, Testing, Coverage, UML/OCL.

1   Introduction

According to the conceptual modeling quality framework proposed by [1], a conceptual schema of an information system has semantic quality when it is valid and complete. Validity means that the schema is correct and relevant. A conceptual schema is correct if the knowledge it defines is true for the domain, and it is relevant if the knowledge it defines is necessary for the system. Completeness means that the conceptual schema includes all relevant knowledge.

Ensuring that a conceptual schema has semantic quality is a fundamental goal for its validation. This goal can be achieved by checking that the knowledge the system requires to know, in order to perform its functions, is the same as the knowledge defined by the conceptual schema.

Some requirements engineering methodologies define the functions the system must perform by means of use cases. In these methodologies, checking the semantic quality of the conceptual schema can only be done with the participation of the system's stakeholders, using manual validation techniques such as inspections, desk-checks, walkthroughs or prototypes [2,3].

However, in modern requirements engineering methodologies, the functions the system must perform are defined by both use cases and concrete scenarios [2].

S.W. Liddle et al. (Eds.): DEXA 2012, Part I, LNCS 7446, pp. 249–264, 2012.

In these methodologies, checking the semantic quality of a conceptual schema can be automatically done by means of testing techniques. This is possible when three conditions are met: (1) the expected system's functions are captured by a set of concrete scenarios written in a formal language; (2) the conceptual schema is executable; and (3) there is a test processor that allows defining and executing test cases. When this happens, each scenario can be written as a test case. The test processor can execute it and check whether or not the conceptual schema makes the test case pass.

When the system's functions are captured by a complete and correct set of scenarios written as test cases, and their execution produces the expected results, then we can ascertain that the conceptual schema is complete. If the conceptual schema were not complete, then some test case would not succeed. We can also ascertain that the part of the conceptual schema involved in the execution of the scenarios is correct, again because otherwise some test case would fail. However, it may happen that the conceptual schema is not valid, because the schema could include knowledge that is not relevant for the scenarios, and this irrelevant knowledge could even be incorrect.

The main contribution of this paper is a set of four properties for checking the validity of conceptual schemas by testing. These properties are inspired by what is called adequacy criteria in the software testing field [4]. If a test set satisfies an adequacy criterion, then it is considered adequate to test the conceptual schema.

We show that each of these criteria is a necessary condition for conceptual schema validity. The criteria are independent of each other but taken together they ensure the relevance of the defined knowledge. They also have the unifying (and interesting) characteristic that they ensure the satisfiability of the entity types, relationship types, integrity constraints and domain event types defined in a schema, which is a necessary property for its correctness [5].

Moreover, the proposed criteria are independent from the conceptual schema language and from the testing language. We have implemented them in our test processor prototype [6] for test sets written in CSTL [7] and conceptual schemas written in UML/OCL. The proposed adequacy criteria have been applied in the testing of a condensed version of the *osCommerce* conceptual schema [8], a popular and widely used e-commerce system. Moreover, they have been used in the development of the complete conceptual schema of the popular *osTicket* system [9].

As far as we know, in the literature there have not been proposals of adequacy criteria for checking conceptual schema validity by testing. The work most related to ours is the study of desirable properties of conceptual schemas and the development of automated or semi-automated reasoning procedures for checking them (a representative set of recent papers is [10-16]). The most studied general properties are satisfiability and non-redundancy of integrity constraints, satisfiability of an entity or relationship type, and operation executability (or, in our terms, domain event occurrence). However, it is well known that the problem of reasoning with integrity constraints and derivation rules in its full generality is undecidable. Therefore, these procedures are restricted to certain kinds of constraints and derivation rules or domains, or they may not terminate in some circumstances [12]. In contrast, testing techniques can be applied to conceptual schemas with any kind of constraint or derivation rule and they determine the relevance of the defined knowledge according to the expected functionalities.

The structure of the paper is as follows. In the next section, we briefly review the main concepts and the notation used to define the conceptual schemas under test and the test sets. A conceptual schema of a civil registry domain is used as a running example throughout the paper. In Section 3 we present the four adequacy criteria. Section 4 comments on our implementation of the test criteria and reports two case study applications. Section 5 summarizes the conclusions and points out future work.

2   Basic Concepts and Notation

In this section, we briefly review the concepts and the notation we use to define the conceptual schemas. We also review the main characteristics of the Conceptual Schema Testing Language (CSTL) used in this paper in order to specify the example test sets. A complete definition of the CSTL language can be found in [7].

2.1   Conceptual Schema under Test

A conceptual schema consists of a structural (sub)schema and a behavioral (sub)schema. The structural schema consists of a taxonomy of entity types (a set of entity types with their generalization/specialization relationships and the taxonomic constraints), a set of relationship types (attributes and associations), the cardinality constraints and a set of other constraints formally defined in OCL [17].

We adopt UML/OCL as the conceptual modeling language, but the ideas presented here can also be applied to schemas in other languages [18,19]. Figure 1 shows the structural schema of a civil registry domain example that will be used throughout the paper. The civil registry records information about the birth and death of the people registered in municipalities. The marital status and the marriage relationships of the inhabitants are also maintained. The main purpose of civil registration systems is computing demographic information such as the population, the life expectancy, etc.

The conceptual schema of Fig. 1 includes the specification of an OCL integrity constraint (*Country::identifiesInhabitantsByCitizenId*) in order to ensure that the inhabitants of a country are identifiable by a unique citizen identifier.

Entity and relationship types may be base or derived. The population of the base entity and relationship types is explicitly represented in the Information Base (IB). If they are derived, there is a formal derivation rule in OCL that defines their population in terms of the population of other types. Figure 1 includes the derivation rules of five derived attributes (*DeadPerson::ageAtDeath*, *Municipality::population*, *Municipality::lifeExpectancy, Country::population* and *Country::lifeExpectancy*).

The behavioral schema consists of a set of event types. We take the view that an event can be modeled as a special kind of entity, which we call event entity [20]. An event entity is an instance of an event type. Event types have characteristics, constraints and an effect. The characteristics of an event are the set of relationships (attributes or associations) in which it participates. The constraints are the conditions that events must satisfy in order to occur. An event constraint involves the characteristics and the state of the IB before the event occurrence.

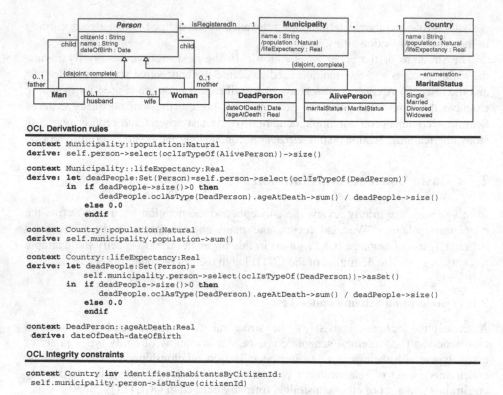

OCL Derivation rules

```
context Municipality::population:Natural
derive: self.person->select(oclIsTypeOf(AlivePerson))->size()

context Municipality::lifeExpectancy:Real
derive: let deadPeople:Set(Person)=self.person->select(oclIsTypeOf(DeadPerson))
        in if deadPeople->size()>0 then
             deadPeople.oclAsType(DeadPerson).ageAtDeath->sum() / deadPeople->size()
           else 0.0
           endif

context Country::population:Natural
derive: self.municipality.population->sum()

context Country::lifeExpectancy:Real
derive: let deadPeople:Set(Person)=
           self.municipality.person->select(oclIsTypeOf(DeadPerson))->asSet()
        in if deadPeople->size()>0 then
             deadPeople.oclAsType(DeadPerson).ageAtDeath->sum() / deadPeople->size()
           else 0.0
           endif

context DeadPerson::ageAtDeath:Real
  derive: dateOfDeath-dateOfBirth
```

OCL Integrity constraints

```
context Country inv identifiesInhabitantsByCitizenId:
  self.municipality.person->isUnique(citizenId)
```

Fig. 1. Structural schema fragment of the civil registry example

An event may occur in the state S of the IB if S satisfies all constraints and the event satisfies its event constraints. Each event type has an operation called *effect()* that gives the effect of an event occurrence. The effect is declaratively defined by the postcondition of the operation. We define both the event constraints and the postcondition in OCL.

For domain event types, the postcondition defines the state of the IB after the event occurrence. It is assumed that the state of the IB after the event occurrence also satisfies all constraints defined over the IB. We deal with executable conceptual schemas, and therefore we need a procedural specification of the method of the *effect()* operation. A method is correctly specified if the IB state after its execution satisfies the postcondition and the IB constraints. In the work reported here, we write those methods using a subset of the CSTL, although we envision the use of standard languages for writing actions in UML schemas, such as the recent *Action Language for Foundational UML (Alf)* [21] proposed by the OMG, as soon as they become mature and associated compilers are developed.

The example used throughout this paper considers the ordinary domain events *Birth*, *Death*, *Marriage* and *Divorce*. Events that create countries and municipalities of a country are also considered (*MunicipalityCreation* and *CountryCreation*).

Figure 2 shows the complete and formal specification in UML/OCL of the domain event *Marriage* including its initial integrity constraint (*marriageIsAuthorized*), its postcondition and the method of its *effect()* operation.

Event constraint

```
context Marriage::marriageIsAuthorized():Boolean
  body: self.husband.oclIsTypeOf(AlivePerson)and self.wife.oclIsTypeOf(AlivePerson)and
        self.husband.oclAsType(AlivePerson).maritalStatus <> MaritalStatus::Married and
        self.wife.oclAsType(AlivePerson).maritalStatus <> MaritalStatus::Married
```

«DomainEvent»
Marriage
effect()
«iniIC»marriageIsAuthorized()

husband 1   1 wife

Man	**Woman**

Event postcondition

```
context Marriage::effect()
  post:   self.husband.wife=self.wife and
          self.husband.oclAsType(AlivePerson).
          maritalStatus=MaritalStatus::Married and
          self.wife.oclAsType(AlivePerson).
          maritalStatus=MaritalStatus::Married
```

Event method

```
method Marriage::effect(){
self.husband.wife := self.wife;
self.husband.maritalStatus := MaritalStatus::Married;
self.wife.maritalStatus := MaritalStatus::Married;
}
```

Fig. 2. Marriage domain event specification

2.2   The CSTL Language

A CSTL program consists of a fixture (may be empty) and a set of one or more test cases. It is assumed that the execution of each test case of a CSTL program starts with an empty IB state. With this assumption, the test cases of a program are independent each other, and therefore the order of their execution is irrelevant. The fixture is a set of statements that create an IB state and define the values of the common program variables. The execution of a test case starts with the execution of the fixture.

The basic construct of CSTL is the concrete test case. Figure 3 shows a test program that consists of a fixture and two test cases.

The last statement of a concrete test case is an assertion, but in general there may be several assertions in the same test case. The verdict of a concrete test case is *Pass* if the verdict of all of its assertions is *Pass*. The objective of the conceptual modeler is to write test cases whose final verdict is *Pass*.

In CSTL there are five kinds of assertions, but in this paper only two are used: asserting the occurrence of domain events and asserting the contents of an IB state, which we briefly describe in the following.

In CSTL, an instance of an event type $EventType_1$ is created with the statement:

```
eventId := new EventType₁(att₁:= value₁,...,attₙ:= valueₙ,
      r₁:= participants₁,...,rₘ:= participantsₘ);
```

The statement creates the instance *eventId* of $EventType_1$, and assigns a value to its characteristics (attributes $att_1,...,att_n$ and binary links with roles $r_1,...,r_m$). Figure 3 shows several examples of statements that create an instance of a domain event type. Once the concrete event *eventId* has been created in a test case, in order to assert that it may occur in the current state of the IB the conceptual modeler writes:

```
assert occurrence eventId;
```

The verdict of this assertion is determined as follows:

- Check that the current IB state is consistent. The verdict is *Error* if that check fails (events may not occur in inconsistent IB states).
- Check that the constraints of the event are satisfied. The verdict is *Fail* if any of the event constraints is not satisfied.
- Execute the method of the corresponding *effect()* operation.

```
testprogram PeopleRegistration{

    belgiumCreation := new CountryCreation(name:='Belgium');
    assert occurrence belgiumCreation;
    belgium := belgiumCreation.createdCountry;
    brusselsCreation := new MunicipalityCreation(name:='City of Brussels', country:=belgium);
    assert occurrence brusselsCreation;
    brussels := brusselsCreation.createdMunicipality;
    antwerpCreation := new MunicipalityCreation(name:='Antwerp', country:=belgium);
    assert occurrence antwerpCreation;
    antwerp := antwerpCreation.createdMunicipality;

    test familyWithoutChildren{
        audreyBirth := new Birth(citizenId:='AUU', name:='Audrey', sex:=Sex::Woman,
            dateOfBirth:='10-10-1934', municipality:=brussels);
        assert occurrence audreyBirth;
        audrey:= audreyBirth.createdPerson;
        assert true audrey.oclAsType(AlivePerson).maritalStatus = MaritalStatus::Single;
        alexBirth := new Birth(citizenId:='ALL', name:='Alex', sex:=Sex::Man,
                               dateOfBirth:='02-31-1936', municipality:= antwerp);
        assert occurrence alexBirth;
        alex:= alexBirth.createdPerson;
        assert true alex.oclAsType(AlivePerson).maritalStatus = MaritalStatus::Single;
        assert equals brussels.population 1;
        assert equals antwerp.population 1;
        assert equals belgium.population 2;
        m := new Marriage(husband:=alex, wife:=audrey);
        assert occurrence m;
        assert true alex.oclAsType(AlivePerson).maritalStatus = MaritalStatus::Married;
        assert true audrey.oclAsType(AlivePerson).maritalStatus = MaritalStatus::Married;
        alexDeath := new Death(person:=alex, dateOfDeath:='06-11-2003');
        assert occurrence alexDeath;
        assert equals belgium.population 1;
        assert true audrey.oclAsType(AlivePerson).maritalStatus = MaritalStatus::Widowed;
    }

    test familyWithADaughter{
        vincentBirth := new Birth(citizenId:='VVV', name:='Vincent', sex:=Sex::Man,
                                  dateOfBirth='01-01-1918', municipality:= brussels);
        assert occurrence vincentBirth;
        vincent:= vincentBirth.createdPerson;
        emmaBirth := new Birth(citizenId:='EEE', name:='Emma', sex:=Sex::Woman,
                               dateOfBirth='01-01-1922', municipality:= brussels);
        assert occurrence emmaBirth;
        emma:= emmaBirth.createdPerson;
        m := new Marriage(husband:=vincent, wife:=emma);
        assert occurrence m;
        julieBirth := new Birth(citizenId:='JJJ', name:='Julie', sex:=Sex::Woman,
                                dateOfBirth:='01-01-1953',
                                father:=vincent, mother:=emma, municipality:= brussels);
        assert occurrence julieBirth;
        div := new Divorce(husband:=vincent, wife:=emma);
        assert occurrence div;
        vincentDeath := new Death(person:=vincent, dateOfDeath:='01-01-1996');
        assert occurrence vincentDeath;
        emmaDeath := new Death(person:=emma, dateOfDeath:='01-01-2007');
        assert occurrence emmaDeath;
        assert equals brussels.lifeExpectancy 81.5;
        assert equals belgium.lifeExpectancy 81.5;
    }
}
```

Fig. 3. CSTL program about people registration in the civil registry example

- Check that the new IB state is consistent. The verdict is *Fail* if any of the constraints is not satisfied.
- Check that the event postcondition is satisfied. The verdict is *Fail* if the postcondition is not satisfied; otherwise the verdict of the whole assertion is *Pass*.

It is often useful to include in a test case an assertion on the current state of the IB. Its purpose may be to check that derivation rules, navigational expressions or domain events behave as expected. In CSTL, to assert that the current state of the IB satisfies a boolean condition defined in OCL, the conceptual modeler writes:

```
assert true booleanExpression;
```

where *booleanExpression* is an OCL expression over the types of the IB and the variables of the test case. The verdict of the assertion is *Error* if the current state is inconsistent. The verdict is *Pass* if *booleanExpression* is true, and *Fail* otherwise. Additionally, CSTL includes similar assertions such as *assert false, assert equals, assert not equals*, etc.

The test program of Fig. 3 contains several assertions about the state of the IB. The statement **assert equals** brussels.lifeExpectancy 81.5, for example, asserts that the life expectancy of the City of Brussels is 81.5. The verdict of this assertion is *Pass* if the conceptual schema derives the life expectancy as expected. The life expectancy in this example is expected to be the average of the ages at death (measured in years) of all the registered dead people within the municipality.

3   Test Adequacy Criteria

The main purpose of conceptual schema testing is exercising the schema in order to trigger failures [22]. In the previous section, we have seen how to test an executable conceptual schema *CS* by writing a set of tests *TS* and making them pass. However, not all possible test sets are equally adequate in order to increase the confidence about the semantic quality of the conceptual schema.

For example, given a conceptual schema *CS* and having a test set *TS* whose verdict is *Pass* is not sufficient to check the schema validity. The reason is that the *CS* may be incomplete or may contain elements whose relevance has not been proved by *TS*.

Then, the following question arises: Which are the basic properties that these tests should have? The key concept developed in the software testing field for this purpose is that of adequacy criterion [4]. A typical example, in program testing, is the criterion that requires that each statement of a program is executed at least once by a test set. Of course, many other criteria are possible. In the context of conceptual schema testing, we can say that an adequacy criterion *C* is a requirement on a test set *TS* of a conceptual schema *CS* such that if *TS* satisfies *C* then *TS* is considered adequate to test *CS* according to *C*.

In this section, we present the main contribution of this paper: a basic set of four adequacy criteria for checking the validity of conceptual schemas by testing. The overall goal of this set is threefold: 1) determining which parts of the conceptual schema have been exercised by a test case, 2) determining which elements of the schema are potentially irrelevant (or even incorrect) and 3) ensuring the satisfiability of the entity types, relationship types, integrity constraints and domain event types, which is a necessary property for correctness.

If the four proposed criteria are simultaneously satisfied, then all the defined elements of the schema are relevant and satisfiable.

In the following, we formally define the four criteria. We denote by TS a test set that consists of a set of one or more test cases TC_i. The execution of a test case implies the execution of one or more test assertions TA_k. TA denotes the set of all the test assertions whose verdict is *Pass*.

3.1   Base Type Coverage

The base types (entity types, attributes and associations) defined in a conceptual schema are valid if they are relevant and correct [1,2,20]. We denote by T_{base} the set of base types. The relevance of each base type $T_i \in T_{base}$ can be ensured by means of testing. The test set TS should include at least one test case TC_j such that it:

- builds a state of the IB having at least one instance of T_i, and
- makes an assertion TA_k that can only *Pass* if the above IB state is consistent (that is, it satisfies all constraints).

If the test set includes such test case TC_j, and the execution of TA_k gives the verdict *Pass*, then it is experimentally proved that T_i is relevant according to the expectations formalized as test cases.

This is the rationale for the test adequacy criterion that we call *base type coverage*, which can be formally stated as follows. Let:

$BaseTypes(TA_k) = \{ T_i | T_i \in T_{base}$ and there are one or more instances of T_i in at least one of the IB states found consistent during the evaluation of $TA_k\}$

$$BaseTypes(TA) = \bigcup_{TA_k \in TA} BaseTypes(TA_k).$$

We say that a test set TS satisfies the *base type coverage* criterion if and only if $T_{base} = BaseTypes(TA)$. Then, it is experimentally proved that all types $T_i \in T_{base}$ defined in a conceptual schema are relevant. It is important to remark, that the accomplishment of this criterion has the interesting property of ensuring the satisfiability of T_{base} (which is a necessary condition for the correctness of T_{base}).

The analysis of the set of uncovered base types ($T_{base} - BaseTypes(TA)$) allows us to identify which base types of the schema have not been exercised in any consistent scenario. Either they need more testing in order to satisfy the *base type coverage* criterion or they are irrelevant or incorrect.

In the test program of Fig. 3, the fixture initializes two municipalities (the *City of Brussels* and *Antwerp*) located in a country (*Belgium*). In a test program, the execution of any of its test cases implies the execution of its fixture. Therefore, the entity types *Municipality* and *Country*, its basic attributes and the relationship type between them are covered according to this criterion.

Moreover, the test case *familyWithoutChildren* registers the births of a woman (*audrey*) and a man (*alex*), the marriage between them and the death of *alex*. The execution of this test case implies that the entity types *Man, Woman, AlivePerson, DeadPerson* (and *Person* due to the taxonomy), their basic attributes, and the relationship types *IsRegisteredIn* and *husband-wife* are also covered.

However, by taking into account only the test case *familyWithoutChildren*, the coverage analysis identifies that the relationship types *father-child* and *mother-child* are not covered. If we consider that the test set also includes the test case *familyWithADaughter*, then all the basic types of the example become covered.

3.2   Derived Type Coverage

Entity and relationship types may be derived. For each derived type, the conceptual schema includes a derivation rule that defines the population of that type in terms of the population of other types. In UML, derivation rules are formally written in OCL. Derived types defined in a schema are valid if they are relevant and correct [1,20].

We denote by T_{der} the set of derived types defined in a conceptual schema. The relevance of a derived type can be checked by means of testing. The test set TS should include a test case that makes an assertion TA_k whose evaluation requires the derivation of at least one instance of that type.

We denote by $DerTypes(TA_k)$ the set of derived entity types such that TA_k has derived one or more instances of them during its evaluation, and by $DerTypes(TA)$ the set of derived entity types that have instances derived during the evaluation of TA. Formally:

$DerTypes(TA_k) = \{T_i \mid T_i \in T_{der}$ and the evaluation of TA_k in a state found consistent has required the derivation of one or more instances of $T_i\}$

$$DerTypes(TA) = \bigcup_{TA_k \in TA} DerTypes(TA_k)$$

We say that a test set TS satisfies the *derived type coverage* criterion if and only if $T_{der} = DerTypes(TA)$. Then, it is experimentally proved that all types $T_i \in T_{der}$ defined in a conceptual schema are relevant. Moreover, the accomplishment of this criterion has the interesting property of ensuring the satisfiability of T_{der} (which is, in turn, a necessary condition for the correctness of T_{der}).

The analysis of the set of uncovered derived types ($T_{der} - DerTypes(TA)$) allows us to identify which derived types have not been exercised in any consistent scenario. Either they need more testing in order to satisfy the *derived type coverage* criterion or they are irrelevant or incorrect.

The first test case (*familyWithoutChildren*) of the test program shown in Fig. 3, makes assertions about the population of the municipalities and the country initialized in the fixture.

```
assert equals brussels.population 1;
assert equals antwerp.population 1;
assert equals belgium.population 2;
```

The verdict of these assertions is *Pass* because the conceptual schema has the knowledge to derive the population as expected. Consequently, this test case ensures that the derived attributes *Municipality::population* and *Country::population* have been correctly derived in a consistent state.

In contrast, the derived attributes *Municipality::lifeExpectancy*, *Country::lifeExpectancy* and *DeadPerson::ageAtDeath* are not covered if we only consider the test case *familyWithoutChildren*. However, if we add the test case *familyWithADaughter* all derived types become covered.

3.3   Valid Type Configuration Coverage

In conceptual models that admit multiple classification (like the UML), an entity may be an instance of two entity types, E_1 and E_2, such that (1) E_1 does not subsume E_2; (2) E_2 does not subsume E_1; and (3) no E_3 is subsumed by both E_1 and E_2. In multiple-classification models, correctness and relevance do not only apply to the individual entity types, but also to the set of valid configurations of entity types [20]. These configurations are completely determined by the entity types and the taxonomic constraints of the conceptual schema.

The example of Fig. 1 assumes multiple classification. There are six valid type configurations: {*Person, Man, AlivePerson*}, {*Person, Woman, AlivePerson*}, {*Person, Man, DeadPerson*}, {*Person, Woman, DeadPerson*}, {*Municipality*} and {*Country*}.

The relevance of a valid type configuration $VTC_i = \{E_1, ..., E_n\}$ can be checked by means of testing. The test set *TS* should include a test case TC_j such that it:

- builds a state of the IB having at least one entity that is an instance of VTC_i, and
- makes an assertion TA_k that can only *Pass* if the above IB state is consistent.

Therefore, if we want to experimentally prove that all valid type configurations $VTC_i \in VTC$ defined in a conceptual schema are relevant, we must require that for each of them there is at least one test assertion that checks the consistency of one or more IB states having at least one instance of VTC_i.

This is the rationale for the test adequacy criterion that we call *valid type configuration coverage*, which can be formally stated as follows.

Let $VTC(TA_k) = \{VTC_i \mid VTC_i \in VTC$ and there are one or more instances of VTC_i in at least one of the IB states found consistent during the evaluation of TA_k }

$$VTC(TA) = \bigcup_{TA_k \in TA} VTC(TA_k)$$

We say that a test set *TS* satisfies the *valid type configuration coverage* criterion if and only if $VTC = VTC(TA)$. Then, it is experimentally proved that all types $VTC_i \in VTC$ defined in a conceptual schema are relevant. Additionally, the accomplishment of this criterion has the interesting property of ensuring the satisfiability of *VTC* (which is, in turn, a necessary condition for the correctness of *VTC*) [20].

When $VTC \neq VTC(TA)$ then there is at least one $VTC_i \in VTC$ but $VTC_i \notin VTC(TA)$. The analysis of the set of uncovered valid type configurations ($VTC - VTC(TA)$)

allows us to identify which type configurations may not be valid. This means that a valid type configuration VTC_i allowed by the conceptual schema has not been tested. If the domain experts confirm that VTC_i is valid in the domain, then the conceptual modeler must write more test cases. Otherwise, if VTC_i is invalid in the domain, then the conceptual modeler must change the taxonomy to prevent it.

The test case *familyWithoutChildren* proves that all the entity types are covered according to the *Base Type Coverage* (see Section 3.1). However, the CSUT example (Fig. 1) assumes multiple classification. Therefore, when analyzing the *Valid Type Configuration Coverage* satisfaction, we realize that {*Person, Woman, DeadPerson*} is not covered (no valid instances of a dead woman participate in the test case). If we also consider the test case *familyWithADaughter*, then all VTCs become covered.

In single-classification schemas, the satisfaction of the *base type coverage* criterion implies the satisfaction of the *valid type configuration coverage* criterion.

3.4   Domain Event Type Coverage

Domain event types must be relevant and correct [20]. We denote by *Dev* the set of domain event types. The relevance of a domain event type $Dev_i \in Dev$ can be checked by means of testing. The test set *TS* should include a test case TC_j such that it 1) builds a state of the IB, 2) creates an instance d of Dev_i, and 3) asserts the occurrence of d.

If the test set includes such test case TC_j, and its execution gives the verdict *Pass*, then it is experimentally proved that Dev_i is relevant. If Dev_i is not relevant, then the test set should not include any assertion stating the occurrence of Dev_i.

This is the rationale for the test adequacy criterion that we call *domain event type coverage*, which can be formally stated as follows. Let TA_k be the assertion of a domain event occurrence. We denote by $DevTypes(TA_k)$ the type of the domain event whose occurrence is asserted and by $DevTypes(TA)$ the set of domain event types that have instances whose occurrence has been asserted during the evaluation of *TA*. Formally:

$$DevTypes(\text{TA}) = \bigcup_{TA_k \in TA} DevTypes(TA_k)$$

We say that a test set *TS* satisfies the *domain event type coverage* criterion if and only if $Dev = DevTypes(TA)$. Then, it is experimentally proved that all types $Dev_i \in Dev$ defined in a conceptual schema are relevant. Again, note that the accomplishment of this criterion has also the interesting property of ensuring the satisfiability of *Dev* (which is, in turn, a necessary condition for the correctness of *Dev*). Satisfiability comprises applicability (the initial IB state has been found consistent and the event constraints have been satisfied) and executability (the new IB state has been found consistent and the event postconditions have been satisfied).

The set of uncovered event types ($Dev - DevTypes(TA)$) allows us to identify which event types do not have valid occurrences in any test case. Either they need more testing to satisfy the criterion or they are irrelevant (or even incorrect).

The example test case *familyWithoutChildren* (Figure 3) exercises the valid execution of all the domain events considered in the example (see Section 2.1), with the exception of the event *Divorce* that becomes covered if we also consider the test case *familyWithADaughter*. The satisfaction of the domain event type coverage criterion ensures that these domain events are satisfiable.

The structural and the behavioral subschema should be consistent [23,24] between them. If the *domain event type coverage* criterion is satisfied but the *base type coverage* criterion is not satisfied, then either the uncovered base types are not relevant, or some event types are missing in the schema, or the existing ones must be instantiated in other test cases.

3.5   Coverage Criteria Satisfaction and Schema Validity

If there exists a test set *TS* that satisfies the four coverage criteria defined in sections 3.1, 3.2, 3.3 and 3.4, then we can ensure that all base and derived types, type configurations and domain event types are relevant and satisfiable, which is a necessary condition for correctness.

Formally, if we denote the relevance by *Rel* and the satisfiability by *Sat*, then:

Base type coverage:	$T_{base} = BaseTypes(TA) \rightarrow Rel(T_{base},TS) \land Sat(T_{base})$
Derived type coverage:	$T_{der} = DerTypes(TA) \rightarrow Rel(T_{der},TS) \land Sat(T_{der})$
Valid type configuration coverage:	$VTC = VTC(TA) \rightarrow Rel(VTC,TS) \land Sat(VTC)$
Domain event type coverage:	$Dev = DevTypes(TA) \rightarrow Rel(Dev,TS) \land Sat(Dev)$

The test program *PeopleRegistration* of Fig. 3 completely satisfies the proposed basic set of test adequacy criteria. Therefore, we can conclude that all the schema elements have been exercised by a test case in at least one consistent scenario, that there are no potentially irrelevant elements, and that these elements are satisfiable. Additionally, detailed steps of the coverage checking process for a condensed version of the *osCommerce* case study are reported in [8].

4   Implementation

Figure 4 shows the relationships between the definition of the conceptual schema and its tests, and their execution by the *information processor* and the *test processor*. Details about the implementation of the tool that supports this testing environment are given in [6]. We have extended our *test processor* in order to provide automatic coverage analysis (Section 4.1.). We have also applied the proposed criteria in two case studies of real-sized information systems (section 4.2).

4.1   The Coverage Processor

The *Coverage Processor* (grey box of the test processor in Fig. 4) automatically checks the satisfaction of test adequacy criteria proposed in this paper.

The *preprocessor* initializes the *coverage database*. This database maintains the set of covered elements for each test adequacy criterion. For each adequacy criterion, the *preprocessor* requests to the *information processor* the set of elements to be covered. Each element is registered in the database and marked as uncovered.

When the execution of a collection of test cases is requested, the *test manager* selects the test programs and requests to the *test interpreter* their execution. The *test interpreter* communicates information about the tests execution to the *adequacy criteria analyzer*, which updates the *coverage database* as follows:

- Every time the *test interpreter* asserts the consistency of the IB, it communicates to the *analyzer* the set *VTC(TA)* of valid type configurations and the set *BaseTypes(TA)* that have valid instances in the current state of the IB. The *analyzer* marks as covered all the uncovered valid type configurations included in *VTC(TA)*. It also marks as covered the entity types included in *BaseTypes(TA)*.
- Every time the test interpreter requests the evaluation of a derivation rule to the *information processor*, the *test interpreter* communicates it to the *analyzer*. The corresponding derived type is marked as covered. A derivation rule may be implicitly evaluated when an integrity constraint is checked, when evaluating other derivation rules, when asserting the contents of the IB or when executing an event.

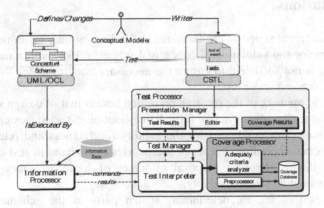

Fig. 4. The coverage processor in the testing environment

- The *test interpreter* informs the *analyzer* every time a valid domain event occurs The asserted domain event types are marked as covered.

After the execution of all test programs, the *adequacy criteria analyzer* queries the *coverage database* in order to obtain the sets of covered and uncovered elements for each criterion. The *analyzer* also computes some statistical information about the coverage results. This information is used by the *coverage results* to show the results. The set of uncovered elements helps the tester to define new relevant test cases.

More information, source files, screenshots and a video demonstration about the use of the CSTL processor for analyzing the coverage of the conceptual schema example explained in this paper may be found in the website of the project [25].

4.2   Case Studies

The basic set of adequacy criteria has been checked in the testing that has been done during the development of (1) a condensed version of the conceptual schema of an e-commerce system (*osCommerce*) [8]; and (2) the complete conceptual schema of a real-sized and widely-used customer support system (*osTicket*) [9].

Fig. 5 summarizes the properties of the conceptual schema of each case study, the executed test cases, and its computation time (including test case execution and coverage analysis) by using a 3.40 GHz processor.

| Case Study | Conceptual Schema Under Test | | | | | | Test Set | | Time |
	Classes	Attributes	Associations	Event types	Derived types	Constraints	Test Cases	CSTL lines	Test execution & coverage
osCommerce	24	49	17	16	19	23	7	185	0.6 sec
osTicket	28	92	44	24	3	54	101	2002	4.3 sec

Fig. 5. Case studies summary

5   Conclusions

Executable conceptual schemas can be tested [6,7]. Testing may be an important and practical means for the validation of conceptual schemas. However, a test set whose verdict is *Pass* is not sufficient to ensure the necessary conditions for the validity of the conceptual schema under test.

The overall framework of the research presented here is that of design science [26]. The problem we try to solve is that of formalizing a set of test adequacy criteria aimed to ensure the necessary conditions for the validity (correctness and relevance) of a conceptual schema, according to a set of user stories formalized as test cases. As far as we know, this is the first proposal of a set of basic adequacy criteria for ascertaining conceptual schema validity.

The goals of this set are determining which parts of the schema have been exercised by a test case; analyzing which elements of the schema are potentially irrelevant (or even incorrect); and ensuring the satisfiability of the entity types, relationship types, integrity constraints and domain event types, which is a necessary property for correctness.

These basic criteria are independent from the conceptual schema and the testing languages and can be automatically checked. We have shown an implementation of the four adequacy criteria in a prototype test processor, which handles the testing of conceptual schemas written in UML/OCL with test programs written in the CSTL language. The criteria have been illustrated by a running example, but we also refer the evaluation of its utility in the incremental definition of the schema of the *osTicket* system [9] and a comprehensive application in the *osCommerce* case study [8].

We have defined the four criteria for conceptual schemas that include both the structural and the behavioral subschemas. However, it is possible to define a variant of these criteria which is applicable when only the structural subschema is available. The idea in this context is that the test cases do not build the IB states by means of the occurrence of domain events, but by means of explicit insertion, deletion and update (CSTL) statements. This variant is useful in projects that aim at developing only the structural schema, or in the initial phases of the development of a complete schema.

All test sets should satisfy the proposed criteria given that they ensure the necessary conditions for conceptual schema validity. However, several additional criteria may be envisaged in order to enhance the confidence about the conceptual schema correctness. As further work, we mention two of them here. The first is similar to the branch coverage criterion in program testing [4], aiming at ensuring that all branches of the OCL integrity constraints have been tested. The second is a criterion that ensures that all integrity constraints that must be enforced by the system have at least one domain event precondition that prevents the occurrence of a domain event that could lead to its violation.

Acknowledgments. This work has been partly supported by the Ministerio de Ciencia y Tecnología and FEDER under the project TIN2008-00444, Grupo Consolidado.

References

1. Lindland, O.I., Sindre, G., Solvberg, A.: Understanding Quality in Conceptual Modeling. IEEE Software 11(2), 42–49 (1994)
2. Pohl, K.: Requirements Engineering. Fundamentals, principles, and techniques. Springer, Berlin (2010)
3. Van Lamsweerde, A.: Requirements Engineering: From System Goals to UML Models to Software Specifications. Wiley (2009)
4. Zhu, H., Hall, P.A.V., May, J.H.R.: Software unit test coverage and adequacy. ACM Computing Surveys 29(4), 366–427 (1997)
5. Thalheim, B.: Entity-Relationship Modeling: Foundations of Database Technology. Springer (2000)
6. Tort, A., Olivé, A., Sancho, M.-R.: The CSTL Processor: A Tool for Automated Conceptual Schema Testing. In: De Troyer, O., Bauzer Medeiros, C., Billen, R., Hallot, P., Simitsis, A., Van Mingroot, H. (eds.) ER Workshops 2011. LNCS, vol. 6999, pp. 349–352. Springer, Heidelberg (2011)
7. Tort, A., Olivé, A.: An approach to testing conceptual schemas. Data Knowl. Eng. 69(6), 598–618 (2010)
8. Tort, A.: A basic set of test cases for a fragment of the osCommerce conceptual schema. Research Report UPC (2009), http://hdl.handle.net/2117/6130
9. Tort, A.: Development of the conceptual schema of the osTicket system by applying TDCM. Research Report UPC (2011), http://hdl.handle.net/2117/12369
10. Berardi, D., Calvanese, D., De Giacomo, G.: Reasoning on UML class diagrams. Artificial Intelligence 168(1-2), 70–118 (2005)
11. Brambilla, M., Tziviskou, C.: An Online Platform for Semantic Validation of UML Models. In: Gaedke, M., Grossniklaus, M., Díaz, O. (eds.) ICWE 2009. LNCS, vol. 5648, pp. 477–480. Springer, Heidelberg (2009)

12. Queralt, A., Teniente, E.: Reasoning on UML Conceptual Schemas with Operations. In: van Eck, P., Gordijn, J., Wieringa, R. (eds.) CAiSE 2009. LNCS, vol. 5565, pp. 47–62. Springer, Heidelberg (2009)
13. Gogolla, M., Kuhlmann, M., Hamann, L.: Consistency, Independence and Consequences in UML and OCL Models. In: Dubois, C. (ed.) TAP 2009. LNCS, vol. 5668, pp. 90–104. Springer, Heidelberg (2009)
14. Kalyanpur, A., Parsia, B., Sirin, E., Hendler, J.: Debugging unsatisfiable classes in OWL ontologies. Web Semantics: Science, Services and Agents on the World Wide Web 3(4), 268–293 (2005)
15. Jarrar, M.: Towards Automated Reasoning on ORM Schemes. In: Parent, C., Schewe, K.-D., Storey, V.C., Thalheim, B. (eds.) ER 2007. LNCS, vol. 4801, pp. 181–197. Springer, Heidelberg (2007)
16. Gogolla, M., Bohling, J., Richters, M.: Validating UML and OCL Models in USE by Automatic Snapshot Generation. Software & Systems Modeling 4(4), 386–398 (2005)
17. Object Management Group (OMG). Object Constraint Language Specification. Version 2.2., formal/2010-02-01, http://www.omg.org/spec/OCL/2.2/
18. Pastor, O., Molina, J.C.: Model-Driven Architecture in Practice. Springer (2007)
19. Halpin, T.A.: Information modeling and relational databases. Morgan Kaufmann (2001)
20. Olivé, A.: Conceptual Modeling of Information Systems. Springer, Berlin (2007)
21. Object Management Group (OMG). Action Language for Foundational UML (Alf). FTF-Beta 1, ptc/2010-10-05, http://www.omg.org/spec/ALF/1.0/Beta1/
22. Meyer, B.: Seven Principles of Software Testing. IEEE Computer 41(8), 99–101 (2008)
23. Salay, R., Mylopoulos, J.: Improving Model Quality Using Diagram Coverage Criteria. In: van Eck, P., Gordijn, J., Wieringa, R. (eds.) CAiSE 2009. LNCS, vol. 5565, pp. 186–200. Springer, Heidelberg (2009)
24. Pilskalns, O., Andrews, A., Knight, A., Ghosh, S., France, R.: Testing UML designs. Information and Software Technology 49(8), 892–912 (2007)
25. Tort, A.: The CSTL Processor website, http://www.essi.upc.edu/~atort/cstlprocessor
26. Hevner, A.R., March, S.T., Park, J., Ram, S.: Design science in information systems research. MIS Quarterly 28(1), 75–105 (2004)

Querying Transaction–Time Databases under Branched Schema Evolution

Wenyu Huo and Vassilis J. Tsotras

Department of Computer Science and Engineering,
University of California, Riverside, USA
{whuo,tsotras}@cs.ucr.edu

Abstract. Transaction-time databases have been proposed for storing and querying the history of a database. While past work concentrated on managing the
data evolution assuming a static schema, recent research has considered data
changes under a linearly evolving schema. An ordered sequence of schema versions is maintained and the database can restore/query its data under the appropriate past schema. There are however many applications leading to a *branched*
schema evolution where data can evolve in parallel, under different concurrent
schemas. In this work, we consider the issues involved in managing the history
of a database that follows a branched schema evolution. To maintain easy
access to any past schema, we use an XML-based approach with an optimized
sharing strategy. As for accessing the data, we explore branched temporal indexing techniques and present efficient algorithms for evaluating two important
queries made possible by our novel branching environment: the vertical historical query and the horizontal historical query. Moreover, we show that our methods can support branched schema evolution which allows version *merging*.
Experimental evaluations show the efficiency of our storing, indexing, and
query processing methodologies.

1 Introduction

Due to the collaborative nature of web applications, information systems experience
evolution not only on their data content but also under different schema versions. For
example, Wikipedia has experienced more than 170 schema changes in its 4.5 years
of lifetime [5]. Schema evolution has been addressed for traditional (single-state)
database systems and issues on how data is efficiently transferred to the latest schema
have been examined [4]. Consider however the case where the application maintains
its past data (typically for archiving, auditing reasons etc.) which may have followed
different schemas. A temporal database can be facilitated to manage the historical
data, but issues related to how data can be queried under different schemas arise. The
pioneering work in PRIMA system [8] addresses the issues of maintaining a transaction-time database under schema evolution by introducing: (i) an XML-based model
for archiving historical data with evolving schemas, (ii) a language of atomic schema
modification operators (SMOs), and (iii) query answering and rewriting algorithms
for complex temporal queries spanning over multiple schema versions. Nevertheless,

S.W. Liddle et al. (Eds.): DEXA 2012, Part I, LNCS 7446, pp. 265–280, 2012.

PRIMA considers only a linear evolution: a new schema is derived from the latest schema and at each time there is only one current schema.

In many applications, however, the schema may change in a more complex way. For instance, in a collaborative design environment, an initial schema may be branched into a number of parallel schemas whose data can evolve concurrently. Another common case of non-linear evolution is in software development management. Revision control enables the modifications and developments happening in parallel along multiple branches. The release history of Mozilla Firefox [1] shows that 10 branches of versions have been developed and 4 more branches are on the way.

In this paper we address the issues involved in archiving, managing and querying a branched schema evolution. In particular, we maintain the branched schema versions in an XML-based document (*BMV-document*) using schema sharing. This choice was made because the number of schema changes is relatively smaller than data changes and the hierarchal structure of XML allows for easy schema querying. The data level changes are stored in column-like tables (*BC-Tables*), one table for each temporal attribute, with the support of applicable temporal indexing. To the best of our knowledge, this is the first work to examine both data and schema evolution in a branched environment. Our contributions can be summarized as:

1. We utilize a *sharing* strategy with *lazy-mark* updating, to save space and update time when maintaining the schema branching.
2. We employ branched temporal indexing structures and link-based algorithms to improve temporal query processing over the data. Moreover, we propose various *optimizations* for two novel temporal queries involving multiple branches, the *vertical* and *horizontal* queries.
3. We further examine how to support *version merging* within the branched schema evolution environment.
4. Our experiments show the space effectiveness of our sharing strategy while the optimized query processing algorithms achieve great data access efficiency.

The rest of the paper is organized as follows. Section 2 summarizes work on linear schema evolution (PRIMA). Section 3 introduces branched schema evolution while section 4 presents the BMV-Document for storing schema versions and the BC-Tables for storing the underlying data changes (with the support of branched temporal indexing). Section 5 provides algorithms and optimizations for efficient processing of temporal queries. The merging challenges are discussed in section 6 and the experimental evaluations are presented in section 7. Finally, conclusions appear in section 8.

2   Preliminaries

2.1   A Linear Evolution Example

Consider the linear schema evolution shown in Table 1 and Fig. 1(a), of an employee database, which is used as the basic running example in this paper. When the database was first created at T_1, using schema version $V_{1.1}$, it contains three tables: **engineer-personnel**, **otherpersonnel** and **job**. As the company seeks to uniformly manage the personnel information, the DBA applies first schema modification at T_2, which merges two tables **engineerpersonnel** and **otherpersonnel**, producing schema $V_{1.2}$. Each schema version is valid for all times between its start-time T_s and its end-time T_e (the time it was updated to a new schema). The rest schema versions and their respective time intervals appear as well until the latest schema $V_{1.5}$. A special value "now" is used to represent the always increasing current time.

Table 1. A linearly evolving employee database

VID	Schema Versions	T_s	T_e
$V_{1.1}$	engineerpersonnel (id, name, title, deptname) otherpersonnel (id, name, title, deptname) job (title, salary)	T_1	T_2
$V_{1.2}$	employee (id, name, title, deptname) job (title, salary)	T_2	T_3
$V_{1.3}$	employee (id, name, title, deptno) job (title, salary) dept (deptno, deptname, managerid)	T_3	T_4
$V_{1.4}$	employee (id, title, deptno) job (title, salary) dept (deptno, deptname, managerid) empbio (id, name, sex)	T_4	T_5
$V_{1.5}$	employee (id, title, deptno, salary) dept (deptno, deptname, managerid) empbio (id, name, sex)	T_5	now

Schema changes are represented by Schema Modification Operators (SMOs) [4]; each operator performs an atomic action on both the schema and the underlying data, like CREATE/MERGE/PARTITION TABLE, ADD/DROP/RENAME COLUMN. For example, two tables in $V_{1.1}$ were merged to one table by a MERGE TABLE operation in $V_{1.2}$. In the following discussion we will use the term SMO to denote a change operator applied to one schema without detailing which SMO was actually used.

2.2   XML Representation of a Linear Schema Evolution

The history of the relational database content and its schema evolution can be published in the form of XML, and viewed under a temporally grouped representation whereby complex temporal queries can be easily expressed in standard XQuery [8, 9]. The MV-Document [8] intuitively represents both schema versions and data tuples using XPath notation, as: **/db/table-name/row/column-name**. Each of the nodes, representing respectively: databases, tables, tuples, and attributes, has two more attributes, start-time (ts) and end-time (te), respectively representing the (transaction-) time in which the element was added to and removed from the database.

Consider our running example: when the three-table schema in version $V_{1.1}$ was created, three table nodes with names **engineerpersonnel**, **otherpersonnel** and **job** were created in the MV-Document, each with interval [T_1, "now"). Similarly, the nodes for their attributes etc., were added in the XML document. In $V_{1.2}$ the schema evolved into the two tables **employee** and **job**; these changes were updated in the MV-Document by changing the end-time of **engineerpersonnel** and **otherpersonnel** to T_2 (as well as the intervals of their attribute and tuple nodes). Meanwhile, a new table node for **employee** is added with interval [T_2, "now"). Since the **job** relation continues in the new version, there is no update on that table node.

To make the storage and querying of MV-Documents more scalable, [9] uses relational databases and mappings between the XML views and the underlying database system. This is facilitated by the use of H-Tables, firstly introduced in [12]. Consider the **employee** (**id, title, deptno, salary**) relation of schema $V_{1.5}$ in Table 1. Its history is stored in four H-Tables, namely: (i) a key table, **employee_key** (**id, ts, te**), that stores the interval (ts, te) during which tuple with key id was stored in the corresponding relation. (ii) three attribute history tables: **employee_title** (**id, title, ts, te**), **employee_deptno** (**id, deptno, ts, te**) and **employee_salary** (**id, salary, ts, te**) that maintain how the individual attributes of a tuple (identified by id) changed over time, and (iii) an entry in the global relation table **relations** (**relationname, ts, te**) which records the time spans covered by the various relations in the database.

3   Branched Schema Evolution

Many modern complex applications need to support schema branching; examples include scientific databases, collaborative design environment, web-based information systems, etc. With branched schema evolution enabled, a new branch can be created by updating the schema of a *parent* version V_p. If version V_p is a current schema version and the data populating the first schema of the new branch is adapted from the currently alive data of V_p, we have a current branching (*c-branching*). An example of c-branching appears in Fig. 1(b) where the most current version of branch B_1 is $V_{1.5}$. At the current time T_6 branch B_2 is created out of $V_{1.5}$ (i.e., the B_2 creation time is T_6) by applying SMOs on the relations that $V_{1.5}$ has at T_6. For example, under branch B_2 a new attribute *status* was added in **empbio** to describe the marital status of employees. As a result, data can start evolving concurrently under two parallel schemas, $V_{1.5}$ and $V_{2.1}$. A real life scenario leading to c-branching is the case when a company establishes a subsidiary. These two companies share the same historical database (branch B_1 from T_1 to T_6) but in the future their schema and data evolve independently. Note that a version can start from any past version (*h-branching*). In this paper however, we concentrate on c-branching due to the challenges of the parallel evolving it imposes.

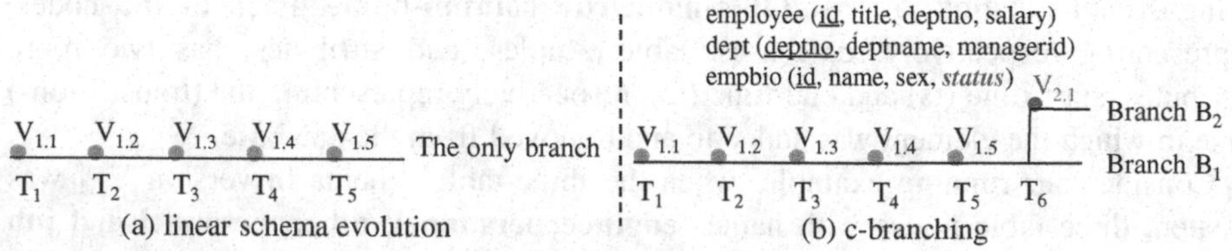

Fig. 1. Linear evolution and branching

As more branches occur, effectively the different schema versions create a *Version Tree*; an example (assuming c-branching) with six branches is shown in Fig. 2, which is an extension of the branched employee DB example from Fig. 1(b). Such version tree can easily display the parent-child relationship among versions and branches; this relationship information is very useful for further optimizations.

The novel problems in supporting c-branching are emanated from its sharing of data: the same original data can evolve in parallel under different branches. To provide efficient access and storage in a branched environment, we use different structures to maintain the evolution of schema versions and their underlying data. Since schema changes are much less frequent, we adopt an XML-based model that enables complex querying (BMV-Document). In contrast, the data evolution over time creates large amounts of historical, disk-resident data, so our focus is on branched column tables (BC-Tables) and efficient index methods.

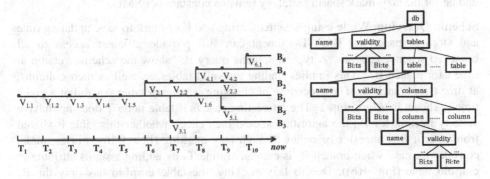

Fig. 2. Example of Version Tree   **Fig. 3.** Illustration of BMV-Document

4   BMV-Document and BC-Tables

4.1   BMV-Document

The BMV-Document is an extension of the MV-Document for storing the branched evolving schema versions in an XML-based representation. The main upgrades are: (i) branch identifier *bid* is needed, because a single timestamp cannot uniquely identify the appropriate schema version. (ii) The BMV-Document refers only to the schema-level storage, and does not detail the data level. (iii) The BMV-Document uses a sharing strategy between versions with various update options and a validity interval (**bid:ts, bid:te**) is thus required, as shown in Fig. 3. When a c-branching is created, the child branch may only modify a relatively small part of its parent schema. Simply copying the schemas of all live tables and their columns from the parent version would incur storage overhead.

Schema Sharing. Consider the c-branching on B_1 that creates a new branch B_2 in Fig. 1(b). B_2's creation time is the start time of its first version, namely $V_{2.1}$, which emanated from $V_{1.5}$ by applying some SMOs.

One approach for schema sharing is *full-mark* which adds new (B_2:ts, B_2:te) interval to all corresponding tables and their columns explicitly for the new branch. While this is better than copying all tables and columns, it still requires update work, especially when there are many current tables and columns. To archive better efficiency, we develop a *lazy-mark* approach, which adds a new (B_2:ts, B_2:te) interval to the db node only, and leaves all shared tables and columns unchanged. If the c-branching partially updated the parent schema, besides adding a validity interval on the db node, the lazy-mark approach updates only the modified tables and columns (based on the corresponding table-level and column-level SMOs).

Therefore, the lazy-mark approach can be summarized as: For each update the path to the corresponding level (db, table or column) is visited and the related nodes are updated. Later on, SMOs can update the BMV-Document within a branch as well, and we re-mark those lazy-marked nodes. As a result, the complexity of each schema update for the lazy-mark sharing strategy remains constant per SMO.

Schema Querying. While using schema sharing and lazy-mark to save updating time and storage space, the BMV-Document can still provide efficient access to all branched schema versions. A typical schema query is: "show the schema version at time t for branch B_i". This implies finding the valid tables, as well as their columns, at time t for branch B_i. The procedure of checking whether a table is valid at a given time is shown in Algorithm 1. The interesting case is if table node T does not have a validity interval for B_i; the algorithm should then check whether this table is shared from one of B_i's ancestor branches through lazy marking (line 7-16). For example, consider the case when branch B_2 is created at time T_6 by adding a status attribute in **empbio** table (Fig. 1(b)). Due to lazy-marking, the table **empbio** has only the B_1 branch id in its interval. However, when we check it for branch B_2, following Algorithm 1, we determine that it has been inherited from B_1 and shared by B_2 at time T_6.

Algorithm 1:   CheckTable (T, t, B_i)
Check whether table node T is valid at time t for branch B_i, **where t is later than B_i's start time.**
1   **if** T has a validity interval for B_i **then**
2   **if** B_i:ts = null **then** return false;
3   **else**
4   **if** B_i:ts <= t < B_i:te **then** return true;
5   **else** return false;
6   **else**
7          $B_h = B_i$'s parent; $B_g = B_i$;
8   **while** (B_h != null)
9   **if** T has a validity interval for B_h **then**
10   **if** B_h:ts = null **then** return false;
11   **else**
12   tt = B_g's start time;
13   **if** B_h:ts<tt<B_h:te **then** return true;
14   **else** return false;
15                    $B_g = B_h$; $B_h = B_g$'s parent;
16   **end while**

4.2   BC-Tables

While the BMV-Document maintains the branched schema versions, the BC-Tables are used to store the underlying evolving data changes. Like H-Table [12], each BC-Table stores the (history of) values for a certain attribute of a base relation.

A BC-Table starts from a particular time and may span over multiple schema versions. However, there are considerable improvements: (i) a BC-Table can be shared by multiple branches; (ii) each data record carries only the start time of its time interval; (iii) suitable branched temporal indexing methods are built on top of BC-Tables.

For indexing a BC-Table we facilitate the branched temporal index ([6, 10]) which is a directed acyclic graph over data and index pages. Data pages (which are at the leaf level) contain temporal data, while index pages contain the searching information to lower level pages. In data pages, due to data sharing, a compact data representation <key, data, ts> is used, where ts corresponds to the record's start time (which will be a bid:time in our BC-Tables) of the original record. In an index page, an entry referencing a child page C is of the form <KR(C), TI(C), address(C)>, where KR is the key-range of the child page, and TI is a list of temporal interval(s) for the shared multiple branches of C.

Splitting occurs when a page becomes full. However, unlike in B+-tree page splitting, when a temporal split happens, the data records currently valid, are copied to a new page. Thus data records are in both the old page and the new page. The motivation for copying valid data from the full page is to make the temporal query efficient. Splits (temporal-split, key-split, and consolidation) cluster data in pages so that when a data page is accessed, a large fraction of its data records will satisfy the query.

Index page splits and consolidations are similar to those of data pages. Since in index page temporal splits, children entries can be copied, this creates multiple parents for these children. As a result, the branched-temporal index is a DAG, not a tree [6].

When the search for a given key k, branch B_i and time t, is directed to a particular data page P through the index page(s), the algorithm checks all the records in P with key k, and finds the record with the largest start time ts, such that $ts <= B_i:t$.

Nevertheless, page P may have been shared by branch B_i, in which case some of its B_i related entries may not contain the B_i interval. Those entries are inherited from B_i's ancestor branches. Therefore, we need to extend the search algorithm of the branched-temporal index [6,10]. In particular, we extend the meaning of the "<" comparison when comparing bid:time tokens. Given two tokens $B_i:T_i$ and $B_j:T_j$ the comparison $B_i:T_i < B_j:T_j$ is satisfied whether $(B_i=B_j \wedge T_i<T_j)$ or $(B_i:T_i < Par(B_j):Ts(B_j)")$, where $Par(B_j)$ is the parent branch of B_j in the version tree, and $Ts(B_j)$ is the start time of B_j.

For example, assume that a data page is shared by branch B_1 and B_2, having entries: <a, v_1, $B_1:t_1$>, <b, v_2, $B_1:t_2$>, <c, v_3, $B_1:t_3$>, <b, v_4, $B_2:t_{14}$>, <c, v_5, $B_1:t_{15}$>, and let branch B_2 be created from B_1 at time t_{10}. So the valid data entries for B_1 at time t_{15} are <a, v_1, $B_1:t_1$>, <b, v_2, $B_1:t_2$>, <c, v_5, $B_1:t_5$>; while the valid data entries for B_2 at time t_{15} are <a, v_1, $B_1:t_1$>, <b, v_4, $B_2:t_{14}$>, <c, v_3, $B_1:t_3$>.

5   Query Processing

Data queries are temporal queries on the data records (stored in the BC-Tables and indexed by the branched temporal index). As with traditional temporal queries [11], a user may ask for: (i) a *snapshot* query, or (ii) a time *interval* query. In a linear schema evolution, snapshot or interval queries deal with a single branch. In a branched sche-ma evolution, the following *multiple*-branch queries (first introduced in [7]) are also of interest: (i) *vertical* query and (ii) *horizontal* query. We first discuss how to process temporal snapshot and interval queries within one branch, and then proceed to vertical and horizontal queries over multiple branches.

5.1   Queries within a Single Branch

In this case, the temporal constraint (time snapshot or interval) falls within the life-time of branch B_i. For a snapshot query, the target schema version that stores the queried data is unique and can be identified easily (from the BMV-Document). The corresponding BC-Tables are then accessed through their branched temporal indices.

Processing a time interval query is more complicated because of two challenges: (i) the time interval may have multiple target schema versions (thus even for a single attribute, multiple BC-Tables may be accessed); (ii) in one BC-Table, many data pag-es may intersect with the time interval, so the search algorithm needs to avoid dupli-cations. The first challenge also appeared in PRIMA [8]: the original temporal query should be reformulated by query rewriting into different sub temporal interval queries for each related BC-Table and the final results are merged from those BC-Tables.

For the second challenge, even in one BC-Table with branched temporal indexing, the naïve depth-first traversal strategy leads to two problems: first, the response set can contain duplicates (due to page splitting copies); second, the same directory entry can be accessed more than once while a query is evaluated. This effect is illustrated in Fig. 4 where the gray-colored rectangles display the pages of the branched temporal index visited for a time-interval query. The naïve algorithm would visit pages 1, 2, 5 once, pages 3, 4, 7 twice, page 8 thrice and page 6 four times.

Traditional duplicate elimination methods such as hashing or sorting may require storage/time overhead, and they are not easy to solve index entry duplication. There-fore, we adopt the Link$_{based}$ algorithm proposed in [3] for (linear) multi-version index structures. The BC-Tables' data pages are equipped with external links pointing to their temporal predecessors.

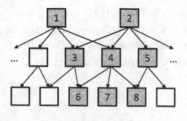   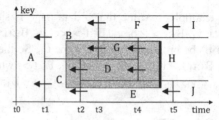

Fig. 4. Visited pages   **Fig. 5.** Data pages with links

An example is presented in Fig. 5 where each page is viewed as the time-key rec-tangle of the records it contains. A key-range time-interval query (the grey rectangle) intersects pages B, C, D, E, G and H. The Link$_{based}$ algorithm consists of two steps.

First, the right border of the query rectangle is used to perform a key-range snapshot query. In Fig 5, this snapshot query will access data pages H and E. Second, for each qualifying page obtained in step 1, its temporal predecessor pages are checked to see whether they contain an answer. If they do, the corresponding pages are put into the buffer, answers are reported and the process is repeated. If the left border of the page is already earlier than the left border of the query rectangle, then we do not proceed further. The worst-case performance of $Link_{Based}$ is $O(log_B n + a/B + u/B)$ where B is the page capacity, n is the number of records at right-border time t, a is the number of answers, and u denotes the number of updates in the query time period.

5.2   Data Queries Over Multiple Branches

Vertical Query. The vertical query is an extension of a single branch query, seeking information for a given branch and its ancestors. An example of a vertical query is: "find the data within a key range KR for a given branch B_i and its ancestor branches, at a time stamp t" (or "during a time interval I"). The time stamp t or interval I must be no later than the end time of branch B_i.

For a *vertical snapshot query* of branch B_i and at time t, if t is earlier than the start time of B_i, then the result conceptually lies in one of B_i's ancestors B_j, whose lifetime covers time t. For a *vertical interval query*, the time interval may span multiple branches along a path in the version tree. For example, in Fig 6, to find titles of employees within a range KR for branch B_4 and its ancestors in a time interval $[T_5, T_{10})$, we need to access data from branches B_4, B_2 and B_1.

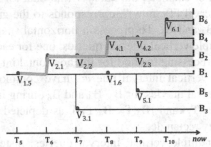

Fig. 6. A part of Version Tree fron Fig. 2.

To process a vertical interval query, we first divide the whole query interval I for branch Bi into multiple smaller adjacent sub-intervals $\{I_1, I_2,..., I_k\}$, one for each ancestor branch along the path $\{B_{i1}, B_{i2},..., B_{ik}\}$ (where $B_{i1} = B_i$, $B_{i2} = B_i$'s parent and so on). In the above example, querying for B_4 with a time interval $I = [T_5, T_{10})$, I should be divided to $[T_8, T_{10})$ for B_4, $[T_6, T_8)$ for B_2 and $[T_5, T_6)$ for B_1 (depicted as the thick orange line in Fig. 6). Then we process the vertical interval query by answering multiple interval queries for each branch and merge the results together.   □

However, certain sub-intervals from different branches may be sharing the same BC-Tables, hence a BC-Table could be processed multiple times by different sub-queries. Notice that the sub-intervals are adjacent and the shared data pages are connected by backward links (Link$_{based}$ approach). Therefore, an optimized processing on vertical interval query is to unite the multiple adjacent sub-queries for the same BC-Table into one "super-query". This optimization, called *reunion*, can guarantee that each BC-Table is processed only once for any vertical interval query.

In the above query example, "find the title of employees within a KR for B_4 and its ancestors during $[T_5, T_{10})$", we assume that the **employee_title** table schema is never changed by any branches after it was created at T_5. With the naïve method, we need to process this table three times for three branches with three time intervals as $[B_4:T_8, B_4:T_{10})$, $[B_2:T_6, B_2:T_8)$ and $[B_1:T_5, B_1:T_6)$. When utilizing the optimized method, the three sub-queries are united into one super-query with an interval $[B_1:T_5, B_4:T_{10})$.

Horizontal Query. The Horizontal query accesses temporal information for a given branch and its descendants. An example is: "find data within a key range KR for a given branch B_i and its descendants, at time point t" (or during "a time period I"). The time stamp t or interval I must be no earlier than the start time of branch B_i.

A *horizontal snapshot query* can be visualized as a snapshot of multiple relevant branches from a sub-tree of the version tree. For example, the query: "find data for branch B_2 and its descendants at time *now*", corresponds to the vertical dash line in Fig 6, involving branches B_2, B_4 and B_6. To process a horizontal snapshot query on time t, we first determine which descendants of branch B_i (including itself) are valid at t, and then issue multiple vertical snapshot queries, one for each branch. □

A *horizontal interval query* can be visualized as a branch-time rectangle on a sub-tree of the version tree. For example, the query: "find data for branch B_2 and its descendants during time interval $[T_7, now)$", corresponds to the grey rectangle in Fig 6, involving branches B_2, B_4 and B_6. To process a horizontal snapshot query on time t, we again first issue multiple vertical interval queries, one for each descendant branch.

However, this naïve processing method for the horizontal interval query will not be efficient if the multiple vertical interval queries have common parts. In the above example, the vertical interval queries for B_2, B_4 and B_6 during interval $[T_7, now)$ have common parts: $[B_2:T_7, B_2:T_8)$ and $[B_4:T_8, B_4:T_{10})$, as depicted in Fig. 6 by the thick orange line inside the grey rectangle.

As a result, for the multiple vertical interval queries, instead of using the same original query time interval I, we should use different intervals for those descendant branches. For each descendant branch B_j, the new query time interval I_j is the intersection of $[ST_j, SE_j)$ with I, where ST_j and SE_j is the start time and end time of branch B_j. For the above example, the optimized vertical interval queries are: $[B_6:T_{10}, B_6:now)$, $[B_4:T_8, B_4:now)$, and $[B_2:T_7, B_2:now)$. This *rearrange* optimization can improve horizontal interval querying by preventing multiple visits of common parts.

6   Merging of Branches

Since branching is allowed for schema evolution, it is quite natural for us to consider the possibility of merging multiple branches. Branching and merging are two key aspects in many modern environments, such as web-based information systems, collaborative framework, and software development managing tools. Branching provides isolation and parallelism, while merging provides subsequent integration. In this section, we consider how to support current version merging (c-merging).

With c-branching, any currently alive version can create a branch; for a c-merging, the currently alive version of branch B_i can merge to another currently alive version from a different branch B_j by creating a new common schema version. In the example shown in Fig. 7, both branching and merging are applied. Such schema evolution will form a *Version Graph* instead of a version tree.

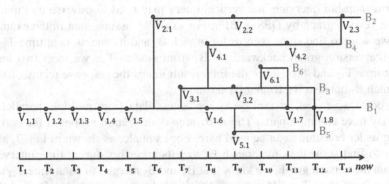

Fig. 7. Schema evolution with branching and merging

6.1   Merging in BMV-Documents

When branch B_i's latest version $B_{i.x}$ merges to branch B_j's latest version $B_{j.y}$ at time t, the branch B_i and version $B_{i.x}$ should be ended and a new version $B_{j.y+1}$ should be created for branch B_j. The branch and version termination can be achieved by updating the end time for corresponding nodes and the lazy-mark process can be utilized for only updating the db and table nodes without reaching to column nodes. After figuring out which elements are discarded from $B_{j.y}$ to $B_{j.y+1}$, and which are added from $B_{i.x}$ to $B_{j.y+1}$, we apply the updates for the corresponding tables and columns. Suitable schema duplication elimination and conflict resolution are applied.

6.2   Merging in BC-Tables

When merging is applied in BC-Tables at the data level records, we still can use the same sharing strategy with the branched temporal index but with special extensions. Assume branch B_i merges to B_j at time t. For both branches, some data records have remained while others are removed (especially when there are conflicts). In BC-Tables, we only delete the removed records by adding null values and keep the remained records unchanged, which is consistent with our sharing method in section 5. Data duplication elimination and conflict resolution are applied as well.

For data accessing, certain extensions should be implemented for merging, since merging integrates data records from two branches into one. Exploring of a branch's ancestors due to lazy mark is extended from one single path to multiple paths with depth-first or breath-first search along the version graph. Meanwhile, the branched temporal indexing can be adapted for merging with certain modifications.

6.3 Query Processing

Here we concentrate on data querying within multiple branches. For vertical queries seeking temporal information for a given branch and its ancestor branches, the ancestors include not only the ones formed by branching but also those by merging. So even for a snapshot querying, the vertical query may need to traverse multiple paths along the version graph by DFS or BFS. For example, assume that in the example of Fig. 8, we want to find some records for branch B_1 and its ancestors at time T_{10}. Traversing the version graph backward for B_1 from *now* to T_{10}, we meet two merging points at time T_{12} and T_{11}. Hence the final result unites the response records from not only branch B_1 but also B_5, B_6 and B_3 at time T_{10}.

To process a vertical interval query we access data from multiple parallel paths which may have common parts. The *rearrange* optimization proposed for horizontal querying under branching can be used here. For example, as shown in Fig. 7, assume we want to find some data for branch B_1 and its ancestors during time interval [T_9, T_{13}). From the version graph, we know that B_3 and B_5 merged to B_1 at time T_{12} and B_6 merged to B_1 at time T_{11}. We can avoid visiting the common paths [B_1:T_{12}, B_1:T_{13}) four times and [B_1:T_{11}, B_1:T_{12}) twice by utilizing *rearrange* to make querying intervals as [B_1:T_9, B_1:T_{13}), [B_3:T_9, B_3:T_{12}), [$B5$:T_9, B_5:T_{12}), and [B_6:T_9, B_6:T_{11}).

7   Experimental Evaluation

To illustrate the efficiency of our framework we present several experiments based on the running example of the employee DB in Fig 2. First, we extend it with more schema versions and branches. The first ten schema changing points (from T_1 to T_{10}) are shown in Fig 2. After that, we make another ten schema changing points (from T_{11} to T_{20}) in two rounds. In each round, there are five schema changes: the first two are linear schema evolutions followed by one schema version branching and two linear schema evolutions. For each linear schema evolution, we choose 50% of the existing branches and make new schema versions for them updating 20% tables and 20% columns in those tables. For each schema branching, we chose all existing branches and make a new branch for each by updating 20% tables and columns. In the end, we have 20 schema changing points with 24 branches of 104 schema versions.

In addition to linear and branched schema evolution, we also create content-level data changes. From T_1 to T20, after each schema changing point, we update the record-level data value 500 times. For each time, we update all existing branches, and for each branch we update 0.2% of all employees for salary, title, and some other randomly chosen attributes. In the end, we have 10,000 time instants of content-level data updates. The Employee DB schema is initialized with 1,000 tables and average 5 columns in each table. We also produce 10,000 employees with 100 titles and other relevant information. For both schema changes and data changes, the tables, attributes and tuples are chosen randomly with a uniform distribution. The page size of our system is 4KB and we set the data page capacity as B = 100 records.

7.1   BMV-Documents

The sharing strategies among multiple branches and the lazy-mark approach are advantageous in space saving for the BMV-Document without sacrificing querying efficiency. We store the branched schema versions, in XML-based BMV-Documents with three different options when branching occurs: (i) copy the schema without any sharing (Non-Shared); (ii) use the sharing strategy and full-mark approach (Shared); (iii) use the sharing strategy and lazy-mark approach (Lazy-mark). Fig. 8 depicts the size per branch (total size / number of versions) of the documents under certain schema changing points: T_{10} (6 branches), T_{15} (12 branches), and T_{20} (24 branches). The options using sharing strategies use much less space than the non-shared option. Compared to the full-mark, the lazy-mark approach is more efficient.

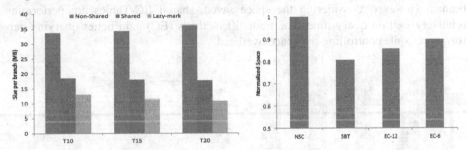

Fig. 8. Space saving in BMV-Documents   **Fig. 9.** Space saving in employee_title table

7.2   BC-Tables

Space Saving. In addition to the shared BC-Tables (SBT), we use a non-shared method which simply copies alive records from the parent branch when a c-branching happens. The non-shared copying method (NSC) utilizes the MVBT ([2]) to store data in each branch separately, so that each single branch has its own data pages and index structure. The total sizes of data pages and index pages for all tables of all 24 branches are: NSC 71.4 GB and SBT 54.9 GB; clearly, the shared BC-Tables provide significant space saving. Nevertheless, the querying performance of the non-shared method will be better than the fully shared BC-Tables since data has been fully materialized at each branch. Therefore, we consider a trade-off between space and querying performance by applying an enforced copying method (EC), which only allows at most p branches that can be shared in one BC-Table. If a shared BC-Table already reaches this number p, then for a later c-branching, we enforce copying (make a new BC-Table for the newly branch) instead of sharing. The fully shared BC-Tables and non-shared method are two extreme situations for this enforced copying (p = 1 corresponds to the non-shared method). In our experiments we implemented an enforced copying method EC with p = 12 (EC-12) and p = 6 (EC-6).

In order to factor out the query reformulating, we choose one particular BC-Table **employee_title**, whose schema never changes from the beginning and is shared by all branches. To compare space usage of the **employee_title** table by the four methods (NSC, SBT, EC-12 and EC-6), we depict a normalized space usage. Since NSC has the largest storage usage (data pages + index pages), the normalized space is computed by (method$_i$'s space) / (NSC's space). As shown in the Fig. 9, the shared **employee_title** BC-Table provides the best space savings followed by EC-12 and EC-6.

Snapshot querying. We use the following query: "find titles of all employees whose ids are within a key range of size 100, for branch B$_i$ at time t" and test on all 24 branches. For each branch, we randomly pick 100 time instants which are in the lifespan of that branch and measure the average snapshot querying time. The average results of all 24 branches are calculated and depicted as normalized page I/O (Fig. 10). The SBT method has the largest I/O usage, so the normalized page I/O is computed by (method$_i$'s I/O) / (SBT's I/O). The non-shared copying method has a better snapshot querying performance because data records are stored separately for each branch. However, considering the space saved, shared BC-Tables are performing relatively well on query time. The trade-off methods (EC) gain better querying performance while controlling the space overhead.

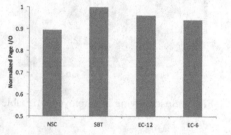 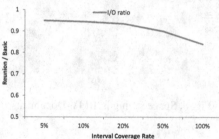

Fig. 10. Snapshot Querying   **Fig. 11.** Vertical interval querying

Interval Query Processing. For interval query processing we implement the $Link_{Based}$ algorithm along with the *reunion* and *rearrange* optimizations in shared BC-Tables. First, we test vertical interval queries involving multiple branches: "find titles of employees whose ids are within a key range of size 100 for branch B$_{24}$ and its ancestors in the time interval I". Five different time intervals are used and their coverage rates with respect to the whole temporal data lifetime are 5%, 10%, 20%, 50%, and 100% correspondingly. Two methods are implemented here: one is the basic solution (Basic) which divides the query interval into multiple sub-intervals for each branch. The other is the optimized *reunion* method (Reunion) that unites the sub-intervals into one super-interval if they are sharing the same BC-Table. The I/O ratio of these two methods (Reunion's I/O) / (Basic's I/O) is shown in Fig. 11. Clearly the *reunion* optimization can improve the vertical interval querying, and the improvements are more significant when the query interval covers more ancestor branches.

Then we consider horizontal interval queries involving multiple branches: "find titles of employees whose ids are within a key range of size 100 for branch B$_1$ and its descendants in the time interval I". The different interval I coverage rates are used as same as above. We again implement two methods: one is the basic solution (Basic) that issues multiple vertical queries with the same query interval for each descendant branch, and the other is the optimized *rearrange* method (Rearrange) that arranges different query intervals for each descendant branch to achieve querying efficiency. The I/O ratio of these two methods (Reunion's I/O) / (Basic's I/O) is shown in Fig. 12. As seen, the *rearrange* optimization can effectively improve the horizontal interval querying especially when the query interval covers more common parts.

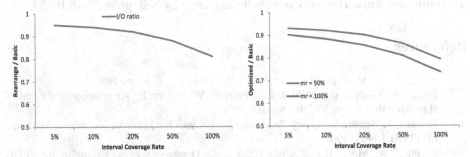

Fig. 12. Horizontal interval querying   **Fig. 13.** Querying with merging added

7.3   Branched Schema Evolution with Merging

Finally, we employ schema merging into the branched system as well. The branched schema versions and datasets are extended as follows: We randomly insert 5 schema merging points into the 20 schema changing points, and for each such schema merging point, we randomly pick some existed branches to do the merges. A parameter mr (0 ~ 1) is used to control the merging rate. For example, if mr = 50%, we randomly pick half of existed branches to do the merges. The content-level data changes are generated as before: the data is updated 500 times after each schema changing point (evolving, branching and merging). The total number of time instants with data updates is increased from 10,000 to 12,500.

Below we only show results for the horizontal interval querying for branch B_1. We set up five different querying interval coverage rates as same as above with two different merging rates as mr = 50% and mr = 100%. The methods we test include (i) the basic method (Basic) without avoiding the common sub-paths and (ii) the optimized method (Optimized) with both *reunion* and *rearrange* implemented. The I/O ratio of these two methods (Optimized's I/O) / (Basic's I/O) is shown in Fig 13 for the two mr rates. The optimized method has an advantage in interval querying processing, and this becomes more apparent for larger merging rates and longer query intervals.

8   Conclusion

We addressed branched schema evolution for transaction-time databases. To the best of our knowledge, this is the first attempt to examine both data and schema evolution in a branched environment. Efficient schema sharing strategies with smart lazy-mark updates are used. Schema versions are stored in XML-based documents for ease of querying. Data records are stored in relational column tables with branched and temporal indexing. We also explored temporal querying optimizations, especially for vertical and horizontal interval queries. The feasibility of supporting schema merging was also examined. In future research, we will investigate temporal joins and aggregations under schema evolution with branching and merging.

Acknowledgements. This work is partially supported by NSF grant IIS-0910859.

References

[1]   http://en.wikipedia.org/wiki/History_of_Firefox
[2]   Becker, B., Gschwind, S., Ohler, T., Seeger, B., Widmayer, P.: An asymptotically optimal multiversion b-tree. VLDB Journal (1996)
[3]   Bercken, J., Seeger, B.: Query Processing Techniques for Multiversion Access Methods. In: VLDB 1996 (1996)
[4]   Curino, C.A., Moon, H.J., Zaniolo, C.: Graceful Database Schema Evolution. In: VLDB (2008)
[5]   Curino, C.A., Moon, H.J., Zaniolo, C.: Managing the history of metadata in support for db archiving and schema evolution. In: ECDM (2008)
[6]   Jiang, L., Salzberg, B., Lomet, D., Barrena, M.: The BT-Tree: A branched and temporal access method. In: VLDB (2000)
[7]   Landau, G.M., Schmidt, J.P., Tsotras, V.J.: Historical Queries along Multiple Lines of Time Evolution. VLDB Journal (1995)
[8]   Moon, H.J., Curino, C.A., Deutsch, A., Hou, C.-Y., Zaniolo, C.: Managing and Querying Transaction-time Databases under Schema Evolution. In: VLDB (2008)
[9]   Moon, H.J., Curino, C.A., Zaniolo, C.: Scalable Architecture and Query Optimization for Transaction-time DBs with Evolving Schemas. In: SIGMOD (2010)
[10] Salzberg, B., Jiang, L., Lomet, D., Barrena, M., Shan, J., Kanoulas, E.: A Framework for Access Methods for Versioned Data. In: Bertino, E., Christodoulakis, S., Plexousakis, D., Christophides, V., Koubarakis, M., Böhm, K. (eds.) EDBT 2004. LNCS, vol. 2992, pp. 730–747. Springer, Heidelberg (2004)
[11] Tsotras, V.J., Jensen, C.S., Snodgrass, R.T.: An Extensible Notation for Spatiotemporal Index Queries. SIGMOD Record 27(1) (1998)
[12] Wang, F., Zaniolo, C., Zhou, X.: Archis: An xml-based approach to transaction-time temporal database systems. VLDB Journal (2008)

Fast Identity Anonymization on Graphs

Xuesong Lu, Yi Song, and Stéphane Bressan

School of Computing, National University of Singapore
{xuesong,songyi,steph}@nus.edu.sg

Abstract. Liu and Terzi proposed the notion of k-degree anonymity to address the problem of identity anonymization in graphs. A graph is k-degree anonymous if and only if each of its vertices has the same degree as that of, at least, k-1 other vertices. The anonymization problem is to transform a non-k-degree anonymous graph into a k-degree anonymous graph by adding or deleting a minimum number of edges.

Liu and Terzi proposed an algorithm that remains a reference for k-degree anonymization. The algorithm consists of two phases. The first phase anonymizes the degree sequence of the original graph. The second phase constructs a k degree anonymous graph with the anonymized degree sequence by adding edges to the original graph. In this work, we propose a greedy algorithm that anonymizes the original graph by simultaneously adding edges to the original graph and anonymizing its degree sequence. We thereby avoid testing the realizability of the degree sequence, which is a time consuming operation. We empirically and comparatively evaluate our new algorithm. The experimental results show that our algorithm is indeed more efficient and more effective than the algorithm proposed by Liu and Terzi on large real graphs.

1 Introduction

The data contained in social media raise the interest of marketers, politicians and sociology researchers, as well as hackers and terrorists. The mining and analysis of the graphs formed by entities and connections in online social networks, messaging systems and the like, should only benefit legitimate users while no one, and, more critically, no malicious user should be able to access or infer private information.

Researchers, such as the authors of [1], quickly observed that simply hiding the identities of the users in a network may not suffice to protect privacy. Indeed, the structure of the graph itself may leak sufficient information for an adversary with minimal external knowledge to infer identity of users, for instance. Consequently several graph anonymization algorithms have been proposed that not only remove identity but also perturb the graph content and structure while trying to preserve utility for the sake of mining and analysis.

1.1 The K-Degree Anonymization Algorithm

Liu and Terzi [13] address the issue of identity disclosure of network users by adversaries with the background knowledge of nodes degree. To prevent such

S.W. Liddle et al. (Eds.): DEXA 2012, Part I, LNCS 7446, pp. 281–295, 2012.
© Springer-Verlag Berlin Heidelberg 2012

attacks they propose the problem of *k-degree anonymity*. A graph is said to be *k-degree anonymous* when each vertex in the graph has the same degree as at least $k - 1$ other vertices. In other words, any vertex cannot be identified with probability higher than $1/k$ if the adversary has the degree information of the graph. The degree sequence of such a graph is said to be *k-anonymous*. Then the problem is to transform a non-*k*-degree anonymous graph into a *k*-degree anonymous graph by adding or deleting a minimum number of edges. For the sake of simplicity, we consider only the addition of edges. Liu and Terzi [13] propose a two-phase algorithm. The first phase (degree anonymization) anonymizes the degree sequence of the original graph to be *k*-anonymous. They propose a dynamic programming algorithm which reproduces the algorithm in [9]. The second phase (graph construction) constructs a *k*-degree anonymous graph with the anonymized degree sequence based on the original graph. We call this algorithm *K-Degree Anonymization* (KDA).

Typically, the degree distribution of large real world graphs follows a power-law or exponential distribution (see [2,6]). Consequently, there are few vertices with very large degrees and many vertices with the same small degrees. Moreover, the difference between consecutive large degrees is great.

The dynamic programming in the degree anonymization phase of KDA is designed to minimize the residual degrees, namely the difference between the original degrees and the degrees in the anonymized degree sequence. On large real world graphs, it generates a sequence at the expense of large residual degrees for large original degrees, as the differences between these large original degrees are great. It also generates the sequence with a small number of changes from the original degree sequence, as many vertices with small original degrees are already *k*-anonymous. It may then be impossible to compensate the large residual degrees. The sequence is then unrealizable. Our experience suggests that, unlike what is claimed by Liu and Terzi, this situation is frequent. For instance, as illustrated in the example below, their dynamic programming in the degree anonymization phase does not generate a realizable degree sequence from the given data set.

Example 1. Email-Enron is the network of Enron employees who have communicated by the Enron email. It is an undirected graph with 36692 vertices and 367662 edges. Each vertex represents an email address. An edge connects a pair of vertices if there is at least one email communication between the corresponding email users. The dataset is available at http://snap.stanford.edu/data/email-Enron.html. The first 10 degrees of its degree sequence in descending order are 1383, 1367, 1261, 1245, 1244, 1143, 1099, 1068, 1026, 924. After the degree sequence is anonymized for $k = 5$, the 10 degrees become 1383, 1383, 1383, 1383, 1383, 1143, 1143, 1143, 1143, 1143. We see that the degree of the last vertex is increased by $1143 - 924 = 219$. This means that 219 vertices with residual degree are required in order to compensate the residual degree of 219. However, during the anonymization the number of vertices that have their degrees increased is 212. Moreover, most of these vertices are those with small original degrees which are already connected to that vertex. Thus there are no enough vertices with residual degrees to be wired to the last vertex. The k-anonymous degree sequence is unrealizable.

Moreover, even if the anonymized degree sequence is realizable, the graph construction phase of the algorithm may not succeed.

Liu and Terzi cater for these two situations by proposing a Probing scheme that operates small random changes on the degree sequence until it is realizable and the graph is constructed. Our experience shows that a large number of Probing steps are in effect necessary to obtain a realizable sequence for practical graphs. After each Probing is invoked, the realizability-testing is conducted. The testing has a time complexity $O(n^2)$ where n is the number of vertices. As Probing is invoked for a large number of repetitions, the complete algorithm is very inefficient.

1.2 Our Contributions

Motivated by the above observations, we study fast k-degree anonymization on graphs at the risk of marginally increasing the cost of degree anonymization, i.e., the edit distance between the anonymized graph and the original graph.

We propose a greedy algorithm that anonymizes the original graph by simultaneously adding edges to the original graph and anonymizing its degree sequence. We thereby avoid realizability testing by effectively interleaving the anonymization of the degree sequence with the construction of the anonymized graph in groups of vertices.

Our algorithm results in larger edit distance on small graphs but smaller edit distance on large graphs compared with the algorithm of Liu and Terzi. Our algorithm is much more efficient than the algorithm of Liu and Terzi.

The rest of the paper is organized as follows. Section 2 discusses background and related work on graph anonymization. Section 3 presents our novel algorithm. Section 4 empirically and comparatively evaluates the performance of our algorithm. Finally, we conclude in Section 5.

2   Related Work

The need for more involved graph anonymization stems from the shortcoming of naive anonymization [1]. Naive anonymization replaces identities of vertices with synthetic identifiers before publishing the graph. With minimal external knowledge, adversaries may be able to recover these identities from the graph structure. Hay et al. [10] further study the problem and quantify the risk of re-identification via graph structural queries.

Several graph anonymization techniques have then been proposed [13,23,5,18,17].

Generally speaking these techniques consists in modifying the graph structure so as to prevent re-identification while preserving sufficient utility. Most of these works have built upon the concept of k-anonymity [16], which was first introduced to anonymize relational micro-data.

Several works consider rather general structural attacks. Namely, they try and protect against attacks from adversaries with diverse background structural knowledge. Hay et al. [10] propose the *k-candidate anonymity* model that requires that, for any structural query, there be at least k candidate vertices. Liu and Terzi [13] suggest to make the degree sequence *k-anonymous* so that each vertex in the graph has the same degree as at least $k-1$ other vertices. Vertices in *k-degree anonymous* graph cannot be identified with probability higher than $1/k$. Zou et al. [23] propose to modify the graph to be *k-automorphic* before releasing. Any vertex in such a graph cannot be distinguished from other at least $k-1$ vertices via graph structure, thus all kinds of structure attacks are prevented. The modifications are achieved by addition and deletion of edges and, occasionally, addition of vertices. Similarly, Wu et al. [18] propose the *k-symmetry* model to prevent identity disclosure. In a *k-symmetric* graph every vertex is structurally indistinguishable from at least $k-1$ other vertices. Cheng et al. [5] consider the same problem as Zou et al. [23], as they also try to prevent general structural attacks on published graphs and protect against not only identity but also link and attribute disclosure. They propose the *k-isomorphism* model that forms k pairwise isomorphic subgraphs, to provide sufficient privacy guarantee.

Although these approaches take all kinds of structural attacks into consideration, and thus provide strong privacy guarantee, they often incur many changes and therefore potentially a loss of utility. Song et al. [15] make a variety of graph structure measurements on social networks both before and after anonymization. They examine the state-of-the-art anonymization algorithm, *k-automorphism* algorithm [23]. The significant changes on degree distribution, diameter, density, algebraic connectivity and other metrics indicate the anonymization can perturb graph structure to a large degree and thus greatly impair data utility, even though strong privacy guarantee is provided.

K-degree anonymity [13] specifically focuses on attacks leveraging an adversary's background knowledge of degree. By not being concerned with other structural attacks, it can achieve privacy with fewer modification and, therefore, at a lesser utility cost. Stronger privacy guarantees than those of *k-degree anonymity* are provided by models such as k^2-*degree anonymity* by Tai et al. [17]. A k^2-*degree anonymous* graph prevents re-identification by adversaries with the background knowledge of the degrees of two vertices connected by an edge.

While the above works and some others focus on identity disclosure [22,3,8] some research studied link discosure [21,19,11,14]. Several works [4,7,12,20] works also look at graph models other than simple graph, for instance bipartite graphs.

3   The Algorithm

The algorithm that we propose simultaneously adds edges to the original graph and anonymizes its degree sequence in groups of vertices.

The main idea of the algorithm is to cluster and anonymize the vertices of the original graph into several anonymization groups. Each group contains at least k vertices. The graph is transformed so as that vertices in each group have the same degree. In order to achieve small local degree anonymization cost, the vertices in each group should have similar degrees. For this reason, our algorithm sorts, examines and groups the vertices in the descending order of their degrees in the original graph. This choice is motivated by the observation that practical graphs often follow a power or exponential law with a long tail according to which many vertices have and share a small degree. We therefore wire vertices with larger degree to vertices with smaller degree in groups until the degree sequence is k-anonymous, if it can be achieved.

Let v be the sorted vertex sequence. The `greedy_examination` algorithm clusters vertices into an anonymization group. An anonymization group is the smallest subset of v that has at least k members and whose members have a degree strictly higher than the remaining vertices. The cost of the subsequent anonymization of such a group is necessarily the sum of residual degrees after anonymization, namely, for an anonymization group $(v_i, \cdots, v_j)$ in descending order of degrees, $\sum_{l=i}^{j}(d_i - d_l)$, where d_l is the degree of vertex v_l.

The `edge_creation` algorithm adds edges in order to anonymize the vertices in a group. It wires vertices with insufficient degree in the anonymization group to vertices with lesser degree in v until all vertices in the group have the same degree d_i for an anonymization group $(v_i, \cdots, v_j)$ in descending order of degrees. However, we constrain the algorithm never to increase the degrees of vertices in and outside the group beyond that of the highest degree in the anonymization group, namely, d_i, for an anonymization group $(v_i, \cdots, v_j)$ in descending order of degrees. After adding edges, v is reordered according to the new degrees. At the next iteration, vertices outside the group may be further added to the newly anonymized group by `greedy_examination`, if their degree is d_i.

The anonymization group is now k-anonymous, because it contains at least k vertices with degree d_i.

The design choices in the algorithms above, in particular the wiring constraint, have been made in order to minimize the need for reordering v and to allow as sequential as possible a processing of vertices and groups.

Because of the wiring constraint, it is however possible that the above deterministic process does not find enough vertices to wire. Therefore it does not construct a graph with an anonymized degree sequence. The `relaxed_edge_creation` algorithm caters for such possible failures. It relaxes the wiring constraint.

The complete algorithm, *Fast K-Degree Anonymization* (FKDA), combines the above three algorithms. FKDA always constructs a k-degree anonymous graph.

3.1   The `greedy_examination` Algorithm

At each iteration, the input to `greedy_examination` is a sequence of vertices v of length n sorted in the descending order of their degrees, an index i such that the vertex sequence $(v_1, v_2, \ldots, v_{i-1})$ has been k-anonymous and the value of k. The output is a number n_a such that the vertices $v_i, v_{i+1}, \ldots, v_{i+n_a-1}$ are selected to be clustered into an anonymization group. Then `greedy_examination` passes v, i and n_a to `edge_creation`.

Algorithm 1. The greedy_examination algorithm

Input: v: a sequence of n vertices sorted in the descending order of their
degrees, i: an index, k: the value of anonymity.
Output: n_a: the number of consecutive vertices that are going to be
anonymized.

1   *Find the first vertex v_j such that $d_j < d_i$;*
2   **if** v_j *is not found* **then**
3 | $n_a = n - i + 1$;
4   **else**
5 | **if** $d_i = d_{i-1}$ **then**
6 | | **if** $n - j + 1 < k$ **then** $n_a = n - i + 1$;
7 | | **else** $n_a = j - i$;
8 | **else**
9 | | **if** $n - i + 1 < 2k$ **or** $n - j + 1 < k$ **then** $n_a = n - i + 1$;
10 | | **else** $n_a = max(k, j - i)$;
11 | **end**
12 **end**
13 *Return n_a;*

The algorithm begins with an sequential examination of v starting from v_i, until v_j such that $d_j < d_i$. If there is no such v_j found, $v_i, v_{i+1}, \ldots, v_n$ have the same degree already. Below we show that there are at least k vertices from v_i to v_n. Thereby v is already k-anonymous. n_a is set to be $n - i + 1$, i.e., the number of all the remaining vertices. If v_j is found, there are two different cases depending on the result of comparison between d_i and d_{i-1}[1]. If $d_i = d_{i-1}$[2] which means that v_i has the same degree as the degree of the last anonymization group, greedy_examination clusters $v_i, v_{i+1}, \ldots, v_{j-1}$ in a group and merges them into the last anonymization group. Then n_a is set to be $j - i$. However, there is an exception when $n - j + 1 < k$. This means that there are less than k vertices after the current group. These vertices cannot be transformed to be k-anonymous in a separated group. Thus greedy_examination has to cluster $v_i, v_{i+1}, \ldots, v_n$ into a group. n_a is set to be $n - i + 1$. In the other case where $d_i < d_{i-1}$, greedy_examination forms a new anonymization group starting from v_i. If $j - i \geq k$, which means there are at least k vertices having the same degree, greedy_examination clusters $v_i, v_{i+1}, \ldots, v_{j-1}$ into the new group. n_a is set to be $j - i$. Otherwise, there are less than k vertices in the sequence $(v_i, v_{i+1}, \ldots, v_{j-1})$. Thereby greedy_examination clusters $v_i, v_{i+1}, \ldots, v_{i+k-1}$ in the new anonymization group. n_a is set to be k. However, there are also two exceptions when $n - i + 1 < 2k$ or $n - j + 1 < k$. The former means that $v_i, v_{i+1}, \ldots, v_n$ cannot form two anonymization groups. The latter means that $v_j, v_{j+1}, \ldots, v_n$ cannot be clustered into a separated group. In either exception, greedy_examination has to cluster $v_i, v_{i+1}, \ldots, v_n$ into an anonymization group. Then n_a is set to be $n - i + 1$.

The algorithm is described in Algorithm 1.

---

[1] If $i = 1$, the comparison is between d_1 with n.
[2] This is caused by edge_creation.

3.2   The edge_creation Algorithm

At each iteration, the input to edge_creation is a sequence of vertices v of length n sorted in the descending order of their degrees, an index i and a number n_a. The goal is to anonymize the vertices $v_i, v_{i+1}, \ldots, v_{i+n_a-1}$ to degree d_i by adding edges to the original graph. The output is an index, which equals $i + n_a$ if the anonymization succeeds, or equals j if v_j cannot be anonymized, where $i < j \leq i + n_a - 1$.

For each v_j in the vertex sequence $(v_i, v_{i+1}, \ldots, v_{i+n_a-1})$, edge_creation wires it to v_l for $j < l \leq n$, such that the edge (j, l) does not previously exist and $d_l < d_i$, until $d_j = d_i$. The former condition avoids creating multiple edges. The latter condition minimizes the need for reordering v. If in the end edge_creation successfully anonymizes these n_a vertices, it reorders the new vertex sequence v in the descending order of their degrees. Otherwise, it returns the index j such that v_j cannot be anonymized with the wiring constraint. Then the repairing algorithm relaxed_edge_creation is invoked.

The algorithm is described in Algorithm 2.

Algorithm 2. The edge_creation algorithm

Input: v : a sequence of n vertices sorted in the descending order of their
   degrees, i : an index, n_a : the number of vertices that are going to be
   anonymized starting from v_i.
Output: j : an index.

1 for $j \in (i + 1, i + n_a - 1)$ do
2 | while $d_j < d_i$ do
3 | | *Create an edge (j, l) where $j < l \leq n$ such that (j, l) does not previously*
   | | *exist and $d_l < d_i$.*;
4 | | if *The edge cannot be created* then   *Return j*;
5 | end
6 end
7 *Sort v in the descending order of degree*;
8 *Return j*;

We consider three heuristics to examine the candidate vertices in v for the creation of edges.

The first heuristics examines v from v_{j+1} to v_n, that is, in the decreasing order of their degrees, and creates the edge (j, l) whenever the constraint is satisfied. The second heuristics examines v from v_n to v_{j+1}. The last heuristics randomly selects a candidature v_l and creates the edge (j, l). Below we denote by 1, 2, and 3, respectively, the variants of the complete algorithm with these three heuristics.

Intuitively, the first heuristics incurs larger anonymization cost than the second heuristics does. This is because the first heuristics increases the degree of vertices with large original degree, so that the largest degrees in the some anonymization groups might be increase. In order to anonymize these groups, more edges will be added. The third heuristics should behavior in between. On the other hand, the first two heuristics construct deterministic anonymized graphs whereas the third heuristics can generate random anonymized graphs, which, as we shall see, has consequences on the preservation of utility.

3.3 The relaxed_edge_creation Algorithm

The edge_creation algorithm is not certain to output a k-degree anonymous graph. The failure occurs when an edge (j, l) with the wiring constraint cannot be created for some j. In this case, relaxed_edge_creation is invoked. It relaxes the wiring constraint.

The algorithm examines v from v_n to v_1 and iteratively creates an edge (j, l) if only the edge does not previously exist, until $d_j = d_i$. Then relaxed_edge_creation returns the index l. Notice that this iteration can always stop because in the worst case v_j will be wired to all the other vertices. Finally relaxed_edge_creation sorts the new vertex sequence v in the descending order of degree and feeds it as the input of greedy_examination in the next iteration.

The algorithm is described in Algorithm 3.

Algorithm 3. The relaxed_edge_creation algorithm

Input: v : a sequence of n vertices sorted in the descending order of their
      degrees, i, j : two indices.
Output: l : an index.

1 **for** $l = n$ to 1 **do**
2   **if** v_j *and* v_l *are not connected* **then**
3   *Create an edge* (j, l);
4   **if** $d_j = d_i$ **then**
5   *Sort* v *in the descending order of degrees*;
6   *Return* l;
7   **end**
8   **end**
9 **end**

Notice that this process may compromise the k-degree anonymity of the vertex sequence $(v_1, v_2, \ldots, v_{i-1})$ if the returned l is less than i, i.e., v_j is wired to some vertex that has been anonymized. In this case, greedy_examination needs to examine v from the beginning in the next iteration, i.e., i is set to be 0. In the other case where $l > i$, greedy_examination still examines v starting from v_i in the next iteration. However, as relaxed_edge_creation examines v from small degree to large degree, there is a high probability that $(v_1, v_2, \ldots, v_{i-1})$ is still k-anonymous.

3.4 The Fast K-degree Anonymization Algorithm

The FKDA algorithm combines the greedy_examination, edge_creation and relaxed_edge_creation algorithms. The input to FKDA is a graph G with n vertices and the value of k. The output is a k-degree anonymous graph G'.

FKDA first computes the vertex sequence v of G in the descending order of degree. Then at each iteration, it invokes greedy_examination to compute the number n_a and passes it with i to edge_creation. If edge_creation successfully anonymizes the n_a vertices, FKDA updates the value of i as $i + n_a$. Then FKDA outputs the anonymized graph G' if $i > n$, or enters the next iteration otherwise. If edge_creation fails to construct the graph, FKDA invokes relaxed_edge_creation and updates the value of i according to the value of l returned by relaxed_edge_creation. Notice that FKDA can always output a valid k-degree anonymous graph because in the worst case a complete graph is constructed.

The complete algorithm is described in Algorithm 4.

Algorithm 4. The Fast K-Degree Anonymization algorithm

Input: G : a graph of n vertices, k : the value of anonymity.
Output: G' : a k-degree anonymous graph constructed from G.

1   v=*the vertex sequence of G in the descending order of degree*;
2   $i = 1$;
3   **while** $i \leq n$ **do**
4   n_a =greedy_examination(v, i, k);
5   j =edge_creation(v, i, n_a);
6   **if** $j = i + n_a$ **then**
7         $i - i + n_a$;
8   **else**
9   l =relaxed_edge_creation(v, i, j);
10   **if** $l < i$ **then** $i = 0$;
11   **end**
12 **end**
13 *Return G'*;

We provide the approximate bounds of the edit distance to the original graph produced by FKDA. Suppose ideally the original vertex sequence v is clustered as follows. The sequence $(v_1, v_2, \ldots, v_{ik})$ is clustered into i groups, each of which contains k vertices, i.e., the $(j+1)^{th}$ group contains the vertices $v_{jk+1}, v_{jk+2}, \ldots,$ $v_{(j+1)k}, 0 \leq j \leq i-1$. The sequence $(v_{ik+1}, v_{ik+2}, \ldots, v_n)$ is already k-anonymous[3]. In the best case (which is encountered in the second heuristics of edge_creation), the vertices in the sequence $(v_1, v_2, \ldots, v_{ik})$ are only wired to the vertices in the sequence $(v_{ik+1}, v_{ik+2}, \ldots, v_n)$ by edge_creation. Suppose the latter sequence is still k-anonymous after anonymization. Then we get the lower bound which is $bound_l = \sum_{j=0}^{i-1} \sum_{l=1}^{k} (d_{jk+1} - d_{jk+l})$. In the worst case (which is encountered in the first heuristics of edge_creation), each vertex in the sequence $(v_1, v_2, \ldots, v_{ik})$ is wired to all of its antecedent vertices. Then the largest degree of the $(j+1)^{th}$ group becomes $d_{jk+1} + jk$. Therefore the upper bound is $bound_u = \sum_{j=0}^{i-1} \sum_{l=1}^{k} (d_{jk+1} + jk - d_{jk+l}) = \frac{i \times (i-1)}{2} k^2 \times bound_l$.

---

[3] This is the usual case for large graphs.

4   Performance Evaluation

4.1   Experimental Setup

We implement KDA and three variants of FKDA, FKDA 1, FKDA 2 and FKDA 3, corresponding to the three heuristics in C++. We run all the experiments on a cluster of 54 nodes, each of which has a 2.4GHz 16-core CPU and 24 GB memory.

4.2   Datasets

We use three datasets, namely, **Email-Urv**, **Wiki-Vote** and **Email-Enron**.

Email-urv contains the email communication among faculty and graduate students at Rovira i Virgili University of Tarragona, Spain. It is an undirected graph with 1133 vertices and 10902 edges. Each vertex represents an email address. An edge connects a pair of vertices if there is at least one email communication between the corresponding email users. The dataset is available at http://deim.urv.cat/ aarenas/data/welcome.htm.

Wiki-Vote is the network of votes for the administrator election for Wikipedia pages. It is a directed graph with 7115 vertices and 103689 edges. Each vertex represents a user of Wikipedia. An edge links vertex i to vertex j if user i votes on user j. The dataset is available at http://snap.stanford.edu/data/wiki-Vote.html.

Email-Enron has been described in Section 1.1.

We conducts experiments on these three graphs. The different sizes of the three graphs illustrate the performance of KDA and FKDA on small (1133 vertices), medium (7115 vertices) and relatively large (36692 vertices) graphs.

4.3   Effectiveness Evaluation

We compare the effectiveness of the algorithms by evaluating the variation of several utility metrics: edit distance (ED), clustering coefficient (CC) and average shortest path length (ASPL) (following [13]).

We calculate the edit distance between the anonymized graph and the original graph. The edit distance is the number of edges Δm added to the original graph. For the sake of convenience for comparison, we normalize it to the number of edges in the original graph and calculate $\Delta m/m$.

We also calculate the clustering coefficient. The clustering coefficient of a vertex is defined as the fraction of the edges existing between the neighbors of the vertex. Then the clustering coefficient of a graph is defined as the average clustering coefficient of all the vertices.

We also calculate the average shortest path length. The shortest path length between a pair of vertices in a simple graph is defined as the number of hops from one vertex to the other. Then the average shortest path length of a graph is defined as the average of the shortest path lengths between all reachable pairs of vertices.

We vary the value of k in the range $\{5, 10, 15, 20, 25, 50, 100\}$. For each value of k, we run each algorithm 10 times on each dataset and compute the average value of the metrics.

Figure 1-3, 4-6 and 7-9 show the results on Email-Urv, Wiki-Vote and Email-Enron, respectively.

Figure 1, 4 and 7 show the evaluation results of the normalized edit distance on the three graphs.

We see that FKDA adds more edges to Email-Urv but less edges to Wiki-Vote and Email-Enron compared to KDA. In Email-Urv which is a small graph with 1133 vertices, the differences between large degrees are not large. By using KDA the residual degrees of the anonymized vertices with large original degrees can be compensated by enough number of anonymized vertices with residual degrees, that is, the anonymized degree sequence is realizable, with only a small number of repetition of Probing. Thus the minimum edit distance found by dynamic programming is still less than the edit distance produced by FKDA. To the contrary, Wiki-Vote and Email-Enron are two relatively larger graphs with 7115 and 36692 vertices, respectively. The differences betweens large degrees of either graph are considerably large. Therefore by using KDA, Probing is invoked a significant number of times before a k-degree anonymous graph is constructed, as explained in Section 1.1. Moreover, by comparing our relaxed_edge_creation algorithm with Probing, we find that relaxed_edge_creation increases a small degree only if the corresponding vertex can be wired to an anonymized vertex with residual degree. To the contrary, Probing randomly increases a small degree regardless the actual structure of the graph. The corresponding vertex may not be able to be wired to an anonymized vertex with residual degree (There might exist already an edge between the two vertices.) Consequently, more repetitions of Probing are invoked. Thus we believe that eventually Probing adds more noise than relaxed_edge_creation does to the degree sequences of the two large graphs. Therefore FKDA adds less edges than KDA does to the two graphs.

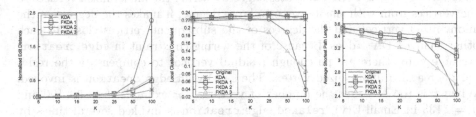

Fig. 1. ED: Email-Urv   **Fig. 2.** CC: Email-Urv   **Fig. 3.** ASPL: Email-Urv

Figure 2, 5, 8 and Figure 3, 6, 9 show the evaluation results of clustering coefficient and average shortest path length, respectively. The constant line shows the value of corresponding metric in the original graph.

We see that FKDA produces less similar results with that in the original graphs on Email-Urv and more similar results on Wiki-Vote and Email-Enron than KDA does. This is generally consistent with the evaluation results of edit distance, since FKDA adds more edges to Email-Urv and less edges to Wiki-Vote and Email-Enron than KDA does.

We further compare the performances of the three variants of FKDA.

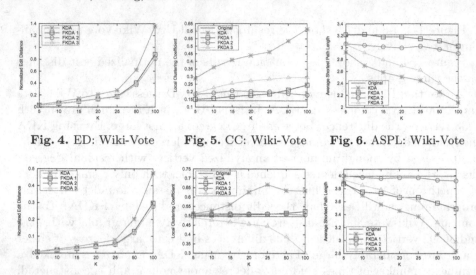

Fig. 4. ED: Wiki-Vote   **Fig. 5.** CC: Wiki-Vote   **Fig. 6.** ASPL: Wiki-Vote

Fig. 7. ED: Email-Enron   **Fig. 8.** CC: Email-Enron   **Fig. 9.** ASPL: Email-Enron

In Section 3.2 we say the that first heuristics incurs larger anonymization cost, i.e. edit distance, than the second heuristics does, and the third heuristics performs in between. The results in Figure 4 and 7 support this claim, although the differences are small. However, in the small graph Email-Urv, we observe that FKDA 2 incurs much larger edit distance than the other two variants and FKDA 1 incurs the smallest edit distance, for $k = 50$ and $k = 100$. The reason is as follow. When k increases, after anonymization the residual degrees of the vertices with large original degrees become larger. Therefore more residual vertices with smaller original degrees are required to compensate these large residual degrees. As FKDA 2 creates edges by wiring the anonymized vertices to the vertices from with small degree to large degree, it makes the degrees of the anonymized vertices and the degrees of the subsequent vertices closer to each other than FKDA 1 does. Because of the wiring constraint in edge_creation, at some point there are no enough residual vertices to compensate the residual degree of a anonymized vertex. Then relaxed_edge_creation is invoked. When k is too large for the number of vertices (for example, $k = 50, 100$ and $n = 1133$ in Email-Urv), relaxed_edge_creation is invoked several times by FKDA 2. Then the edit distance to the original graph is enlarged. To the contrary, FKDA 1 creates edges by wiring the anonymized vertex with large residual degree to the vertices from with large degree to small degree. It maintains a sufficient gap between the degrees of the anonymized vertices and the degrees of the subsequent vertices. The residual degree of the anonymized vertices can be compensated under the wiring constraint in edge_creation, without invoking relaxed_edge_creation. Therefore the edit distance is small. FKDA 3 creates edges by wiring the anonymized vertices to random residual vertices, so that it incurs the edit distance to the original graph in between.

The abilities of the three heuristics on the preservation of utility of the original graph differ from each other, depending on the structure of the original graph.

For example, Figure 6 shows that FKDA 1 incurs larger average shortest path length in the anonymized Wiki-Vote than FKDA 2 does. This suggests that the vertices in Wiki-Vote with similar degrees are more connected than the vertices with very different degrees. So creating edges by wiring an anonymized vertex to the vertices from with large degree to small degree (similar degree to different degree) in edge_creation of FKDA 1 does not reduce the average shortest path length much. To the contrary, FKDA 2 links vertices with very different degrees in edge_creation, which results in a significant reduction in the average shortest path length. However, Figure 9 shows the reverse result in the anonymized Email-Enron, which suggests that the vertices in Email-Enron with similar degrees are less connected than the vertices with very different degrees. The overall results show that FKDA 1 and FKDA 2 preserve the utilities of the original graph better than FKDA 3 does. Nevertheless, FKDA 3 has an interesting property that it can generate a random k-degree anonymous graph.

4.4 Efficiency Evaluation

We compare the efficiency of the algorithms by measuring their execution time.

We vary the value of k in the range $\{5, 10, 15, 20, 25, 50, 100\}$. For each value of k, we run each algorithm 10 times on each dataset and compute the average execution time. We also compute the speedup of FKDA versus KDA for each parameter setting.

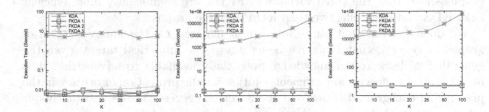

Fig. 10. Execution time on Email-Urv   **Fig. 11.** Execution time on Wiki-Vote   **Fig. 12.** Execution time on Email-Enron

Figure 10, 11 and 12 show the execution times on Email-Urv, Wiki-Vote and Email-Enron, respectively. Figure 13, 14 and 15 show the corresponding speedups.

We see that FKDA is significantly more efficient than KDA. The speedup varies from hundreds to one million on different graphs. The inefficiency of KDA is due to the decoupling of the checking of realizability of the anonymized degree sequences from the construction of graph.

The efficiency of the three FKDA variants is similar. FKDA 1 and FKDA 2 are slightly faster than FKDA 3. This is because FKDA 3 maintains additional a list of candidate residual vertices in edge_creation.

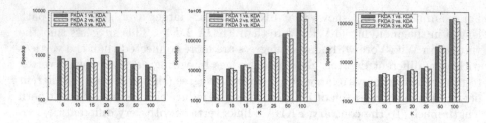

Fig. 13. Speedup of FKDA vs. KDA on Email-Urv   **Fig. 14.** Speedup of FKDA vs. KDA on Wiki-Vote   **Fig. 15.** Speedup of FKDA vs. KDA on Email-Enron

5   Conclusion

In this paper, we propose a greedy k-degree anonymization algorithm that anonymizes a graph by simultaneously adding edges and anonymizing its degree sequence in groups of vertices.

The algorithm is designed to overcome the shortcomings of the KDA algorithm proposed by [13]. The simultaneity of degree anonymization and graph construction in the new FKDA algorithm eliminates the need for realizability testing, which, as confirmed by our experiments, is a significant factor in the poor efficiency of the KDA algorithm.

We propose three variants of the algorithm, corresponding to three wiring heuristics. The comparative empirical performance evaluation on three real world graphs shows that the three variants of FKDA are significantly more efficient than KDA and more effective than KDA on large graphs.

We do not claim that our solution is a panacea for the anonymization of graphs in general, that objective being anyway a chimerical target given the generality of background knowledge potentially available to adversaries. It is however a very effective and efficient solution for the protection of privacy in the presence of background knowledge about vertex degrees. More importantly our solution shows that it is possible to tightly knit realizability and construction into one anonymization process and therefore paves the way to the development of algorithms catering for a variety of background structural knowledge.

References

1. Backstrom, L., Dwork, C., Kleinberg, J.M.: Wherefore art thou R3579X?: Anonymized social networks, hidden patterns, and structural steganography. Commun. ACM 54(12) (2011)
2. Barabási, A.-L., Albert, R.: Emergence of Scaling in Random Networks. Science 286, 509–512 (1999)
3. Bhagat, S., Cormode, G., Krishnamurthy, B., Srivastava, D.: Class-based graph anonymization for social network data. PVLDB 2(1) (2009)
4. Campan, A., Truta, T.M.: A clustering approach for data and structural anonymity in social networks. In: PinKDD (2008)

5. Cheng, J., Fu, A.W.-C., Liu, J.: K-isomorphism: privacy-preserving network publication against structural attacks. In: SIGMOD (2010)
6. Clauset, A., Shalizi, C.R., Newman, M.E.J.: Power-law distributions in empirical data. SIAM Reviews (2007)
7. Cormode, G., Srivastava, D., Yu, T., Zhang, Q.: Anonymizing bipartite graph data using safe groupings. PVLDB 19(1) (2010)
8. Francesco Bonchi, A.G., Tassa, T.: Identity obfuscation in graphs through the information theoretic lens. In: ICDE (2011)
9. Ghinita, G., Karras, P., Kalnis, P., Mamoulis, N.: Fast data anonymization with low information loss. In: VLDB, pp. 758–769 (2007)
10. Hay, M., Miklau, G., Jensen, D., Towsley, D., Weis, P.: Resisting structural re-identification in anonymized social networks. PVLDB 1(1), 102–114 (2008)
11. Korolova, A., Motwani, R., Nabar, S.U., Xu, Y.: Link privacy in social networks. In: CIKM (2008)
12. Li, Y., Shen, H.: Anonymizing graphs against weight-based attacks. In: ICDM Workshops (2010)
13. Liu, K., Terzi, E.: Towards identity anonymization on graphs. In: SIGMOD Conference, pp. 93–106 (2008)
14. Liu, L., Wang, J., Liu, J., Zhang, J.: Privacy preserving in social networks against sensitive edge disclosure. In: SIAM International Conference on Data Mining (2009)
15. Song, Y., Nobari, S., Lu, X., Karras, P., Bressan, S.: On the privacy and utility of anonymized social networks. In: iiWAS (2011)
16. Sweeney, L.: K-anonymity: a model for protecting privacy. International Journal of Uncertainty, Fuzziness and Knowledge-Based Systems 10(5) (2002)
17. Tai, C.-H., Yu, P.S., Yang, D.-N., Chen, M.-S.: Privacy-preserving social network publication against friendship attacks. In: SIGKDD (2011)
18. Wu, W., Xiao, Y., Wang, W., He, Z., Wang, Z.: K-symmetry model for identity anonymization in social networks. In: EDBT (2010)
19. Ying, X., Wu, X.: Randomizing social networks: a spectrum perserving approach. In: SDM (2008)
20. Yuan, M., Chen, L., Yu, P.S.: Personalized privacy protection in social networks. PVLDB 4(2) (2010)
21. Zheleva, E., Getoor, L.: Preserving the Privacy of Sensitive Relationships in Graph Data. In: Bonchi, F., Malin, B., Saygın, Y. (eds.) PInKDD 2007. LNCS, vol. 4890, pp. 153–171. Springer, Heidelberg (2008)
22. Zhou, B., Pei, J.: Preserving privacy in social networks against neighborhood attacks. In: ICDE (2008)
23. Zou, L., Chen, L., Özsu, M.T.: K-automorphism: a general framework for privacy-preserving network publication. PVLDB 2(1) (2009)

Probabilistic Inference
of Fine-Grained Data Provenance

Mohammad Rezwanul Huq, Peter M.G. Apers, and Andreas Wombacher

University of Twente, 7500 AE Enschede, The Netherlands
{m.r.huq,p.m.g.apers,a.wombacher}@utwente.nl

Abstract. Decision making, process control and e-science applications process stream data, mostly produced by sensors. To control and monitor these applications, reproducibility of result is a vital requirement. However, it requires massive amount of storage space to store fine-grained provenance data especially for those transformations with overlapping sliding windows. In this paper, we propose a probabilistic technique to infer fine-grained provenance which can also estimate the accuracy beforehand. Our evaluation shows that the probabilistic inference technique achieves same level of accuracy as the other approaches do, with minimal prior knowledge.

1 Introduction

Sensors produce data tuples in form of streaming and these tuples are used by the applications to take decisions as well as to control operations. In case of any wrong decision, it is important to have reproducibility to validate the previous outcome. Reproducibility refers to the ability of producing the same output after having applied the same transformation process on the same set of input data, irrespective of the process execution time. To be able to reproduce results, we need to store provenance data, a kind of metadata relevant to the transformation process and associated input and output dataset.

Data provenance refers to the derivation history of data from its original sources [15]. It can be defined either at the tuple-level or at the relation-level [6] also known as fine-grained and coarse-grained data provenance respectively. Fine-grained data provenance can achieve reproducibility because it documents the used set of input tuples for each output tuple and the transformation process as well. On the other hand, coarse-grained data provenance cannot achieve reproducibility because of the updates and delayed arrival of tuples. However, maintaining fine-grained data provenance in stream data processing is challenging. In stream data processing, a transformation process is continuously executed on a subset of the data stream known as a window. Executing a transformation process on a window requires to document fine-grained provenance data for this processing step to enable reproducibility. If a window is large and subsequent windows overlap significantly, the size of provenance data becomes a multiple of the actual sensor data. Since provenance data is 'just' metadata and less often used by the end users, this approach seems to be infeasible and too expensive [11].

S.W. Liddle et al. (Eds.): DEXA 2012, Part I, LNCS 7446, pp. 296–310, 2012.

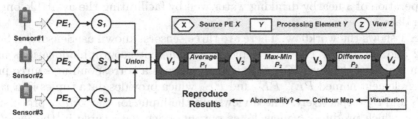

Fig. 1. Example workflow

The storage requirement can be significantly reduced if the fine-grained data provenance is not explicitly documented, but inferred based on coarse-grained data provenance and reproducible states of the database enabled by a temporal data model [12], known as *basic provenance inference*. Since the characteristics of a stream data processing often varies over time, the inference mechanism has to account for this dynamics. In particular, two parameters are important:

- Processing delay or δ refers to the time required to execute the transformation process on the current window.
- Sampling time or λ refers to the time between the arrival of the current tuple and the subsequent one.

The inference algorithm proposed in this paper uses the given processing delay, δ and sampling time, λ distribution to improve the basic inference algorithm. In particular, the input window is shifted such that the achievable accuracy of the inferred fine-grained data provenance is optimized. The distance of the shift is determined by the relationship among δ, λ distribution and tuples arrival within a window.

The proposed probabilistic approach has an advantage over the approach discussed in [10], which requires to observe specific distributions deduced from the sampling time distribution at runtime. As a consequence, estimating the accuracy of the inference algorithm is not possible at the design time of the processing, since the special distributions are not known in prior. The probabilistic method can estimate the accuracy of the inference at design time since the method has no requirement of observing any distribution. Inference of tuple-based windows is independent of these special distributions, thus the results are not repeated here and we only focus on time-based windows in this paper.

2   Motivating Scenario

RECORD[1] is one of the projects in the context of the Swiss Experiment[2], which is a platform to enable real-time environmental experiments. Several sensors have been deployed to monitor river restoration effects. Some of them measure electric conductivity of water which indicates the number of ions in the water. Increasing conductivity refers to higher level of salt in the water. We are interested to control

---

[1] http://www.swiss-experiment.ch/index.php/Record:Home
[2] http://www.swiss-experiment.ch/

the operation of a nearby drinking water well by facilitating the available online sensor data.

Fig. 1 shows the workflow. There are three sensors, known as: Sensor#1, Sensor#2 and Sensor#3. They are deployed in different geographic locations in a known region of the river. For each sensor, there is a corresponding source processing element named PE_1, PE_2 and PE_3 which provides data tuples in a *view* S_1, S_2 and S_3 respectively. These views are the input for the *Union* processing element which produces a view V_1 as output. Each data tuple in the view V_1 is attached with an explicit timestamp referring to the point in time when it is inserted into the database (also known as *transaction time*). Next, the view V_1 is fed to the processing element P_1 which calculates the average value per window and then generates a new view V_2. The task of P_2 is to calculate the maximum and minimum value per input window of view V_2 and store the aggregated value in view V_3. Next, V_3 is used by P_3 which calculates the difference between the maximum and minimum electric conductivity over the selected region at a particular point in time. The view V_4 holds these output data tuples along with the *transaction time* and gives significant information about the fluctuation of electric conductivity. Later, *Visualization* processing element facilitates V_4 to produce a contour map of the fluctuation of the electric conductivity in that selected region of the river. If the map shows any abnormality, researchers may want to reproduce results to validate their model. We consider the shaded part in Fig. 1 to discuss and evaluate our proposed solution later in this paper.

3   Basic Provenance Inference

The *basic provenance inference* algorithm has been reported in [12]. Since our proposed *probabilistic provenance inference* algorithm is based on the fundamental principle of the basic algorithm, we discuss this algorithm first and then explain its limitations and propose the *probabilistic provenance inference* algorithm. To explain this algorithm, we consider the processing element P_1 shown in Fig.1, that takes view V_1 as input and produces view V_2. Moreover, we assume that, sampling time is 2 time units, window size is 5 time units and the window triggers after every 5 time units.

3.1   Document Coarse-Grained Provenance

At first, we document coarse-grained provenance of P_1 which is a one-time action, and performed during the setup of this processing element. The stored provenance information is quite similar to *process provenance* reported in [16]. Inspired from this, we keep the following information of a processing element specification based on [17] as coarse-grained data provenance.

- Number of sources: indicates the total number of source views.
- Source names: a set of source view names.
- Window types: a set of window types; one element for each source. The value can be either *tuple* or *time*.

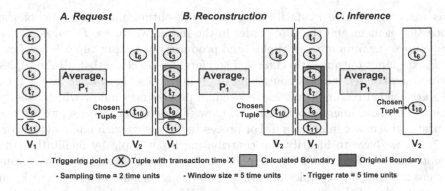

Fig. 2. Request, Reconstruction & Inference of Provenance Algorithm

- Window predicates: a set of window predicates; one element for each source. The value actually represents the size of the window.
- Trigger type: specifies how the *processing element* will be triggered for execution (e.g. *tuple* or *time* based)
- Trigger predicate: specifies when a *processing element* will be triggered for execution.

3.2 Reconstruct Processing Window

This phase will be only executed if the provenance information is requested for a particular output tuple T generated by P_1 and it returns the set of tuples which reconstruct the processing window. Here, the tuple T is referred to as *chosen tuple* for which provenance information is requested and the horizontal dashed line indicates the time when that particular window triggers, known as triggering point (see Fig. 2.A).

We apply a temporal data model on streaming sensor data to retrieve appropriate data tuples based on the given timestamp. The temporal attributes are: i) **valid time** represents the point in time a tuple was created by a sensor and ii) **transaction time** is the point in time a tuple is inserted into a database. The *valid* and *transaction* time is also known as application and system timestamp. While *valid time* is anyway maintained in sensor data, *transaction time* attribute requires extra storage space.

Fig. 2.B shows the reconstruction phase. The *transaction time* of the chosen tuple is t_{10} which is the *reference point* to reconstruct the processing window. Since window size is 5 time units, we retrieve the tuples having *transaction time* within the boundary $[t_5, t_{10})$ from the view V_1. This set of tuples reconstruct the processing window which is shown by the tuples surrounded by a light shaded rectangle in Fig. 2.B.

3.3 Provenance Inference

The last phase of the *basic provenance inference* establishes the relationship among the chosen output tuple with the set of contributing input tuples.

This mapping is done by facilitating the input-output mapping ratio of the processing element and the tuple order in the respective views. P_1 takes all the input tuples (i.e. n number of tuples) and produces one output tuple. Therefore, for P_1, the input-output ratio is $n : 1$. Therefore, we conclude that all the tuples in the reconstructed window contribute to produce the *chosen tuple*. In Fig.2.C, the dark shaded rectangle shows the original processing window which exactly coincides with our inferred processing window. Therefore, in this case, we achieve accurate provenance information. For processing elements with input-output ratio $1 : 1$, we have to identify the contributing input tuple by facilitating the *monotonicity in tuple ordering* property in both views V_1 and V_2. This property ensures that input tuples of view V_1 producing output tuples of view V_2 in the same order of their *transaction time* and this order is also preserved in the output view V_2.

3.4   Discussion

The *basic provenance inference* algorithm has few requirements to be satisfied. Most of the requirements are already introduced to process streaming data in literature. In [13], authors propose to use transaction time on incoming stream data. We assume that the windows are defined and evaluated based on *transaction time*, i.e. system timestamp. However, our inference-based methods are also applicable if the window is built on *valid time* or application timestamp. In this case, if an input tuple arrives after the window execution, we can ignore that tuple since it's transaction time is greater than the transaction time of the output tuple. Ensuring temporal ordering of data tuples is another requirement for provenance inference.

The basic inference method performs well if the processing delay is not significant, i.e. processing is infinitely fast. However, in case of a significant processing delay and variable sampling time it cannot infer accurate provenance. The next section demonstrates few cases where inaccurate provenance is provided by the basic inference method.

4   Inaccuracy in Time-Based Windows

To explain different cases where inaccurate provenance is inferred, we introduce few basic concepts of our inference model first. For the processing element P_j, λ_j refers to the sampling time of the input view of P_j. The windows are defined over the input view of P_j and assuming that W be the set of processing windows where $W = \{w_i \mid w_i \in W\}$ where $i = 1, 2, ..., n$. There might be a small time gap between the starting of the window w_i and appearance of the first tuple in w_i. This time gap is denoted by $\alpha(w_i)$. Accordingly, the time between the last tuple in w_i and the triggering point is denoted by $\beta(w_i)$. Then, each w_i needs some time to finish the processing, i.e. processing delay, which is denoted as $\delta(w_i)$.

Fig. 3 shows different cases in a time-based window of 5 time units which triggers after every 5 time units, defined over the input view V_1 of the processing

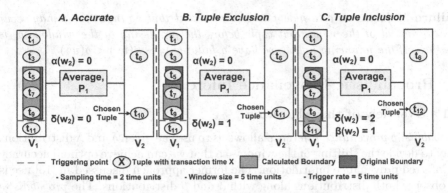

Fig. 3. Inaccuracy in time-based windows

element P_1 with $\lambda_1 = 2$ time units. The first case shown in Fig. 3.A, is the case described in Section 3. The window w_2 triggers at t_{10} shown by the dashed line and the output tuple is also produced at t_{10}. Therefore, processing delay $\delta(w_2) = 0$ time unit. Since the processing is infinitely fast, both original and inferred processing window have the same boundary $[t_5, t_{10})$. Therefore, the basic provenance inference provides accurate provenance in this case.

Fig. 3.B shows another case where the same window w_2 triggers at t_{10} and the output tuple is produced at t_{11}. Therefore, $\delta(w_2) = 1$ time unit. Earlier, the window w_2 began at t_5 and the transaction time of the first tuple within w_2 is also t_5. Therefore, $\alpha(w_2) = 0$ time unit. Based on the basic provenance inference technique, the reconstructed processing window contains tuples having transaction time within $[t_6, t_{11})$ shown by the light shaded rectangle. However, the original window w_2 has the boundary $[t_5, t_{10})$ shown by the dark shaded rectangle. Therefore, the inferred provenance is inaccurate since the input tuple with transaction time t_5 is not included in the reconstructed window. This failure of providing accurate provenance can be defined as follows.

Failure 1. *Exclusion of a contributing tuple from the lower end of the window w_i may occur if the processing delay $\delta(w_i)$ is longer than the difference between the first input tuple in w_i and the time at which w_i starts. If the following condition holds, we have a failure: $\alpha(w_i) < \delta(w_i)$*

Fig. 3.C shows the last case where the window w_2 triggers at the same time as in the previous cases and the output tuple is produced at t_{12}. Therefore, $\delta(w_2) = 2$ time units. As described in the previous case, $\alpha(w_2)$ remains the same which is 0 time unit. The transaction time of the last tuple within w_2 is t_9. Therefore, $\beta(w_2) = 1$ time unit. The basic algorithm returns the reconstructed window with the boundary $[t_7, t_{12})$. However, the original window w_2 has the boundary $[t_5, t_{10})$. Therefore, the inferred provenance is inaccurate and one of the reasons for that is the input tuple with transaction time t_{11} is included in the reconstructed window which was not contributing to produce the *chosen tuple* during the original processing. This failure can be defined as follows.

Failure 2. *Inclusion of a newly arrived non-contributing input tuple may occur due to arrival of the new input tuple before the processing of the window w_i is finished. If the following holds, we have a failure:* $\lambda_j - \beta(w_i) < \delta(w_i)$

5   Probabilistic Provenance Inference

5.1   Overview of the Algorithm

Probabilistic provenance inference allows us to use the given δ and λ distributions only to decide the shifting of the window so that we can achieve optimal accuracy of inferred provenance information. The former approach discussed in [10] needs to observe both distributions along with α and β distributions. The *probabilistic* approach facilitates Markov chain modeling on the arrival of data tuples within a window to calculate both α and β distributions which are then used in the process of adapting the window size. A Markov chain is a mathematical system that represents the undergoing transitions from one state to another in a chain-like manner [4].

The major advantage of using the *probabilistic* method is that it can estimate the accuracy at the design time since it depends on the given distributions. This accuracy estimation provides users useful hint about the applicability of the inference mechanism beforehand. Furthermore, our evaluation shows that the actual accuracy achieved using probabilistic inference is comparable to the accuracy of the *adaptive* approach [10] although less prior knowledge is required for the *probabilistic* approach to achieve this level of accuracy.

5.2   Required Parameters

We propose a novel *tuple-state* graph based on the principle of a Markov chain to calculate both α and β distributions which eventually help us to infer fine-grained data provenance. To do so, different parameters are required. The number of vertices in the *tuple-state* graph depends on the given *window size* of the processing element. The transitions from one vertex to another depend on the λ distribution and the *trigger rate*. We use the example described in Section 3 where a time-based window is defined over the input view of P_1 with window size $= 5$ time units and trigger rate $= 5$ time units.

Furthermore, to build the *tuple-state* graph, the given λ and δ distributions are used. The λ distribution of the input view V_1, i.e. λ_1, follows poisson distribution with the following values: $P(\lambda_1 = 1) = 0.37$, $P(\lambda_1 = 2) = 0.39$ and $P(\lambda_1 = 3) = 0.24$ where $mean = 2$. The δ distribution of P_1 also follows poisson distribution with $mean = 1$. The values of the δ distribution are: $P(\delta(w_i) = 1) = 0.68$, $P(\delta(w_i) = 2) = 0.32$.

5.3   Building Tuple-State Graph to Calculate α Distribution

Based on the given λ_1 distribution, it is possible to construct a Markov model for determining the α distribution, i.e., the probability for a tuple arriving with a specific distance from the start of the window.

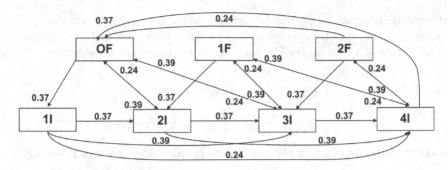

Fig. 4. Tuple-state graph to calculate α distribution

For each processing element P_i, a *tuple-state* graph G_α has to be built to compute the corresponding α distribution. Each vertex in the *tuple-state* graph represents a state, which identifies the position of a tuple within a processing window w.r.t. the start of the window. There are two different types of states in a *tuple-state* graph. These are:

1. *First states*: These states represent that the current tuple is the first tuple of a particular window. These are denoted as the arrival timestamp of the tuple in the window w.r.t the start of the window followed by a letter 'F' (e.g. 0F, 1F, 2F).
2. *Intermediate states*: These states represent the arrival of tuples within a window without being the first tuple. The states are represented by the arrival timestamp of the new tuple in the window w.r.t the start of the window followed by a letter 'I' (1I, 2I, 3I, 4I).

The construction of the *tuple-state* graph for processing element P_1 mentioned in Fig. 3 is described below. First, a set of first and intermediate states as vertices are added to $G_\alpha(V, E)$. The number of vertices in both states is bounded by the window size. It can be expressed as: $V = \bigcup_{j=0}^{WS_1} \{jF, jI\}$ where WS_1 be the *window size* of V_1 which is the input view of P_1.

Next, we add edges from all vertices based on the value of the tuple arrival distribution λ_1. An edge is defined via the start vertex (*from vertex*), the end vertex (*to vertex*), and the probability of this edge occurring (*weight*).

A directed edge can be defined from every point in the window to a later point in the window without crossing the window boundary. The start vertex could be a first or intermediate state, while the end vertex is an intermediate state. Assume that, TR_1 be the *trigger rate* of P_1, the formula below represents these edges, where the weight associated to an edge corresponds to the probability of two subsequent tuples arriving with a distance of $k - j$ time units.

$$E_1 = \bigcup_{j=0}^{WS_1-1} \bigcup_{k=j+1}^{j+max(\lambda_1)} \{\{(jF, kI, P(\lambda_1 = k - j)),$$

$$(jI, kI, P(\lambda_1 = k - j))\} \mid k < TR_1\}$$

Furthermore, directed edges can be defined which are crossing window boundaries. In this case, the start vertex is either a first or an intermediate state, while the end vertex is a first state. The formula below represent these edges.

$$E_2 = \bigcup_{j=0}^{WS_1-1} \bigcup_{k=j+1}^{j+max(\lambda_1)} \{\,\{\,(\,jF, k'F, P(\lambda_1 = k - j)\,),$$

$$(\,jI, k'F, P(\lambda_1 = k - j))\}|k \geq TR_1 \wedge k' = k \mod TR_1\}$$

The complete set of edges in the *tuple-state* graph is the union of E1 and E2.

$$E = E_1 \cup E_2$$

Fig. 4 depicts a *tuple-state* graph to calculate α distribution for the processing element P_1. Given, $P(\lambda_1 = 1) = 0.37$, $P(\lambda_1 = 2) = 0.39$ and $P(\lambda_1 = 3) = 0.24$. Starting from the vertex $0F$, edges are added to $1I$, $2I$ and $3I$ with weight 0.37, 0.39 and 0.24 respectively. These edges are the elements of set E_1. As another example, consider starting from vertex $4I$, we add edges to $0F$, $1F$ and $2F$ with weight 0.37, 0.39 and 0.24 respectively. These edges are elements of set E_2. This process will be continued for all the veritces to get a complete G_α.

5.4   Steady-State Distribution Vector

The long-term behavior of a Markov chain enters a steady state, i.e. the probability of being in a state will not change with time [9]. In the steady state, the vector s represents the average probability of being in a particular state. To optimize the steady state calculation, vertices with no incoming edges can be ignored. Since the steady state analysis of the Markov model considers these states irrelevant those vertices and associated edges are removed.

Assuming uniformly distributed initial probabilities, the steady state of the Markov model can be derived. The probabilities of states with suffix 'F' form the α distribution for processing element P_1, i.e., the probabilities of the first tuple in a window arrives after a specific number of time units. The steady state distribution vector s_α for the *tuple-state* graph, G_α (see Fig. 4) is:

$$s_\alpha = \left(\frac{0F \quad 1F \quad 2F \quad 1I \quad 2I \quad 3I \quad 4I}{0.20\,0.13\,0.05\,0.07\,0.15\,0.20\,0.20}\right)$$

The components of the states 0F, 1F, 2F represent the probability of the value of $\alpha = 0$, 1 and 2 respectively. After normalizing the probability of these values, we get the model-given distribution of α. Table 1.a shows that the α distribution achieved facilitating the tuple-state graph G_α is comparable with the observed α distribution.

5.5   Calculating β Distribution

Along the lines of the previous two subsections, also the β distribution indicating the probability distribution on the distance between the last tuple in a window and the end of the window can be calculated. Due to a lack of space we do not describe this construction in detail.

Table 1. Comparison of α distributions and Joint prob. distribution of α & δ

(a) Observed vs. model-given α distribution

α	observed	model-given
0	0.532	0.535
1	0.347	0.337
2	0.121	0.128

(b) Joint probability distribution of model-given α and δ

$\alpha = x$	$\delta = y$	$P(\alpha = x, \delta = y)$
0	1	0.364 (a)
0	2	0.171 (b)
1	1	0.229 (c)
1	2	0.108 (d)
2	1	0.087 (e)
2	2	0.041 (f)

5.6 Accuracy Estimation and Shifting of the Window

The proposed *probabilistic* technique shifts the window based on the model-given α and the given δ distribution. The original window should be shifted in such a way that we can avoid both failures described in Section 4. Before shifting the window, the transaction time of the *chosen tuple* is referred to as the *current reference* point which also indicates the upper end of the window beyond which no more tuples would be considered. At first, the upper end of the window is adjusted and the point in time after the adjustment is known as the *new reference* point. The time gap between current and new reference point is called as *offset*. Therefore, the formula to calculate the upper end is: $UpperEnd = TransactionTime - offset$. To calculate the lower end of the window, we subtract the window size from the upper end, i.e. $LowerEnd = TransactionTime - windowSize - offset$. The value of *offset* is determined using the joint probability distribution of model-given α and the given δ distribution. Both of these distributions are related to the processing element P_1 which is mentioned in Fig. 3.

Since α and δ are two independent variables, the joint probability distribution of these two variables can be calculated and is shown in Table 1.b. The δ distribution given in Section 5.2 is used for the calculation. If the *offset* is set to 0, the value of δ remains the same. Based on the definitions of failures in Section 4 and from Table 1.b, it is clear that if the *offset* = 0, only about 36% (c+e+f) accurate provenance could be achieved. However, if the *offset* is set to 1, δ is also subtracted by 1 which greatly reduces the chance of inaccuracy. According to our failure conditions discussed in Section 4 and from Table 1.b, the chance of inaccuracy is 17% (b). Therefore, setting the *offset* value 1 would achieve around 83% accuracy. If the *offset* is 2 then the percentage of inaccuracy again increases mainly due to the inclusion of non-contributing tuples from the lower end of the window. Therefore, based on the joint probability distribution, we choose *offset* = 1 which gives the optimal estimated accuracy of 83%.

In Fig. 3, we discuss three different cases by altering the δ value. Since $\delta = 0$ in the first case, it falls outside the scope of our proposed algorithm because the proposed *probabilistic* approach assumes that there exists some processing delay. Applying our probabilistic inference algorithm with *offset*= 1 for the other two cases, case B would return the inferred window $w_2 = [t_5, t_{10})$ which exactly coincides with the actual window w_2. Thus, we infer accurate provenance. For case C, the inferred window w_2 contains tuples within the range $[t_6, t_{11})$ which differs from the actual window w_2. This is one of the examples where *probabilistic* inference provides inaccurate provenance due to the bigger processing delay.

6　Evaluation

6.1　Evaluating Criteria, Test Cases and Datasets

We evaluate our proposed *probabilistic provenance inference* algorithm using i) accuracy and ii) storage consumption. To compare accuracy, the traditional fine-grained provenance information, also known as *explicit* method, is used as a ground truth. We compare the accuracy among *basic* [12], *adaptive* [10] and *probabilistic* approach proposed in this paper through a simulation.

The simulation is executed for 10000 time units for the processing element P_1 mentioned in Section 2. Based on queuing theory, we assume that both sampling time λ and processing delay δ distribution follows poisson distribution. The 6 test cases shown in Table 2 are chosen carefully. Test case 1 is used through out this paper to explain our method. Test case 2 and 3 is almost similar to each other except the trigger rate. Test case 4 and 5 are the example of non-overlapping, tumbling windows with the only difference in the processing delay. Test case 6 is similar to test case 4 except the deviation in the sampling time.

Furthermore, we compare the storage requirement of *inference-based* approaches with *explicit* method of maintaining provenance. A real dataset[3] reporting electric conductivity of the water, collected by the RECORD project is used for this purpose. The input dataset contains 3000 tuples consuming 720kB.

6.2　Accuracy

Table 2 shows the accuracy achieved using the different algorithms for the aforesaid test cases. Test case 1 is the one which is used as the example through

Table 2. Different test cases used for evaluation and Evaluation result

			Test Cases					Accuracy		
Test case	Window size	Trigger rate	$avg(\lambda)$	$max(\lambda)$	$avg(\delta)$	$max(\delta)$	Basic	Adaptive	Probabilistic Estimated	Achieved
1	5	5	2	3	1	2	36%	83%	84%	83%
2	10	5	2	3	1	2	40%	83%	82%	83%
3	10	10	2	3	1	2	39%	85%	82%	83%
4	10	10	3	5	1	2	53%	87%	87%	87%
5	10	10	3	5	2	3	41%	75%	75%	74%
6	10	10	4	6	1	2	61%	92%	91%	92%

---

[3] http://data.permasense.ch/topology.html\#topology

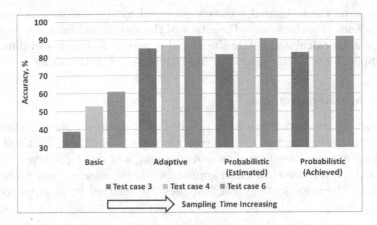

Fig. 5. Influence of Sampling Time over the accuracy

out this paper. In test case 1 & 2, only the window size is changed with other parameters remain unchanged. In both cases, we achieve almost the same level of accuracy for all algorithms. Therefore, it seems that *window size does not influence the accuracy*.

Next, we discuss the accuracy achieved comparing test case 2 and 3. These two cases have the same parameters except the trigger rate. Nevertheless, the result is again almost identical for all algorithms. This might indicate that *trigger rate has very little influence to the accuracy*.

The difference in parameters between test case 3 and 4 is $avg(\lambda)$ and $max(\lambda)$. The accuracy achieved in test case 4 for all the approaches is higher than those of case 3. The reason is that *increasing the sampling time of tuples and keeping the processing delay the same, may lower the chance of inaccuracy*.

Test case 4 and 5 differ in $avg(\lambda)$ and $max(\delta)$ parameters. The processing takes longer in test case 5 which influences the level of accuracy. The accuracy achieved in test case 5 is around 74% for our probabilistic approach where as it is 87% in test case 4. Therefore, *keeping the sampling time equal and increasing the processing delay might cause to achieve lower accuracy*.

Lastly, we introduce another test case for better understanding of the influence of sampling time on the accuracy. Test case 6 has the same parameters like test case 3 and 4 except $avg(\lambda)$ and $max(\lambda)$. The value of $avg(\lambda)$ is 4 and $max(\lambda)$ is 6 time units. Figure 5 shows the accuracy achieved for test case 3, 4 and 6 for the different approaches. From Fig. 5, we observe that increasing sampling time with other parameters unchanged, might provide more accurate inferred provenance information. Therefore, it might give a useful hint that *the higher the sampling time, the higher the accuracy*.

Our *probabilistic* algorithm uses minimal prior knowledge to infer fine-grained provenance data. However, the proposed *probabilistic* algorithm provides the same level of accuracy compared to the *adaptive* approach. The reason is that the α-distribution given by our *tuple-state* graph is very similar to the observed α-distribution produced for all test cases (see Table 1.a for test case 1).

Furthermore, the estimated accuracy provided by the *probabilistic* algorithm is almost identical to the achieved accuracy of the algorithm. Since, the estimated accuracy can be calculated before the actual experiment, it is a useful indicator for the applicability of the algorithm for a given set of distributions.

6.3   Storage Requirement

We measure the storage overhead to maintain fine-grained data provenance for the same processing element P_1. The result is reported in Table 3 for test case 1 and 2 which are the examples of non-overlapping and overlapping windows respectively. All three inference based approaches: *basic, adaptive* and *probabilistic* method have the same storage cost and they are referred to as *inference-based* methods.

Table 3. Provenance data storage consumption (in KB)

Method	Non-Overlapping (Test Case 1)		Overlapping (Test Case 2)	
	Space consumed	Ratio	Space consumed	Ratio
Explicit method	950	5.5:1	1925	11:1
Inference-based methods	175		175	

Table 3 shows the storage cost to maintain fine-grained provenance data for different methods. In case of test case 1, *inference*-based methods take almost 6 times less space than the *explicit* method. Since the trigger rate is the same, test case 2 also produces as many output tuples as produced in test case 1. The storage cost of inference-based methods only depends on the number of input and output tuples. Therefore, the storage consumed in test case 2 by the *inference-based* methods remains the same. However, the consumed storage space for the explicit method gets bigger due to the larger window size and overlapping windows. Therefore, in test case 2, the *inference-based* methods take 11 times less space than the explicit method. This ratio of course will vary based on the window size, overlapping between windows and number of output tuples. *The bigger the window and overlapping between windows, the higher the ratio of space consumption between explicit and inference-based approaches.*

7   Related Work

The work reported in [2] and [1] discuss the projects which facilitate the execution of continuous queries and stream data processing . All these techniques proposed optimization for storage space consumed by sensor data. However, none of these systems offer fine-grained data provenance in stream data processing.

In [5], authors described a data model to compute provenance on both relation and tuple level. This data model follows a graph pattern and shows case studies for traditional data but it does not address how to handle streaming data and associated overlapping windows.

In [8], authors have presented an algorithm for lineage tracing in a data warehouse environment. They have provided data provenance on tuple level. LIVE [14] is an offshoot of this approach which supports streaming data. It is a complete DBMS which preserves explicitly the lineage of derived data items in form of boolean algebra. However, these techniques incur extra storage overhead to maintain fine-grained data provenance.

In sensornet republishing [13], the system documents the transformation of online sensor data to allow users to understand how processed results are derived and support to detect and correct anomalies. They used an annotation-based approach to represent data provenance explicitly. However, our proposed method does not store fine-grained provenance data rather infer provenance data.

In [7], authors proposed approaches to reduce the amount of storage required for provenance data. To minimize provenance storage, they remove common provenance records; only one copy is stored. Then, using an extra provenance pointer, data tuples can be associated with their appropriate provenance records. Their approach seems to have less storage consumption than traditional fine-grained provenance in case of sliding overlapping windows.

A layered model to represent workflow provenance is introduced in [3]. The layers presented in the model are responsible to satisfy different types of provenance queries including queries about a specific activity in the workflow. A relational DBMS has been used to store captured provenance data. The authors have not introduced any inference mechanism for provenance data.

Our earlier work described in [12] can infer fine-grained provenance information for one processing step only. This technique is known as *basic provenance inference*. However, it did not take system dynamics into account. *Adaptive inference* technique provides inferred provenance considering the changes in system characteristics [10]. However, it requires to have some additional knowledge about different specific distributions which must be observed during runtime. Proposed *probabilistic provenance inference* can infer provenance and estimate the accuracy without observing those specific distributions.

8   Conclusion and Future Work

The proposed *probabilistic* approach is capable of addressing the dynamics of a streaming system because of it's adaptivity based on tuple arrival patterns and processing delay. Further, it provides highly accurate provenance. We compare the *probabilistic* method with other inference-based methods and the results show that it gives the same accuracy as the *adaptive inference* method. However, the advantage of using the *probabilistic* method is to have a guaranteed accuracy level based on the given distributions. Furthermore, it also reduces storage costs to maintain provenance data like any other inference-based methods. In future, we will extend this technique to infer provenance for a chain of processing elements.

References

1. Abadi, D., et al.: The design of the borealis stream processing engine. In: CIDR 2005, Asilomar, CA, pp. 277–289 (2005)
2. Babcock, B., et al.: Models and issues in data stream systems. In: ACM SIGMOD-SIGACT-SIGART Symposium, pp. 1–16. ACM (2002)
3. Barga, R., Digiampietri, L.: Automatic capture and efficient storage of e-science experiment provenance. Concurrency and Computation: Practice and Experience 20(5), 419–429 (2008)
4. Bishop, C.M.: Patter Recognition and Machine Learning. Springer Science+Business Media LLC (2006)
5. Buneman, P., Khanna, S., Tan, W.-C.: Why and Where: A Characterization of Data Provenance. In: Van den Bussche, J., Vianu, V. (eds.) ICDT 2001. LNCS, vol. 1973, pp. 316–330. Springer, Heidelberg (2001)
6. Buneman, P., Tan, W.C.: Provenance in databases. In: SIGMOD, pp. 1171–1173. ACM (2007)
7. Chapman, A., et al.: Efficient provenance storage. In: SIGMOD, pp. 993–1006. ACM (2008)
8. Cui, Y., Widom, J.: Lineage tracing for general data warehouse transformations. VLDB Journal 12(1), 41–58 (2003)
9. Gebali, F.: Analysis of Computer and Communication Networks. Springer Science+Business Media LLC (2008)
10. Huq, M.R., Wombacher, A., Apers, P.M.G.: Adaptive inference of fine-grained data provenance to achieve high accuracy at lower storage costs. In: 7th IEEE International Conference on e-Science, pp. 202–209. IEEE Computer Society Press (2011)
11. Huq, M.R., Wombacher, A., Apers, P.M.G.: Facilitating fine grained data provenance using temporal data model. In: Proceedings of the 7th Workshop on Data Management for Sensor Networks (DMSN), pp. 8–13 (2010)
12. Huq, M.R., Wombacher, A., Apers, P.M.G.: Inferring Fine-Grained Data Provenance in Stream Data Processing: Reduced Storage Cost, High Accuracy. In: Hameurlain, A., Liddle, S.W., Schewe, K.-D., Zhou, X. (eds.) DEXA 2011, Part II. LNCS, vol. 6861, pp. 118–127. Springer, Heidelberg (2011)
13. Park, U., Heidemann, J.: Provenance in Sensornet Republishing. In: Freire, J., Koop, D., Moreau, L. (eds.) IPAW 2008. LNCS, vol. 5272, pp. 280–292. Springer, Heidelberg (2008)
14. Das Sarma, A., Theobald, M., Widom, J.: LIVE: A Lineage-Supported Versioned DBMS. In: Gertz, M., Ludäscher, B. (eds.) SSDBM 2010. LNCS, vol. 6187, pp. 416–433. Springer, Heidelberg (2010)
15. Simmhan, Y.L., et al.: A survey of data provenance in e-science. SIGMOD Rec. 34(3), 31–36 (2005)
16. Simmhan, Y.L., et al.: Karma2: Provenance management for data driven workflows. International Journal of Web Services Research 5, 1–23 (2008)
17. Wombacher, A.: Data workflow - a workflow model for continuous data processing. Technical Report TR-CTIT-10-12, CTIT, University of Twente, Enschede (2010)

Enhancing Utility and Privacy-Safety via Semi-homogenous Generalization

Xianmang He[1], Wei Wang[2], HuaHui Chen[1],
Guang Jin[1], Yefang Chen[1,*], and Yihong Dong[1]

[1] School of Information Science and Technology, NingBo University
No.818, Fenghua Road, Ning Bo, 315122, P.R. China
{hexianmang,chenhuahui,jinguang,chenyefang,dongyihong}@nbu.edu.cn
[2] School of Computer Science and Technology, Fudan University,
No.220, Handan Road, Shanghai, 200433, P.R. China
weiwang1@fudan.edu.cn

Abstract. The existing solutions to privacy preserving publication can be classified into the homogenous and non-homogenous generalization. The generalization of data increases the uncertainty of attribute values, and leads to the loss of information to some extent. The non-homogenous algorithm which is based on ring generalization, can reduce the information loss, and in the meanwhile, offering strong privacy preservation. This paper studies the cardinality of the assignments based on the ring generalization, and proved that its cardinality is $\alpha^n (\alpha > 1)$. In addition, we propose a semi-homogenous algorithm which can meet the requirement of preserving anonymity of sensitive attributes in data sharing, and reduce greatly the amount of information loss resulting from data generalization for implementing data anonymization.

Keywords: non-homogenous algorithm, privacy preservation, ring generalization, semi-homogenous algorithm.

1 Introduction

Disseminating aggregate statistics of private data has much benefit to the public. Organizations may need to release private data for the purposes of facilitating data analysis and research. For example, medical records of patients may be released by a hospital to aid the medical study. Assume that a hospital wants to publish records of Table 1, which is called microdata (T). Since attribute *Disease* is sensitive, we need to ensure that no adversary can accurately infer the disease of any patient from the published data. For this purpose, any unique identifier of patients, such as *Name* should be anonymized or excluded from the published data. However, it is still possible for the privacy leakage if adversaries have certain background knowledge about patients. For example, if an adversary knows that Alex is of age 19, Zipcode 14k and Sex F, s/he can infer that Bob's disease is emphysema since the combination of Age, Zipcode and Sex uniquely

---

* Corresponding author.

S.W. Liddle et al. (Eds.): DEXA 2012, Part I, LNCS 7446, pp. 311–325, 2012.

identify each patient in Table 1. The attribute set that uniquely identify each record in a table is usually referred to as a quasi-identifier(QI for short) of the table.

To overcome the privacy threat under the attack guided by background knowledge, k-anonymity ($k \geq 2$) was firstly proposed. A data set is k-anonymous if each record in the data set is indistinguishable from at least $k - 1$ other records with the same data set. The larger the value of k, the better the privacy is protected. A plethora of of k-anonymization algorithms have been developed, some of them but not limited to these are: Incognito [1], Mondrian [2], Utility anonymization [3], Spatial indexing techniques [4], Fast algorithm [5],etc. All these algorithms share a common framework: *partition the table into many QI-groups such that the size of each QI-group is not smaller than k, and then generalize each QI-group.* Such an approach, to which we refer as homogeneous generalization.

1.1   Homogeneous and Non-homogeneous Generalization

Homogeneous generalization has been widely used in many anonymization solutions [3, 6–10]. In a typical homogenous-based solution, tuples are first divided into subsets (each subset is referred to as a QI-group). Then, QI-values of each QI-group are generalized to the same generalized value in all quasi-identifiers within a partition. As an example, in order to achieve 4-anonymity, we generalize Table 1 into a QI-group. In other words, the homogenous generalization results into 1 equivalent classes with QI-values ($< 19 - 25, 12 - 20k, F/M >$). That is, assign the same generalized QI-values to tuples in the same group.

In contrast, non-homogeneous approach was proposed in recently years [11]. This approach allows tuples within a QI-group to take different generalized quasi-identifier values for groups of size larger than k. After a random assignment was posed on the sensitive attributes(an assignment is essentially a permutation, see Definition 1), it can provide better privacy protection and security, be able to resist various attacks.

Definition 1 (Math, Assignment [11]). *Given a table T and its anonymized table T^*, a match m is a 2-tuple $< t_i, t'_j >$, where $t_i \in T$, $t'_j \in T^*$ and the QI-values of t_i is included in that of t'_j. An assignment a is a set of matches $m_i = < t_{x_i}, t'_{y_j} >$, where $t_{x_i} \in T$, $t'_{y_j} \in T^*$, and for each pair of matches $m_i, m_j \in a$, then $x_i \neq x_j$ and $y_i \neq y_j$.*

Example 1. Table 1 is anonymized into Table 2 by a strategy called ring generalization: to assign the k consecutive tuples gen($t_i, t_{i+1}, \cdots, t_{i+k-1}$) to t_i. Note that if $i + j > |T_1|$, we use $i + j - |T_1|$ instead($|T_1|$ is the size of table T_1). Thus, tuples Alex, Lucy and Lily have a different generalized QI-values. Let *gen* be a generalization function that takes as input a set of tuples and returns a generalized domain. Lucy=gen(Lucy,Lily, Jane, Bob) ={[19-25], [12k-18k], [F/M]}. Similarly, Lily=gen(Lily, Jane, Bob, Sarah),Jane=gen (Jane, Bob, Sarah, Alex), Bob=gen (Bob, Sarah, Alex, Lucy), Sarah=gen(Sarah, Alex, Lucy,Lily). Finally,

Table 1. Microdata T

Name	Age	Zip	Sex	Disease
Alex	19	14k	F	Emphysema
Lucy	21	12k	F	Dyspepsia
Lily	19	12k	F	Bronchitis
Jane	21	13k	F	Pneumonia
Bob	25	18k	M	Gastritis
Sarah	21	20k	M	Flu

Table 2. Non-Generalization T^*

Name	Age	Zip	Sex	Disease
Alex	[19-21]	[12k-14k]	F	Emphysema
Lucy	[19-25]	[12k-18k]	F/M	Dyspepsia
Lily	[19-25]	[12k-20k]	F/M	Bronchitis
Jane	[19-25]	[12k-20k]	F/M	Pneumonia
Bob	[19-25]	[12k-20k]	F/M	Gastritis
Sarah	[19-21]	[12k-20k]	F/M	Flu

a random assignment of sensitive attributes (tuples can only be replacement to those covered by its quasi-identifier values) is: {Emphysema, Dyspepsia, and Bronchitis, Pneumonia, Gastritis, Flu} was replaced by {Emphysema, Dyspepsia, Pneumonia, Gastritis, Flu, and Bronchitis}.

By intuition, the smaller the sizes of intervals in the generalized tuples, the less information loss in the anonymization. Non-homogeneous generalization can achieve less information loss than homogeneous generalization. In our example, the NCP of homogenous generalized table is 18, while the non-homogenous generalized table is $15\frac{1}{3}$. (See the NCP in definition 2).

Although non-homogeneous-based or homogeneous-based algorithms have successfully achieved the privacy protection objective, as well recognized in many data analysis applications, another issue *utility* still needs to be carefully addressed. One of the direct measures of the utility of the generalized data is information loss. In order to make the anonymized data as useful as possible, it is required to reduce the information loss as much as possible. Consistent efforts have been dedicated to developing algorithms that improve utility of anonymized data while ensuring enough privacy-preservation.

Then, we may wonder, does generalization have to be homogeneous or non-homogeneous? In this paper, we propose a cross approach between homogenous and non-homogenous generalization to anonymize data. We will give the basic idea of our approach in the next subsection.

1.2 Semi-homogeneous Generalization

We propose *Semi-homogeneous anonymization* as our solution to improve the utility of generalization approaches while ensuring privacy preservation. The framework consists of the following major steps. First, we partition the tuples of the microdata into several QI-groups based on certain strategies, such that each QI-group has at least k tuples. Then, we generalize QI-values of each current QI-group such that for each tuple in the table the accumulated size of all QI-groups satisfying the tuple is not less than k by taking advantage of overlaps of generalized groups. To resist various attacks, the third step is to assign a random permutation on the sensitive attribute.

As an example, we illustrate the details to generalize Table 1 by our approach. Table 1 is partitioned into 3 sub-groups, denoted by G_1, G_2, G_3, respectively.

Table 3. Small QI-groups Table T_3

G_ID	Age	Zip	Sex	Disease
G1	[19-21]	[12k-13k]	F	Bronchitis
	[19-21]	[12k-13k]	F	Pneumonia
G2	[19-21]	[12k-14k]	F	Emphysema
	[19-21]	[12k-14k]	F	Dyspepsia
G3	[21-25]	[18k-20k]	M	Gastritis
	[21-25]	[18k-20k]	M	Flu

Table 4. Semi-homogenous Table T_4

Name	Age	Zip	Sex	Disease
Lily	[19-25]	[12k-20k]	F/M	Gastritis
Jane	[19-25]	[12k-20k]	F/M	Flu
Alex	[19-21]	[12k-14k]	F	Bronchitis
Lucy	[19-21]	[12k-14k]	F	Pneumonia
Bob	[19-25]	[12k-20k]	F/M	Emphysema
Sarah	[19-25]	[12k-20k]	F/M	Dyspepsia

Secondly, by exchanging the position of G_1 and G_2, we reorder $\{G_1, G_2, G_3\}$ to $\{G_2, G_1, G_3\}$ as indicated by the group-ID shown in Table 3. Order of other QI-groups is not significant. We construct 4-anonymized table to assign $gen(G_1, G_2)$ to G_2, $gen(G_2, G_3)$ to G_3, $gen(G_3, G_1)$ to G_1.

Let us take this example a little further. For tuples in G_2, we add tuples of Lily and Jane into the input of the procedure generalizing G_2, as a result we have the generalized QI-values of G_2 as $gen(\{$Alex, Lucy, Lily, Jane$\})=\{$19-21, 12k-14k, F$\}$. Similarly, for tuples in G_3, G_3's QI-values will be generalized from G_2's tuples as well as tuples of Sarah and Bob. As a result, G_2's QI-values are updated to be $gen(\{$Sarah, Bob, Alex, Lucy$\})=\{$19-25, 12k-20k, F/M,$\}$. The QI-values of G_1 are updated from its own tuples as well as tuples from the third group. Eventually, each tuple in the Table 4 satisfies the 4-anonymity requirement, that is, the projection of each tuple in Table 4 on Age, Zip, Sex will contain at least 4 tuples.

The final step is to assign a random assignment to the sensitive attribute of $\{G_1, G_2, G_3\}$. In our example, { {Bronchitis, Pneumonia}, {Emphysema, Dyspepsia}, { Gastritis, Flu} } was replaced by {{Gastritis, Flu}, {Bronchitis, Pneumonia}, {Emphysema, Dyspepsia}}. The number of such replacement is $O(\alpha^n)$, $\alpha > 1$, which is proved in this paper. The cardinality is far larger than the privacy level k therefore, the attacker will be very difficult to success.

From the prospective of data security, we can see that the semi-homogeneous method provides stronger privacy preservation. Assume that the attacker has some background knowledge about Lucy. He/She found that Lucy belongs to the second tuple in Table 2, and Lucy's sensitive value has not been replaced, then the attacker is easy to infer her sensitive value is Dyspepsia. As for our method, in the face of the same attacker, the attacker only come to learn that Lucy belongs the second QI-group in the table 4, even if the attacker captures that her sensitive values has been replaced to the third QI-group, the final success probability is only 50% (Emphysema or Dyspepsia). (Please note that these small groups in Table 4 is homogenous.)

To test the utility improvement of our approach, we calculate NCP (See the definition 2) for Table 2 and Table 4 are $15\frac{1}{3}$, $12\frac{1}{6}$, respectively. The significant decrease of NCP of Table 4 strongly suggests that information loss has been successfully reduced by our semi-homogeneous anonymization approach.

In this paper, our contribution consists of two parts: first, we study the cardinality of the random assignments, and prove the cardinality theorem; Then, we provide a technique to which we refer to *semi-homogeneous* that generates a cross between homogenous and non-homogenous generalization to anonymize the microdata. We conduct extensive experiments on real data sets to show the performance and utility improvement of our model.

The rest of the paper is organized as follows. In Section 2, we give the basic definitions and problem definition. In Section 3, the cardinality theorem is introduced. In Section 4, we present the details of our generalization algorithm. We review the previously related research in Section 5. In Section 6, we experimentally evaluate the efficiency and effectiveness of our techniques. Finally, the paper is concluded in Section 7.

2 Preliminaries

In this section, we will first discuss several fundamental concepts, then give the formal definition of major problem that will be studied in this paper. At last, the complexity of semi-homogenous anonymization is analyzed.

2.1 Basic Notations

Let T be a microdata table that contains the private information of a set of individuals and has d QI-attributes $A_1, ..., A_d$, and a sensitive attribute (SA) S. We consider that S is categorical, and every QI-attribute $A_i(1 \leq i \leq d)$ can be either numerical or categorical. All attributes have finite and positive domains. For each tuple $t \in T, t.A_i(1 \leq i \leq d)$ denotes its value on A_i, and $t.A_s$ represents its SA value. Now, we are ready to clarify several fundamental concepts.

A *quasi-identifier* $QI = \{A_1, A_2, \cdots, A_d\} \subseteq \{A_1, A_2, \cdots, A_n\}$ is a minimal set of attributes, which can be joined with external information in order to reveal the personal identity of individual records.

A *partition* P consists of several subsets $G_i(1 \leq i \leq m)$ of T, such that each tuple in T belongs to exactly one subset and $T = \bigcup_i^m G_i$. We refer to each subset G_i as a QI-group.

Example 2. Table 3 stems from the following partition of table 1. $P_1=\{\{$Lucy, Alex$\}, \{$Jane, Lily$\}, \{$Jame, Linda$\}, \{$Sarah, Bob$\}\}$.

2.2 Problem Definition

In our anonymization framework, K-anonymity of T^* is achieved by QI-groups with size of k. In general, we use a divisor of K as k such that each tuple in T^* satisfies $\frac{K}{k}$ QI-groups through linking operation simultaneously. The number of QI-groups satisfying a tuple is called the *order* of our anonymization approach. It is better but not necessary that k is a divisor of K. Assume that a publisher plans to anonymize a table to be 100-anonymity, which can be achieved by either 20-anonymity with *order* 5 or 34-anonymity with *order* 3.

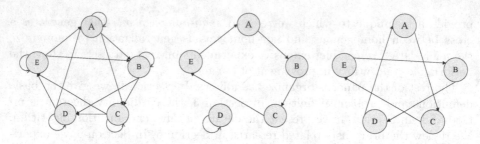

Fig. 1. Assignment Graph G **Fig. 2.** Assignment 3   **Fig. 3.** Assignment 13

Some methods have been developed to measure the information loss in anonymization. In this paper, we adopt the normalized certainty penalty to measure the information loss. We aim to produce a utility-friendly version anonymization for a microdata such that the privacy can be guaranteed and the information loss quantified by NCP is minimized.

Definition 2 (Normalized Certainty Penalty [3]). *Suppose a table T is anonymized to T^*. In the domain of each attribute in T, suppose there exists a global order on all possible values in the domain. If a tuple t in T^* has range $[x_i, y_i]$ on attribute $A_i (1 \leq i \leq d)$, then the normalized certainty penalty in t on A_i is $NCP_{A_i}(t) = \frac{|y_i - x_i|}{|A_i|}$, where $|A_i|$ is the domain of the attribute A_i. For tuple t, the normalized certainty penalty in t is $NCP(t) = \sum_i^d w_i \cdot NCP_{A_i}(t)$, where w_i are weights of attributes. The normalized certainty penalty in T is $\sum_{t \in T^*} NCP(t)$.*

Now, we are ready to give the formal definition about the problem that will be addressed in this paper.

Definition 3 (Problem Definition). *Given a table T and an integer K, anonymize it to be T^* such that T^* is K-anonymity by semi-homogenous generalization and information loss is minimized.*

2.3 Complexity

The studies [2,12,13] show that the problem of optimal k-anonymity is NP-hard even a simple quality metric is employed. Paper [3] proved that under the metric of NCP, the K-anonymity is NP-hard for $K \geq 2$. We have the following results on the complexity for linking-based model proposed in this paper.

Theorem 1. *(Complexity) The problem of optimal semi-homogenous anonymization is NP-hard.*

Proof. We can show that the suppression model used in [14] is a special case of the k-anonymity model defined here, where all weight are set to 1. It is clearly that the theorem follows from the result in [14].

3   The Cardinality of Assignments

We start this section by an example, which is to help us to understand the cardinality theorem.

Example 3. Consider that a table with five tuples {A, B, C, D, E} as illustrated in Figure 1. The Assignment Graph was constructed from the ring-generalization with $k = 3$: $A = gen(A, B, C), B = gen(B, C, D), C = gen(C, D, E), D = gen(D, E, A), E = gen(E, A, B)$. Figure 2-3 show 2 assignments, while Table 5 provides all the assignments.

Table 5. All assignments

1, A->B->C->D->E->A	2, A->B->C->D->A, E->E	3, A->B->C->E->A, D->D,
4, A->B->D->E->A, C->C	5, A->B->D->A, C->C, E->E	6, A->C->D->E->A, B->B
7, A->C->D->A, B->B, E->E	8, A->C->E->A, B->B, D->D	9, A->A, B->C->D->E->B
10, A->A, B->C->E->B, D->D	11, A->A, B->D->E->B, C->C	12, A->A, B->B, C->C, D->D, E->E
13, A->C->E->B->D->A		

Please note that the larger the cardinality of assignments, the better privacy protection achieved. In order to determine the cardinality of assignments, we use an assignment graph which is used to visualize the assignments.

Definition 1 (Assignment Graph). *Consider a table T and its anonymized table T*. An assignment graph $G(V, E)$ is a directed graph with n vertices, which represent the tuples(for simplify, we use n to denote the number of tuples in T). For $i = 1$ to n, vertex $v_i \in V$ represents G_i and G'_i. An edge $v_i \longrightarrow v_j$ is present if and only if the generalized QI-values of t_i is included in that of t'_j, that is, $t_i \in t'_j$.*

Now, we are ready to give our main theorem.

Theorem 2 (Cardinality Theorem). *Let $G(V, E)$ be an assignment graph. Assume that G have n vertices, and each vertex has one self-circuit and $e-1$ out-edges ($e > 2$). Denote by φ^n_e the cardinality of all assignments which satisfying Definition 4, then $\varphi^n_e = O(\alpha^n)$, where $\alpha > 1$.*

Proof. It's a trivial case for $e = 2$, thus we assume $e > 2$ in the following text.

1) According to the length of the circuit containing A (is equal to or greater than 1, all the assignments can be divided into two categories:

a) The length is greater than 1, i.e., the circuits start from A and then come back A through other points. Denote by E^e_n the cardinality of all such circuits in the category.

b) The length is equal to 1. Delete point A and associated edges, and compare the remainder graph to G, then it can be inferred that the cardinality of all circuits in this category is less than that in the above category. Thus, $E^e_n < \varphi^e_n < 2 \cdot E^e_n$.

2) Now we consider the point next to A in the circuits of Category a), the points can be $B, C, \cdots$ etc. In other words, there are $e - 1$ choices. According to the addition rule, the following recursion holds:

$$E^e_n \geq E^e_{n-1} + E^e_{n-2} + \cdots, +E^e_{n-e+1}$$

By comparing the above inequality to the famous Fibonacci Sequence ($F_n = F_{n-1} + F_{n-2}$), the theorem holds.

The cardinality theorem demonstrates that given the the assignment graph, the cardinality is bounded in $O(\alpha^n)$, where ($\alpha > 1$). Here are some examples: $\varphi_4^5 = 53, \varphi_4^{10} = 870, \varphi_4^{12} = 2922, \varphi_5^{12} = 28450$. It is worth noting that only two trivial assignments when $e = 2$.

4   Generalization Algorithm

In this section, we will elaborate the details of our semi-homogeneous anonymiza- tion as our major solution to the problem defined in Definition 3. This algorithm is an improved version of ring generalization proposed in [11]. The example in the section 1.2 illustrates the general process of the algorithm.

To prepare the semi-homogenous algorithm, we need to define the relationship between two multi-attribute domain. We use $D_i \leq D_j$ to denote the fact that domain D_j is either identical to or one of its generalization of D_i. When $D_i \leq D_j$ and $D_i \neq D_j$, we denote it by $D_i < D_j$. For quasi-identifer consisting of multi- ple attributes $(A_1, A_2, \cdots, A_d)$, we can define corresponding d-dimension vector $V_A = \langle D_{A_1}, D_{A_2}, \cdots, D_{A_d} \rangle$ with each D_{A_i} being the domain of A_i. Such kind of d-dimension vector for the set of d attributes is referred to as *multi-attribute domain*. The multi-attribute domain of the G_1 in Table 4 is $\langle$[19-25], [12k-20k], $F/M \rangle$. Given two d-dimension attribute domains $V_A = \langle D_{A_1}, D_{A_2}, \cdots, D_{A_d} \rangle$ and $V_B = \langle D_{B_1}, D_{B_2}, \cdots, D_{B_d} \rangle$, V_B is a *multi-attribute domain generalization* of V_A (this relationship is also denoted by $\leq$) if for each j, $D_{A_j} \leq D_{B_j}$.

Now, we elaborate the detailed procedure. In general, we first divide the mi- crodata table T into a partition $P = \{G_1, G_2, \cdots, G_m\}$ by certain off-the-shelf partitioning algorithms, such as the partition algorithm used in [11]. The algo- rithm framework is shown in Figure 4. At the beginning, if the size of QI-group G_i in P is equal to k exactly, we generalized this group directly. To reduce in- formation loss, we sort the tuples in G_i into the ascending order of their values of $\sum_{i=1}^{d} \frac{t.A_i}{|A_i|}$ (step 2) when the size of $G_i > k$. Then, varying the order from 2 to k, for each order ord, distribute the tuples into c different sub-groups, where $c = \frac{|G_i| * ord}{k}$ (step 5-6). Sort these sub-groups $\{g_1, g_2, \cdots, g_c\}$ by the above defi- nition of multi-attribute domain generalization. The two steps (step 2 and step 7) of sorting is to distribute the tuples sharing the same or quite similar QI- attributes into the same sub-groups. Step 9-10 is to generalize each sub-group to satisfy k-anonymous. We can get total information loss $IL = \sum_{i=1}^{c} NCP(g_i)$. Step 4 to step 11, for different order, we can get (k-1) different partitions of G_i. Among all the $k - 1$ partitions, we pick the one that minimizes the sum of $\sum_{i=1}^{c} NCP(g_i)$ as the final partition. The final step is to assign a random assignment to the recorded partition(step 14).

In addition, we can easily show that it can give equal or better utility com- pared to the non-homogeneous presented in [11].

Input: a partition $P=\{G_1, G_2, \cdots, G_m\}$ **of microdata** T, **Privacy Level** k
Output: generalization table T^*;
Method:
For each QI-group $G_i \in P$, DO
1. IF $|G_i| = k$, RETURN;
2. sort the tuples in G_i;
3. FOR $ord = 2$ to k
4.   let $gsize = \frac{k}{ord}$, $c = \frac{|G_i|}{gsize}$, $bestIL = MAX\_FLOAT$;
5.   distribute tuples into c groups $\{g_1, g_2, \cdots, g_c\}$:
6. $g_1 = \{t_1, t_2, \cdots, t_{gsize}\}, g_2 = \{t_{gsize+1}, \cdots, t_{2 \cdot gsize}\}, \cdots$,
           $g_c = \{t_{gsize \cdot (c-1)+1}, \cdots, t_{|G_i|}\}$
7.   sort the groups $\{g_1, g_2, \cdots, g_c\}$;
8.   ring generalization for each group g_i:
9.   for $i = 1$ to ord
10. $g_i = gen(g_i, g_{i+1}, \cdots, g_{i+ord \bmod c})$;
11.   compute the the total information loss $IL = \sum_{i=1}^{c} NCP(g_i)$
12.   IF the current partition is better than previous tries: $bestIL > IL$,
13.   record $\{g_1, g_2, \cdots, g_c\}$ and $bestIL = IL$;
14. assign a random assignment to the recorded partition $\{g_1, \cdots, g_c\}$;

Fig. 4. The framework of semi-homogenous algorithm

Theorem 3. *Semi-homogeneous generalization gives a k-anonymized table of equal or better utility than that given by a non-homogeneous generalization under the same partition* P.

Proof. Note that the step 3 in Figure 4, the value of order varies from 2 up to k. When order equals k, it is the case of the non-homogenous algorithm. For the different order, step 3 to step 11 makes the $k-1$ different partitions of G_i. Among all the $k-1$ different partitions, the step 12 picks the one that minimizes the sum of $\sum_{i=1}^{c} NCP(g_i)$ as the final partition, thus, the theorem holds.

5   Related Work

In this section, previous related work will be surveyed. In this section, previous related work will be surveyed. All the privacy-preserving transformation of the microdata is referred to as recoding, which can be classified into two of classes of models: global recoding and local recoding. In global recoding, a particular detailed value must be mapped to the same generalized value in all records, which implies that the data space will be partitioned into a set of non-overlapping regions. While local recoding allows the same detailed values being mapped to different generalized values of different QI-groups. Obviously, global recoding is a more special case of local recoding and local recoding is more flexible and has the potential to achieve lower information loss. Efficient greedy solutions following certain heuristics have been proposed [3,11,15,16] to obtain a near optimal solution. Generally, these heuristics are general enough to be used in many anonymization models. Incognito [1] provides a practical framework for implementing full-domain generalization, borrowing ideas from frequent item set mining, while Mondrian [2] takes a partitioning approach reminiscent of KD-trees. To achieve K-anonymity, [5] presents a framework mapping the multi-dimensional quasi-identifiers to 1-Dimensional(1-D) space. For 1-D quasi-identifiers, an algorithm of $O(K \cdot N)$ time complexity for optimal solution is also developed. It is discovered that K-anonymizing a data set is strikingly similar to building a spatial index over the data set, so that classical spatial indexing techniques can be used for anonymization [4].

The idea of non-homogeneous generalization was first introduced in [17], which studies techniques with a guarantee that an adversary cannot associate a generalized tuple to less than K individuals, but suffering additional types of attack. Authors of paper [11] proposed a randomization method that prevents such type of attack and showed that k-anonymity is not compromised by it, but its partitioning algorithm is only a special of the top-down algorithm presented in [3]. The anonymity model of the paper [11,17] is different to us. In their model, the size of QI-groups is fixed as 1, while in our model it is varying from 1 to K.

6   Empirical Evaluation

In this section, we will experimentally evaluate the effectiveness and efficiency of the proposed techniques. Specifically, we will show that by our technique (presented in Section 4) have significantly improved the utility of the anonymized data with quite small computation cost.

In the following experiments, we compare our semi-generalization anonymity algorithm (denoted by HG) with the existing state-of-the-art technique: the Non-homogeneous generalization [11](NH for short). For a fair comparison, the two algorithms are processed under the same partition, which is generated by certain off-the-shelf partitioning algorithm used in [11]. Two widely-used real databases SAL and INCOME(downloadable from http://ipums.org) with 500k and 600k tuples, respectively, will be used in following experiments. Each tuple describes the personal information of an American. The two data sets are summarized in Table 6. In order to examine the influence of dimensionality, we create two sets of micro-data tables from SAL and INCOME. The first set has 4 tables, denoted as SAL-4, $\cdots$, SAL-7, respectively. Each SAL-d ($4 \leq d \leq 7$) has the first d attributes in Table 6 as its QI-attributes and Occupation as its sensitive attribute(SA). For example, SAL-4 is 5-Dimensional, and contains QI-attributes: Age, Gender, and Education, Marital. The second set also has 4 tables INC-4, $\cdots$, INC-7, where each INC-d ($4 \leq d \leq 7$) has the first d attributes as QI-attributes and income as the SA.

In the experiments, we investigate the influence of the following parameters on information loss of our approach: (i) number of tuples n; (ii) number of attributes d in the QI-attributes; (iii) value of K in K-anonymity. Table 7 summarizes the parameters of our experiments, as well as their values examined. Default values are in bold font. Data sets with different cardinalities n are also generated by randomly sampling n tuples from the *full* SAL-d or INC-d ($4 \leq d \leq 7$). All experiments are conducted on a PC with 1.9 GHz AMD Dual Core CPU and 1 gigabytes memory. All the algorithms are implemented with Microsoft VC++

Table 6. Summary of attributes

Attribute	Number of distinct values	Types
Age	78	Numerical
Gender	2	Categorical
Education	17	Numerical
Marital	6	Categorical
Race	9	Numerical
Work-class	10	Categorical
Country	83	Numerical
Occupation	50	Sensitive

Attribute	Number of distinct values	Types
Age	78	Numerical
Occupation	711	Numerical
Birthplace	983	Numerical
Gender	2	Categorical
Education	17	Categorical
Race	9	Categorical
Work-class	9	Categorical
Marital	6	Categorical
Income	[1k,10k]	Sensitive

(a) SAL   (b) INCOME

Table 7. parameters and tested values

Parameter	Values
k	250,200,150,**100**,50
cardinality n	**100k**,200k,300k,400k,500k
number of QI-attributes d	3,4,5,**6**

2008. In all following experiments, without explicit statements, default values in Table 7 will be used for all other parameters.

6.1 Privacy Level K

In order to study the influence of k on data utility, we observe the evolution of GCP that has been widely used to measure the information loss of the generalized tables by varying k from 50 to 250 with the increment of 50. The results on SAL-d and INC-d ($4 \leq d \leq 7$) data are shown in Figure 4 (a)-4(h). From the results, we can clearly see that for all the tested data, information loss of our model is quite significantly smaller than that of NH-anonymization. Another advantage of our model over NH is that the utility achieved by our model is less sensitive to domain size than NH. From the figures, we can see that data sets generated by NH has a lower GCP on SAL-d than that on INC-d ($4 \leq d \leq 7$) due to the fact that domain size of SAL is smaller than that of INC. Such a fact implies that the information loss of NH is positively correlated to the domain size. However, in our model, domain size of different data set has less influence on the information loss of the anonymized data. Results of this experiment also suggest that for almost all tested data sets the GCP of two algorithms grows linearly with K. This can be reasonably explained since larger K will lead to more generalized QI-groups, which inevitably will sacrifice data utility.

6.2 QI-Attributes Dimensionality d

Experiments of this subsection is designed to show the relation between the information loss of two algorithms and data dimensions d. In general, the information loss will increase with d, since data sparsity or more specifically the

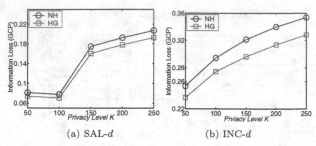

(a) SAL-d (b) INC-d

Fig. 5. The Global Certainty Penalty vs. Privacy Level K

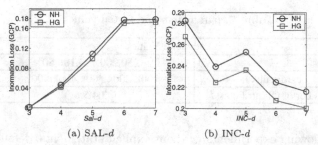

(a) SAL-d (b) INC-d

Fig. 6. The Global Certainty Penalty vs. QI-Dimensionality d

data space characterized by a set of attributes exponentially increases with the
number of attributes in the set, i,e, dimensions of the table. Figure 6(a) and
6(b) compare the information loss of the anonymization generated by the two
methods with respect to different values of d on SAL-d and INC-d, respectively.
It is clear that the anonymization generated by the linking-based method has
a lower global certainty penalty compared to that of NH. The advantage of LB
over NH is obvious, and such an advantage of LB can be consistently achieved
when d lies between 4 to 7.

6.3   Cardinality of Data Set n

In this subsection, we investigate the influence of the the table size n on information loss of LB and NH. The results of experiments on two data sets SAL-7
and INC-7 are shown in Figure 7(a) and 7(b), respectively. We can see that
the information loss of two methods on both two data sets decreases with the
growth of n. This observation can be attributed to the fact that when the table
size increases more tuples will share the same or quite similar QI-attributes. As
a result, it is easier for the partitioning strategies to find very similar tuples to
generalize. Similar to previously experimental results, our method is the clear
winner since information loss of LB is significantly small than that of NH, which
is consistently observed for various database size.

6.4   Efficiency

Finally, we evaluate the overhead of performing anonymization. Figure 8(a) and
8(b) show the computation cost of the two anonymization methods on two data

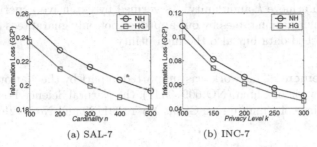

(a) SAL-7   (b) INC-7

Fig. 7. The Global Certainty Penalty vs. Cardinality n

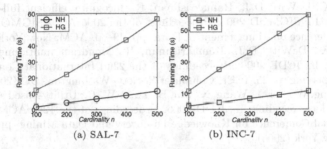

(a) SAL-7   (b) INC-7

Fig. 8. Running time vs. Cardinality n

sets, respectively. The running time of both two algorithms increases linearly when n grows from 100k to 500k, which is expected since more tuples that need to be anonymized will consume longer time to finish the anonymization procedure. Comparison results show that the advantages of our method in anonymization quality do not come for free. However, in the worst case, our algorithm can be finished in 60 seconds, which is acceptable. In most real applications, quality is more important than running time, which justifies the strategy to sacrifice certain degree of time performance to achieve higher data utility.

Summary. Through these sets of experiments(Figure 5-7) , we can conclude that the HG algorithm is indeed reduce the infromation loss, comparing with NH algorithm, as proved in Theorem 3 and illustrated in section 1.2. Calculations show that on the SAL data sets the HG algorithm compared with the NH algorithm, reduces about 3.01% and 8.98% probably, while on INC data set, probably reduced by between 5.28% and 7.83%.

The trad-off of high anonymization quality is the runtime and the NH method is more efficient. However, the runtime of the HG method is not far away from that of the NH in practice. Moreover, for anonymization, the computation time is often a secondary consideration yielding to the quality.

7   Conclusion

In this paper, we systematically investigate the cardinality of the assignments. We propose a technique called semi-generalization which uses QI-groups with varying size to achieve k-anonymity. As verified by extensive experiments, the produced anonymization tables by our approach not only guarantees the privacy safety of published data but also the high utility.

Acknowledgement. This work was supported in part by the National Natural Science Foundation of China (NO.60973047), the Natural Science Foundation of Zhejiang Province of China under Grant No.Y1091189 and No.Y12F020065.

References

1. LeFevre, K., DeWitt, D.J., Ramakrishnan, R.: Incognito: efficient full-domain k-anonymity. In: SIGMOD 2005: Proceedings of the 2005 ACM SIGMOD International Conference on Management of Data, pp. 49–60. ACM, New York (2005)
2. LeFevre, K., DeWitt, D.J., Ramakrishnan, R.: Mondrian multidimensional k-anonymity. In: ICDE 2006: Proceedings of the 22nd International Conference on Data Engineering, p. 25. IEEE Computer Society, Washington, DC (2006)
3. Xu, J., Wang, W., Pei, J., Wang, X., Shi, B., Fu, A.W.-C.: Utility-based anonymization using local recoding. In: KDD 2006: Proceedings of the 12th ACM SIGKDD International Conference on Knowledge Discovery and Data Mining, pp. 785–790. ACM, New York (2006)
4. Iwuchukwu, T., Naughton, J.F.: K-anonymization as spatial indexing: toward scalable and incremental anonymization. In: VLDB 2007: Proceedings of the 33rd International Conference on Very Large Data Bases, pp. 746–757. VLDB Endowment (2007)
5. Ghinita, G., Karras, P., Kalnis, P., Mamoulis, N.: Fast data anonymization with low information loss. In: VLDB 2007: Proceedings of the 33rd International Conference on Very Large Data Bases, pp. 758–769. VLDB Endowment (2007)
6. Samarati, P., Sweeney, L.: Generalizing data to provide anonymity when disclosing information (abstract). In: PODS 1998: Proceedings of the Seventeenth ACM SIGACT-SIGMOD-SIGART Symposium on Principles of Database Systems, p. 188. ACM, New York (1998)
7. Sweeney, L.: k-anonymity: a model for protecting privacy. Int. J. Uncertain. Fuzziness Knowl.-Based Syst. 10(5), 557–570 (2002)
8. Samarati, P.: Protecting respondents' identities in microdata release. IEEE Trans. on Knowl. and Data Eng. 13(6), 1010–1027 (2001)
9. Xiao, X., Tao, Y.: M-invariance: towards privacy preserving re-publication of dynamic datasets. In: SIGMOD 2007: Proceedings of the 2007 ACM SIGMOD International Conference on Management of Data, pp. 689–700. ACM, New York (2007)
10. Tao, Y., Chen, H., Xiao, X., Zhou, S., Zhang, D.: Angel: Enhancing the utility of generalization for privacy preserving publication. IEEE Transactions on Knowledge and Data Engineering 21, 1073–1087 (2009)
11. Wong, W.K., Mamoulis, N., Cheung, D.W.L.: Non-homogeneous generalization in privacy preserving data publishing. In: SIGMOD 2010: Proceedings of the 2010 International Conference on Management of Data, pp. 747–758. ACM, New York (2010)
12. Meyerson, A., Williams, R.: On the complexity of optimal k-anonymity. In: PODS 2004: Proceedings of the 23nd ACM SIGMOD-SIGACT-SIGART Symposium on Principles of Database Systems, pp. 223–228. ACM, New York (2004)

13. Bayardo, R.J., Agrawal, R.: Data privacy through optimal k-anonymization. In: ICDE 2005: Proceedings of the 21st International Conference on Data Engineering, pp. 217–228. IEEE Computer Society, Washington, DC (2005)
14. Aggarwal, G., Feder, T., Kenthapadi, K., Motwani, R., Panigrahy, R., Thomas, D., Zhu, A.: Anonymizing Tables. In: Eiter, T., Libkin, L. (eds.) ICDT 2005. LNCS, vol. 3363, pp. 246–258. Springer, Heidelberg (2005)
15. Fung, B.C.M., Wang, K., Yu, P.S.: Top-down specialization for information and privacy preservation. In: International Conference on Data Engineering, pp. 205–216 (2005)
16. LeFevre, K., DeWitt, D.J., Ramakrishnan, R.: Workload-aware anonymization. In: KDD 2006: Proceedings of the 12th ACM SIGKDD International Conference on Knowledge Discovery and Data Mining, pp. 277–286. ACM, New York (2006)
17. Gionis, A., Mazza, A., Tassa, T.: k-anonymization revisited. In: ICDE 2008: Proceedings of the 2008 IEEE 24th International Conference on Data Engineering, pp. 744–753. IEEE Computer Society, Washington, DC (2008)

Processing XML Twig Pattern Query
with Wildcards*

Huayu Wu[1], Chunbin Lin[2], Tok Wang Ling[3], and Jiaheng Lu[2]

[1] Institute for Infocomm Research, Singapore
huwu@i2r.a-star.edu.sg
[2] School of Information, Renmin University of China
{jiahenglu,chunbinlin}@ruc.edu.cn
[3] School of Computing, National University of Singapore
lingtw@comp.nus.edu.sg

Abstract. In this paper, we present a novel and complementary technique to optimize XML twig pattern queries with wildcards(*). Our approach is based on utilizing a new axis called *AD-dis*, to equivalently rewrite a query with wildcards (non-branching as well as branching wildcards) into a single query *without* any wildcards. We present efficient rewriting algorithms and also twig pattern matching algorithms to process the rewritten queries with AD-dis, which is proven to be I/O and CPU optimal. In addition, the experimental results not only verify the scalability and efficiency of our extended matching algorithms, but also demonstrate the effectiveness of our rewriting algorithms.

1 Introduction

With the growing importance of XML in information exchange, how to efficiently query an XML document becomes a hot research topic. XML employs a tree-structured model for representing data. In most XML query languages (e.g., XPath and XQuery), the queries are expressed by twig (i.e., small tree) patterns. Finding all occurrences of a twig pattern in an XML database is a core operation in XML query processing.

The twig pattern queries solved by existing algorithms require each query node to be either a tag name or a value comparison, and the relationship between each pair of adjacent query nodes to be either a parent-child edge ("/") or an ancestor-descendant edge ("//"). However XPath is more expressive than the twig pattern representation solved by these algorithms. Although some works try to extend twig pattern matching for more practical features, e.g., NOT-predicate [11], OR-predicate [5] and advanced content search in predicate [10], there are still important XPath features that are not carefully considered by the existing works, such as wildcards.

Wildcard(*) is widely used in many practical queries when the element names are unknown or do not matter. Consider the football team document (fragment)

---

* We may use *wildcard node* and * *node* exchangeably in this paper.

S.W. Liddle et al. (Eds.): DEXA 2012, Part I, LNCS 7446, pp. 326–341, 2012.
© Springer-Verlag Berlin Heidelberg 2012

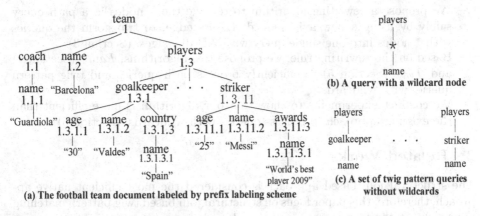

(a) The football team document labeled by prefix labeling scheme

(b) A query with a wildcard node

(c) A set of twig pattern queries without wildcards

Fig. 1. Example XML tree and twig pattern query with wildcard node

shown in Fig. 1(a). Assume a user wants to find all players' names, no matter what position they play. She/He may issue a query as //players//name. However, this query cannot get the correct answer, since the country names and the award names are also returned. In order to meet the requirement, a query with wildcard (i.e., //players/*/name, as shown in Fig. 1(b)) should be issued. Otherwise, the user needs to create all the possible queries (e.g., //players/goalkeeper/name and //players/striker/name, as shown in Fig. 1(b)) and merge the results from them. This example illustrates the usefulness of wildcard in XML query expression.

Wildcard brings in convenience for users to issue queries, on the other hand, it also poses challenges in query evaluation. For example, in the structural join based XML query processing algorithms, processing a query "//*//b" would require accessing all the document nodes to match the wildcard query node. The most common way to resolve wildcards is to replace * nodes by all possible element nodes under the help of schematic information (e.g., DTD) before pattern matching [7]. Then the query in Fig. 1(b) can be equivalently transformed into the union of a set of queries that replace the * node with all the possible player positions (e.g., striker), which are shown in Fig. 1(c). Obviously this approach is tightly bound to document schema. Also, combining all possible elements for a wildcard during pattern matching is very inefficient. In addition, a query may contain more than one * nodes, and a certain * node may appear as a branching node (whose out-degree is at least 2). The cost to identify possible elements for all * nodes may increase exponentially w.r.t. the number of * nodes.

In this paper, we propose native rewriting algorithms to equivalently rewrite the queries with wildcards to wildcard-free queries by using a new axis called *AD-dis*. Further, we extend the most popular twig pattern matching approach, i.e., the structural join based approach to process twig pattern queries with wildcards more efficiently without enumerating all the possible queries to reduce the I/O and CPU cost. In addition, we also discuss the optimality of the our algorithms. The contributions can be summarized as follows:

- We propose a rewriting algorithm to remove the * nodes in a path query safely by using a new axis, named *AD-dis* edge, to transform the queries with * nodes into one single query with AD-dis edges. (Section 3)
- Based on the rewriting rule, we propose two algorithms, *Path** (Section 4) and *Twig**(Section 5) to efficiently process path queries and twig pattern queries with wildcards.
- We conduct experiments to show that our algorithm is more efficient than the existing approach to process queries with wildcards. (Section 6)

2   Related Works

The structural join based approach is considered the most efficient native approach, therefore, this paper focus on structural join based twig pattern matching approach. At the beginning, binary join based approaches were proposed, e.g., [12]. However, such prior works would produce a large size of intermediate result during twig pattern matching. *TwigStack* [2] is the first work on *holistic* structural join, which significantly improves the space usage and is proven optimal for queries containing "//"-axis only. There were many subsequent works to optimize *TwigStack* in terms of I/O, or extend it for different kinds of problems. For example, [6] introduced a *list* structure to make it optimal for queries containing parent-child relationships under non-branching nodes. There are also some extensions to enrich twig pattern matching for general XPath queries. E.g., [11] and [5] made twig pattern matching support NOT and OR predicates, [10] enhanced twig pattern matching to handle advanced content search. However, all these structural join based twig pattern matching algorithms are not suitable for queries containing wildcards.

[3] proposed an optimization to wildcards in XPath queries. However, it failed to build a connection between their optimized XPath queries and the state-of-the-art query processing algorithms. We implemented their optimization strategy and compare it with our algorithm in Section 6. The only work we identified to process twig pattern queries with wildcards is the extension proposed by [7], which replaces * nodes by element nodes based on schema, to process twig pattern queries. As mentioned in Section 1, this approach is inefficient.

More related works can be found in the full version [9] of our work.

3   Path Rewriting

Matching a path query is essential to twig pattern matching. Existing algorithms have already made it efficient and optimal to perform structural joins in matching a path query without wildcards. However, when we consider the * node within a path query, the existing structural join algorithms will fail to work. In this section, we propose a rewriting algorithm to rewrite a path query by removing all * nodes. Then we extend the existing path matching algorithms to solve the path query.

In a twig pattern query, a query edge can be either a parent-child (PC) edge, denoted by "/", or an ancestor-descendant (AD) edge, denoted by "//". We propose a new type of query edge to aid our query rewriting.

Definition 1. *(AD-dis edge) An AD-dis edge (*dis *is an abbreviation for* distance*) in a path query is an extension to an AD edge by assigning a distance condition to constraint the ancestor and descendent query nodes. It is denoted as "//^{range}", where* range *is an interval to specify the distance range of the corresponding AD edge. In the twig pattern representation of a path query, we put a range beside an AD edge to declare it as an AD-dis edge.*

For example, a path query A//$^{[2,4]}$B contains an AD-dis edge. It aims to find an A-typed node and a B-typed node which are in AD relationship, and the distance between them is at least 2 and at most 4.

Algorithm 1. Path rewriting

1: tokenize the input path query by the delimiters of non-* nodes
2: **while** there are more tokens **do**
3:    let e to be the next token
4:    **if** e contains * **then**
5:      replace e by an AD-dis edge //^{range}
6:      let n = total number of PC and AD edges in e
7:      **if** e contains only PC edges **then**
8:         set *range* to [n, n]
9:      **else**
10:        set *range* to [n, ∞)
11: merge the tokens by the corresponding delimiters to form a new path query

---

The general idea to rewrite a path query with * nodes is to replace the PC and AD edges that connect several consecutive * nodes to non-* element nodes by a single AD-dis edge, in which the range is calculated based on the number PC and AD edges in the original path query. We first tokenize a path query by the non-* nodes. There are three cases of each token: (1) a PC edge, (2) an AD edge, and (3) a set of PC/AD edges connected by * nodes. We do not care the first two cases, but only concentrate on the last case. We replace all the PC/AD edges with * nodes by an AD-dis edge. To decide the distance range of the AD-dis edge, we rely on the type of original edges. If all edges in the original path are PC edges, then the distance of the AD-dis edge is exactly the number of edges in the original path. If the edges in the original path contains both PC edges and AD edges, the distance will be at least the number of total edges. The detailed path rewriting algorithm is presented in Algorithm 1.

The rewriting rule is based on the semantic meaning of PC and AD edges, i.e., the PC edge means the distance of exact 1 and the AD edge means the distance is at least 1 between the two connected nodes.

Example 1. Suppose we have a path query */A//B/C /*//*/D//*. After applying the path writing algorithm on it, the original path query is rewritten as follows:

$$*/A//B/C/*//*/D//* \equiv //^{[1,1]}A//B/C//^{[3,\infty)}D//^{[1,\infty)}$$

4 Algorithm Path*

In this section, we present our algorithm **Path*** to process path queries with wildcards. We follow the idea of the PathStack [2], which is a well known algorithm to process path query, to maintain multiple stacks to achieve optimality of path matching. In Path*, we rewrite a path query with wildcards by replacing all * nodes and relevant AD and PC edges by AD-dis edges, as presented in Section 3. After that the existing algorithms can be extended to optimally process the rewritten queries. We do not repeat their work on how to use stacks to achieve optimality in structural joins. The only difference between our rewritten path query and the normal path query they solve is the additional AD-dis edge requiring structural join. Thus we only need to specially handle the structural join between two nodes connected by an AD-dis edge.

In existing algorithms, they use the labels of two document nodes to perform structural join for PC edge and AD edge. In particular, using Dewey IDs in the prefix labeling scheme (as shown under each labeled node in the document in Fig. 1), two nodes v and u are in AD relationship if the Dewey ID of v is the prefix of the Dewey ID of u; furthermore, if the length gap of their Dewey IDs is 1, they are also in PC relationship. In order to check the AD-dis, Dewey IDs are also utilized in our paper. We can use the prefix property to check the AD relationship, and the gap of the length of the IDs to identify the distance range in the AD-dis edge. For example, the query in Fig. 1(b) can be rewritten to be $players//^{[2,2]}$ $name$, and the nodes "players (1.3)" and "name (1.3.1.2)" in Fig. 1 satisfy the query, since 1.3 is the prefix of 1.3.1.2 and the length gap between them is 2, which meet the distance constraint. Therefore, the path "players/goalkeeper/name=Valdes" is an answer. Note that, there are two special cases: 1) the AD-dis edge is the prefix of a path query; 2) the AD-dis edge is the postfix of a path query. We will discuss this two cases as follows.

4.1   AD-dis Edge as Prefix

In this case, we firstly execute the path query without the prefix AD-dis edge. Then we check whether the Dewey ID of the data node matching the first node in the path query satisfies the recorded range in its length. For example, to process a rewritten path query $//^{[m,\infty)}A//B$, we first process the path $A//B$, and then get the final results by checking whether the length of the Dewey ID of each A-typed node is longer than m.

4.2   AD-dis Edge as Postfix

Similar to the prefix processing, we first execute the partial path query without the postfix edge, and then get the answers by checking whether the length of the leaf node satisfies the range specified in the AD-dis edge. The original Dewey ID of a certain node can only identify the distance to its ancestor nodes, but here we want to know the distance to the bottom node of the path. Therefore, we define "Gap label" to record the distance between a node and the deepest node below it.

Definition 2. *(Gap label) A gap label is assigned to each document node, beside the Dewey label during document labeling. A gap label indicates the distance between the current node to its deepest descendant node.*

For an example query $A//B//^{[m,\infty)}$, after matching $A//B$ to the document, we check whether each matched B-typed node has a gap value greater than m.

Since the core technique of path matching, i.e., using stack to achieve optimality, is proposed in previous work [2] and our extension is easy to understand, we do not present the pseudo-code of *Path**. Compared with the naive extension of existing path matching algorithms which replaces the * nodes in a path query, our approach by query rewriting (1) does not need schematic information for query replacing, and (2) reduces the number of query nodes and the number of structural joins in a path query, thus achieve a better performance. Note that if a path query does not contain any * node, *Path** will omit the rewriting step, and thus it is exactly the same as *PathStack*.

5   Algorithm Twig*

In this section, we propose **Twig*** to process twig pattern queries with wildcards as the following four steps:(i) Optimize a twig pattern query by simplifying the all-* subtwigs (defined later). (ii) Decompose the optimized twig pattern query into paths and rewrite the paths as we discussed above. (iii) Match the rewritten path queries and merge the results to get actual twig answers. (iv) Browse the matches to the all-* subtwigs, if necessary. In this paper, we only focus on the first three steps and omit the last step, because it is less important and rarely necessary. Step 4 can be fond in our full report [9].

5.1   [Step 1]: All-* Subtwig Optimization

The cost of performing structural joins among wildcards is very expensive, In addition, multiple wildcard nodes are usually not the return node in the query, but only play the role of structural constraint. Thus before processing a query, we first optimize the all-* subtwig.

Definition 3. *(All-* subtwig) In a twig pattern query with wildcards, if a subtwig is rooted at an element query node whose descendant query nodes are all * nodes, we call it an all-* subtwig.*

For example, in Fig. 2(a), the subtree rooted at "B" is an all-* subtwig.

Lemma 1. *If a document node matches the root of an all-* subtwig T in a query Q, it must match the path in T which starts at the root and ends at the lowest * node in T, and vice versa.*

Proof [sketch]. When a node n satisfies the deepest path with level of l in an all-* subtwig T rooted at n, i.e., n has a descendant l levels below, then n satisfies any other paths with depth less than l in T, i.e., n also has descendants less than l levels below.   □

Based on Lemma 1, the longest path in an all-* subtwig can fully represent the subtwig, however, there might be more than one such paths. For example, in Fig. 2(b), path $P_a = B/*_2//*_3$ and $P_b = B/*_2/*_4$ have the same length, since they both contain two * nodes, and they both are the longest path in the all-* subtwig. Therefore, we need to discuss whether the two paths having the same length in the same all-* subtwig are equivalent.

Lemma 2. *For a given node N in a query, the path queries N/* and N//* are equivalent to qualify an N-typed node.*

Lemma 3. *For a given all-* subtwig, the two paths having the same length are equivalent to qualify the root of the all-* subtwig.*

The proof to Lemma 2 and 3 can be found in [9]. According to Lemma 1-3, we can reduce all-* subtwig to any longest path in the subtwig. For example, the all-* subtwig rooted at B in the twig pattern query shown in Fig. 2(a) can be reduced as shown in Fig. 2(b) (when we optimize the subtwig rooted at $*_2$, we can keep either $*_3$ or $*_4$, based on Lemma 3.).

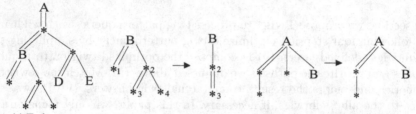

(a) Twig query   (b) All-* subtwig optimization   (c) Optimizing sibling wildcard nodes

Fig. 2. All-* subtwig and sibling wildcard nodes optimization

We have discussed how to reduce the all-* subtwig. However, in practice, a subtwig usually contains both element nodes and * nodes. The above lemmas actually can also be applied to simplify the subtwig which is not an all-* subtwig but contains sibling * nodes. Fig. 2(c) shows such an example.

We design a recursive algorithm to optimize an all-* subtwig. We also discuss how to browse all matches to an all-* subtwig, as the last step of query processing. Due to the space limitation, these are omitted in this paper, and can be found in our full report [9].

5.2 [Step 2]: Twig Pattern Query Decomposition and Path Rewriting

In the second step, we decompose a twig pattern query, resulted from Step 1, into path queries. When we do this, we need to record down the branching nodes so that after matching every path query we can join the path results based on the correct branching node. Consider the twig pattern query shown in Fig. 3(a), which is a simplified pattern of the query in Fig. 2(a), after Step 1. There are three leaf nodes in this query, thus the query is intuitively decomposed into three path queries, as shown in Fig. 3(b). In each path query, we need to mark out the branching node appearing in the original twig pattern query. For example, we can mark that $*_1$ in P1 is a branching node in the twig pattern query with respect to P2 and P3. Such a branching node is called the lowest common node (LCN) of the corresponding decomposed paths.

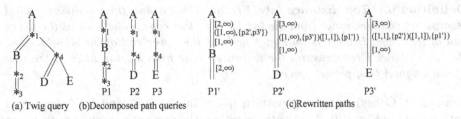

(a) Twig query   (b)Decomposed path queries   (c)Rewritten paths

Fig. 3. Twig pattern query with wildcards decomposition examples

Definition 4. *Lowest Common Node (LCN) In a twig pattern, no matter a query or a matched data pattern, if a branching node* v *contains* m *child nodes, then after decomposing the twig into paths,* v *is called the lowest common node (LCN for short) of the paths that contain these* m *nodes.*

For example, $*_1$ is a branching node in the twig pattern query shown in Fig. 3(a), and it is also the LCN of paths P1, P2 and P3 in Fig. 3(b). Knowing the LCN of a set of decomposed path queries, we can easily merge these path queries to the original twig pattern query.

When a path query contains * nodes, we rewrite the query by removing the * nodes based on Algorithm 1 in Section 3. After performing path rewriting, all non-* LCNs remain in rewritten path queries. Thus we can simply add a label beside each non-* LCN node to indicate all other path queries that branch at this LCN from the current path query, just like what they do in existing approaches for twig pattern matching without wildcards. However, after path rewriting, the * LCNs of original path queries cannot be identified any more, as they are hidden within relevant AD-dis edges.

In our approach, for each rewritten decomposed path query, we record all LCNs along it, together with participating path queries and location information.

Definition 5. *(LCN label) When some unknown node within an AD-dis edge is an LCN, we assign an LCN label to the AD-dis edge. An LCN label contains two components: (1) a range to indicate the relevant distance from the LCN to its lower LCN, or to the non-* query node at the lower end of the AD-dis edge, and (2) a list of path queries (IDs) which share this LCN with the current path.*

Since an AD-dis edge may contain multiple LCN labels w.r.t. different sets of query paths, we use an *LCN label list* to store all LCN labels within each AD-dis edge in bottom-up order. Suppose the AD-dis edge has a non-* node at its lower end, the range of each LCN label is the distance from this LCN to the non-* query node at the lower end of the AD-dis edge, or to its child LCN, depending on whether it is the lowest LCN in the edge. To find the range in an LCN label, we can adopt the similar approach in finding the range of an AD-dis edge, as shown in Section 3. If an AD-dis edge does not have a lower-end non-* node, we keep a *null* for the range of the lowest LCN.

The last label we record for an AD-dis edge involving LCNs is the *top distance label*. Using the top distance label and the LCN label list, we can determine the location of all LCNs along an AD-dis edge.

Definition 6. *(Top distance label) In an AD-dis edge, the top distance label records the distance range from the top LCN in this edge to the node at the upper end of this edge. If there is no non-* node at the upper end of the AD-dis edge, the top distance label measures the distance range from the top LCN to the root of the original twig pattern query.*

Example 2. Consider the twig pattern query in Fig. 3(a) and its decomposed path queries in Fig. 3(b). After path rewriting, the new path queries are shown in Fig. 3(c). There are three lines of labels beside each AD-dis edge containing LCNs in Fig. 3(c). The first line is the range label of the AD-dis edge, as introduced in Section 3. The second line is the LCN label list. In the first AD-dis edge in P1', the LCN label list contains only one LCN label, i.e., $([1,\infty),\{P2',P3'\})$. The first component $[1,\infty)$ indicates the corresponding LCN node is at least 1 level above the node B. The second component means it is the LCN of the current path and the paths P2' and P3'. The LCN label list in the AD-dis edge in P2' contains more than one LCN labels. Especially, the second LCN label $([1,1],\{P1'\})$ indicates the second LCN in this edge is exactly 1 level above the previous LCN, and it is the LCN of the current path and path P1'. The third line in each AD-dis edge is the corresponding top distance label, indicating that the top LCN in every AD-dis edge is at least 1 level lower than the node A.

5.3 [Step 3]: Matching Rewritten Paths and Merging Results

This is the main step of Twig*, we match each rewritten decomposed path query, and then merge the results to get twig answers. To reduce redundant path matches, we follow the idea of holistic join first proposed in *TwigStack* [2], and extend it by adding the function to perform structural join for AD-dis edges, to match each rewritten query. For example, a twig pattern query is decomposed into three rewritten path queries in Fig. 4. Using the holistic joins in TwigStack, we can guarantee that the A-typed nodes in all path matches to P1 must have a descendant C-typed node and descendant a D-typed node satisfying the constraints in P2 and P3. This cannot be guaranteed by PathStack. Thus by extending TwigStack, our algorithm will also significantly reduce the number of useless intermediate results. Furthermore, other extended algorithms to improve TwigStack, e.g., [6], are also feasible to extend our approach during pattern matching.

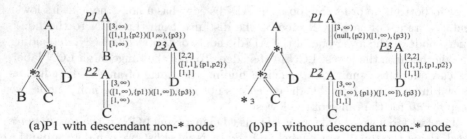

(a)P1 with descendant non-* node   (b)P1 without descendant non-* node

Fig. 4. Path query merging example

After matching each decomposed path queries in a holistic manner, now we put more focus on merging the matched path results. This process can be easily done by joining sorted path results based on labels of relevant branching nodes in TwigStack, for queries without wildcards. However if a query contains branching wildcard nodes as in our examples, it is not so trivial. Because when a branching node is a * node, it will hide in a certain AD-dis edge in the corresponding rewritten query. We cannot simply merge two path answers by the equivalence of any two node labels, but have to rely on the positional relationship of answer nodes. We present how Twig* merges path results based on wildcard LCNs.

Merge Order. From the LCN label list of each AD-dis edge in the decomposed path queries, we can merge the path queries into original twig pattern query in any order. However, in our approach, we need to adopt a bottom-up fashion to merge paths, i.e., merging the paths share the lowest LCN first. This is because the LCN labels along each AD-dis edge are sorted in bottom-up order. Thus we always start merging from a path which contains the lowest LCN and also has a non-* node as a leaf.

Example 3. Consider a twig pattern query and its rewritten decomposed path queries shown in Fig. 4(a). Suppose we have three path instances, P1-matched path, P2-matched path and P3-matched path. We merge P1-matched path and the P2-matched first (assumed mergeable) because the LCN of these two paths is lower. After deciding the position of the node matching the LCN of the two paths, i.e., $*_2$, we then check whether the intermediate answer can be merged with the P3-matched path at an LCN at least 1 level above the $*_2$-matched node.

Path Result Merging. When merging two path instances from two path queries, if the LCN is an element node, we simply check whether the labels of the two nodes matching the LCN are equivalent or not. This is what they do in existing approaches. However, if the LCN is a * node, during path matching we can only find the nodes matching the two non-* nodes at the ends of the corresponding AD-dis edge, but cannot decide where exactly the LCN is between the two non-* nodes. To solve this, we use the property of prefix label to find the LCN match.

Proposition 1. *The Dewey ID of a common ancestor of two nodes u and v must be a common prefix of the Dewey IDs of the two nodes, and vice versa.*

According to Proposition 1, we can merge two path answers based on the labels of two nodes below the LCN and their distances to the LCN. We consider two cases when merging a new path to a merged intermediate result w.r.t. the AD-dis edge involving the LCN: (1) both the AD-dis edges have non-* query nodes at lower ends, and (2) one such AD-dis edge does not have a non-* query node at its lower end. It is not possible that neither of the AD-dis edges has a non-* query node at lower end, because this case can be avoided in Step 1. The general idea is that for the first case, we find each common prefix of the two node labels that match the lower-end query nodes or the previously decided LCN nodes in the two AD-dis edges. Then check whether each common prefix is in the right level to satisfy the range constraints in the LCN labels, as well as the top distance labels if the LCN is the top LCN in each AD-dis edge. For the second case, we do not need to compute the common prefix. Instead we simply check each ancestor of the node matching the only lower-end query node by also considering its gap label, to see whether its position satisfies the range constraints in the LCN labels or the top distance labels. There are some special cases such as an AD-dis edge involving LCNs does not have an upper-end node. Basically, with the LCN labels and the top distance label in the AD-dis edge we can easily determine the location of branching wildcard nodes. Then by checking the position and the common prefix of relevant nodes, we can merge the path answers. The detailed algorithm of Twig* is omitted here due to the space limitation and can be found in the full version [9].

Example 4. Consider the query and decomposed paths in Fig. 4(a). Suppose two paths *1//1.1.2.3.2* and *1//1.1.2.3.5.1* match P1 and P2 respectively. When we try to merge these two paths, we first check whether they have the same ancestor node. Then we find the common prefixes of the two descendant node, including 1, 1.1, 1.1.2 and 1.1.2.3. We check that only 1.1.2.3 satisfies the range specified in the first LCN label of the AD-dis edge in P1, i.e., the distance between 1.1.2.3 and 1.1.2.3.2 in P1 is exactly 1. Similarly, the constraint in the first LCN label in P2 is also satisfied by 1.1.2.3, so we can output the merged result *1//1.1.2.3[//1.1.2.3.2]//1.1.2.3.5.1*. Next, we match this intermediate result to a P3 path *1//1.1.4*. The common prefixes of 1.1.2.3 and 1.1.4 are 1 and 1.1. Now only 1.1 satisfies the range constraint in the two LCN labels and the top distance label, i.e., 1.1 is exactly 1 level below the A-typed node 1 in P3 and exactly 1 level higher than the D-typed node 1.1.4 in P3. Then we output the final matched pattern *1//1.1[//1.1.2.3[//1.1.2.3.2]//1.1.2.3.5.1] //1.1.4*.

For the query in Fig. 4(b), suppose we merge a P1 path 1 (actually a single A-typed node) and a P2 path *1//1.1.2.3.5.1* first. P1 does not have a lower-end node, so we simply check all the ancestor nodes of 1.1.2.3.5.1 in the P2 path. There are four possible LCNs to merge P1 and P2, i.e., 1.1, 1.1.2, 1.1.2.3, and 1.1.2.3.5. Then we merge four intermediate results to a P3 path 1.1.4, to get the only LCN for the three paths, which is 1.1.   □

Theorem 1. *Consider a query twig pattern q with n nodes (among which there are m (m <= n) wildcard nodes), and only ancestor-descendant edges, and an XML database D. Algorithm Twig* has worst-case I/O and CPU time complexities linear in the sum of size of the (n − m) input lists and the output list. Further, the worst-case space complexity of Algorithm Twig* is the minimum of (i) the sum of sizes of n-m input lists, and(ii) n-m times the maximum length of a root-to-leaf path in D.*

If $m = 0$, Twig* equals to TwigStack, which has been proven to be both I/O and CPU optimal in the wildcard-free queries with only ancestor-descendant edges. If $m > 0$, according to our rewritten strategy, all the wildcards are removed. Thus, Twig* selectively loads only the data nodes whose labels are explicitly referenced in the query (i.e.,$n − m$ nodes) instead of loading the entire input data into main memory to evaluate the query, which is independent of the number of wildcards. In addition, the non-wildcard nodes are processed in the way of TwigStack, as desired.

5.4 Discussion

In this section, we would like to show that all the existing twig pattern matching algorithms based on Dewey-class labeling (e.g.,Extended Dewey [8], JDewey[4]) could be extended to answer the queries with AD-dis axis. The above four steps are available for any other matching algorithms. And the key operation is to (i) identify the LCN nodes and (ii) calculate the distance between two nodes connected with an AD-dis axis. For (i), the above approaches could be adopted, while for(ii), any two Dewey-class IDs could efficiently calculate the distance by checking whether they share the common prefix and the size gap of their Ids.

6 Experiments

In this section, we report an extensive experimental evaluation of our algorithms, using three real-life datasets. Our experiments were conducted to verify the efficiency and scalability of our algorithms.

Implementation and Environment. All the algorithms were implemented in Java and the experiments were performed on a dual-core Intel Xeon CPU 2.0GHz running Windows XP operating system with 2GB RAM and a 320GB hard disk.

Datasets. We use three datasets including $DBLP^1$, $XMark^2$ and $TreeBank^3$ to test the efficacy of our algorithms in the real world. The characters of the three datasets are as follows: DBLP is 127MB with max/average depth of 6/2.9, XMark is 110MB with max/average depth of 12/5.5 and TreeBank is 83MB with max/average depth of 36/7.8.

DataSize	ID	Xpath Expression	DataSize	ID	Xpath Expression
DBLP	Q1	//*//book//author="Jim Gray"	TreeBank	Q9	//*/S/VP/NP/NN
	Q2	//inproceedings//*//author="Stephen F. Smith"		Q10	//EMPTY/*/VP/S/*//NP/*/PP/NP/NNPS
	Q3	//proceedings[//number]//title//*		Q11	/FILE/EMPTY/S/VP/*/*
	Q4	//dblp//*/*[/sub]/i		Q12	//*/S/VP/S[/TO]/NP/_NONE_
XMark	Q5	//*//item/description/parlist/listitem		Q13	//EMPTY/*[/*/PP]//VP[//IN]/VBZ
	Q6	/site/*//*/item//description/*		Q14	//EMPTY//S/*[/TO][/NP]//PP[/IN]//*/_COMMA_
	Q7	//*/regions[//*/location][/*//shipping]//mail/*		Q15	//*//S[/NP/*]/VP[/VBZ]/PP[/*]/*
	Q8	/site/*/europe/item[./incategory][//*/from]//*/listitem			

Fig. 5. Experimental queries for performance comparison

Queries. Regarding to the real-world user queries, the most recent 100 queries are selected from the query logs, e.g., from a DBLP online demo [1]. Then we create the queries with wildcards by replacing some nodes in those queries or

---

[1] http://dblp.uni-trier.de/xml/

[2] http://www.xml-benchmark.org/

[3] http://www.cs.washington.edu/research/xmldatasets/www/repository.html

merging some queries with a wildcard as the branching nodes. Figure 5 reports some of the queries in our experiments.

Compared Algorithms and DataBases. In order to verify the efficiency and scalability of our algorithms, we compared our algorithms with three different types of approaches: (i) DataBases, including (1) famous commercial relational databases that supporting XML queries (e.g., SQL/XML), here we hide the name of the database and call it **RDB**; (2) pure XML database, here we choose BaseX 7.1.1. We execute the XML queries in BaseX with and without indexes, respectively, called **BaseX-I** and **BaseX** respectively. (ii) The approaches in [3], we implemented the wildcard-step elimination strategy (i.e., **layer+eval**) as well as the selective-loading evaluation strategy (i.e., **layer+optEval**) proposed in [3]. (iii) The existing twig pattern matching method for the queries with wildcards proposed in [7], called **TwigMS**, i.e., replacing * nodes by possible element nodes according to XML schemes (e.g.,DTD).

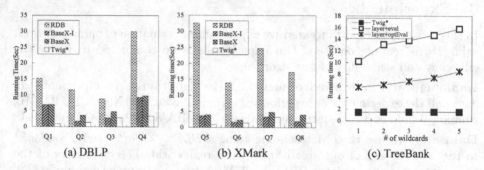

(a) DBLP   (b) XMark   (c) TreeBank

Fig. 6. The running time on DBLP and XMark datasets are reported for comparing the efficiency of Twig* and relational databases and pure XML databases, while the running time on TreeBank dataset is drawn to show the comparison results of Twig* and the approaches proposed in [3].

6.1 Experimental Results

We inspected all the results returned from all the tested algorithms and found that their results are all the same, which verifies the validity of our algorithms. Each experiment was repeated over 10 times and the average numbers are reported here.

Compared with Databases. In Figure 6(a)(b), we test the running time for Twig*, RDB, BaseX-I and BaseX by executing the queries Q_1 Q_8 in the DBLP and XMark respectively. As shown, Twig* outperforms the other three approaches, sine twig* only needs to read data nodes whose labels are explicitly referenced in the query which is independent to the number of wildcards in the query. RDB achieves the worst performance since SQL/XML language does not optimize queries with wildcards. We observed that BaseX-I also achieves a good performance and less running time than BaseX due to the indexes and both of them outperform RDB, the reason is that pure XML databases carefully optimize the queries with wildcards.

Compared with Algorithms in [3]. To test the efficiency of *twig** and the algorithms proposed in [3], i.e., *layer+eval* and *layer+optEval*. We vary the number of wildcards from 1 to 5 in the TreeBank dataaset. As shown in Figure 6(c), Twig* outperforms the other two algorithms, more precisely, Twig* is about one order of magnitude more efficient than *layer+eval* method and saves 4.6 times of running time than *layer+optEval*, which indicates the effects of our algorithm.

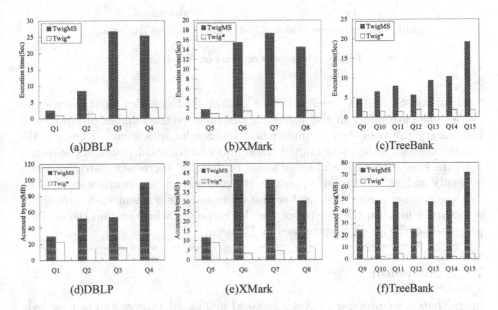

Fig. 7. The execution time and space usage on DBLP, XMark and TreeBank

Compared with Twig Pattern Join Algorithm. Figure 7(a)(b)(c) report the execution time of the Twig* and TwigMS. As shown, *Twig** is faster than *TwigMS*. The first reason is that *Twig** does not need to consider different possible element nodes for each * node, thus it avoids reading many useless labels . We further validate this point by showing the data size read by the two approaches during query processing in Figure 7(d)(e)(f). The second reason is that *Twig** performs structural join across each * node. This attempt will have less round of structural joins compared to that in *TwigMS*.

However, when the possible element nodes for the * nodes is quite limited and the structural joins saved is not expensive, the two approaches should perform similarly. For example, in the first query to the DBLP and XMark documents (Q1 and Q5) there is only one possible element node to replace the * node in *TwigMS*. Moreover in these two queries the * node corresponds to an element with very few labels, which means the structural joins between the * node and the adjacent nodes are not very expensive. Thus the execution time of the two approaches are quite similar.

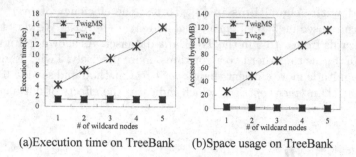

(a)Execution time on TreeBank   (b)Space usage on TreeBank

Fig. 8. The execution time on TreeBank

Scalability. Since the TreeBank data is deep and complex in structure, we choose one path query in the TreeBank and increase the number of * nodes to test the scalability of the two approaches. The results are shown in Figure 8. We can see that with the increasing of the number of wildcard nodes in the query, both the execution time and space usage of *TwigMS* increase rapidly. This can be easily understood. In our approach, the execution time is rather stable, and the space usage even decreases because there are less element nodes requiring structural join when the number of * nodes increases, which verifies the optimal properties of Twig* as we proposed in Theorem 1.

7   Conclusion

In this paper, we propose a novel structural join based approach to process twig pattern queries with wildcards effectively. Different from the existing structural join algorithms, we do not need to replace the * query nodes by element query nodes based on schemas. Instead, we propose a new query edge (i.e., AD-dis edge) and a rewriting algorithms to remove the * nodes in a query. In addition, we experimentally verify the effectiveness and efficiency of approaches.

References

1. Bao, Z., Ling, T.W., Chen, B., Lu, J.: Effective xml keyword search with relevance oriented ranking. In: ICDE, pp. 517–528 (2009)
2. Bruno, N., Koudas, N., Srivastava, D.: Holistic twig joins: optimal xml pattern matching. In: SIGMOD Conference, pp. 310–321 (2002)
3. Chan, C.Y., Fan, W., Zeng, Y.: Taming xpath queries by minimizing wildcard steps. In: VLDB, pp. 156–167 (2004)
4. Chen, L.J., Papakonstantinou, Y.: Supporting top-k keyword search in xml databases. In: ICDE, pp. 689–700 (2010)
5. Jiang, H., Lu, H., Wang, W.: Efficient processing of XML twig queries with OR-predicates. In: SIGMOD, pp. 59–70 (2004)
6. Lu, J., Chen, T., Ling, T.W.: Efficient processing of XML twig patterns with parent child edges: a look-ahead approach. In: CIKM, pp. 533–542 (2004)

7. Lu, J., Ling, T.W., Bao, Z., Wang, C.: Extended XML tree pattern matching: theories and algorithms. IEEE Trans. Knowl. Data Eng. (2010)
8. Lu, J., Ling, T.W., Chan, C.Y., Chen, T.: From region encoding to extended Dewey: On efficient processing of XML twig pattern matching. In: VLDB, pp. 193–204 (2005)
9. Wu, H., Lin, C., Ling, T.W., Lu, J.: Processing xml twig pattern queries with wildcards. Technical report, http://datasearch.ruc.edu.cn/full
10. Wu, H., Ling, T.-W., Chen, B.: VERT: A Semantic Approach for Content Search and Content Extraction in XML Query Processing. In: Parent, C., Schewe, K.-D., Storey, V.C., Thalheim, B. (eds.) ER 2007. LNCS, vol. 4801, pp. 534–549. Springer, Heidelberg (2007)
11. Yu, T., Ling, T.-W., Lu, J.: TwigStackList¬: A Holistic Twig Join Algorithm for Twig Query with Not-Predicates on XML Data. In: Lee, M.L., Tan, K.-L., Wuwongse, V. (eds.) DASFAA 2006. LNCS, vol. 3882, pp. 249–263. Springer, Heidelberg (2006)
12. Zhang, C., Naughton, J.F., DeWitt, D.J., Luo, Q., Lohman, G.M.: On supporting containment queries in relational database management systems. In: SIGMOD, pp. 425–436 (2001)

A Direct Approach
to Holistic Boolean-Twig Pattern Evaluation

Dabin Ding, Dunren Che, and Wen-Chi Hou

Department of Computer Science
Southern Illinois University, Carbondale, IL 62901, USA
{dding,dche,hou}@cs.siu.edu

Abstract. XML has emerged as a popular formatting and exchanging language for nearly all kinds of data, including scientific data. Efficient query processing in XML databases is of great importance for numerous applications. Trees or twigs are the core structural elements in XML data and queries. Recently, a holistic computing approach has been proposed for extended XML twig patterns, i.e., B-Twigs (Boolean Twigs), which allows presence of AND, OR, and NOT logical predicates. This holistic approach, however, resorts to pre-normalization on input B-Twig queries, and therefore causes extra processing time and possible expansion on input queries. In this paper, we propose a direct, holistic approach to B-Twig query evaluation without using any preprocessing or normalization, and present our algorithm and experimental results.

Keywords: Query processing, XML Query, Boolean Twig, Twig join, Tree Query, Twig Query, Tree Pattern, Logical Predicate.

1 Introduction

XML (Extensible Markup Language) as a de facto standard for data exchange and integration is ubiquitous over the Internet. Many scientific datasets are represented in XML, such as the Protein Sequence Database, which is an integrated collection of functionally annotated protein sequences [1] and the scientific datasets at NASA Goddard Astronomical Data Center [2]. XML is frequently adopted for representing meta data for scientific and other computing tasks. In addition, numerous domain-specific XML markup languages are defined, such as the Chemical Markup Language (CML), the Mathematics Markup Language (MathML) and the Geography Markup Language (GML). Efficiently querying XML data is a fundamental request to fulfill these scientific applications. In addition to examine the contents and values, an XML query requires matching implied twig patterns against XML datasets. Twig pattern matching is a core operation in XML query processing. In the past few years, many algorithms have been proposed to solve the XML twig pattern matching problem. Holistic twig pattern matching has been demonstrated as so far the most efficient approach to XML twig pattern computation. Well-known holistic twig join/matching algorithms include [5], [10], [6], [11], [12], [15], [8], [14], [7].

S.W. Liddle et al. (Eds.): DEXA 2012, Part I, LNCS 7446, pp. 342–356, 2012.

These algorithms, however, can only deal with twig queries which contain limited types of predicates. However, queries in practical applications may contain all three boolean predicates (AND, OR, NOT). Such kinds of twigs are called Boolean-Twigs or B-Twigs for short. Some example B-Twig XML queries are given below (in XPath-like format):

Q1: /department/employee [age > 30 OR city = "NYC"]/name

Q2: /vehicle/car/[[made = "ford" AND year < 2005] OR [NOT[type = "coupe"] AND color = "white"]]

Q1 selects the employees who are either older than 30 or live in NYC. Q2 (involving all three types of logical predicates) selects the cars that are either "made by FORD and before 2005" or "white but not a coupe".

OR and NOT predicates are very important not only in theory but also in practical applications. It is hence a very natural requirement to support these two types of predicates, in addiction to ANDs, in twig pattern matching algorithms. We have seen several efforts being made toward holistically computing B-Twig pattern matches. Jiang et al. [6] proposed a solution to AND/OR-twigs; Yu et al. [15] proposed an algorithm for holistically evaluating AND/NOT-twigs; most recently, Che et al. (in our own group) [7] proposed the first algorithm, called *BTwigMerge*, for holistic computing of full B-Twigs (i.e., AND/OR/NOT-twigs), but requiring significant preprocessing (normalization) on input B-Twigs first. Normalization helps control the complexity in holistic B-Twig pattern matching, but inevitably introduces an extra processing step and may cause query expansion especially when NOTs are pushed down to lower levels in the B-Twigs. In this paper, we propose the first, *direct* holistic B-Twig pattern matching algorithm, called *DBTwigMerge*, without relying on any preprocessing or normalization on the input B-Twig queries. Compared with our prior algorithm, *BTwigMerge* [7], our new algorithm, *DBTwigMerge*, relies on a new mechanism, called *status* which is associated with every query node, to unify the processing needed for different types of query nodes (including AND, OR, and NOT nodes) into a coherent holistic processing framework. Our new algorithm thus can very *elegantly* process any input B-Twig without the need to normalize [7] it first, and yet our new algorithm remains runtime optimal, i.e., linear to the total size of input and output. We believe what we achieved and reported in this paper represents one major advance in the area of XML query processing as our algorithm is unique and is the first of its kind – as a holistic twig join algorithm designed to be directly applied to arbitrary input B-Twig queries.

The remainder of the paper is organized as follows. Section 2 discusses related works. Section 3 provides preliminary knowledge, including our data model and the tree representation used in this paper. Section 4 elaborates on the novel *DBTwigMerge* algorithm that directly applies to arbitrary B-Twig queries without relying on any preprocessing or normalization on the input queries. Section 5 presents the result of our performance study. Finally, Section 6 concludes this paper.

2   Related Work

Twig pattern matching is a core operation in XML query processing. Naive navigation (or pointer-chasing), structural joins, and holistic twig joins have all been studied for twig pattern matching. In the following, we review representative works on structural joins and particularly on holistic twig joins.

The first structural join (or containment join) algorithm was proposed by Zhang *et al.* [16], which extends the traditional merge join to multi-predicate merge join (MPMGJN). Al-Khalifa *et al.* [4] later proposed two families of structural join algorithms, i.e., tree-merge and stack-based structural joins, as primitives for XML twig query processing. In 2002, Bruno *et al.* [5] first proposed the so-called holistic twig join approach to XML twig queries, of which the main goal was to overcome the drawback of structural joins that usually generate large sets of unused intermediate results. Bruno *et al.* designed the holistic twig join algorithm, named *TwigStack*, which is optimal for twigs with only AD edges (but not with PC edges). The work of Lu *et al.* [11] aimed at making up this flaw and they presented a new holistic twig join algorithm called *TwigStackList*, in which a list structure is used to cache limited elements in order to identify a larger optimal query class. Chen *et al.* [8] studied the relationship between different data partition strategies and the optimal query classes for holistic twig joins. Lu *et al.* [12] proposed a new labeling scheme, called *extended Dewey*, and an interesting algorithm, named *TJFast*, for efficient processing of XML twig patterns. Unlike all previous algorithms based on region encoding, to answer a twig query, *TJFast* only needs to access the labels of the *leaf* query nodes. The result of Lu *et al.* [12] includes enhanced functionality (can process limited wildcard), reduced disk access, and increased total query performance. The same group [13] also studied efficient processing for *ordered* XML twig patterns using their proposed encoding scheme. S.K. Izadi *et al.* [17] proposed a series of algorithm called S3. Twig queries are first evaluated against structural summary of documents to avoid document access as far as possible. Later on, they published newer version of their algorithms to support B-Twig queries [18]. NOT and OR operators are processed by special techniques in their algorithms. Overall, their series of S3 algorithms are based on structural joins and do not produce matching result holistically. Additional techniques are introduced in [18] to reduce intermediate results and redundant I/O.

In an ordinary twig, the multiple sibling nodes under a common parent node automatically signify the AND logic relationship among them, and all holistic twig join algorithms discussed above already support this implied AND logic in their implementation schemes. Users would take all the three commonly used logical predicates, AND, OR, and NOT, as *granted facilities* in formulating their XML queries and thus would expect full support from a query engine for unlimited use of all these predicates in their XML queries. Jiang *et al.* [6] made the first effort toward incorporating support for OR predicates into the holistic twig join approach pioneered by Bruno *et al.* [5]. Jiang *et al.* [6] presented an interesting framework for holistic processing of AND/OR-twigs based on the concept of OR-blocks. With resort to the mechanism of OR-blocks, an AND/OR-twig is transformed to an AND-only twig carrying OR-blocks. Yu *et al.* [15] made effort for supporting NOT predicates in XML twig queries. The recent publication of Xu *et al.* [14] proposed another interesting algorithm that claims to be able to efficiently compute the answers to XML queries without holistically computing the twig patterns — the answers obtained contain *individual elements* corresponding to designated *output* query nodes. So basically this work does not belong to the category of holistic twig join algorithms. But what is interesting of their work [14] is the proposed *path-partitioned element encoding scheme*, which bears efficiency potential and may be adopted in the future course of seeking improved performance on holistic B-Twig pattern matching.

Most recently, to support query that contains AND/NOT/OR predicates, we [7] proposed normalization procedure to regulate the arbitrary combination of AND, NOT, OR predicates in B-Twig. Normalization transforms the original B-Twig into an equivalent one that resembles the DNF form of Boolean logic, and which is then evaluated by an efficient computing scheme. However normalization comes with a cost- extra processing steps and possible query expand.

Our goal in this paper is to design a more powerful, holistic computing scheme that can directly apply to arbitrary input B-Twigs without normalization. Our approach and new algorithms are to be detailed in the subsequent sections.

3   Preliminaries

In this section, we address the data model and tree representation of XML data and XML queries.

We adopt the general perspective [5] that an XML database is a forest of rooted, ordered, and labeled trees, each node corresponds to a data element or a value, and each edge represents an element-subelement or element-value relation. The order among sibling nodes implicitly defines a total order on the tree nodes. Node labels are important for efficient processing of twig pattern queries, as properly designed node labels may leave out the necessity of accessing the node contents during query evaluation. This is especially true with twig pattern matching, which is at the core of XML query processing. Node labels typically encode the *region* information of data elements and reflect the relative positional relationships among the elements in the source data file. We assume a simple encoding scheme — a triplet region code $(start, end, level)$ — which is assigned to each data element in a XML document. When multiple documents are present, the *document-id* is added to the labels to differentiate the documents. Region code can be conveniently obtained through preorder document-tree traversing. For example, region codes for each XML data element are given in Fig. 1(a). Region code can preserve the tree structure among nodes. For twig pattern evaluation, our algorithm only need to store and access the region code of each data element. Each element type node in input twig pattern is associated with a stream of data elements (represented in region code) of that type. As in Fig. 1(a), the stream of element type *Author* is (3,6,3)(7,10,3)(18,21,3)(22,25,3). Each XML query implies a twig pattern, small or large. The smallest twig may contain just a single node, but a typical twig usually comprises a number of nodes. In this paper, we will use upper case letters to denote the query nodes in a twig query, and the same lower case letters to denote the current data elements in their associated streams. The target of our investigation is B-Twigs that allow arbitrary combination of ANDs, ORs, and NOTs. In general, a B-Twig may consist of two general categories of nodes: ordinary *query nodes* standing for element types (or tags) and the special *connective nodes* denoting logical predicates — ANDs, ORs, and NOTs. More specifically, we introduce the following types of nodes that a B-Twig may contain:

- QNode: An ordinary query node, associates with an element type (or tag name) in an XML database, and is further associated with an input stream (of which the data elements are represented by their region code, respectively).
- LgNode: A logical node, can be either an AND node, an OR node, or a NOT node. An LgNode does not have any stream associates with it. We further introduce the following specific types of LgNodes:
  - ANode: An AND logical node, always takes the text 'AND' as its content. It connects two or more child subtrees through the AND logic.
  - ONode: An OR logical node, always takes the text 'OR' as its content. It connects two or more child subtrees through the OR logic.
  - NNode: The NOT logical node, always takes the text 'NOT' as its content. It negates the node below it;
- DAQ: a Direct Ancestor Query node. Every node (either QNode or LgNode) has a DAQ node. We define the DAQ of a LgNode to be the first QNode met when traversed up in query tree; and the DAQ of a QNode to be itself.

For the convenience of presentation, we use "a query node" to generally refer to any node (QNode or LgNode) that appears in the B-Twig query under discussion. In contrast, an ordinary node refers to a QNode in the B-Twig. There are two kinds of relationship between two connected query nodes, parent-child(PC) or ancestor-descendant(AD) relationship.

Every B-Twig XML query can be represented as tree (as shown in figure 1(b)) or in an XPath-like expression. The answer to a B-Twig query is a set of qualified twig instances (i.e., the embedding of the twig pattern into the XML database). We assume the following *output model* for B-Twigs: each *output twig instance* of a B-Twig query comprises of elements from only those QNodes that are not inside of any predicate. The sub-twig resulted from the original input B-Twig after pruning all predicate branches (which are subtrees rooted at a predicate node) is called the *output twig* of the B-Twig query. Each remaining QNode on the output twig is called an *output node*, and each leaf on the output twig is called an *output leaf*.

4   Direct B-Twig Evaluation

As pointed out earlier, precious efforts have been made toward holistic B-Twig pattern matching: *GTwigMerge* for AND/OR-twigs, *TwigStackList¬*

for AND/NOT-twigs, and *BTwigMerge* for normalized B-Twigs, and none for general B-Twigs. Developing a holistic approach for general B-Twigs involves great challenges, and requires new, creative supporting mechanism.

4.1   *Status* Mechanism

A general B-Twig query may contain arbitrary combination of logical nodes. For arbitrary combination of logical predicates, we mean that there is no limitation on the usage of logical predicates as long as they are meaningful and bear no redundancy. All the meaningless queries, e.g. NOT as a leaf node, and redundant queries, e.g. NOT/NOT branch can be easily eliminated by a simple preprocessing process. For arbitrary B-Twigs, what most troublesome is the involvement of multiple NOT nodes; in that case, the repeated negations implied in the B-Twig can be very hard to interpreted and programmatically dealt with.

The real challenge in B-Twig pattern matching is centered around the LgNodes in the B-Twig and their evaluation. It is possible for a B-Twig query to have more than four levels of LgNodes between any two QNodes. To deal with complex combination of logical predicates, we need to take the LgNodes into consideration during path-match and subtree match. The nodes in B-Twig need to bear a status of its subtree matching result and LgNodes need to inherit a fake region code from its DAQ node. To facilitate the evaluation process, we introduce an explicit *status* flag (true or false) for each LgNode and each QNode. Our algorithm continuously detects the status of the current node. More formally, we give the following definition for the status mechanism of every node in a B-Twig:

Definition 1 (Status of Node). *Every node in a B-Twig is associated with a boolean flag, called the status of the node; the status of a node takes the value of either true or false, indicating whether a match for the sub-tree rooted at this node is found.*

To motivate the new *status* mechanism which is fundamental to our algorithm, we look at the twig query in Fig. 1(b) and the sample XML data in Fig. 1(a). To save up space, we will use I# to denote the data element, e.g., I1 denotes Refinfo(1,15,1) and I3 denotes Author(3,6,3). Initially, I12/'1962' is read in, then the status of QNode Year is set as true since it is a match. LgNode NOT is set as false since the status of its child is true, which should be negated. When I3 is processed, the status of QNode Author is false, then later on the stream of Author is advanced and I7 is processed and the status of Author is updated to be true. After all two child branches of LgNode OR are updated, then the status of OR can be determined as true since one of the child branch, Author, is true. Finally we can find out that the status of QNode Refinfo is true for I1, and I1/I12/'1962' is a matching instance and be outputted. Similarly, I16/I27/'1988' can also be found as a matching instance.

To summarize, we introduced the status for each query node and each logical node in a B-Twig; based on this mechanism, we are able to uniformly deal with all the nodes (logical and query nodes) in the B-Twig during the matching process. Now we can envision our holistic B-Twig matching process as follows: we simultaneously keep an eye on the input B-Twig and an eye on the input streams; we start from the root of the input B-Twig and probe into the associated streams to see if we can find a match; often, whether an element is a match is not only decided by the element itself, but also by whether the lower level nodes in the B-Twig can find a match from their input streams; stream heads need be promptly advanced during this process when the head element is found disqualified by probing the range covering relationship between the elements; the whole process consists of repeatedly deciding the status of all query nodes in the input B-Twig, outputting successful matches (resulting in a 'true' status value on the root query node), advancing stream heads, and updating the statuses of query nodes (including logical nodes). Our direct, holistic B-Twig join algorithm is designed based on the above process.

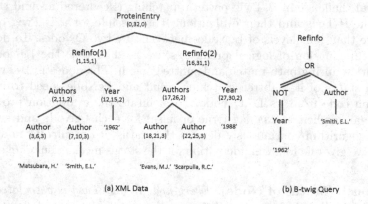

Fig. 1. Illustrative XML data and query

4.2 *DBTwigMerge* Algorithm Design

Status updating is going to be a primary supporting mechanism in our direct, holistic B-Twig join algorithm. We will discuss how the status of different types of query node is going to be updated. First, we need to define some terms and prerequisites. As mentioned earlier, we use upper case letters (e.g., Q) to denote query nodes, and same lower case letters (e.g., q) to denote the current data instances in the corresponding streams. R(Q,S) denotes the region covering relationship between query nodes Q and S, which is supported by and equals to R(q,s), where q and s refer to the two corresponding, stream head instances. A special case is when S is a LgNode, then R(Q,S) is always set as true. For an AD relationship of Q//S, R(q,s) is true (i.e., q covers s) iff q.Start≤s.Start, q.End≥s.End, and q.Level<s.Level. For a PC relationship of Q/S, R(q,s) is true (i.e., q covers s) iff q.Start≤s.Start, q.End≥s.End, and q.Level=s.Level-1.

In this paper, we further denote the *status* of a query node Q as S_Q. The following equations are defined for the status updating with regard to different types of query node.

A QNode Q may connect to zero or more child subtrees Q_i, $i \in [0, n]$. The *status* of Q is decided by the following equation:

$$S_Q = \underset{i \in [0,n]}{\wedge} \{S_{Q_i} \wedge R(Q, Q_i)\}$$

In the beginning, the statuses of all QNodes are set as 'true'.

The *statuses* of different LgNodes are respectively calculated as follows:

ANode: the status updating of an AND node is basically the same as a QNode, except that an AND node does not have data stream associated with it; instead, it inherit a fake region code from its DAQ's to determine its status. An ANode connects to zero or more child subtrees, AND_i, $i \in [1, n]$. The *status* of an AND node is defined as:

$$S_{AND} = \underset{i \in [1,n]}{\wedge} \{S_{AND_i} \wedge R(DAQ_{AND}, AND_i)\}$$

DAQ_{AND} denotes the DAQ of an AND node, similarly for DAQ_{OR} and DAQ_{NOT}.

ONode: an OR node connects to one or more subtrees OR_i, $i \in [1, n]$. The *status* of an ONode is defined as:

$$S_{OR} = \underset{i \in [1,n]}{\vee} \{S_{OR_i} \wedge R(DAQ_{OR}, OR_i)\}$$

NNode: a NOT node connects to only one subtree, NOT_c. The *status* of an NNode is defined as:

$$S_{NOT} = \neg(S_{NOT_c} \wedge R(DAQ_{NOT}, NOT_c))$$

In our implementation, the initial statuses of all ANodes and NNodes are set as true, but the initial statuses of all ONodes are set as false; status initialization and updating are implemented through the key supporting function, *hasExtension* which is given in Algorithm 1.

In Algorithm 1, line 1 sets up initial *statuses* of query nodes. All nodes' *statuses* are set to true, except for ONodes; lines 2-12 update the *statuses* of different kinds of nodes according to the equations given above; line 13 returns the *status* of the current node.

Another key support function is *getNextNode*, which is given in Algorithm 2. It recursively calls itself to find and return the next query node that has been confirmed to have a matching instance in the input streams.

In Algorithm 2, lines 1-3 return current q if q is a leaf node. Lines 4 sets local variable DAQ to the DAQ of q if q is a logical node, otherwise to q itself. Lines 5-6 reset qMax and qMin of q to be null. Each node keeps track of max (qMax) and min (qMin) child nodes in the subtrees. In lines 7-35, the function recursively call itself to return q or qMin depending on which is associated with

---

Algorithm 1. hasExtension(QueryNode Q)

1: setInitialStatus();
2: **for all** Q_i : Q.getChildrenList() **do**
3:     **if** Q.isQNode **then**
4:         Q.status = $(Q.status \land Q_i.status \land R(Q, Q_i))$;
5:     **else if** Q.isANode **then**
6:         Q.status=$(Q.status \land Q_i.status \land R(Q.DAQ, Q_i))$;
7:     **else if** Q.isONode **then**
8:         Q.status=$(Q.status \lor (Q_i.status \land R(Q.DAQ, Q_i)))$;
9:     **else if** Q.isONode **then**
10:        Q.status=!$(Q_i.status \land R(Q.DAQ, Q_i))$;
11:    **end if**
12: **end for**
13: RETURN Q.status;

---

a stream element with a smaller *start* value, or return the query node q with status equals to true. In lines 8, it recursively calls itself for child node q_i to find a matching for the subtree at q_i. If the matching node is not q_i (root of subtree) then return this node directly, as this node might be part of previous solutions. In lines 12-14, if q_i is *older* than the current DAQ (e.g., beyond left boundary of DAQ), then return q_i for the same reason as explained above. In lines 15-18, update qMin and qMax of q by q_i, except when q_i is a NNode — in that case, ancestor node q does not need to update its qMax and qMin, as q does not need to advance its stream according to children NNodes later on. In lines 20-21, if q is connected to a OR-NOT branch, then we do not need to advance q to qMax because instances between current q and qMax might satisfy the NOT branch. This will prevent some solutions being missed. In lines 23-25, the stream of q is advanced according to qMax. Lines 27-35 deal with the return of this function. If $hasExtention(q)$ evaluates to be true, then the current q is a match and returned, otherwise, q or qMin is returned depending on which associates with an element with a smaller *start* .

Our main algorithm, *DBTwigMerge*, is slightly different from the original *TwigStack* algorithm. We doest not give it here for space reason. One special case is added in the main loop as q.stream can be advanced immediately if q.status is false after q is returned from *getNextNode* function.

4.3   Cost Analysis

In this subsection, we analyze the I/O and CPU cost of our algorithm, *DBTwigMerge*. For ease of presentation, the following parameters are defined:

- $|Q|$: the total number of nodes in queries Q, including QNodes and LgNodes.
- $|Input|$: the total size of all the input streams relevant to query Q.
- $|Output|$: the total count of the data elements included in all output B-Twig instances produced for query Q.

For a query with only AD edges, our algorithm guarantees that all the input streams are read in once and no backtrack. So the I/O cost of our algorithm is proportional to $|Input| + |Output|$, i.e., linear to input and output. Due to the holistic processing feature, our $DBTwigMerge$ does not generate useless intermediate paths on stacks, the total CPU cost is linear to $|Q| * |Input| + |Output|$. In summary, our $DBTwigMege$ remains optimal in CPU cost and have linear I/O cost. Please note that the selection ratio of query have direct relation with $|Output|$, so we did not take selection ratio as a parameter here.

For queries containing PC edges, as our algorithm does not make strict checking on these edges, some unused intermediate results may occur. So $DBTwigMerge$ is no longer optimal in CPU cost when an input B-Twig contains PC edges, but it is still linear in I/O cost.

5   Experiments

In this section we present the experiment results of our algorithm and compare its performance with relevant algorithms on the aspects that they are related. We select the following algorithms to compare with: $TwigStack$ [5] ,$GTwigMerge$ [6], $TwigStackList^\neg$ [15] and $BTwigMerge$ [7]. All those algorithms used the same region encoding schemes for XML documents. We focus our comparison on running time of each algorithm.

5.1   Experimental Setup

Before proceeding to the details of our experiment study, we address a few related issues regarding this experimental study.

Platform Setup. The platform of our experiments contains an Intel Core 2 DUO 2.2 GHz running Windows 7 System with 4GB memory and a 250GB hard disk. Java SE is the software platform on which these algorithms are implemented and tested. A simple Java code based on SAX 2.0 [3] is implemented to parse XML datasets and generates region code for all elements. Region codes of elements are stored as external files on hard disk. Test queries are kept in an external query file that is parsed and transformed into in-memory query trees before sent for execution.

Dataset. To avoid potential bias of using a single dataset, we did experiments with multiple popular XML datasets, and obtained basically consistent results. But for space reason, herein we choose to present only the results with the Protein Sequence Database (PSD) [1]. PSD is an integrated collection of functionally annotated protein sequences. It is published by Georgetown Protein Information Resource, and is freely downloadable [1]. The size of this dataset is 683MB. PSD contains 21,305,818 elements, 1,290,647 attributes, with max depth of 7, average depth of 5.15147. PSD is the largest XML datasets available at [1], which facilitates scalability and robustness test of our algorithms. For experimentally "verifying" the correctness of our implementation, we transformed and stored a

Algorithm 2. getNextNode(QueryNode q)

```
 1: if q.isLeaf() then
 2:     return q;
 3: end if
 4: DAQ= q.isLogicalNode()? q.DAQ : q;
 5: q.updateQmax(null);
 6: q.updateQmin(null);
 7: for all qᵢ ∈ q.getChildrenList() do
 8:     nᵢ = getNextNode(qᵢ);
 9:     if nᵢ ≠ qᵢ then
10:         return nᵢ;
11:     end if
12:     if qᵢ.nextL() < DAQ.nextL() then
13:         return qᵢ;
14:     end if
15:     if ¬qᵢ.isNotNode() then
16:         q.updateQmin(qᵢ);
17:         q.updateQmax(qᵢ);
18:     end if
19: end for
20: if q.hasORNOTBranch() then
21:     break;
22: else
23:     while (q.nextR() < q.getQmax().nextL()) do
24:         q.advanceStream();
25:     end while
26: end if
27: if hasExtension(q) then
28:     return q;
29: else
30:     if q.nextL() < q.getQmin().nextL() then
31:         return q;
32:     else
33:         return q.getQmin();
34:     end if
35: end if
```

portion of this dataset into an Oracle 11g Database that is queried through SQL and led the same query results as the holistic algorithms over the input streams with region encoded elements.

Query Set. We design our test queries with consideration of the following aspects: topology of twigs, distribution of LgNodes, and selection ratio of queries. With regard to topology, attention is directed to the number of nodes, the depth and fanout of twigs. With regard to distribution of LgNodes, we design queries that have LgNodes at different depths and with different combinations. With regard to selection ratio, we require our queries to have a full range (0% to 100%) varying selection ratio. We define selection ratio as the ratio between the count of total query results and the count of all data elements in the data set. We designed totally four sets of queries, each representing a distinct class of B-Twigs. Query set 1 (T1) is a set of plain (or AND-only) twig queries. Query set 2 (T2) is a set of AND/OR-twigs. Query Set 3 is a set of AND/NOT-twigs. Query set 4 (T4) is a set of AND/NOT/OR-twigs.

Performance Metric. This study focused on the key performance metric – CPU cost, and accordingly, we only compare the algorithms in terms of their execution times on the tested queries.

5.2   Experiment Results

We now present the performance results of the four sets of queries, respectively.

Experiment 1: Plain Twig Queries

The performance result of this set of queries is plotted in Fig. 2(a). With this set of plain twig queries, all five tested algorithms show similar performance. On all five test queries, *DBTwigMerge* performs slightly worse than the others. The reason is that the status updating mechanism that consumes extra time (but necessary for handling B-Twigs). However, this disadvantage can be eliminated by pre-probing into the query in the future, and if it is plain query, existing algorithms like *TwigStack* can be called instead.

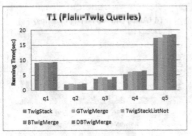
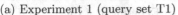
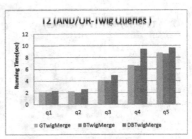

(a) Experiment 1 (query set T1)   (b) Experiment 2 (query set T2)

Fig. 2. Experiment 1 and 2

Experiment 2: AND/OR-twig Queries

For AND/OR-twigs, *TwigStack* and *TwigStackListNot* are not applicable. The performance results are shown in Fig. 2(b). We notice that *GTwigMerge* and *BTwigMerge* slightly outperform *DBTwigMege*. Especially with q4, which has 3 OR nodes in it, extra iterations are rendered by *DBTwigMege* because of its status updating mechanism.

Experiment 3: AND/NOT-twig queries

With AND/NOT-twig queries, *TwigStack* and *GTwigMerge* are not applicable and the performance varies according to different queries according to Fig. 3(a). For the first three queries, *TwigStackList¬* and *BTwigMerge* perform almost the same, and both preforms slightly better than *DBTwigMerge*. For q4,

TwigStackList¬ and *BTwigMerge* take a big advantage over *DBTwigMerge*. In this set of queries, q4, with 12 QNodes and 4 LgNodes, has the most nodes among all five queries, and the status updating mechanism of *DBTwigMerge* slows its performance to the most (but still comparable). For q5, *BTwigMerge* performs worse than the other two because of the extra query nodes introduced by *BTwigMerge* after normalization. More specifically, on q5, the normalization step of *BTwigMerge* introduces 50% of new nodes into the query, which explains the degradation its performance (The total number of nodes in the query increased from 8 to 12).

Experiment 4: AND/OR/NOT Twig Queries
For this set of queries, all three boolean nodes(AND/OR/NOT) are introduced and the query patterns get more complex and only two algorithms, *BTwigMerge*, and *DBTwigMerge* are applicable. As seen from Fig. 3(b) *DBTwigMerge* showed similar performance as *BTwigMerge* on the first four queries, and outperformed on q5. This result is somehow counter-expectation – we expected *DBTwigMerge* (which waived the normalization) to outperform *BTwigMerge* (which requires normalization). The reason is that *DBTwigMerge* iterates over all query nodes (including logical nodes) because of its status updating mechanism, while *BTwigMerge* iterates only over regular query nodes (i.e., QNodes), which saves processing time for *BTwigMerge*.

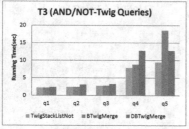

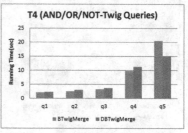

(a) Experiment 3 (query set T3)   (b) Experiment 4 (query set T4)

Fig. 3. Experiment 3 and 4

In summary, the overall performance of *DBTwigMerge* is not as good as we and anyone else would expect, actually more or less a surprise, but we gained precious insight into the complex issue of holistic B-Twig pattern matching. Faced a great challenge, *DBTwigMerge* was intentionally designed to get rid of the normalization step of *BTwigMerge* [7], and logically it should outperform *BTwigMerge* in all aspects. However, instead of win, it lose in all cases except for one case – i.e., when an input B-Twig has a NOT node at a high lever (close to the root), which is the worst case that normalization causes B-Twig expansion in *BTwigMerge*. In all (or most) other cases, the overhead saved from normalization is offset by the extra iterations introduced in *DBTwigMerge* for uniformly dealing with all kinds of query nodes in a B-Twig. From *BTwigMerge* to *DBTwigMerge*, we switched from one extreme (normalization-based) to another extreme (without any normalization), and neither beats out the other completely. The conclusion we drew from our experience with *BTwigMerge* and *DBTwigMerge* can be summarized as follows. Normalization indeed helps control the logical complexity of B-Twigs and leads to an effective, holistic B-Twig pattern matching algorithm, *BTwigMerge* [7], but has to pay the cost of potential expansion on B-Twig queries; *DBTwigMerge* on the other hand (in addition to its elegance), has to rely on a more complex processing scheme (i.e., iterating over all query nodes to check and update the statuses of all the nodes) that degrades its performance. We herein envision a new approach that would outperform all previous approaches in all aspects – this must be one that can combine the advantages of both *DBTwigMerge* and *BTwigMerge*. This envisioned a new approach that shall retain the normalization step but reduce it to the minimum, get the complexity of arbitrary B-Twigs under control, and avoid the rather complex processing in *DBTwigMerge*.

6   Summary

Holistic twig join is a critical operation in XML query processing. All the three types of logical predicates, AND, OR, and NOT, are equally important and needed for general XML queries. However, nearly all previously proposed holistic twig join algorithms failed to provide an integral solution for efficient processing of all these predicates in a single algorithmic framework. In this paper, we presented a direct approach to holistic computing of B-Twigs. We developed a novel algorithm, *DBTwigMerge*, that directly evaluates arbitrary B-Twig queries. A key supporting mechanism in *DBTwigMerge* is its novel *status* updating mechanism that aids the holistic evaluation of B-Twigs. We reported analytical and experimental results of our algorithm with regard to its validity and performance, and drew inspiring conclusions.

As future work, we plan to do the following:

(1) Design a new holistic B-Twig join algorithm, as envisioned in Section 5, which is going to be a compromise, combining the advantages of both *BTwigMerge* and *DBTwigMerge*.

(2) Investigate the potential introduction of new label encoding schemes such as *Extended Dewey* [12] and *path partitioned* [14] into the framework of our algorithms to further boost their performance as these encoding schemes have the potential of extracting ancestor elements' labels without accessing them.

References

1. University of Washington XML repository,
   http://www.cs.washington.edu/research/xmldatasets/
2. NASA Goddard Astronomical Data Center (ADC) 'Scientific Dataset' in XML,
   http://xml.coverpages.org/nasa-adc.html
3. Simple API for XML(SAX), http://www.saxproject.org/about.html

4. Al-Khalifa, S., Jagadish, H.V., Patel, J.M., et al.: Structural joins: A primitive for efficient XML query pattern matching. In: ICDE 2002 Conf. Proc., pp. 141–152 (2002)
5. Bruno, N., Koudas, N., Srivastava, D.: Holistic twig joins: Optimal XML pattern matching. In: SIGMOD 2002 Conf. Proc., pp. 310–321. ACM (June 2002)
6. Jiang, H., Lu, H., Wang, W.: Efficient processing of twig queries with OR-predicates. In: SIGMOD 2004 Conf. Proc., pp. 59–70 (2004)
7. Che, D., Ling, T.W., Hou, W.-C.: Holistic boolean twig pattern matching for efficient XML query processing. IEEE Transactions on Knowledge and Data Engineering, preprint available:
    http://www.computer.org/portal/web/csdl/doi/10.1109/TKDE.2011.128
8. Chen, T., Lu, J., Ling, T.W.: On boosting holism in XML twig pattern matching using structural indexing techniques. In: SIGMOD 2005 Conf. Proc., pp. 455–466 (June 2005)
9. Jagadish, H.V., Al-Khalifa, S., Chapman, A., et al.: Timber: A native XML database. The VLDB Journal 11(4), 274–291 (2002)
10. Jiang, H., Wang, W., Lu, H., Yu, J.X.: Holistic twig joins on indexed XML documents. In: VLDB 2003 Conf. Proc., pp. 273–84 (September 2003)
11. Lu, J., Chen, T., Ling, T.W.: Efficient processing of XML twig patterns with parent child edges: A look-ahead approach. In: CIKM 2004 Conf. Proc., pp. 533–542 (November 2004)
12. Lu, J., Ling, T.W., Chan, C.-Y., Chen, T.: From region encoding to Extended Dewey: On efficient processing of XML twig pattern matching. In: VLDB 2005 Conf. Proc., pp. 193–204 (August 2005)
13. Lu, J., Ling, T.-W., Yu, T., Li, C., Ni, W.: Efficient Processing of Ordered XML Twig Pattern. In: Andersen, K.V., Debenham, J., Wagner, R. (eds.) DEXA 2005. LNCS, vol. 3588, pp. 300–309. Springer, Heidelberg (2005)
14. Xu, X., Feng, Y., Wang, F.: Efficient processing of XML twig queries with all predicates. In: ICIS 2009 Proc., pp. 457–462. IEEE/ACIS (June 2009)
15. Yu, T., Ling, T.-W., Lu, J.: TwigStackList¬: A Holistic Twig Join Algorithm for Twig Query with Not-Predicates on XML Data. In: Li Lee, M., Tan, K.-L., Wuwongse, V. (eds.) DASFAA 2006. LNCS, vol. 3882, pp. 249–263. Springer, Heidelberg (2006)
16. Zhang, C., Naughton, J., DeWitt, D., et al.: On supporting containment queries in relational database management systems. In: SIGMOD 2001 Conf. Proc, pp. 425–436 (May 2001)
17. Izadi, S.K., Harder, T., Haghjoo, M.S.: S3: Evaluation of tree pattern XML queries supported by structural summaries. Data and Knowledge Engineering 68(1), 126–145 (2009)
18. Izadi, S.K., Haghjoo, M.S.: Theo Harder, S3: Processing tree-pattern XML queries with all logical operators. Data Knowledge Engineering 72, 31–62 (2012)

Full Tree-Based Encoding Technique
for Dynamic XML Labeling Schemes

Canwei Zhuang and Shaorong Feng[*]

Department of Computer Science, Xiamen University, 361005 Xiamen, China
cwzhuang0229@163.com, shaorong@xmu.edu.cn

Abstract. It is important to design dynamic labeling schemes which can process the updates when nodes are inserted into or deleted from the XML tree. One class of these schemes is based on lexicographical order. Lexicographical order allows dynamic insertions and thus supports updates in dynamic XML. However, these schemes are of inefficient memory usage when initial labeling. They all require creating an encoding table to produce dynamic labels. As the size of the encoding table can be prohibitively large for large XML documents and main memory remains the limiting resource, having encoding algorithms of memory efficiency is desirable. In this paper, we propose an encoding technique which can be applied broadly to lexicographical-based schemes to produce dynamic labels with high memory usage. Meanwhile, since we don't need the costly table creation, the labeling time of our encoding technique is also efficient. The experimental results confirm that our proposed techniques substantially surpass previous dynamic label schemes.

Keywords: XML Data, Dynamic Labeling Scheme, Lexicographical order.

1   Introduction

Documents obeying XML standard are typically modeled as a tree. Labeling schemes encode structural information so that queries can exploit them without accessing the original XML file. If XML data are static, the labeling schemes, such as containment scheme[1] and prefix scheme[2], can determine the ancestor–descendant, etc. relationships efficiently. However, if XML data become dynamic, it incurs costly relabeling of large amounts of nodes. Designing dynamic labeling schemes to efficiently update the labels is an important research topic.

Several schemes [4, 5, 6, 7, 8, 9, 10] have been proposed to support dynamic XML. One class of these schemes is assigning labels based on lexicographical order. Lexicographical order allows dynamic insertions and thus without re-encoding the existing labels. Such schemes include CDBS [7, 8] and QED [9, 10] which transform the original labels to binary strings and quaternary strings respectively. The following example illustrates the application of CDBS to containment scheme, which is the representative of encoding schemes based on lexicographical order.

---

[*] Corresponding author.

S.W. Liddle et al. (Eds.): DEXA 2012, Part I, LNCS 7446, pp. 357–368, 2012.
© Springer-Verlag Berlin Heidelberg 2012

Example 1. Fig.1(a) shows the original containment scheme using integers to label every node in a XML tree. When CDBS is applied, the integers are transformed into CDBS Codes based on the encoding table in Fig.1(b). Fig.1(c) shows the CDBS-containment labels. CDBS supports that a new code can be inserted between any two consecutive codes with the orders kept and hence avoid re-labeling.

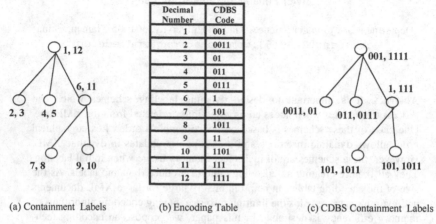

Decimal Number	CDBS Code
1	001
2	0011
3	01
4	011
5	0111
6	1
7	101
8	1011
9	11
10	1101
11	111
12	1111

(a) Containment Labels   (b) Encoding Table   (c) CDBS Containment Labels

Fig. 1. Applying CDBS encoding scheme to containment labeling scheme

The ranges in a set of containment labels come from a sequence of integers from 1 to $2n$ for an XML tree with n elements. Application of lexicographical order is through the process of order-preserving transformation. The transformation is defined as a mapping f from the original integer codes to the target codes such that the target codes preserve the order of the original labels and the total size of the target codes are minimized for a given range. The mapping may be different for different ranges. The following example illustrates how this mapping is derived based on CDBS encoding scheme. QED encoding schemes adopt similar algorithm.

Example 2. The CDBS encoding algorithm is a recursive procedure to create the encoding table (Fig.1.(b) is an example). Given an encoding range m (when labeling for an XML tree with n elements, $m=2n$), CDBS assigns empty string to the 0th position and the $(m+1)^{th}$ position. CDBS encodes a middle position P_M by applying an insertion algorithm takes the encoded codes of the start position P_L and that of the last position P_R. The insertion algorithm takes two CDBS codes as input and output another CDBS code that are lexicographically between them with smallest size. Then the encoding algorithm is recursively applied to $[P_L, P_M]$ and $[P_M, P_R]$ until all the positions are assigned CDBS codes. Since CDBS encodes the numbers randomly--not sequentially, the encoding algorithm needs the temporary array with size $O(m)$ but cannot drop it when labeling XML documents with m elements, which is memory costly.

We classify both CDBS and QED as memory-based schemes since they create an encoding table with size $O(n)$ for labeling XML documents with n elements. It may fail to process a large-scale XML when memory is limited. Since XML documents can be prohibitively large but main memory remains the limiting resource, it is desirable to have a both memory and computation efficient encoding algorithm. In this paper, we present a Full Tree-based (FT) encoding technique which is not only high memory usage but also computationally efficient. FT encoding technique make it possible to generate the codes in the encoding table sequentially and thus no table is needed when XML initial labeling. Importantly, our memory efficiency is not at the sacrifice of labeling time. The advantages of our FT encoding are more significant when applied to improve the performance of QED.

2   Preliminary

In this section, we review the related work on labeling schemes.

Containment Labeling Schemes. In containment scheme, each element node is assigned two values: "start" and "end" where "start" and "end" define a range that contains all its descendant's ranges. Although containment labeling schemes work well for static XML, node insertions may led to costly re-labeling due to the limitation of continuous integers.

Dynamic codes based on lexicographical order are proposed to avoid the re-labeling when XML updating, which include CDBS code[7, 8] and QED code[9, 10]. CDBS Code is a binary string that ends with "1". QED code is a quaternary string that ends with "2" and "3". Only "1", "2" and "3" are used in the QED code itself and "0" is used as the separator. Both CDBS Code and QED Code are compared based on lexicographical order and robust enough to allow insertions without re-labeling.

Example 3. Let "1", "11" be two CDBS codes satisfying "1 $\prec$ "11"lexicographically, we can insert "101" which is another CDBS code between them and we have "1" $\prec$ "101" $\prec$"11". To continue to insert between"1" and "101", we can use"1001", satisfying "1" $\prec$"1001"$\prec$"101". In fact, given any two CDBS codes S_L and S_R satisfying $S_L \prec S_R$, we can always find a middle code S_M such that $S_L \prec S_M \prec S_R$. And the same for QED.

A dynamic encoding scheme can be defined as a mapping from the original containment labels to the dynamic codes. CDBS scheme and QED scheme firstly create an encoding table with preserving order and optimal size to realize the mapping, which are memory inefficiencies.

Search Tree-Based(ST) Encoding Technique[13]. ST encoding provide a novel order-preserving and size-optimal mapping from integer codes in the containment labels to the dynamic codes. The mapping is derived based on ST tree. To encode a range m with ST technique is to realize the mappings represented by an ST table of size m which can be achieved by traversing the ST tree of size m in inorder. ST encoding technique can be applied to CDBS codes and QED codes, and are called STB and STQ encoding schemes respectively. STB tree and STQ tree are two basic data structures for STB and STQ encoding schemes respectively.

STB tree. An STB tree is a complete binary tree. The STB code of root is "1".Given a code n in STB tree, the codes of its left child lc and right child rc are derived as follows: $Clc=Cn$ with the last "1"replaced with "01"; $Crc=Cn \oplus$ "1" ($\oplus$ means concatenation).

STQ tree. An STQ tree is a complete ternary tree of each node associated with two STQ codes: left code (L) and right code (R) where $R = L$ with the last number "2" change to "3". The codes of root are "2" and "3". Given a node n in the STQ tree, the left code of its left child (lc), middle child (mc) and right child (rc) can be derived as follows: $Llc=Ln$ with the last number "2" change to "12";$Lmc= Ln \oplus$ "2"; $Lrc=Rn \oplus$ "2". Additionally, for every node, we have $R=L$ with the last number "2" change to "3". Fig. 2 shows an STB tree and an STQ tree of the same size 12.

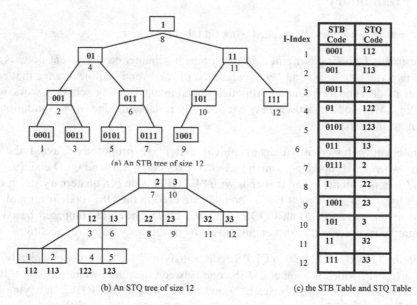

(a) An STB tree of size 12

(b) An STQ tree of size 12

(c) the STB Table and STQ Table

I-Index	STB Code	STQ Code
1	0001	112
2	001	113
3	0011	12
4	01	122
5	0101	123
6	011	13
7	0111	2
8	1	22
9	1001	23
10	101	3
11	11	32
12	111	33

Fig. 2. STB and STQ encoding of ranges 12

Theorem 1. *An inorder traversal of the STB tree visits the STB codes in increasing lexicographical order and of optimal size. And the same for the STQ tree.*

Proof [Sketch] In an STB tree, the left subtree of a node n contains STB codes lexicographically less than Cn; The right subtree of n contains STB codes lexicographically greater than Cn. Thus, inorder traversal sequence is of order preserving. Moreover, an STB tree has all the possible STB codes of length i at level i (except possibly the lowest level). STB codes with length i are always used up before STB codes with length $i+1$ are used. Therefore the codes in STB tree are of optimal total size. The proof of STQ follows similarly.   □

Based on Theorem 1, encoding a range m can be achieved by inorder traversing the ST tree of size m. Our Full Tree-based(FT) encoding provides a technique to avoid the construction of the encoding table but achieve the codes in ST tree sequentially. FT-based encoding technique can be applied to both CDBS and QED, and we call them BFT Encoding Technique and QFT Encoding Technique respectively.

3   BFT Encoding Technique

The ranges in a set of containment labels come from a sequence of integers from 1 to $2n$ for an XML tree with n elements. Application of dynamic encoding is through the process of order-preserving transformation. As we have shown in sec.2, STB encoding provide an order-preserving and size-optimal mapping from integer codes in the containment labels to the target codes. For encoding a range m, the STB mapping f can be represented by an STB table of size m, i.e. $f(i) = STB\text{-}TABLE[i]$. STB encoding cannot avoid table creation to realize the mapping. Our BFT encodings provide a technique that doesn't need to create the encoding table, but can calculate the codes in *STB-TABLE* on the fly. BFT encoding technique is based by the data structure we call BFT tree(Binary Full Tree). We first introduce the BFT tree and then show how BFT technique realize the mappings $f(i) = STB\text{-}TABLE[i]$ with no encoding table needed.

3.1   Binary Full Tree(BFT)

The STB codes are the binary strings that ended with "1". We construct the BFT tree of level l to cover all the STB codes of which the size is not larger than l.

Data Structure. A BFT tree is a full binary tree where each node is associated with a STB code. The rule assigning codes to BFT is the same as that of STB tree(see Sec.2). Fig.3(a) shows a BFT tree with 4 levels.

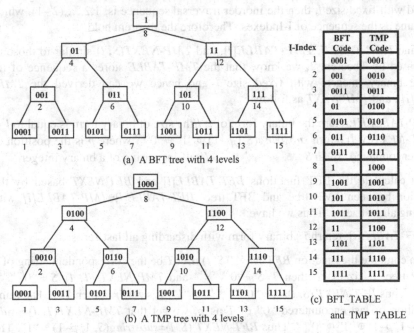

(a) A BFT tree with 4 levels

(b) A TMP tree with 4 levels

I-Index	BFT Code	TMP Code
1	0001	0001
2	001	0010
3	0011	0011
4	01	0100
5	0101	0101
6	011	0110
7	0111	0111
8	1	1000
9	1001	1001
10	101	1010
11	1011	1011
12	11	1100
13	1101	1101
14	111	1110
15	1111	1111

(c) BFT_TABLE and TMP TABLE

Fig. 3. A BFT tree and a TMP tree with 4 levels

A BFT tree of l levels has 2^l-1 STB codes. Since there are 2^l-1 possible STB codes of size less than or equal to l, thus a BFT tree of l levels covers all the possible STB codes of which the size is not larger than l. In additional, BFT tree is order preserving as STB tree(see Theorem 1). An inorder traversal of the BFT tree visits STB codes lexicographically.

Basic Functions. We use a table **BFT-TABLE** to stores the codes of a BFT tree in order of inorder traversal and use **I-Index** to denote the index in *BFT-TABLE*. We define **BFT-TABLE[i]** to denote the ith code in the table. Additionally, given the BFT tree level l and the code S in *BFT-TABLE*, we define **BFT-NEXT(S, l)** to produce the code after S in the table.

To compute the functions defined above, we construct a temp full tree **TMP** which has the same level as BFT. The strings in TMP are of same size l where l is equal to the tree level. Fill "0"s to the end of each string in BFT if its size less than l, and then assign the new string to correspondent node in TMP. Fig.3(b) shows a TMP tree with 4 levels. The strings in TMP are of great regularity as is shown in Theorem 2.

Theorem 2.   *The string in TMP tree is equal to the binary form of its I-Index.*

Proof. A TMP tree of l levels has 2^l-1 codes. Excluding the binary string in which all the bits are '0', there are 2^l-1 possible string of which the size is equal to l, and hence a TMP tree of l levels covers all possible binary strings of size l excluding the string "0^l"("0^l" means l '0's). Furthermore, an inorder traversal of the tree visits these strings in increasing lexicographical order. Thus if we view a string as a binary integer stored with fixed size l, then the inorder traversal sequence is: $1,2,\ldots,(2^l-1)$, which is the same as the sequence of I-Indexes. Therefore the theorem hold.   □

We define **TMP-TABLE**, **TMP-TABLE[i]**, and **TMP-NEXT(S, l)** similar to those of BFT. Based by Theorem 2, we know that the *TMP-TABLE* stores a sequence of increasing binary integers with fixed size l and hence we can derived the *TMP-TABLE[i]* , and *TMP-NEXT* as follows:

- *TMP-TABLE[i]*:return the binary form of i; //the size of binary form is equal to l
- *TMP-NEXT(S, l)*:return $substring(S,1,p-1) \oplus$ "1" $\oplus$ "0^{l-p}",where p is the position of last encountered "0" in S.   //self-increasing of a binary integer

We can calculate the BFT functions *BFT-TABLE[i]* and *BFT-NEXT* based by the correlation between TMP tree and BFT tree. *BFT-TABLE[i]=TMP-TABLE[i]* with discarding all last zeros. Thus we have:

- *BFT-TABLE[i]*: return i's binary form with discarding all last zeros;

We then compute the function *BFT-NEXT(S, l)*. Let T be the correspondent string of S in TMP tree, if size$(S)<l$, then $T=S \oplus$ "$0^{l-size(S)}$" and *TMP-NEXT* $(T, l)=S \oplus$ "$0^{l-size(S)-1}$" $\oplus$ "1". Thus *BFT-NEXT* $(S, l)=S \oplus$ "$0^{l-size(S)-1}$" $\oplus$ "1"; If size$(S)=l$, then $T=S$. Denoting the position of last encountered "0" of T as p, we get that *TMP-NEXT(T, l)= substring(S, 1, p-1)* $\oplus$ "1" $\oplus$ "0^{l-p}". Thus *BFT-NEXT* $(S, l)=substring(S, 1, p-1) \oplus$ "1". The implements of *BFT-NEXT* is shown in Algorithm 1.

Algorithm 1. *BFT_NEXT(S, l)* //return the code after *S* in the *BFT-TABLE* of tree level *l*

```
1    if size(S)<l
2        then return   S ⊕ "0^{l-size(S)-1}" ⊕ "1";
3    else
4        denote the position of last encountered "0" in S as p;
5        return substr(S, 1, p − 1) ⊕ "1";
6    endif
```

The implement of *BFT-TABLE[i]* shows that we don't need to store strings in a *BFT-TABLE* but can calculate them on the fly. *BFT-NEXT* provides the other way to get the strings in the *BFT-TABLE*. Initializing *S* to be null, we can get all strings in a *BFT-TABLE* with *l* levels by (2^{l-1}) iterations of *BFT-NEXT*. We will use these two functions to discuss how BFT technique labeling initial XML without table needed.

3.2   Labeling Initial XML without Encoding Table

To encode a range *m* is to realize the mappings represented by an *STB-TABLE* of size *m*. This can be achieved by our BFT tree of the same level as STB tree(see Fig.4 which is an example of encoding a range 12).

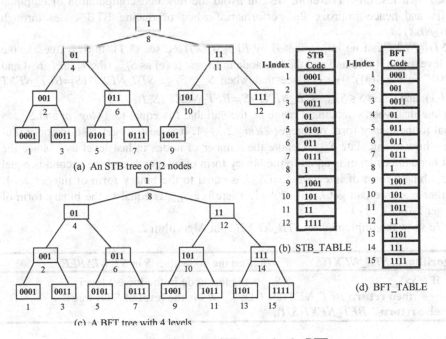

Fig. 4. Achieving the STB codes by the BFT tree

In STB, denote the I-Index of the last node in the lowest level as $INDEX_{Last}$(the value is 9 in Fig.4(a)). We can see that when $i \leq INDEX_{Last}$, the string with I-Index of *i* in STB is equal to that with I-Index of *i* in BFT. And when $i > INDEX_{Last}$, the string with I-Index of i in STB is equal to that with I-Index of $[i+(i-INDEX_{Last}-1)]$ in BFT, which is $(2i-INDEX_{Last}-1)$. Thus we have:

$$STB\_TABLE[i] = \begin{cases} BFT\_TABLE[i] & , when \quad i \leq INDEX_{Last} \\ BFT\_TABLE[2i-1-INDEX_{Last}], when & i > INDEX_{Last} \end{cases} \quad (1)$$

To label XML documents, we just need transforming the integer codes i in the containment labels to STB codes by the mapping: $f(i) = STB\text{-}TABLE[i]$. Since we have discussed how to calculate $BFT\_TABLE[i]$ without table needed(see sec. 3.1), we can calculate $STB\_TABLE[i]$ without table needed as well based on (1). Therefore, we can cast away any encoding table but label the initial document correctly.

However, if we produce STB codes based on (1), we need to compute the binary form of all integers from 1 to m when encoding a range m, which is time-consuming when encoding a large range. We observe that the transformation from the integer codes i to $STB\_TABLE[i]$ is in sequential order when XML labeling. That is, $STB\_TABLE[1]$ is assigned firstly, and then $STB\_TABLE[2]$,...,and so on. Thus, we can implement the function $STB\_NEXT(S)$ which produces the STB code from its previous code S in STB_TABLE to produce all the STB codes. On the other fact, the adjacent STB codes have a long common prefix and have only a little difference between their last bits. Therefore we can avoid the repeated computation of common prefix and hence improve the performance when producing STB codes through STB_NEXT.

STB_NEXT can be implemented by BFT_NEXT(see sec. 3.1). In STB tree, denote the level of STB tree as l and the last node in lowest level as S_{Last} (the values are 4 and "1001" in Fig.4(a)). We can see that when $S > S_{Last}$, $STB\_NEXT(S)=BFT\_NEXT(S, l-1)$; and when $S \leq S_{Last}$, $STB\_NEXT(S)=BFT\_NEXT(S, l)$.

Note that when encoding a range m, the variable l is equal to $1+log_2 m$ and S_{Last} is equal to the binary form of integer $2 \times (m-2^{l-1})+1$. It is easy to get l, so we only illustrate how to calculate S_{Last}. Denote the number of nodes in last level as k, since the first node in last level is equal to the binary form of integer 1, and the second is equal to the binary form of integer 3...so S_{Last} is equal to the binary form of integer $2k-1$. Further more we can get $k=m-2^{l-1}+1$. Therefore S_{Last} is equal to the binary form of integer $2 \times (m-2^{l-1})+1$.

We show the implement of $STB\_NEXT(S)$ in Algorithm 2.

Algorithm 2. $STB\_NEXT(S)$ //return the code after S in $STB\text{-}TABLE$

1   **if** $S>S_{Last}$ // S_{Last} is the last node in the lowest level
2   **then return** $BFT\_NEXT(S, l-1)$;   // l is the level of STB tree
3   **else return** $BFT\_NEXT(S, l)$;

Initializing S to be null, we can produce all the m STB codes in $STB\text{-}TABLE$ sequentially by m iterations of $STB\text{-}NEXT$, which has both computational and memory efficiencies when applied to transform the integer codes in the containment labels to STB codes in order to label the initial XML documents.

4   QFT Encoding Technique

As we have shown in sec.2, STQ encoding provide an order-preserving and size-optimal mapping from integer codes in the containment labels to the STQ codes. For encoding a range m, the STQ mapping f can be represented by an STQ table of size m, i.e. $f(i) = STQ\text{-}TABLE[i]$. Similar to BFT, our QFT encodings provide a technique that doesn't need to create the encoding table, but can calculate the codes in $STQ\text{-}TABLE$ on the fly. QFT encoding technique is based by the data structure we call QFT tree(Quaternary Full Tree). We first introduce the QFT tree and then show how QFT technique realize the mappings $f(i) = STQ\text{-}TABLE[i]$ with no encoding table needed.

4.1   Quaternary Full Tree(QFT)

The STQ codes are the quaternary strings that ended with "2" or "3". We construct QFT tree of l levels to cover all the STQ codes of which the size are not larger than l. A QFT tree is a full ternary tree where each node is associated with two STQ codes. The rule assigning codes to QFT is the same as that of STQ(see sec.2). Additionally, we construct a temp ternary full tree **QTMP** which has the same level as QFT. The strings in QTMP are of the same size l. Fill "1"s to the end of each string in QFT if its size is less than l, and then assign the new string with every "symbol–1" to the correspondent node in QTMP, where "symbol–1" means if the symbol is "3", then return "2", if "2", then "1", and if "1", then "0".A QFT tree and a QTMP tree with 2 levels are shown in Fig.5. The strings in QTMP tree are of great regularity as is shown in Theorem 4. We omit the proof since it is of similar details to that of BFT.

Theorem 4. *The string in* QTMP *tree is equal to the* ternary *form of its I-Index.*

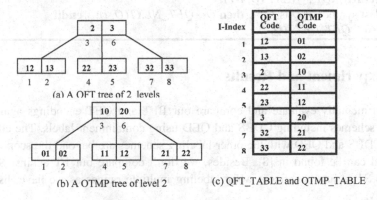

(a) A QFT tree of 2 levels

(b) A QTMP tree of level 2

(c) QFT_TABLE and QTMP_TABLE

I-Index	QFT Code	QTMP Code
1	12	01
2	13	02
3	2	10
4	22	11
5	23	12
6	3	20
7	32	21
8	33	22

Fig. 5. A QFT tree and a QTMP tree with 2 levels

We define **QFT-NEXT(Q, l)** similar to that of *BFT*. We implement it(see Algorithm 3) based by Theorem 4 and the correlation between QFT and QTMP. The details are ignore here as they are similar to those of BFT.

Algorithm 3. $QFT\_NEXT(Q, l)$//return the code after Q in $QFT\text{-}TABLE$ of tree level l

1   **if** size $(Q) < l$

2   **then return** $Q \oplus "1^{\,l\text{-size}(Q)-1}2"$;

3   **else**

4   denote the position of last encountered "1" or "2" as p;

5   **if** $substring(Q, p, p) = "1"$ **then return** $substr(Q, 1, p-1) \oplus "2"$; **endif**

6   **if** $substring(Q, p, p) = "2"$ **then return** $substr(Q, 1, p-1) \oplus "3"$; **endif**

7   **endif**;

4.2   Labeling Initial XML without Encoding Table

How QFT technique labeling for initial XML is similar to that of BFT. Algorithm 4 shows the implement of STQ_NEXT which are used in QFT technique to label the initial XML without encoding table. Compared to that of BFT, one difference is the values of l(equal to $1 + log_3 m$ in QFT)and the value of Q_{Last} (equal to the ternary form of integer $(3 \times (m-3^l) + 2)/2$ in QFT). The other difference is when Q is equal to Q_{Last} and the last symbol of Q_{Last} is "2", then we skip the next code of Q as it is not a code we need and $STQ\_NEXT(Q) = QFT\_NEXT\ (QFT\_NEXT(Q, l)\ ))$(line 5).

Algorithm 4 $STQ\_NEXT(Q)$ //return the code after Q in $STQ\text{-}TABLE$

1   **if** $Q < Q_{Last}$ // Q_{Last} is the last node in lowest level

2   **then return** $QFT\_NEXT(Q, l)$; // l is the level of STQ tree

3   **else if** $Q > Q_{Last}$

4   **return** $QFT\_NEXT(Q, l\text{-}1)$;

5   **if**   last symbol of Q is "2"   **then** $Q \leftarrow QFT\_NEXT(Q, l)$;   **endif**

6   **return** $QFT\_NEXT(Q, l)$;

---

5   Experiment and Results

We experimentally evaluate and compare our BFT and QFT encodings against the previous schemes including CDBS and QED using containment labels. The comparisons of CDBS and QED with the other labeling schemes are beyond the scope of this paper and can be found in [8]. Besides, we don't compare our FT against ST [13] since the ST is just beneficent when labeling multiple ranges as we have discussed previously.

We use Dev-C++ for our implementation and all the experiments are carried out on AMD 2.8GHZ with 2G of RAM running on windows XP. The test datasets choose from real world and their characteristics are shown in table 2.

Table 1. Test data sets

Data set	Document Name	No. of nodes	Max depth	Average depth
D1	Hamlet	6,636	6	4.79
D2	All_shakes	179,690	7	5.58
D3	Nasa	476,646	8	3.16
D4	Lineitem	1,022,976	3	2.94
D5	Treebank	2,437,666	36	7.87

When labeling initial XML documents, both FT encoding techniques and memory-based labeling schemes produce labels of optimal size. Thus we just evaluate their different encoding times and different memory usages which are dominated by the encoding table. The results are shown in Fig.6. We observe clear difference of memory usages between FT encodings and memory-based schemes: Our BFT and QFT both need no table while both CDBS and QED need encoding tables of size $O(N)$. Furthermore, we observe the encoding time of BFT is approximately similar to that of CDBS, and QFT needs significantly fewer encoding time over QED since QED has very time-consuming division operation by "3" to create the encoding table. We can draw a conclusion that the high memory usage for our encoding is not at the sacrifice of labeling time but can reduce the computation. The advantages are more significant for our QFT encoding technique. The results confirm that our FT techniques substantially surpass the memory-based encoding especially when processing large XML data sets with limited memory available.

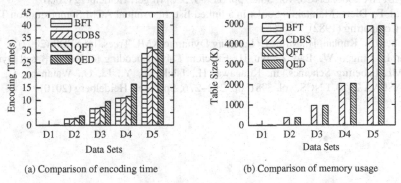

(a) Comparison of encoding time   (b) Comparison of memory usage

Fig. 6. Performance study on initial labeling

6   Conclusion

In this paper, we propose FT technique to optimize the performance of both CDBS and QED. Compared with former encodings which are memory-based, our FT encoding technique labels initial XML with no encoding table needed and is therefore able to process very large XML with limited memory. Moreover, the advantage of high memory usage of our encoding is not at the sacrifice of the encoding time but   computationally efficient. The experimental results have also demonstrated the benefits of our encoding techniques compared to previous approaches.

References

1. Zhang, C., Naughton, J.F., DeWitt, D.J., Luo, Q., Lohman, G.M.: On Supporting Containment Queries in Relational Database Management Systems. In: SIGMOD (2001)
2. Tatarinov, I., Viglas, S., Beyer, K.S., Shanmugasundaram, J., Shekita, E.J., Zhang, C.: Storing and Querying Ordered XML Using a Relational Database System. In: SIGMOD (2002)
3. Wu, X., Lee, M., Hsu, W.: A prime number labeling scheme for dynamic order XML tree. In: ICDE (2004)
4. O'Neil, P., O'Neil, E., Pal, S., Cseri, I., Schaller, G., Westbury, N.: ORDPATHs: Insert-friendly XML Node Labels. In: SIGMOD (2004)
5. Xu, L., Bao, Z., Ling, T.-W.: A Dynamic Labeling Scheme Using Vectors. In: Wagner, R., Revell, N., Pernul, G. (eds.) DEXA 2007. LNCS, vol. 4653, pp. 130–140. Springer, Heidelberg (2007)
6. Xu, L., Ling, T.W., Wu, H., Bao, Z.: DDE: from Dewey to a fully dynamic XML labeling scheme. In: SIGMOD (2009)
7. Li, C., Ling, T.W., Hu, M.: Efficient Processing of Updates in Dynamic XML Data. In: ICDE (2006)
8. Li, C., Ling, T.W., Hu, M.: Efficient Updates in Dynamic XML Data: from Binary String to Quaternary String. In: VLDB J. (2008)
9. Li, C., Ling, T.W.: QED: A Novel Quaternary Encoding to Completely Avoid Re-labeling in XML Updates. In: CIKM (2005)
10. Li, C., Ling, T.-W., Hu, M.: Reuse or Never Reuse the Deleted Labels in XML Query Processing Based on Labeling Schemes. In: Lee, M.L., Tan, K.-L., Wuwongse, V. (eds.) DASFAA 2006. LNCS, vol. 3882, pp. 659–673. Springer, Heidelberg (2006)
11. Paul, F.: Dietz. Maintaining order in a linked list. In: Annual ACM Symposium on Theory of Computing (1982)
12. Cohen, E., Kaplan, H., Milo, T.: Labeling Dynamic XML Trees. In: SPDS (2002)
13. Xu, L., Ling, T.W., Bao, Z., Wu, H.: Efficient Label Encoding for Range-Based Dynamic XML Labeling Schemes. In: Kitagawa, H., Ishikawa, Y., Li, Q., Watanabe, C. (eds.) DASFAA 2010. LNCS, vol. 5981, pp. 262–276. Springer, Heidelberg (2010)

Top-k Maximal Influential Paths in Network Data

Enliang Xu, Wynne Hsu, Mong Li Lee, and Dhaval Patel

School of Computing, National University of Singapore
{xuenliang,whsu,leeml,dhaval}@comp.nus.edu.sg

Abstract. Information diffusion is a fundamental process taking place in networks. It is often possible to observe when nodes get influenced, but it is hard to directly observe the underlying network. Furthermore, in many applications, the underlying networks are implicit or even unknown. Existing works on network inference can only infer influential edges between two nodes. In this paper, we develop a method for inferring top-k maximal influential paths which can capture the dynamics of information diffusion better compared to influential edges. We define a generative influence propagation model based on the Independent Cascade Model and Linear Threshold Model, which mathematically model the spread of certain information through a network. We formalize the top-k maximal influential path inference problem and develop an efficient algorithm, called TIP, to infer the top-k maximal influential paths. TIP makes use of the properties of top k maximal influential paths to dynamically increase the support and prune the projected databases. We evaluate the proposed algorithms on both synthetic and real world data sets. The experimental results demonstrate the effectiveness and efficiency of our method.

1 Introduction

The prevalence of online social media such as Facebook, Twitter, Flickr and YouTube has led to research on social analytics with important applications in online advertising, viral marketing, and recommendation. For example, a company could target a small number of influential users to adopt a new product in the hope that through the "word of mouth" in the social network, these users may persuade their friends and followers to adopt the same product. Early attempts to find the top-k influential users/nodes in a social network assume the existence of a social graph with edges labeled with probabilities of influence between users [9,10,11,17,4,3]. However, this assumption is not realistic as such edges are often implicit or even unknown in the networks.

Recent works aim to infer the "hidden" network from a list of observations of when and where an event occurs [6,16]. The work in [6] infers top-k influential edges in the context of information propagation among blogs and online news sources where bloggers write about newly discovered information without explicitly citing the source. In other words, we can only observe the time when a blog gets influenced but not where it got the influence from.

Figures 1(a) and 1(b) show the top-5 influential nodes and top-5 influential edges obtained from the MemeTracker dataset [12]. Each node in the network is a news website and a directed edge from node a to node b indicates that information has propagated from

S.W. Liddle et al. (Eds.): DEXA 2012, Part I, LNCS 7446, pp. 369–383, 2012.

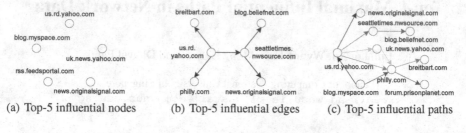

(a) Top-5 influential nodes   (b) Top-5 influential edges   (c) Top-5 influential paths

Fig. 1. MemeTracker

a to *b*. Based on the influential edges, we can only know that when the website *seattle-times.nwsource.com* has new information, it gets propagated to either *blog.beliefnet.com* or *news.originalsignal.com* or both. However, if we have the top-5 influential paths as shown in Figure 1(c), then we see that a new piece of information gets propagated from *us.rd.yahoo.com* to *seattletimes.nwsource.com* to *blog.beliefnet.com*. Further, we observe that if the top-5 influential paths have some node(s) in common, then any disruptions to these common nodes may lead to news blackout. We call these nodes as critical nodes which should have mirror sites.

Another important application of top-*k* influential paths and critical nodes is in the surveillance of computer virus propagation. Inferring the top-*k* influential paths from the list of sites infected by computer virus allows one to better understand how the virus spreads over time. For example, Figure 2(c) shows the top-5 influential paths generated from Code-Red Worm[1]. We can identify the critical nodes in these top-5 influential paths and stop the virus propagation by bringing down these sites.

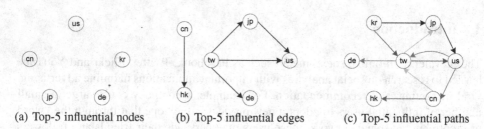

(a) Top-5 influential nodes   (b) Top-5 influential edges   (c) Top-5 influential paths

Fig. 2. Code-Red Worm

In this paper, we develop a method for inferring top-*k* maximal influential paths which can truly capture the dynamics of information diffusion. Given a log of propagation observations of some information over a hidden network, our goal is to infer the top-*k* maximal influential paths that best explain these observations. We define a generative influence propagation model based on the Independent Cascade Model and the Linear Threshold Model [9], which mathematically model the spread of certain information through a network. We design an algorithm called TIP to infer the top-*k* maximal influential paths. TIP utilizes the properties of top-*k* maximal influential paths to dynamically increase the support and prune the projected databases. We evaluate

---

[1] http://www.caida.org/data/passive/codered_worms_dataset.xml

the proposed algorithms on both synthetic and real world data sets. The experimental results demonstrate the effectiveness and efficiency of our method.

2   Influence Propagation Model

An influence network aims to capture the propagation of influence among a set of entities based on a list of observations. We model the network using a directed graph $G = (V, E)$ where V and E are the sets of nodes and edges respectively.

A node u in V denotes an entity and can be *active* or *inactive*. It is considered *active* if it has been influenced. Nodes can switch from being inactive to active, but not vice versa. When a node u gets influenced, it in turns may influence each of its currently inactive neighbors v with some small probability. Node u can only influence its neighbor v if their time difference is within some time threshold τ.

Each directed edge $(u, v) \in E$ has a weight $weight(u, v) \in [0, 1]$ denoting the likelihood of node v being influenced by node u. Suppose t_u and t_v are the times at which nodes u and v get influenced respectively. Then $weight(u, v) = 0$ if $t_v \leq t_u$, i.e., nodes cannot be influenced by nodes from the future time points. Otherwise, $weight(u, v) = e^{-\frac{t_v - t_u}{\alpha}}$ where α is radius of influence.

We associate each node u with an influence measure which is computed from the weights of the edges connecting u to its active neighboring nodes as follows:

$$influence(u, S) = 1 - \prod_{w \in S}(1 - weight(w, u)) \tag{1}$$

where S is the set of active neighbors of u.

One immediate concern is the cost of updating $influence(u, S)$ when the status of nodes change. Since the node status changes frequently, this update cost can be computationally expensive. We derive an expression that allows $influence(u, S)$ to be updated incrementally.

Suppose a new neighboring node w of u becomes active. Then

$$influence(u, S \cup \{w\}) = 1 - (1 - weight(w, u)) * \prod_{u' \in S}(1 - weight(u', u))$$

$$= 1 - (1 - weight(w, u)) * (1 - influence(u, S))$$

$$= influence(u, S) + (1 - influence(u, S)) * weight(w, u) \tag{2}$$

We observe that the influence measure $influence(u, S)$ is both monotonic and submodular.

A function $f(.)$ is *monotonic* if $f(S) \leq f(T)$, for $S \subseteq T$. From Equation 2, we have

$$influence(u, S \cup \{w\}) - influence(u, S) = (1 - influence(u, S)) * weight(w, u) \geq 0$$

A function $f(.)$ is *submodular* if $f(S \cup \{w\}) - f(S) \geq f(T \cup \{w\}) - f(T)$, for $S \subseteq T$. This means that adding a node w to S increases the score more than adding w to T when $S \subseteq T$. We show that $influence(u, S)$ is sub-modular as follows:

$$influence(u, S \cup \{w\}) - influence(u, S) - (influence(u, T \cup \{w\}) + influence(u, T)$$

$$= (1 - influence(u, S)) * weight(w, u) - (1 - influence(u, T)) * weight(w, u)$$

$$= (influence(u, T) - influence(u, S)) * weight(w, u) \tag{3}$$

By monotonicity, $influence(u,T) \geq influence(u,S)$. Hence,

$$(influence(u,T) - influence(u,S)) * weight(w,u) \geq 0$$

Definition 1. *An observation $o = <(u_1,t_1), (u_2,t_2), \cdots, (u_n,t_n)>$ is a sequence of tuples (u_i,t_i) where t_i is the time when node u_i becomes active, and $\forall\, i < j,\, t_i < t_j$. Further, $u_i \neq u_j\, \forall\, i \neq j$. The length of observation o, denoted as $|o|$, is the number of (u_i,t_i) tuples in o.*

Definition 2. *An influential path is a sequence of nodes, denoted as $p = <v_1 \rightarrow v_2 \rightarrow \cdots \rightarrow v_n>$, such that $weight(v_i,v_{i+1})$ is larger than some user defined threshold for all $i,\, 1 \leq i \leq n-1$. The length of p is given by $|p| = n-1$.*

Definition 3. *An observation o supports an influential path p if*

- *$\forall v \in p,\, v \in \{\, u_i \mid (u_i,t_i) \in o \,\}$, and*
- *if u_i and u_j are nodes in o that correspond to $v_{i'}$ and $v_{i'+1}$, then $0 < t_j - t_i < \tau,\, 1 \leq i' \leq n-1$.*

Let D be an observation database. The *support* of an influential path p, denoted as $support(p)$, is the fraction of observations in D that support p.

The *score* of a path $p = <v_1 \rightarrow v_2 \rightarrow \cdots \rightarrow v_n>$ w.r.t. an observation o is defined as

$$score(p,o) = \log\left(influence(v_1,S) \prod_{1 \leq i \leq n-1} weight(v_i,v_{i+1})\right) - \log\varepsilon, \qquad (4)$$

where $\varepsilon \in [0,1]$ is some small value and S is the set of active neighbors of v_1 w.r.t. o.

Let S_p be the set of observations in D that support influential path p. The *total score* of p, denoted as $total\_score(p)$, is defined by

$$total\_score(p) = \sum_{o \in S_p} score(p,o). \qquad (5)$$

An influential path $p = <v_1 \rightarrow v_2 \rightarrow \cdots \rightarrow v_m>$ is a *sub-path* of another influential path $p' = <v'_1 \rightarrow v'_2 \rightarrow \cdots \rightarrow v'_n>$, denoted as $p \sqsubseteq p'$, if and only if $\exists\, i_1, i_2, \cdots, i_m$, such that $1 \leq i_1 < i_2 < \cdots < i_m \leq n$, and $v_1 = v'_{i_1},\, v_2 = v'_{i_2},\, \cdots,$ and $v_m = v'_{i_m}$. We also call p' a *super-path* of p.

An influential path p is **maximal** if there exists no influential path p' such that $p \sqsubseteq p'$ and $support(p) = support(p')$.

Definition 4. *An influential path p is a top-k maximal influential path if p is maximal and there exist no more than $(k-1)$ maximal influential paths whose total score is greater than that of p.*

The following theorem states the relation between the support and total score of two maximal influential paths. This theorem is utilized by our proposed algorithm in Section 3 to effectively prune off the search space.

Theorem 1. *For any two maximal influential paths p and p', if $support(p) > support(p')$ and $\varepsilon < e^{-|D|(|o|+1)\tau}$, then $total\_score(p) > total\_score(p')$ where o is an observation with maximum length in database D.*

Proof. Let p be a maximal influential path with support s and length $|p|$. We can calculate the total score of path p as

$$
\begin{aligned}
total\_score(p) = \sum_{o \in S_p} score(p, o) \\
> (\log e^{-\tau} + |p| * \log e^{-\tau} - \log \varepsilon) * s \\
= -s\tau - s|p|\tau - s * \log \varepsilon \\
= -s\tau - s|p|\tau - s * \log \varepsilon + \log \varepsilon - \log \varepsilon \\
= (-\log \varepsilon) * (s - 1) + (-s\tau - s|p|\tau - \log \varepsilon) \\
> (-\log \varepsilon) * (s - 1) + (-s\tau - s|p|\tau - \log e^{-|D|(|o|+1)\tau}) \\
= (-\log \varepsilon) * (s - 1) + (|D|(|o| + 1) - s(|p| + 1))\tau \\
> (-\log \varepsilon) * (s - 1)
\end{aligned}
$$

Since $(|D|(|o| + 1) - s(|p| + 1)) \geq 0$, we have $(\log e^{-\tau} + |p| * \log e^{-\tau} - \log \varepsilon) * s > (-\log \varepsilon) * (s - 1)$. Note that $(\log e^{-\tau} + |p| * \log e^{-\tau} - \log \varepsilon) * s$ is the lower bound for the *total_score* of any maximal influential path with support s, and $(-\log \varepsilon) * (s - 1)$ is the upper bound for the *total_score* of any maximal influential path with support $(s - 1)$. Further, the value of *total_score* decreases with the length of a path. Hence, $(\log e^{-\tau} + |p| * \log e^{-\tau} - \log \varepsilon) * s > (-\log \varepsilon) * (s - 1)$ implies that the *total_score* of any maximal influential path with support s is greater than all the maximal influential paths whose support is less than s.   □

3   The TIP Algorithm

In this section, we describe our method, TIP, for mining top-k maximal influential paths without the need to specify a minimum support threshold. TIP is a prefix-based influential path mining method. It extends the classical projection-based pattern growth method [13] with time constraint. Instead of projecting observation databases by considering all possible occurrences of prefixes, TIP examines the frequent prefix sub-paths and projects only the corresponding valid observations which satisfy the time constraint into the projected databases. The influential paths are then extended by exploring the valid frequent nodes in the projected databases.

Given an influential path $p = < v_1 \rightarrow v_2 \rightarrow \cdots \rightarrow v_n >$ and a node α, we can extend p by α if the last node of p, i.e. v_n, can *influence* α, that is, the time difference between t_{v_n} and t_α is within the time threshold τ. We denote the extension as $p \rightarrow \alpha = < v_1 \rightarrow v_2 \rightarrow \cdots \rightarrow v_n \rightarrow \alpha >$.

Let $p' = p \rightarrow \alpha$ be an extension of p. we say p is a *prefix* of p' and α is a *suffix* of p'. For example, in our sample observation database D as shown in Table 1, $< a \rightarrow d \rightarrow g >$ is a prefix of path $< a \rightarrow d \rightarrow g \rightarrow i >$ and $< i >$ is its suffix.

Let S_p be the set of observations that support influential path p. Suppose each $o \in S_p$ is of the form $< (u_1, t_1), (u_2, t_2), \cdots, (u_a, t_a), (u_{a+1}, t_{a+1}), \cdots, (u_b, t_b) >$. Then we define the p-*projected* database as $D_p = \{ < (u_{a+1}, t_{a+1}), \cdots, (u_b, t_b) > \}$ if the last node $v_n \in p$ corresponds to $u_a \in o$ and the time difference $t_{a+1} - t_a$ is less than τ.

Table 1. A Sample Observation Database D

ID	Observation
o_0	$<(a,1) (d,5) (g,10) (i,16)>$
o_1	$<(c,8) (e,15) (f,20)>$
o_2	$<(c,4) (d,10) (g,16) (i,20)>$
o_3	$<(c,3) (e,12) (i,36)>$
o_4	$<(c,5) (e,9) (h,20) (i,24)>$

Table 2. Frequent nodes in D

Node	Support
c	4
i	4
e	3
d	2
g	2
a	1
f	1
h	1

Consider the sample observation database in Table 1. Let time threshold $\tau = 20$. The projected database for path $< c \to e >$ is $D_{<c \to e>} = \{< (f, 20) >, < (h, 20), (i, 24) >\}$. Note that for observation o_3, the time stamp of e is 12, while the next time stamp is 36. Since the time difference is 24 which is more than τ, node e cannot influence node i, and hence $< (i, 36) >$ is not included in the projected database.

Algorithm 1. TIPMiner(D, k, τ)

Require: observation database D, an integer k, and time threshold τ
Ensure: Top-k maximal influential path set $PathSet$
1: $V \leftarrow$ nodes in D
2: Initialize $min\_sup = 1$
3: Initialize $PathSet = \emptyset$
4: **for each** node $v \in V$ **do**
5:     $PathSet = \text{TIP}(v, D_{<v>}, k, min\_sup, \tau)$
6: **end for**
7: return $PathSet$

Having defined the concept of path-projected databases, we next describe the framework TIPMiner for mining the top-k maximal influential paths from a given observation database D. Algorithm 1 gives the details. It first finds all the nodes in D and sorts them in decreasing order of their support values. A global variable $PathSet$ is used to keep track of the set of top-k maximal influential paths. This global variable is updated by calling Algorithm TIP (see Algorithm 2) for each node.

Algorithm TIP finds the top-k maximal influential paths by constructing the prefix search tree in a depth-first manner. Inputs to TIP algorithm are an influential path p, the p-projected database D_p, the number of maximal influential paths k, minimum support threshold min_sup, and time threshold τ. The output is the set of top-k maximal influential paths $PathSet$.

Given an influential path p, TIP algorithm attempts to extend p by first obtaining the p-projected database D_p. Initially, the path consists of only one node. Given a path p, we first check if this path is promising (lines 1-3). Line 4 checks whether there exists an influential path $p' \in PathSet$ such that p is a sub-path or super-path of p'. If p' exists, we perform maximal influential path verification (lines 6-16). If p' is a sub-path of p,

---

Algorithm 2. TIP(p, D_p, k, min_sup, τ)

Require: a path p, D_p, an integer k, minimum support threshold min_sup, and time threshold τ

Ensure: Top-k maximal influential path set $PathSet$

1: **if** $support(p) < min\_sup$ **then**
2:    **return**
3: **end if**
4: check whether a discovered influential path p' exists, s.t. either $p \sqsubseteq p'$ or $p' \sqsubseteq p$, and $support(p) = support(p')$
5: **if** such super-path or sub-path exists **then**
6:    **if** $\exists$ node $\alpha \in D_p$, such that $support(p \to \alpha) = support(p)$ **then**
7:       **return**
8:    **end if**
9:    **for each** $p' \in PathSet$ such that $support(p') = support(p)$ **do**
10:      **if** $p \sqsubseteq p'$ **then**
11:         **return**
12:      **end if**
13:      **if** $p' \sqsubseteq p$ **then**
14:         replace p' with p
15:      **end if**
16:   **end for**
17: **else**
18:   **if** $|PathSet| < k$ **then**
19:      $PathSet = PathSet \cup \{p\}$
20:   **else**
21:      let path $q \in PathSet$ such that $\nexists q' \in PathSet, total\_score(q') < total\_score(q)$
22:      **if** $total\_score(p) > total\_score(q)$ **then**
23:         replace q with p
24:      **end if**
25:   **end if**
26: **end if**
27: **if** $|PathSet| = k$ **then**
28:   let path $q \in PathSet$ such that $\nexists q' \in PathSet, total\_score(q') < total\_score(q)$
29:   $min\_sup = support(q)$
30: **end if**
31: $Q \leftarrow$ empty priority queue
32: scan D_p once, find every frequent node α such that p can be extended to $p \to \alpha$
33: Q.insert(α)
34: **if** no valid α available **then**
35:   **return**
36: **end if**
37: **while** !Q.isEmpty() **do**
38:   $\alpha = Q$.pop()
39:   **Call** TIP($p \to \alpha$, $D_{p \to \alpha}$, k, min_sup, τ)
40: **end while**
41: **return**

---

then we replace p' by p in the *PathSet* since p is now the maximal influential path (lines 13-15). However, if p' is a super-path of p, then p is not a maximal influential path and can be discarded (lines 10-12).

If p' does not exist and *PathSet* contains less than k maximal influential paths, then we add p to the *PathSet* (lines 18-19). Otherwise, if *PathSet* already contains k maximal influential paths, we compute the *total_score* of p. If the *total_score* of p is larger than any of the k maximal influential paths in *PathSet*, we replace the path with the smallest *total_score* by p (lines 20-25). By Theorem 1, we raise *min_sup* to the support of the path whose *total_score* is the minimum in *PathSet* (lines 27-30). This allows us to prune off unpromising paths.

Next, the algorithm attempts to extend p by finding all the frequent nodes $\alpha \in D_p$ such that we can extend p to $p \rightarrow \alpha$ (lines 31-40). We scan the p-projected database D_p to find every frequent node α, such that path p can be extended to $p \rightarrow \alpha$, and insert α into a priority queue Q (lines 32-33). If no α can be found, then we stop extending this path (lines 34-36). Otherwise, we recursively call TIP algorithm to extend another path using the next frequent node in Q (lines 37-40). The algorithm terminates when Q is empty.

Table 3. $< c >$-projected database $D_{<c>}$

ID	Observation
o_1	$<(e,15)\ (f,20)>$
o_2	$<(d,10)\ (g,16)\ (i,20)>$
o_3	$<(e,12)\ (i,36)>$
o_4	$<(e,9)\ (h,20)\ (i,24)>$

Table 4. Frequent nodes in $D_{<c>}$

Node	Support
e	3
i	2
d	1
f	1
g	1
h	1

Let us now use the example in Table 1 to illustrate the TIP algorithm. The entity with the highest support value is c (see Table 2). We obtain the projected database $D_{<c>}$ as shown in Table 3. The frequent nodes with their support values are shown in Table 4. We insert these nodes into the priority queue Q and recursively call TIP to extend $< c >$. Since node e has support 3 in Q, we extend $< c >$ to $< c \rightarrow e >$.

Figure 3 shows the prefix search tree constructed. Each node in the tree corresponds to an influential path starting from the root to the node and its support is shown next to the node. The number along each edge denotes the *total_score* of the path from the root to the end node of the edge. We assume that the time threshold $\tau = 20$ and $\varepsilon = e^{-64}$. We observe that $< c \rightarrow e >$ are supported by three observations o_1, o_3 and o_4 in Table 1. The scores with respect to these observations are as follows:

$$score(p,o_1) = \log(influence(c,S) * weight(c,e)) - \log \varepsilon$$
$$= \log e^{-\frac{15-8}{1.0}} - \log e^{-64}$$
$$= 57$$

Similarly, we have $score(p,o_3) = 55$ and $score(p,o_4) = 60$. Thus the *total_score* of the influential path $p = < c \rightarrow e >$ is $total\_score(p) = 57 + 55 + 60 = 172$. In the same manner, we build $< c \rightarrow e >$-projected database and extend $< c \rightarrow e >$ to $< c \rightarrow e \rightarrow f >$.

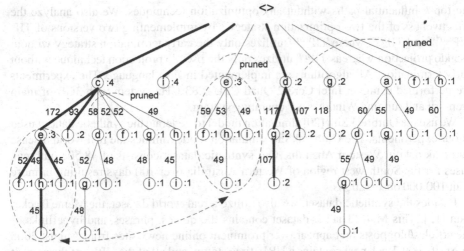

Fig. 3. Prefix search tree for sample database

Suppose we wish to find the top-2 maximal influential paths. After obtaining the paths $< c \rightarrow e >$ and $< c \rightarrow i >$, the min_sup is raised to 2. This implies that all the branches rooted at node a are pruned as their support values are less than 2. Similarly, branches rooted at node e are also pruned as they have already been traversed previously from node c. The bold lines in Figure 3 show the explored paths.

To further improve the efficiency of TIP algorithm, we propose two optimization strategies.

Early Termination by Equivalence. Early termination by equivalence is a search space reduction technique developed in CloSpan [14]. Let $N(D)$ represent the total number of nodes in D. The property of early termination by equivalence shows if two influential paths $p \sqsubseteq p'$ and $N(D_p) = N(D_{p'})$, then $\forall \gamma, support(p \rightarrow \gamma) = support(p' \rightarrow \gamma)$. It means the descendants of p in the prefix search tree cannot be maximal. Furthermore, the descendants of p and p' are exactly the same. We can utilize this property to quickly prune the search space of p.

Pseudo Projection. As with traditional projection-based mining method, the major cost of TIP is the construction of projected databases. To reduce the cost of projection, we apply the pseudo-projection technique [13]. Instead of constructing a *physical projection* by collecting all the postfixes, we use pointers referring to the observations in the database as a *pseudo projection*. Every projection consists of two pieces of information: *pointer* to the observation in database and *offset* of the postfix in the observation. This allows us to avoid physically copying postfixes: only pointers to the projected point are saved for each observation. Thus, it is efficient in terms of both running time and space.

4   Experimental Evaluation

In this section, we conduct experiments to evaluate the effectiveness and efficiency of our TIP algorithm. We compare the TIP algorithm with the Naïve algorithm that finds the top-k influential paths without any optimization techniques. We also analyze the effectiveness of the two optimization strategies by implementing two versions of TIP, TIP^{early} and TIP^{pp}, where TIP^{early} utilizes only the early termination strategy without pseudo projection whereas TIP^{pp} utilizes only the pseudo projection technique without early termination. All algorithms are implemented in Java language. The experiments are performed using an Intel Core 2 Quad CPU 2.83 GHz system with 3GB of main memory and running Windows XP operating system.

We use the OutbreakSim [20] simulation model to generate the synthetic dataset used in our experiments. The OutbreakSim simulation model mimics the real world disease outbreak data in Western Australia. Our synthetic dataset consists of 48,507 outbreak cases for the South-west region of Western Australia over 100 days resulting in more than 100,000 observations.

Besides the synthetic dataset, we also utilize a real-world dataset, the MemeTracker data [12]. This MemeTracker dataset contains the quotes, phrases, and hyperlinks of the articles/blogposts that appear over prominent online news sites from August 2008 to April 2009. Each post contains a URL, time stamp, and all of the URLs of the posts it cites. Nodes are mostly news portals or news blogs and the time stamps in the data capture the time that a quote/phase was used in a post. Finally, there are directed hyperlinks among the posts. We use these hyperlinks to trace the flow of information. A site publishes a piece of information and uses hyperlinks to refer to the same or closely related pieces of information published by other sites. An observation is thus a collection of time-stamped hyperlinks among different sites that refer to the same or closely related pieces of information. We record one observation per piece – or closely related pieces – of information. We extract the most active media sites and blogs with the largest number of posts, and generate 46,352 observations.

Table 5 shows the characteristics of the synthetic and real world datasets used in the experiments including the the number of input observations, average observation length, maximum and minimum observation lengths.

Table 5. Datasets Characteristics

Datasets	Cardinality	Avg Len	Max Len	Min Len
Synthetic	100,000	8.00	20	6
MemeTracker	46,352	13.72	42	3

4.1   Efficiency Experiments

In this set of experiments, we evaluate the efficiency of TIP algorithm on both synthetic and real world datasets. First, we vary the synthetic database size from 10k to 90k. We set k to 10, time threshold $\tau = 100$, and radius of influence $\alpha = 1.0$. Figure 4 shows the results. We observe that TIP algorithm remains efficient as the database size increases. In particular, the early termination optimization strategy is more effective in reducing the runtime compared to the pseudo projection.

Similarly, for the real world MemeTracker dataset, we generate the top-10 (i.e. $k = 10$) maximal influential paths by setting time threshold τ to 1000 and radius of influence α to 1.0. We randomly sample the dataset to vary the database size from 10k to 40k. As

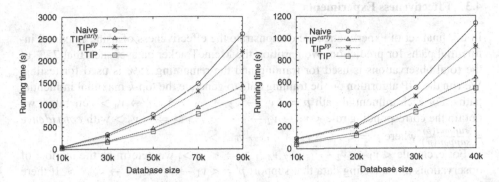

Fig. 4. Performance of varying database size on synthetic dataset

Fig. 5. Performance of varying database size on MemeTracker dataset

can be seen from Figure 5, TIP algorithm outperforms the Naïve algorithm with early termination playing a greater role in reducing the runtime of TIP.

4.2 Sensitivity Experiments

Effect of k. Next, we investigate the effect of the number of maximal influential paths, k, on the performance of the four algorithms. We set the database size to 20k and vary k from 5 to 25. Figure 6 shows the experimental results for the synthetic dataset. As can be seen, the runtime for both TIP and Naïve algorithm increases as k increases. However, the runtime of TIP algorithm is half that of the Naïve algorithm demonstrating that TIP remains efficient even when k increases.

Effect of τ. Here, we examine the effect of varying the time threshold τ on the performance of TIP algorithm. Note that increasing τ is equivalent to increasing the search space, i.e. the number of potential influential paths. We set the number of maximal influential paths k to 5, database size $|D|$ to 20k and vary the time threshold τ from 10 to 50. Figure 7 shows that the runtime for both algorithms increases as τ increases. Similar trend is observed here with the TIP algorithm showing a significant reduction in runtime as compared to the Naïve algorithm.

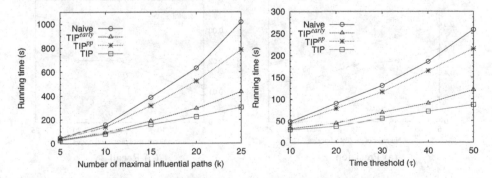

Fig. 6. Performance of varying k

Fig. 7. Performance of varying τ

4.3   Effectiveness Experiments

In the final set of experiments, we demonstrate the effectiveness of using maximal influential paths for prediction. We partition the MemeTracker dataset such that 75% of the total observations is used for training and the remaining 25% is used for testing. We run the TIP algorithm on the training data to generate the top-k maximal influential paths. For each influential path $p = <v_1 \to v_2 \to \cdots \to v_{n-1} \to v_n >$ generated, we obtain the corresponding rule $<v_1 \to v_2 \to \cdots \to v_{n-1}> \Rightarrow <v_n>$ with $confidence$ $= \frac{support(p)}{support(p')}$ where $p' = <v_1 \to v_2 \to \cdots \to v_{n-1}>$.

For each rule $<v_1 \to v_2 \to \cdots \to v_{n-1}> \Rightarrow <v_n>$, we determine the number of observations in the testing data that support $p' = <v_1 \to v_2 \to \cdots \to v_{n-1}>$. If there is at least one support observation in the testing data, we assign the probability of node v_n being influenced to the confidence of the rule, i.e. $\frac{support(p)}{support(p')}$. If we have more than one rule predicting that node v_n will be influenced, we assign the maximum confidence of the rules as the probability of node v_n being influenced.

The set of predicted nodes are sorted in decreasing order of the probability of getting influenced. We consider a node to be the next influenced node if it is among the top-n nodes. Here top-n nodes are the first n non-duplicate nodes with highest probability of being influenced.

Let X be the set of nodes influenced in test data, and Y be the set of nodes predicted to be influenced in test data, then precision and recall are defined by the following equations:

$$precision = \frac{|X \cap Y|}{|Y|} \tag{6}$$

$$recall = \frac{|X \cap Y|}{|X|} \tag{7}$$

We compare the prediction accuracy of TIP algorithm with NetInf algorithm [6], which can only infer influential edge between two nodes. Similarly, we run NetInf algorithm

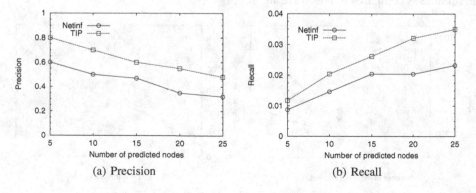

(a) Precision   (b) Recall

Fig. 8. Precision and recall on MemeTracker dataset

on the training data to generate a set of influential edges, say $< i \rightarrow j >$. We assign the probability of node j being influenced as $\frac{support(<i \rightarrow j>)}{support(<i>)}$.

We perform cross validation for evaluating the prediction performance of both algorithms. Figure 8 shows the precision and recall results by varying the number of predicted nodes, n, from 5 to 25. We observe that TIP algorithm significantly outperforms NetInf algorithm for different values of n. This is because influential paths are more informative than influential edges and hence in predicting which node will be influenced next, the TIP algorithm tends to be more accurate than NetInf algorithm.

5   Related Work

Research on information diffusion has focused on validating the existence of influence [5,2], studying the maximization of influence spread in the whole network [9,11,4,3], modeling direct influence in homogeneous networks [19], and mining topic-level influence on heterogeneous networks [15].

The works in [5,18] first study the influence maximization problem as an algorithmic problem. Kempe et al. [9] examine the influence maximization problem for a family of influence models. The authors design approximation algorithms for the independent cascade model. However, a drawback of their work is the efficiency issue of their greedy algorithm. Several recent studies try to address the efficiency issue by using new heuristics [11,10,4,3,17].

Gomez et al. [6] study the diffusion of information among blogs and online news sources. They assume that connections between nodes cannot be observed and use the observed cascades to infer a sparse, "hidden" network of information diffusion. They propose an iterative algorithm called NetInf which is based on submodular function optimization. NetInf first reconstructs the most likely structure of each cascade. Then it selects the most likely edge of the network in each iteration. The algorithm assumes that the weights of all edges have the same values.

Mathioudakis et al. [16] investigate the problem of sparsifying influence networks. Given a social graph and a list of actions propagating through it, they design the SPINE algorithm to find the "backbone" of the network through the use of the independent-cascade model [9]. SPINE has two phases: the first phase selects a set of arcs that yields a finite log-likelihood, while the second phase greedily seeks a solution of maximum log-likelihood. The effectiveness of SPINE came from its ability to reduce computation speed significantly.

In the field of sequence mining, Giannotti et al. [21] introduce a novel form of sequential pattern, called Temporally-Annotated Sequence (*TAS*), representing typical transition times between the events in a frequent sequence. They formalize the novel mining problem of discovering representative frequent *TAS*'s as a combination of frequent sequential pattern mining and density-based clustering.

Information diffusion has also been considered from the view of the blogosphere, since it provides a unique resource for studying information flow. The works in [1,7] model and study the dynamics of diffusion of information in the blogosphere, while [8,2] design algorithms to identify influential blog posts and influential bloggers in a blogosphere.

6 Conclusion

In this paper, we develop a method for inferring top-k maximal influential paths which can truly capture the dynamics of information diffusion. Given a log of propagation observations of some information over a hidden network, our goal is to infer the top-k maximal influential paths that best explain these observations. We define a generative influence propagation model based on the Independent Cascade Model and Linear Threshold Model, which mathematically model the spread of certain information through a network. We design an algorithm called TIP to infer the top-k maximal influential paths. TIP utilizes the properties of top-k maximal influential paths to dynamically increase the support and prune the projected databases. We evaluate the proposed algorithms on both synthetic and real world data sets. The experimental results demonstrate the effectiveness and efficiency of our method.

References

1. Adar, E., Adamic, L.A.: Tracking Information Epidemics in Blogspace. In: Web Intelligence, pp. 207–214 (2005)
2. Agarwal, N., Liu, H., Tang, L., Yu, P.S.: Identifying the Influential Bloggers in a Community. In: WSDM 2008, pp. 207–218 (2008)
3. Chen, W., Wang, C., Wang, Y.: Scalable Influence Maximization for Prevalent Viral Marketing in Large-Scale Social Networks. In: KDD 2010, pp. 1029–1038 (2010)
4. Chen, W., Wang, Y., Yang, S.: Efficient Influence Maximization in Social Networks. In: KDD 2009, pp. 199–208 (2009)
5. Domingos, P., Richardson, M.: Mining the Network Value of Customers. In: KDD 2001, pp. 57–66 (2001)
6. Gomez-Rodriguez, M., Leskovec, J., Krause, A.: Inferring Networks of Diffusion and Influence. In: KDD 2010, pp. 1019–1028 (2010)
7. Gruhl, D., Guha, R., Liben-nowell, D., Tomkins, A.: Information Diffusion through Blogspace. In: WWW 2004, pp. 491–501 (2004)
8. Java, A., Kolari, P., Finin, T., Oates, T.: Modeling the Spread of Influence on the Blogosphere. In: World Wide Web Conference Series (2006)
9. Kempe, D., Kleinberg, J., Tardos, É.: Maximizing the Spread of Influence through a Social Network. In: KDD 2003, pp. 137–146 (2003)
10. Kimura, M., Saito, K.: Tractable Models for Information Diffusion in Social Networks. In: Fürnkranz, J., Scheffer, T., Spiliopoulou, M. (eds.) PKDD 2006. LNCS (LNAI), vol. 4213, pp. 259–271. Springer, Heidelberg (2006)
11. Leskovec, J., Krause, A., Guestrin, C., Faloutsos, C., VanBriesen, J., Glance, N.: Cost-effective Outbreak Detection in Networks. In: KDD 2007, pp. 420–429 (2007)
12. Leskovec, J., Backstrom, L., Kleinberg, J.: Meme-tracking and the Dynamics of the News Cycle. In: KDD 2009, pp. 497–506 (2009)
13. Pei, J., Han, J., Mortazavi-Asl, B., Pinto, H., Chen, Q., Dayal, U., Hsu, M.-C.: Prefixspan: Mining Sequential Patterns Efficiently by Prefix-projected Pattern Growth. In: ICDE 2001, pp. 215–224 (2001)
14. Yan, X., Han, J., Afshar, R.: Clospan: Mining Closed Sequential Patterns in Large Datasets. In: SDM 2003, pp. 166–177 (2003)
15. Liu, L., Tang, J., Han, J., Jiang, M., Yang, S.: Mining Topic-level Influence in Heterogeneous Networks. In: CIKM 2010, pp. 199–208 (2010)

16. Mathioudakis, M., Bonchi, F., Castillo, C., Gionis, A., Ukkonen, A.: Sparsification of Influence Networks. In: KDD 2011, pp. 529–537 (2011)
17. Narayanam, R., Narahari, Y.: A Shapley Value-Based Approach to Discover Influential Nodes in Social Networks. IEEE T. Automation Science and Engineering 8(1), 130–147 (2011)
18. Richardson, M., Domingos, P.: Mining Knowledge-Sharing Sites for Viral Marketing. In: KDD 2002, pp. 61–70 (2002)
19. Tang, J., Sun, J., Wang, C., Yang, Z.: Social Influence Analysis in Large-scale Networks. In: KDD 2009, pp. 807–816 (2009)
20. Watkins, R., Eagleson, S., Beckett, S., Garner, G., Veenendaal, B., Wright, G., Plant, A.: Using GIS to Create Synthetic Disease Outbreaks. BMC Medical Informatics and Decision Making 7(1), 4 (2007)
21. Giannotti, F., Nanni, M., Pedreschi, D., Pinelli, F.: Mining Sequences with Temporal Annotations. In: SAC 2006, pp. 593–597 (2006)

Learning to Rank
from Concept-Drifting Network Data Streams

Lucrezia Macchia, Michelangelo Ceci, and Donato Malerba

Dipartimento di Informatica, Università degli Studi di Bari "Aldo Moro",
via Orabona, 4 - 70126 Bari, Italy
lucrezia.macchia@uniba.it, {ceci,malerba}@di.uniba.it

Abstract. Networked data are, nowadays, collected in various application domains such as social networks, biological networks, sensor networks, spatial networks, peer-to-peer networks etc. Recently, the application of data stream mining to networked data, in order to study their evolution over time, is receiving increasing attention in the research community. Following this main stream of research, we propose an algorithm for mining ranking models from networked data which may evolve over time. In order to properly deal with the concept drift problem, the algorithm exploits an ensemble learning approach which allows us to weight the importance of learned ranking models from past data when ranking new data. Learned models are able to take the network autocorrelation into account, that is, the statistical dependency between the values of the same attribute on related nodes. Empirical results prove the effectiveness of the proposed algorithm and show that it performs better than other approaches proposed in the literature.

1   Introduction

The coming of new technologies has determined the generation, at a rapid rate, of massive structured and complex data that in most of cases can be represented as data networks. The networks have become ubiquitous in several social, economical and scientific fields ranging from the Internet to social sciences, biology, epidemiology, geography, finance and many others. Indeed, researchers in these fields have proven that systems of different nature can be represented as networks [19]. For instance, the Web can be considered as a network of webpages, which may be connected with each other by edges representing various explicit relations, such as hyperlinks. Sensor networks are networks where nodes represent sensors and edges represent the (spatial) distance between two sensors.

In the real world, network data may evolve over time. This evolution can be both in the structure of the network (nodes can be added or removed, edges can be added or removed) and in the distribution of the attribute values associated with the nodes. As an example, consider a sensor network whose nodes collect temperature, humidity, etc. at single positions in a specific environment. In this case, new sensors can be either added to the network or removed from it as well as the underlying data distribution of some variables may change. Moreover, as

S.W. Liddle et al. (Eds.): DEXA 2012, Part I, LNCS 7446, pp. 384–396, 2012.

observed by Swanson [24], in this situation, data can be affected by temporal autocorrelation according to which two values of the some variable are cross correlated over a certain time lag.

Another important aspect that we have to consider when mining networked data is that they are characterized by a particular form of autocorrelation [12] according to which a value observed at a node depends on the values observed at neighboring nodes in the network [22]. The major difficulty due to the autocorrelation is that the independence assumption (i.i.d.), which typically underlies machine learning and data mining methods, is no longer valid. The violation of the instance independence has been identified as the main responsible of poor performance of traditional machine learning methods [18]. To remedy the negative effects of the violation of independence assumptions, autocorrelation has to be explicitly accommodated in the learned models.

By taking into account these two aspects, in this paper we face the problem of mining ranking functions. Although in the recent years learning ranking functions has received increasing attention due to its potential application to problems raised in information retrieval, machine learning, data mining and recommendation systems [10], at the best of our knowledge, very few works in the literature face the problem of learning to rank from network data which may evolve over time.

At this aim, we propose an ensemble learning approach in order to weight the importance of ranking models which are learned from past data. This weighting schema is then used for ranking new data. In this way, individual models can be learned from different time windows and can work as a team to enhance a final model. By appropriately defining the weighting schema, it is possible to give more importance to models learned from recent time windows than to models learned from distant time windows. As in [11][3][13] the ranking problem is boiled down into a regression problem. Individual models are tree-structured and allow us to consider network autocorrelation in the data. This is performed by considering a locally weighted regression function to be associated with the nodes of the tree, where the local weighting explicitly considers the network proximity of single elements of the network.

The paper is organized as follows. The next section and Section 3 report relevant related work. Section 4 describes the proposed algorithm. Section 5 describes the datasets, experimental setup and reports relevant results. Finally, in Section 6, some conclusions are drawn and some future work are outlined.

2   Related Works

In the following, we report some related works which i) mine ranking models able to work on network data and ii) exploit ensemble learning techniques for data stream mining.

2.1   Mining Ranking Models Able to Work on Network Data

The task considered in this work is that of learning to rank. This task has received increasing attention due to its potential application to problems in information retrieval and recommendation systems [1,5,10]. The aim of the methods developed in this field is to learn a ranking model which returns the output predictions in the form of a ranking of the examples given in input. According to [4], it is possible distinguish three types of ranking problems. The first type is *Label ranking*, where the goal is to learn a "label ranker" in the form of an $X \rightarrow S_Y$ mapping, where the output space S_Y is given by the set of all total orders (permutations) of the set of labels Y. The second type is *Instance ranking*, where an instance $x \in X$ belongs to one among a finite set of classes $Y = y_1, y_2, \ldots, y_k$ for which a natural order $y_1 < y_2 < \ldots < y_k$ is defined. The third type is *Object ranking*. In this case, the goal is to learn a ranking function $f(\cdot)$ which, given a subset of an underlying referential set of objects as an input, produces a ranking of these objects.

By focusing on object ranking, studies reported in the literature solve this problem by resorting to two alternative approaches. The first approach, determines a regression function that assigns a numerical value to each element of a set, then the same is used to sort the items. The second approach aims at learning preference functions, which are able to perform pairwise comparisons in order to define a relative order between two objects. The first approach is generally more efficient but it is applicable only when a single total order between objects is acceptable. While, when not all the objects have to necessarily be included in the ranking, the second approach is preferable. Since in this work we do not consider the problem of defining partial orders, we consider the first approach.

According to this approach, Herbrich et al. [11] propose to learn a function which, given an object description, returns an item belonging to an ordered set. The function is determined so that a loss function is minimized. A similar approach was proposed by Crammer et al. [3], in which the learned functions are modelled by perceptrons. Tesauro [25] proposed a symmetric neural network architecture that can be trained with representations of two states and a training signal that indicates which of the two states is preferable. In the framework of *constraint classification*, some works [8,9] exploit linear utility functions to find a way to express a constraint in the form $f_i(x) - f_j(x) > 0$, in order to transform the original ranking problem into a single binary classification problem.

By keeping in mind that we intend to learn ranking functions from network data, we propose to use regression trees which are proved to be easily adapted to work with this kind of data [22]. Regression trees [2] are supposed to be more comprehensible then classical regression models. They are built top-down by recursively partitioning the sample space. An attribute may be of varying importance for different regions of the sample space. A constant is associated with each leaf of a regression tree, so that the prediction performed by a regression tree is the same for all sample data falling in the same leaf. A generalisation of regression trees is represented by model trees, which associate multiple linear models with each leaf. Hence, different values can be predicted for sample data falling in the same leaf. Some of the model tree induction systems are RETIS [14], M5' [29], HTL [26], TSIR [15] and SMOTI [17].

One of the main peculiarities that led us to consider model trees is that their tree structure allows us to deal with the so-called "ecological fallacy" problem [21] according to which individual sub-regions do not have the same data distribution of the entire region. This is coherent with the research reported in [28], where the authors argue that the concept drift is not uniform over the feature space. Moreover, the tree structure allows us to capture different effects of the autocorrelation either at a global (higher levels of the tree) or at a local (lower levels of the tree) granularity level.

In this work we consider the system SMOTI, which is characterized by a tree structure with two types of nodes: regression nodes, which perform only straight-line regression, and splitting nodes, which partition the feature space. The multiple linear model associated with each leaf is then the composition of the straight-line regressions reported along the path from the root to the leaf. A brief description of SMOTI is reported in Section 3.

2.2   Ensemble Learning for Data Streams

The idea of *ensemble learning* is to employ multiple learners and combine their predictions. As observed in several works found in the literature, (see, for example [20]), the resulting ensemble is generally more accurate than any of the individual classifiers that contribute to the ensemble.

In this work we intend to exploit the idea of the ensemble to deal with data whose underlying distribution may evolve over time. Indeed, this idea is not novel and several papers in the literature exploit ensemble learning techniques in order to work with streaming data. For instance, in [23], the authors propose an ensemble learning of individual decision trees learned at different time windows. Decision trees are built through the classical Quinlan's C4.5 learning algorithm and ensemble aims at preserving only the k most accurate trees for future data. New classifies are added to the ensemble if the ensemble size does not exceed k, otherwise, new classifiers are added only if they improve the ensemble performance. Obviously, this leads to sacrifice exiting classifiers. Wang et al. [27] propose weighted classifier ensembles to mine streaming data affected by the concept drift phenomenon. They train an ensemble of classifiers from sequential data chunks in the stream. Subsequently, they associate each classifier with a weight which represents the expected prediction accuracy of the classifier on the current test examples. In [28] the authors propose to capture the non-deterministic nature of concept drifts in a region of the feature space and model the concept drift process as a continuous time Markov chain. In particular, they propose to train k classifiers from data in recent time windows (they use a decay factor) and assume that each classifier partitions the feature space into a set of non-overlapping regions (as decision trees do). When a new example arrives, on the basis of the region it belongs to, a score that represents the drift is computed (on the basis of the continuous time Markov chain). This score is then used to compute the probability that the example belongs to a class.

Although all cited works are proved to be effective in their capability to use models learned in the past for new data, none of them consider the problem of mining ranking models from networked data, which is the main objective of the present work. The only work that, at the best of our knowledge, faces the same problem we consider in this works, is reported in [16], where the authors modify the SVMRank algorithm in order to emphasize the importance of models learned in time periods during which data follow a data distribution that is similar to that observed in the time period for which prediction has to be made. However, their approach (called SVMRankT) does not consider the "ecological fallacy" problem (see section 2.1).

3   Background: Model Tree Induction in SMOTI

SMOTI (Stepwise Model Tree Induction) performs the top-down induction of model trees by considering not only a partitioning procedure, but also by some intermediate prediction functions [17]. This means that there are two types of nodes in the tree: regression nodes and splitting nodes. The former compute straight-line regressions, while the latter partition the sample space. They pass down training data to their children in two different ways. For a splitting node t, only a subgroup of the $N(t)$ training data in t is passed to each child, with no change on training cases. For a regression node t, all the data are passed down to its only child, but the values of both the dependent and independent numeric variables not included in the multiple linear model associated with t are transformed in order to remove the linear effect of those variables already included. Thus, descendants of a regression node will operate on a modified training set. Indeed, according to the statistical theory of linear regression [6], the incremental construction of a multiple linear model is made by removing the linear effect of introduced variables each time a new independent variable is added to the model. For instance, let us consider the problem of building a multiple regression model with two independent variables through a sequence of straight-line regressions:

$$\hat{Y} = a + bX_1 + cX_2 \tag{1}$$

We start regressing Y on X_1, so that the model $\hat{Y} = a_1 + b_1X_1$ is built. This fitted equation does not predict Y exactly. By adding the new variable X_2, the prediction might improve. Instead of starting from scratch and building a model with both X_1 and X_2, we can build a linear model for X_2 given

$$X_1 : \hat{X}_2 = a_2 + b_2X_1 \tag{2}$$

Then we compute the residuals on $X_2 : X_2' = X_2 - (a_2 + b_2X_1)$ and on $Y : Y' = Y - (a_1 + b_1X_1)$. Finally, we regress Y' on X_2' alone:

$$\hat{Y}' = a_3 + b_3X_2' \tag{3}$$

By substituting the equations of X_2' and Y' in the last equation we have:

$$\left(Y - \widehat{(a_1 + b_1X_1)}\right) = a_3 + b_3(X_2 - (a_2 + b_2X_1)) \tag{4}$$

Since $Y - \widehat{(a_1 + b_1 X_1)} = \hat{Y} - (a_1 + b_1 X_1)$, we have:

$$\hat{Y} = (a_3 + a_1 a_2 b_3) + (b_1 - b_2 b_3) X_1 + b_3 X_2 \qquad (5)$$

It can be proven that this last model coincides with (1), that is, $a = a_3 + a_1 a_2 b_3$, $b = b_1 - b_2 b_3$ and $c = b_3$. Therefore, when the first regression line of Y on X_1 is built we pass down both the residuals of Y and the residuals of the regression of X_2 on X_1. This means that we remove the linear effect of the variables already included in the model (X_1) from both the response variable (Y) and those variables to be selected for the next regression step (X_2).

4   Considering Network Autocorrelation in Model Tree Induction and Ranking Models' Ensemble

The original SMOTI algorithm uses the least squares method in order to identify the parameters to be used in all the linear regression functions such as (2) and (3). The least squares method works as follows: Let $\hat{Y} = \alpha + \beta X_j$ $(j = 1, \ldots, m)$ be a generic regression function to be built between variables X_j and Y,

$$\begin{cases} \beta = (\mathbf{X}_j^T \mathbf{X}_j)^{-1} \mathbf{X}_j^T \mathbf{Y} = \dfrac{\displaystyle\sum_{i=1,..,n} (x_{ji} - \overline{x_j})(y_i - \overline{y})}{\displaystyle\sum_{i=1,..,n} (x_{ji} - \overline{x_j})^2} \\ \alpha = \overline{y} - \beta \overline{x_j} \end{cases} \qquad (6)$$

where $\mathbf{X}_j = [x_{j1}, x_{j2}, \ldots, x_{jn}]$ and $\mathbf{Y} = [y_1, y_2, \ldots, y_n]$ are vectors of values of X_j and Y, respectively, and $\overline{y}$ and $\overline{x_j}$ are the averages of values in $\mathbf{X}_j$ and $\mathbf{Y}$, respectively.

However, although standard, this method neglects possible autocorrelation in the network data. In other words, the least squares method would lead to regression models (and so, in our case, piecewise ranking functions) which do not consider the network. In fact, in networks, data are organized according to a graph structure $G = (V, E)$, where $V = \{v_1, v_2, \ldots, v_n\}$ is a set of vertices $v_i = (\mathbf{x}_i, y_i)$, $i = 1, \ldots, n$, $E = \{\langle v_i, v_h, w_{i,v} \rangle | i \neq h\}$ is a set of edges and $w_{i,h}$ represents the strength of the connection among the nodes. As stated before, neglecting the network structure (i.e. autocorrelation), may result in inaccurate prediction models [18].

To better explain how networks can be exploited in order to extract a ranking model that takes autocorrelation into account (i.e. models able to generate ranking labels which are more coherent in the network), we report a simple example.

Example 1. Let Figure 1(a) be a network structure, let $\mathbf{X}_j = [1, 5, 4, 8]$ and $\mathbf{Y} = [2, 3, 4, 6]$ be the values for the nodes d, b, a, c and let e be the example for which a ranking label has to be identified (query point). According to the standard least squares method, the (ranking) regression line is the regression

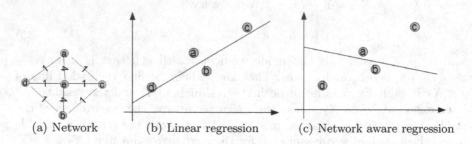

(a) Network   (b) Linear regression   (c) Network aware regression

Fig. 1. A simple network (a) and two (ranking) regression functions. In (c), the regression line gives more importance to vertices a and b since the query point is e.

line reported in Figure 1(b). However, by taking into account the network, the regression line should change (see Figure 1(c)) by giving more importance to the effect of the nodes a and b, which are strongly correlated with e.

To take autocorrelation into account, the alternative solution we intend to exploit in this work is that of locally weighted regression method (LWR). The basic idea in LWR, is that a local model should fit to nearby data, according to a proximity measure defined on the feature space $X_1, \ldots, X_m$. Differently, in this work we exploit the network in order to account the degree of correlation between pairs of vertices.

Formally, let $a_i \in V$ be a vertex, it is possible to build a diagonal matrix

$$\mathbf{W_i} = \begin{pmatrix} w_{i1} & 0 & \cdots \\ 0 & w_{i2} & \cdots \\ \vdots & \vdots & w_{in} \end{pmatrix}$$

where each non-zero element w_{iv}, $v = 1, \ldots, n$ represents the strength of the connection between vertices a_i and a_v and n represents the number of examples falling in the node of the tree. Then, it is possible to build a regression function between Y and X_j associated with each example falling in the node.

$$Y = \alpha_i + \beta_i X_j \tag{7}$$

in this case, $\beta_i = ((\mathbf{W}_i \mathbf{X}_j)^T (\mathbf{W}_i \mathbf{X}_j))^{-1} (\mathbf{W}_i \mathbf{X}_j)^T (\mathbf{W}_i \mathbf{Y})$. This means that

$$\begin{cases} \beta_i = \dfrac{\sum\limits_{t=1,..,n} (w_{it} x_{jt} - r)(w_{it} y_t - s)}{\sum\limits_{t=1,..,n} (w_{it} x_{jt} - r)^2} \\ \alpha_i = s - \beta_i r \end{cases} \tag{8}$$

where $r = \frac{1}{n} \sum_t w_{it} x_{jt}$ and $s = \frac{1}{n} \sum_t w_{it} y_t$.

In the regression step, in order to uniformly modify all the examples that are passed down in the model tree construction, we use the averages of the parameters' values. More, formally, suppose we are introducing a regression node on X_1, then residuals are computed as follows: $X_2' = X_2 - (\overline{\alpha_2} + \overline{\beta_2}X_1)$ and $Y' = Y - (\overline{\alpha_1} + \overline{\beta_1}X_1)$ where $\overline{\alpha} = \frac{1}{n}\sum_i \alpha_i$ and $\overline{\beta} = \frac{1}{n}\sum_i \beta_i$ for each regression problem (on X_1 and on Y).

It is noteworthy that this approximation does not compromise the correct consideration of the autocorrelation since residuals will be explained at lower levels of the model tree. Moreover, autocorrelation is mainly considered during the prediction phase. In fact, during prediction, once the example to be classified reaches the leaf, the system uses the α_i and β_i parameters associated with the training example for which the connection is the strongest (according to the network structure). In case more than one training example satisfies this property, the average of the parameters' values is considered. This approach allows the system to predict values for network nodes that are not available during the learning phase.

It is worth to notice that this approach also solves problems coming from the collinearity phenomenon, that is, when some of the independent variables are related to each other. Indeed, as recognized in [7], when some variables are (approximately) collinear, several problems may occur, such as indeterminacy of regression coefficients, unreliability of the estimates of the regression coefficients, and impossibility of evaluating the relative importance of the independent variables. In our approach, this problem is solved by introducing one variable at a time in the model tree.

By introducing the temporal dimension in the analysis, we resort to a temporal sliding window framework. Let $\langle G_1, \dots G_{t-1} \rangle$ be a sequence of time windows associated with $t - 1$ time points, and $\tau : \mathcal{T} \times V \to \mathbb{R}$ be a ranking labeling function which maps a model tree $T_q \in \mathcal{T}$ and a vertex $v \in V$ in the space of ranking labels ($\mathbb{R}$), then it is possible to define a general framework able to classify a generic vertex v at time t according to all the ranking models $T_1, T_2, \dots, T_{t-1}$ learned from $\langle G_1, \dots G_{t-1} \rangle$:

$$\tau'(T_1, T_2, \dots, T_{t-1}, v) = \frac{1}{t * (t-1)/2} \sum_{q=1,\dots,t-1} \frac{q}{t} * \tau(T_q, v) \qquad (9)$$

where the first factor is only used for normalization purposes and $\frac{q}{t}$ is a decay factor which gives more importance to models learned in recent time windows than to models learned in distant time windows.

It is noteworthy that the method well adapts to cases in which the network structure changes over time. In fact, it is not strictly necessary that the vertex v belongs to the network in the time windows $\{G_1, \dots G_{t-1}\}$. Moreover, our choice to ignore network properties in the splitting nodes of the model tree allows us to use the ranking model in different (but related) networks.

5   Experiments

In order to evaluate the effectiveness of the proposed solution, we performed experiments on three real world datasets, that is, Intel Lab database, Portuguese rivers database and California Truck database. The proposed approach has been compared with results obtained with SVMRankT [16].

As a measure to evaluate the learned ranking models, we use the Spearman's rank correlation coefficient. Let $V^{(t)} = \{v_1^{(t)}, v_2^{(t)}, \ldots, v_n^{(t)}\}$ be the real dataset at time t, $\hat{y}_i^{(t)} = \tau'(T_1, T_2, \ldots, T_{t-1}, v_i)$ be the estimated ranking for the example v_i and $y_i^{(t)}$ be the real ranking of v_i at time t, the Spearman's rank correlation coefficient is defined as:

$$\rho = \frac{\sum\limits_{i=1,\ldots,n_t} \left(y_i^{(t)} - \overline{y^{(t)}}\right)\left(\hat{y}_i^{(t)} - \overline{\hat{y}^{(t)}}\right)}{\sqrt{\sum\limits_{i=1,\ldots,n_t} \left(y_i^{(t)} - \overline{y^{(t)}}\right)^2\left(\hat{y}_i^{(t)} - \overline{\hat{y}^{(t)}}\right)^2}} \tag{10}$$

where $\overline{y^{(t)}}$ and $\overline{\hat{y}^{(t)}}$ are the average true and estimated rankings, respectively. ρ ranges in the interval [-1,1], where -1 means negative correlation in the ranking and 1 means perfect ranking.

The first dataset considered in the experiments is the Intel Lab Database which contains real information collected from 54 sensors deployed in the Intel Berkeley Research lab between February 28th and and March 21st, 2004. The dataset is taken from http://db.csail.mit.edu/labdata/labdata.html. The sensors which we consider in this experiment have collected timestamped temperature, humidity and luminosity values once every 31 seconds. Networks are built by considering the spatial distance between sensors and the target attribute is represented by the temperature (we removed temperature and we used the temperature ranking as y value). In the experiments, we only considered working days and we used 1-day time-intervals, this means that we built 15 networks in all.

In Table 1, results of the Spearman's rank correlation coefficient are reported. In this dataset, the proposed approach is not able to show the same performances as those shown by SVMRankT. This is mainly due to the small number of attributes that prevents the model trees learning approach to learn accurate models. This is different from what happens in the case of SVMRankT that is based on SVMs that are able to create oblique partitions of the feature space. Moreover, in this dataset, autocorrelation phenomenon is quite uniform across the space (ecological fallacy does not hold).

The Portuguese rivers dataset holds water's information of the rivers Douro and Paiva. The dataset may be incomplete because the controls are manually done and are not done systematically. The original dataset is composed of a fact table and six additional relational tables: The fact table (ANALYSIS) contains information on the measures under control (pH, % Coliformi Bacteria, conductivity, turbidity, % Escherichia Coli Bacteria) and the gathering method. Additional tables are directly (or indirectly) connected to ANALYSIS according to a snowflake logic schema. They are: PARAMETERS (that are considered in the analysis), INSTITUTIONS (that collected data), DAY, CONTROL POINTS and PATH (that specifies the position of a control point according to the course of the rivers). From the table PATH we got the course of the river and the

Table 1. Intel lab Dataset: Spearman's rank coefficient

Train	Test	$SVMRankT$	Our approach
1 2	3	**0.9021739**	0.8312905
1 2 3	4	**0.9529370**	0.9062211
1 2 3 4	5	**0.9364014**	0.8566142
1 2 3 4 5	6	**0.9240286**	0.8507169
1 2 3 4 5 6	7	**0.9084181**	0.8576550
1 2 3 4 5 6 7	8	0.5514569	**0.5795560**
1 2 3 4 5 6 7 8	9	**0.8632053**	0.8554579
1 2 3 4 5 6 7 8 9	10	**0.8931544**	0.8331406
1 2 3 4 5 6 7 8 9 10	11	**0.8375346**	0.8252775
1 2 3 4 5 6 7 8 9 10 11	12	**0.8341813**	0.5730804
1 2 3 4 5 6 7 8 9 10 11 12	13	**0.7514743**	0.7318166
1 2 3 4 5 6 7 8 9 10 11 12 13	14	**0.8169229**	0.4641246
1 2 3 4 5 6 7 8 9 10 11 12 13 14	15	**0.9021739**	0.5225197
Avg.		**0.8518510**	0.7451901

Table 2. Portuguese rivers dataset: Spearman's rank coefficient

Train	Test	$SVMRankT$	Our approach
2004-2005	2006	0.4024926686	**0.6146444282**
2004-2005-2006	2007	0.4761730205	**0.6251832845**
2004-2005-2006-2007	2008	0.4769061584	**0.6017155425**
2004-2005-2006-2007-2008	2009	0.4226539589	**0.6462609971**
Avg.		0.44455645161	**0.64445106305**

position of the control points in order to build the network structure. The weights on the edges represent the navigation distance between the control points (in all we have 115 control points). We considered data aggregated by year and for each node, we represented institution, gathering method, pH, % Coliformi Bacteria, conductivity, turbidity, % Escherichia Coli Bacteria. Aggregation is performed by considering mode (average) for discrete (continuous) values. In all, we considered 6 years (from 2004 to 2009). The experiments are performed using the pH feature as target since it is recognized to be a good indicator of river pollution.

Results of the Spearman's rank correlation coefficient are reported in Table 2. Differently from conclusions drawn in the case of Intel lab dataset, in this case our approach outperforms SVMRankT approach of a great margin. This confirms our intuitions since in this dataset, autocorrelation phenomenon is not uniform across the space.

The California traffic dataset concerns the traffic on the highways of California. The dataset is taken from http://traffic-counts.dot.ca.gov/index.htm. The experiments are carried out using the following independent attributes observed by sensors on highways: the percentage of trucks, the percentage of 2-axle vehicles, the percentage of 3-axle vehicles, the percentage of 4-axle vehicles, the percentage of 5-axle vehicles. The goal is to rank sensors' positions on the basis of the sum of the volumes of traffic on a road in both directions. Each sensor

Table 3. California traffic dataset: Spearman's rank coefficient

Train	Test	$SVMRankT$	Our approach
2001-2002	2003	0.7484557	**0.8064138**
2001-2002-2003	2004	0.7417657	**0.8064138**
2001-2002-2003-2004	2005	0.7402272	**0.7970344**
2001-2002-2003-2004-2005	2006	0.7468870	**0.8122890**
2001-2002-2003-2004-2005-2006	2007	0.7447633	**0.8125018**
2001-2002-2003-2004-2005-2006-2007	2008	0.7500496	**0.8247747**
2001-2002-2003-2004-2005-2006-2007-2008	2009	0.7375568	**0.7725733**
Avg.		0.7443340	**0.8045715**

represents a node in the network (in all, we have 969 sensors), while weights on the edges represent the driving distance between two sensors. However, the network is not fully connected and only nodes whose driving distance is less than 25 miles are connected (in all, there are 34093 edges). The dataset refers to the period 2001-2009 and for each year, a network is created.

Results of the Spearman's rank correlation coefficient confirm results obtained on the Portuguese rivers dataset (see Table 3). This result is quite interesting since, in this case, the network edges are quite sparse. This means that the LWR model is able to concentrate the attention only on truly interesting vertices.

6   Conclusions

In this paper we have faced the problem of mining ranking models from networked data whose data distribution may change over time. The first contribution of this paper is to provide a way to learn ranking models based on model trees which are able to consider (different effects of) the autocorrelation phenomenon. This is obtained by resorting to a modified version of the locally weighted regression method. The second contribution of this paper is to provide a time windows framework able to combine models learned in the past for prediction in future time windows.

We evaluate our approach on several real world problems of learning to rank from network data, coming from the area of sensor networks. An empirical comparison with an SVM-based approach shows the superiority of our approach when working with datasets where the effect of the autocorrelation phenomenon is not uniform in the network and with not fully connected networks.

For future works, we intend compare our approach with additional existing regressions/ranking approaches (after adaptation). Moreover, we intend to modify our approach in order to allow it to distinguish between smooth and abrupt changes. Finally, we intend to apply our approach in the context of biological (literature) data analysis for gene prioritization, that is, identification of the most relevant genes that are "connected" to a given disease (e.g. regulate the disease).

References

1. Aiolli, F.: A preference model for structured supervised learning tasks. In: ICDM, pp. 557–560. IEEE Computer Society (2005)
2. Breiman, L., Friedman, J.H., Olshen, R.A., Stone, C.J.: Classification and Regression Trees. Statistics/Probability Series. Wadsworth Publishing Company, Belmont (1984)
3. Crammer, K., Singer, Y.: Pranking with ranking. In: NIPS, pp. 641–647. MIT Press (2001)
4. Dembczyski, K., Kotlowski, W., Slowiski, R., Szelag, M.: Learning of rule ensembles for multiple attribute ranking problems. In: Fürnkranz, J., Hüllermeier, E. (eds.) Preference Learning, pp. 217–247. Springer (2010)
5. Doyle, J.: Prospects for preferences. Computational Intelligence 20(2), 111–136 (2004)
6. Draper, N.R., Smith, H.: Applied regression analysis. Wiley series in probability and mathematical statistics. Wiley, New York (1996)
7. Draper, N.R., Smith, H.: Applied regression analysis. John Wiley & Sons (1982)
8. Har-Peled, S., Roth, D., Zimak, D.: Constraint Classification: A New Approach to Multiclass Classification. In: Cesa-Bianchi, N., Numao, M., Reischuk, R. (eds.) ALT 2002. LNCS (LNAI), vol. 2533, pp. 365–379. Springer, Heidelberg (2002)
9. Har-Peled, S., Roth, D., Zimak, D.: Constraint classification for multiclass classification and ranking. In: Becker, S., Thrun, S., Obermayer, K. (eds.) Advances in Neural Information Processing Systems 15 (NIPS 2002), pp. 785–792 (2003)
10. Herbrich, R., Graepel, T., Bollmann-sdorra, P., Obermayer, K.: Learning preference relations for information retrieval (1998)
11. Herbrich, R., Graepel, T., Obermayer, K.: Large margin rank boundaries for ordinal regression. MIT Press (2000)
12. Jensen, D., Neville, J.: Linkage and autocorrelation cause feature selection bias in relational learning. In: Proc. 9th Intl. Conf. on Machine Learning, pp. 259–266. Morgan Kaufmann (2002)
13. Joachims, T.: Optimizing search engines using clickthrough data. In: Proceedings of the Eighth ACM SIGKDD International Conference on Knowledge Discovery and Data Mining, KDD 2002, pp. 133–142. ACM, New York (2002)
14. Karalic, A.: Linear regression in regression tree leaves. In: Proceedings of ECAI 1992, pp. 440–441. John Wiley & Sons (1992)
15. Lubinsky, D.: Tree structured interpretable regression. In: Fisher, D., Lenz, H.J. (eds.) Learning from Data. Lecture Notes in Statistics. Springer (1994)
16. Macchia, L., Ceci, M., Malerba, D.: Mining Ranking Models from Dynamic Network Data. In: Perner, P. (ed.) MLDM 2012. LNCS, vol. 7376, pp. 566–577. Springer, Heidelberg (2012)
17. Malerba, D., Esposito, F., Ceci, M., Appice, A.: Top-down induction of model trees with regression and splitting nodes. IEEE Trans. Pattern Anal. Mach. Intell. 26(5), 612–625 (2004)
18. Neville, J., Simsek, O., Jensen, D.: Autocorrelation and relational learning: Challenges and opportunities. In: Wshp. Statistical Relational Learning (2004)
19. Newman, M.E.J., Watts, D.J.: The structure and dynamics of networks. Princeton University Press (2006)
20. Opitz, D., Maclin, R.: Popular ensemble methods: An empirical study. Journal of Artificial Intelligence Research 11, 169–198 (1999)

21. Robinson, W.S.: Ecological Correlations and the Behavior of Individuals. American Sociological Review 15(3), 351–357 (1950)
22. Stojanova, D., Ceci, M., Appice, A., Džeroski, S.: Network Regression with Predictive Clustering Trees. In: Gunopulos, D., Hofmann, T., Malerba, D., Vazirgiannis, M. (eds.) ECML PKDD 2011, Part III. LNCS, vol. 6913, pp. 333–348. Springer, Heidelberg (2011)
23. Street, W.N., Kim, Y.: A streaming ensemble algorithm (sea) for large-scale classification. In: Proceedings of the Seventh ACM SIGKDD International Conference on Knowledge Discovery and Data Mining, KDD 2001, pp. 377–382. ACM, New York (2001)
24. Swanson, B.J.: Autocorrelated rates of change in animal populations and their relationship to precipitation. Conservation Biology 12(4), 801–808 (1998)
25. Tesauro, G.: Connectionist learning of expert preferences by comparison training. In: Advances in Neural Information Processing Systems 1, pp. 99–106. Morgan Kaufmann Publishers Inc., San Francisco (1989)
26. Torgo, L.: Functional models for regression tree leaves. In: Fisher, D.H. (ed.) ICML, pp. 385–393. Morgan Kaufmann (1997)
27. Wang, H., Fan, W., Yu, P.S., Han, J.: Mining concept-drifting data streams using ensemble classifiers. In: Proceedings of the Ninth ACM SIGKDD International Conference on Knowledge Discovery and Data Mining, KDD 2003, pp. 226–235. ACM, New York (2003)
28. Wang, H., Yin, J., Pei, J., Yu, P.S., Yu, J.X.: Suppressing model overfitting in mining concept-drifting data streams. In: Proceedings of the 12th ACM SIGKDD International Conference on Knowledge Discovery and Data Mining, KDD 2006, pp. 736–741. ACM, New York (2006)
29. Wang, Y., Witten, I.H.: Induction of model trees for predicting continuous classes. In: Poster papers of the 9th European Conference on Machine Learning. Springer (1997)

Top-k Context-Aware Queries on Streams

Loïc Petit[1,3], Sandra de Amo[2], Claudia Roncancio[3], and Cyril Labbé[3]

[1] Orange Labs, France
name.surname@orange.com
[2] Federal University of Uberlândia, Brazil
deamo@ufu.br
[3] Grenoble University, France
Name.Surname@imag.fr

Abstract. Preference queries have been largely studied for relational systems but few proposals exist for stream data systems. Most of the existing proposals concern the skyline, top-k or top-k dominating queries, coupled with the sliding-window operator. However, user preferences queries on data streams may be more sophisticated than simple skyline or top-k and may involve more expressive operations on streams. This paper improves the existing work on data stream query-answering personalization by proposing a solution to express and handle *contextual* preferences together with a large variety of queries including one-shot and continuous queries. It adopts a more expressive preference model supporting context-based preferences, allowing to capture a wide range of situations. We propose algorithms to implement the new preference operators on stream data and validate their performance on a real-world dataset of stock market streams.

1 Introduction

Query-answering personalization has been attracting much attention in the database community in recent years [3,7]. Such works have been motivated by the need to select the data items that better fit user preferences. This is useful in situations when the number of potential answers is either too high or too small. When it is too high, user preferences are used to restrict the answer set by identifying the subset of the most preferred data items. On the other hand, some queries may involve hard conditions which imply a very small (or even a disappointing empty) answer-set. In this case, user preferences could be used to enhance the set of retrieved data by including answers which could be of user interest even if they do not verify the hard constraints specified in the query.

Numerous application domains such as financial, monitoring and sensor-based applications require now data stream management. Supporting preference queries on evolving data is more challenging than their evaluation on persistent data. *Contextual* preference queries are particularly helpful to users dealing with data streams. For instance, in a stock market scenario, buyers may want to know the most interesting deals so far before making their trading decisions. Some statistical data such as the *volatility rate* of the stock options in the last three days

S.W. Liddle et al. (Eds.): DEXA 2012, Part I, LNCS 7446, pp. 397–411, 2012.
© Springer-Verlag Berlin Heidelberg 2012

or the *economic situation* of the stock options country can influence their decision. So, it is possible to support queries as "What are the most interesting deals
you propose given that for stock options coming from countries in bad economic
situation in the last year I prefer those presenting a lower volatility rate in the
last three days" may be issued. The existing proposals on contextual preferences
query processing [11,4] are designed for conventional DBMS and are not tailored to handle stream data. Besides, few proposals in the literature support
preference queries on data streams [12,13,9]. Most of them concern the skyline,
top-k and top-k dominating queries, coupled with the sliding-window operator.
To the best of our knowledge there are no proposal in the literature dealing with
contextual preference query processing on data streams.

This work goes a step beyond by proposing contextual preference queries
on both conventional and stream databases. We consider the stream algebra
Astral introduced in a previous paper [16] as the core stream query language.
We extend Astral by the introduction of two preference operators Best and
KBest. These operators are adapted from the preference operators of the query
language CPrefSQL originally designed for querying static data [4].

Main Contributions. The main contributions of this paper can be summarized
as follows: (1) The introduction of the top-k contextual preference queries in a
data stream context; (2) The introduction of two new operators in the Astral algebra designed to query relations and data streams; (3) The design and implementation of incremental algorithms for evaluating continuous and instantaneous
queries over streams and relations; (4) The implementation of the preference operators in the original prototype of Astral and their performance evaluation.

This paper is organized as follows: In Section 2, we motivate our proposal by
presenting a real-world scenario where user preferences are naturally influenced
by the user context. Section 3 introduces the main theoretical concepts underlying the preference model and the stream algebra Astral. In Section 4, we present
the Preference Astral algebra incorporating two new preference operators Best
and KBest. In Section 5, we present the incremental algorithms for continuously
evaluating the preference operators Best and KBest. In Section 6 we present and
discuss some experimental results. Related work are resented in section 7. Finally,
in Section 8 we conclude the article and present some research perspectives.

2   A Motivating Example

Tom is a very cautious investor who likes to get as much information as possible
before making his decisions about buying and selling stocks shares. He has free
access to a web site that provides information about real-time quotations and
volatility rates as well as real-time transactions. These data involve the following
data streams and static data stored in a relational DBMS:

• Relation *StockOption(StOpName, Category, Country)*: stores the stock name,
its category (Commodities (*c*), Info-Tech (*it*)), and the country where the company headquarters are located.

• Stream *Transactions(OrderID, TTime, StOpName, Volume, Price)*: a data stream providing real-time information about stock options transactions. It includes the transaction time *(TTime)*, the quantity of shares *(Volume)* and the price *(Price)* of the stock option share.

• Stream *Volatility(StOpName, ETime, Rate, Method)*: a data stream providing real-time information about the estimated volatility *(rate)* of stock options. It includes the time of the estimation *(ETime)* and the estimation method *(Method)*.

Based on his past experience and the information he reads in the papers, Tom has some preferences he wants to be taken into account in order to facilitate and speed up his decisions. His preferences are described by the following statements:

[P1] Concerning commodities stocks, at each moment Tom prefers those with a volatility-rate less than 0.25. On the other hand, concerning IT stocks, Tom is more aggressive and prefers those with a volatility-rate greater than 0.35.

[P2] For stock options with volatility-rate greater than 0.35 at present (calculated according to some method) Tom prefers those from Brazil than those from Venezuela.

[P3] For stock options with volatility-rate greater than 0.35 at present, Tom is interested in transactions carried out during the last 3 days concerning these stock options, preferring those transactions with quantity exceeding 1000 shares than those with a lower amount of shares.

Notice that Tom's preferences are expressed by means of *rules* of form IF *some context is verified* THEN *Tom prefers something to something else*. *Contexts* are conditions involving the values of some data attributes. For instance, in statement [P1] the context is *StockOption.Category = 'Commodities'* and Tom's preference is *Volatility.Rate ≤ 0.25 better than Volatility.Rate > 0.25*. Preference rules may involve streams or relational data on both the context side and the preference side of the rule.

As well as his preferences, Tom's queries may concern relational and stream data and be "one-shot" or continuous queries. Here are some of Tom's:

[Q1] Considering the last 100 transactions with a volume greater than 1000 shares, list my top 10 most preferred ones.

[Q2] Give me the list of quotations during the last 2 days, concerning the stock options which most fulfill my preferences.

[Q3] Give me the list of quotations during the last 2 days, concerning only IT stock options which most fulfill my preferences.

[Q4] Every 30 minutes, give me a complete description of my 10 preferred stock options: country, category, the last transaction concerning the stock option (volume and quotation), the volatility rate with its corresponding estimating method.

It is important to emphasize that, differently from the *hard constraint* expressed by statements like "IT stock options", preferences should be viewed as *soft constraints*: If no database entry fulfills the hard constraints (for instance, there is no IT stock options in the database), the result answer-set is empty. On the other hand, if there are not K tuples in the database which are considered *perfect* according to the preferences, a list of K tuples respecting the preference hierarchy is returned instead.

3 Preliminaires

To achieve our work we build-on two existing proposals (1) the Astral algebra proposing operators to query data stream and relational data together and (2) theoretical foundations on contextual preferences rules. This Section introduces such proposals. Section 4 and the following, present our proposal to integrate preferences in Astral queries.

3.1 The Astral Stream Algebra

We use the Astral algebra [16] which provides a formal definition of operators involved in streams querying. Such a formalism facilitates the expression and understanding of queries. Putting aside preferences and the top-k operator, queries presented in section 2 can be expressed in Astral. This section provides the very few definitions necessary to introduce Astral queries. Our examples refer to queries **Q1, Q2, Q3 and Q4** of section 2. We'll use **Q1', Q2', Q3' and Q4'** which are their counterpart without preferences.

In Astral, streams and relations (denoted by S and R respectively in the following) are two different concepts [1]. A *stream* S is a possibly infinite set of tuples s with a common schema containing two special attributes: a timestamp t and, the position in the stream, p^1. A *temporal relation* R is a step function that maps a time identifier t to a set of tuples $R(t)$ having a common schema. Classical relational operators, as selection σ, projection π and join $\bowtie$, are extended to temporal relations. The extension of π and σ for streams is simple whereas joins are very complex. For example, $\sigma_{Volume>10000}(Transactions)$ is the stream of transactions having $Volume > 10000$, whereas $\sigma_{Category=it}(StockOption)$ is a *temporal relation* containing only tuples for which $Category$ is *it* .

A temporal relation can be extracted from a stream using windows operators. Astral provides an extended model for windows operators including positional, temporal and non standard cross domain windows (e.g. slide n tuples every δ seconds). The following expressions represent some very useful windows: (a) $S[L]$ contains the last tuple of a stream;
(b) $S[A/L]$ is the partitioned window containing the last tuple of each sub-stream identified with the attribute A. For example $Transaction[StOpName/L]$ contains the last known transaction for each stock option.
(c) $S[N \text{ slide } \Delta]$ is a sliding window of size N sliding Δ every Δ. N and Δ are either a time duration or a number of tuples. For instance, **Q1'** is written as $(\sigma_{Volume>10000}(Transactions))[100 \text{ slide } 1]$. The window is 100 tuples large and slides of 1 tuple whenever one new tuple arrives. **Q2'** is $\pi_{Price}(Transactions[2 \text{ days } \text{ slide } 1])$. The window is 2 *days* large and slides of 1 tuple whenever one new tuple appears.

---

[1] These definitions can be extended using the notion of batch [16].

	StOpName	Cat	Country	ETime	Rate	Method
t_1	MS	it	USA	T1	0.30	M1
t_2	AP	it	India	T2	0.55	M1
t_3	USSteel	c	USA	T1	0.20	M2
t_4	Petr4	c	Brazil	T2	0.40	M2
t_5	Bel5	m	Venezuela	T3	0.55	M2

(a)   (b)

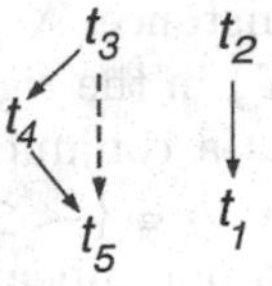

Fig. 1. Instance of *StockOption* ⋈ *Volatility* and Better-than Graph associated to Γ

A stream can be generated from a temporal relation using a *streamer* operator. Among them, $I_S(R)$ produces the stream of tuples inserted in R. Streamers and windows may be composed in order to join two streams or a stream and a relation. Given a window description W, a streamer Sc and a join condition c, the join operator (stream×relation)→stream can be defined as: $S \bowtie_c R = Sc(S[W] \bowtie_c R)$. In the following we will use: $S \bowtie_c R = I_S(S[L] \bowtie_c R)$. The stream $S \bowtie_c R$ contains tuples generated by updates in R. Other types of join operators can be defined. For instance, tuples can be added to the output stream only for new tuples in S and not when R is updated: the semi-sensitive-join operator (stream×relation)→stream produces a stream resulting from a join between the last tuple of the stream and the relation at the time of the last tuple of the stream: $S \bowtie_c R = I_S(S[L] \bowtie_c R(\tau_S(S[L])))$. Here ι_S denotes the function that gives the timestamp of a tuple in the stream S. (see [16] for more details). For instance, query **Q3'** is written as

$$\pi_{Price}((Transactions \bowtie (\sigma_{Category=IT}(StockOption)))[2 \ days \ slide \ 1]).$$

The expression for query **Q4'** provides a stream built from the join between the *Transaction* stream, the last known values of the *Volatility* and the *StockOption* relation. The required windows expression is $[W] = [30 \ min \ slide \ 30 \ min]$.

3.2 The Preference Model

In this section we present the main concepts concerning the logical formalism for *specifying and reasoning* with preferences. Details can be found in [4,5].

Let R be a relational schema with attributes $\text{Attr}(R) = \{A_1, A_2, ..., A_n\}$. R can be a non-temporal or temporal relational schema. If R is temporal then one of its attributes is T (time). For each attribute $A \in Attr(R)$, let $\textbf{dom}(A)$ be the set of values of A (the domain of A). The set $\text{Tup}(Attr(R)) = \textbf{dom}(A_1) \times \textbf{dom}(A_2) \times ... \times \textbf{dom}(A_n)$ is the set of all possible tuples over $Attr(R)$.

Definition 1 (Contextual Preference Rules). A *conditional preference rule* (or *cp-rule* for short) over the relational schema R is a statement φ of the form:
$\varphi: u \rightarrow Q_1(X) \succ Q_2(X) \quad [W]$ where:
- X is a non-temporal attribute of R, $W \subseteq Attr(R)$, $X \notin W$,
- $Q_i(X)$ (for $i = 1, 2$) is a statement of the form $X\theta a$ where $\theta \in \{=, \neq, \leq, \geq, <, <\}$ and $a \in \textbf{dom}(X)$.

- There is no $x \in \mathbf{dom}(X)$ satisfying both $Q_1(X)$ and $Q_2(X)$ simultaneously. For instance, $X > 1$ and $X \leq 3$ cannot be considered as statements $Q_1(X)$ and $Q_2(X)$ in the right side of a cp-rule, since $X > 1 \cap X \leq 3 = (1, 3] \neq \emptyset$.
- u is a conjunction of simple statements of the form: $A_1 \theta_1 a_1 \wedge ... \wedge A_k \theta_k a_k$, where $\theta_i \in \{=, \leq, \geq, <, <\}$ for $i = 1, ..., k$. We assume X and the attributes in W do not appear among the attributes of u.

The formula u in the left side is called the *context* of the rule φ. The statement $Q_1(X) \succ Q_2(X)$ in the right side is called the *preference statement* and the attributes in W are called the *ceteris paribus* attributes. This will be clearer in the sequel. A tuple $t \in Tup(Attr(R))$ is said to be *compatible* with a cp-rule φ if t satisfies its context. For instance, the tuples $t_1 = (1, 2, 4, 1)$ and $t_2 = (0, 2, 5, 6)$ over the relation schema $R(A, B, C, D)$ are compatible with the cp-rule $(A < 2 \wedge B = 2 \wedge C > 3) \rightarrow (D \leq 2 \succ D > 4)$. The context of this cp-rule is the formula $(A < 2 \wedge B = 2 \wedge C > 3)$ and its preference statement is $(D \leq 2 \succ D > 4)$. Intuitively, this cp-rule means that between two tuples compatible with the context $(A < 2 \wedge B = 2 \wedge C > 3)$ I prefer the one with $D \leq 2$ than the one with $D > 4$. So, between the tuples t_1 and t_2, I prefer t_1.

A *contextual preference theory* (*cp-theory* for short) over R is a finite set of cp-rules Γ over R. We denote by $Attr(\Gamma)$ the set of attributes appearing in the cp-rules of Γ. Notice that $Attr(\Gamma) \subseteq Attr(R)$.

Example 1: Let us consider the two preference statements P1 and P2 of our motivating example. They can be expressed by the following cp-theory over the schema $T(StOpName, Cat, Country, ETime, Rate, Method)$:

- φ_1: $Cat = c \rightarrow (Rate < 0.25 \succ Rate \geq 0.25)$, $[M]$
- φ_2: $Cat = it \rightarrow (Rate \geq 0.35 \succ Rate < 0.35)$, $[M]$
- φ_3: $Rate > 0.35 \rightarrow (Country = Brazil \succ Country = Venezuela)$

The attributes between brackets mean that in order to compare two tuples by means of a cp-rule, these tuples must coincide on these attributes. For the other attributes there is no restriction. For instance, in the scenario of Example 1, let t_1 and t_2 as described in Figure 1(a). Then t_1 and t_2 can be compared by using the rule φ_2, since they have the same context ($Cat = it$), the $Rate$ (volatility rate) of t_2 is greater than 0.35 and the $Rate$ of t_1 is lower than 0.35, and the method used to measure $Rate$ is the same for both tuples.

It is clear by now that a cp-rule φ over R induces a *binary relation* (denoted by $\succ_\varphi$ on the set $Tup(R)$: the set of pairs (t, t') such that t is better than t' according to φ. Of course, this binary relation is not necessarily an *order relation*, since it is not always transitive. In the following we define the notion of *Preference Relation* induced by a cp-theory Γ.

Definition 2 (Preference Relation). *Let Γ be a contextual preference theory over a relational schema R (temporal or non-temporal). The Preference Relation associated to Γ (denoted by $\succ_\Gamma$) is defined as: $\succ_\Gamma = (\bigcup_{\varphi \in \Gamma} \succ_\varphi)^*$, where $*$ denotes transitive closure.*

Example 2: Let us consider the cp-theory Γ of Example 1. Let us consider instances I and J of relation schemas $StockOption$ and $Volatility$ respectively, such that the result of $StockOption \bowtie Volatility(I, J)$ is given in Figure 1(a). It is clear that $t_3 \succ_{\varphi_1} t_4$ and $t_4 \succ_{\varphi_3} t_5$. Then, by transitivity, we conclude that $t_3 \succ_\Gamma t_5$. Notice that t_3 and t_5 cannot be compared using only one rule in Γ. However, they can be compared by transitivity using different rules in Γ.

Discussion. We say that a cp-theory Γ is *consistent* if and only if the induced order $>_\Gamma$ is irreflexive and consequently, a *strict partial order* over $Tup(R)$. In [17], a sufficient condition for ensuring consistency of a cp-theory is given. This condition involves testing the acyclicity of the *dependency graph* associated to the cp-theory and its *local consistency*. For lack of space we omit the details. In this paper, we will suppose our cp-theories are consistent, that is the associated Preference Relation $\succ_\Gamma$ is a strict partial order. For more details on the theoretical foundations and consistency test see [5].

4   Introducing Preference Operators into ASTRAL

Let us focus on the integration of contextual preferences in the ASTRAL algebra. This Section presents the syntax and semantics of the preference operators integrated to ASTRAL whereas Section 5 presents the algorithms for implementing them.

4.1   Global Approach

The objective of our proposal is to provide an integrated solution where the full expressivity of both, queries and preferences are available. We propose to capture the semantics of the preference evaluation as algebraic operators that extend the ASTRAL algebra. Such preference operators can be part of instantaneous and continuous queries performing on streams and relations, and using any of the existing operators. Particularly, queries involving data streams can use the wide variety of temporal, positional and hybrid windows[15]. The preference operators calculate user preferred answers according to the available cp-theory. Each user provides the system with his/her preferences (a cp-theory Γ) which become some kind of *user profile*. During querying, these preferences are used for answer customization if the user asks for. Concretely, we will allow powerful contextual *most preferred* and *top-k* queries by the introduction of two operators:

(1) The $Best_\Gamma$ operator selects a subset of *optimal* tuples according to user preferences Γ.
(2) The $KBest_\Gamma$ operator selects the K most preferred tuples respecting the preference *hierarchy* specified by Γ. For the sake of simplifying the presentation we omit the subscript Γ whenever it is implied by the context.

4.2   Best and KBest Operators

The *Best* operator selects from a given temporal or non-temporal relation those tuples which are not dominated by other tuples according to the preference order inferred from Γ (see Definition 2).

Definition 3 (Best). *Let R be a relational schema and Γ be a cp-theory over R. Let $r(t)$ be an instance over R at time t* $\mathbf{Best}(r(t)) = \{u \in r(t) \mid \nexists v \in r(t)$ *such that $v \succ_\Gamma u$ }*

The operator **KBest** selects the *top-k tuples* according to the preference *hier-archy* dictated by Γ. Intuitively, **KBest**(I, k) returns the set of k tuples of I having the minimum number of tuples dominating them in the preference hier-archy. In order to define its semantics, we need first to introduce the notion of *level* of a tuple u (denoted by $l(u)$) according to a cp-theory Γ. The level of a tuple reflects "how far" is the tuple from the most preferred ones (those which best fit the user preferences).

Definition 4 (Level). *Let R be a relational schema and Γ be a cp-theory over R. Let $r(t)$ be a tuple-set or instance of R at time t, and let tuple $u \in r(t)$. The level of u, $l(u)$, according to Γ is inductively defined as follows:*

- *If there is no $u' \in r(t)$ such that $u' \succ_\Gamma u$, then $l(u) = 0$.*
- *Otherwise $l(u) = 1 + max\{l(u') \mid u' \succ_\Gamma u\}$*

It is easy to show that if $u \succ u'$ then $l(u) < l(u')$. The reverse implication does not hold. The semantics of the **KBest** operator is defined as follows.

Definition 5 (KBest). *Let R be a temporal (or non-temporal) relation and Γ be a cp-theory over R. Let $r(t)$ be a tuple-set or instance of R at time t. $\mathbf{KBest}(r(t), k)$ is the set of the k tuples $\in r(t)$ with the lowest levels. The posi-tional order is used to sort tuples at the same level.*

The following examples illustrates the semantics of both operators.

Example 3: Let us consider the cp-theory $\Gamma = \{\varphi_1, \varphi_2, \varphi_3\}$ of Example 1 and the temporal relation I of Figure 1(a). Figure 1(b) shows the *Better-Than Graph* G (*BTG*)associated to cp-theory Γ over I. The nodes are the tuples of I. An edge (t_i, t_j) expresses that t_i is preferred to t_j according to a rule of Γ. A dotted edge (t_i, t_j) means that t_i is preferred to t_j by transitivity. We have that $\mathbf{Best}(I) = \{t_1, t_3\}$, since these are the tuples which are not dominated by others. Note that level$(t_1) =$ level$(t_3) = 0$, level$(t_2) =$ level$(t_4) = 1$ and level$(t_5) = 2$. So, $\mathbf{KBest}(I, 3) = \{t_1, t_3, t_2\}$. As t_2 and t_4 have the same level in the preference hierarchy the positional order is used to decide between them.

As an example of a query in the extended ASTRAL algebra, let us consider the query **Q2** of Section 2. It is expressed in the extended algebra by:
$\pi_{Price}\mathbf{Best}(Transactions[2days$ slide $1])$

5   Best and KBest Algorithms

This section presents the algorithms we propose to evaluate the **Best** and **KBest** operators (see section 5.2). As such operators require the preference hierarchy of tuples which is represented by a *BTG*, the algorithms to create the *BTG* are first introduced in section 5.1. Section 5.3 presents an incremental alternative to manage the *BTG*.

Algorithm 1. Compare(t_1, t_2, φ)	**Algorithm 2.** CompT(t_1, t_2, Γ)
	(without transitive closure)

Algorithm 1. Compare(t_1, t_2, φ)

Data: $\varphi : u \to Q_1(X) > Q_2(X)$ [W]
Result: $\{1, -1, \emptyset\}$,
    resp. $\{t_1 >_\varphi t_2, t_1 <_\varphi t_2, \text{inc.}\}$
if $t_1 \not\models u \parallel t_2 \not\models u$ **then return** $\emptyset$
foreach $V \in W$ **do**
    | **if** $t_1(V) \neq t_2(V)$ **then return** $\emptyset$
if $t_1 \models Q_1(X)$ & $t_2 \models Q_2(X)$ **then**
    | **return** 1
if $t_2 \models Q_1(X)$ & $t_1 \models Q_2(X)$ **then**
    | **return** -1
return $\emptyset$

Algorithm 2. CompT(t_1, t_2, Γ)
(without transitive closure)

Data: $\Gamma = \{\varphi_1, ..., \varphi_k\}$ a
    cp-theory
Result: $\{1, -1, \emptyset\}$,
    resp. $\{t_1 >_\Gamma t_2, t_1 <_\Gamma t_2, \text{inc.}\}$
foreach $\varphi_k \in \Gamma$ **do**
    | $r \leftarrow$ Compare(t_1, t_2, φ_k)
    | **if** $r \neq \emptyset$ **then return** r
return $\emptyset$

5.1 The Preference Hierarchy and the Better-Than Graph

First and foremost, we introduce the algorithm to establish the preference order between two tuples t_1 and t_2 according to a rule φ. The straight forward Algorithm 1 returns $\emptyset$ if t_1 and t_2 are incomparable, 1 if t_1 is more preferred than t_2 and -1 otherwise. Algorithm 2 extends the comparison to a cp-theory Γ. It relies on the set of rules (φ_n) to identify the preference order. However, it does not compute the transitive closure as stated in the cp-theory definition. The transitivity will be computed in the preference algorithms.

Better-Than Graph: The Best and KBest preference operators are applied on a tuple-set TS and require the *BTG* of TS. For its implementation we adopt the *Graph(Next, Prec, Src)* defined as follows: *Next* associates to each tuple the list of its direct dominated tuples. *Prec* associates to each tuple the list of tuples that directly dominate it. *Src*, the no-dominated tuples, that is the sources of the graph. From a formal point of view:

$$Next = s \in TS \mapsto \{s' \in TS, \exists \varphi_n \in \Gamma, s >_{\varphi_n} s'\}$$
$$Prec = s \in TS \mapsto \{s' \in TS, \exists \varphi_n \in \Gamma, s <_{\varphi_n} s'\}$$
$$Src = \{s \in TS, Prec(s) = \emptyset\}$$

To provide good performance, the implementation of the graph uses hash-sets and hash-maps. The *Next* and *Prec* functions also have a *keys()* method defined by: $s \in F.keys() \Leftrightarrow F(s) \neq \emptyset$.

The construction and maintenance of the graph require *Insert* and *Delete* methods. To insert a tuple, Algorithm 3 iterates over the graph to update *Next*, *Prec* and the *Src* set. As the cost of insertion and deletion in a hash structure can be considered as $\mathcal{O}(1)$, the global cost of the insertion of a tuple in the graph is $\mathcal{O}(|G|)$. The deletion of a tuple s, presented in Algorithm 4, is an iteration over the nodes connected to the one we delete[2]. The cost is $\mathcal{O}(\text{degree}(s))$.

---

[2] As a recall, in graph theory, the number of edges connected to a node is called the *degree* of a node.

Algorithm 3. G.Insert ; Insert a tuple in the BTG	**Algorithm 4.** Graph.Delete ; Removes a tuple from the BTG
Input: A tuple s, Γ and the BTG structure $L \leftarrow Prec.\text{keys}() \cup Src$ $Src.\text{add}(s)$ **foreach** $s' \in L$ **do** $\quad$ $r \leftarrow \text{CompT}(s, s', \Gamma)$ $\quad$ **if** $r > 0$ **then** $\quad\quad$ $Src.\text{remove}(s')$ $\quad\quad$ $Prec.\text{put}(s', s)$ $\quad\quad$ $Next.\text{put}(s, s')$ $\quad$ **else if** $r < 0$ **then** $\quad\quad$ $Src.\text{remove}(s)$ $\quad\quad$ $Prec.\text{put}(s, s')$ $\quad\quad$ $Next.\text{put}(s', s)$	**Input**: A tuple s and the BTG structure $Src.\text{remove}(s)$; $P \leftarrow Prec.\text{remove}(s)$; $Dom \leftarrow Next.\text{remove}(s)$ **foreach** $s' \in P$ **do** $\quad$ $Anc \leftarrow Next.\text{get}(s')$ $\quad$ **if** $Anc.\text{size}() = 1$ **then** $Next.\text{remove}(s')$ $\quad$ **else if** $r < 0$ **then** $Anc.\text{remove}(s)$ **foreach** $s' \in Dom$ **do** $\quad$ $Anc \leftarrow Prec.\text{get}(s')$ $\quad$ **if** $Anc.\text{size}() = 1$ **then** $\quad\quad$ $Src.\text{add}(s')$ $\quad\quad$ $Prec.\text{remove}(s')$ $\quad$ **else if** $r < 0$ **then** $Anc.\text{remove}(s)$

Given a known cp-theory Γ and a tuple-set, the construction of the entire BTG relies on the insert method (see Algorithm 6).

5.2 Evaluation of Best and KBest

By definition, the *Graph* includes *Src* which corresponds to the most preferred tuples. *Src* is the answer of the Best operator. However, Algorithm 3 can be optimized by avoiding the entire construction of the *BTG*. The *Prec.put* and *Next.put* sequences can be suppressed from it. This variant will be named *Src* in section 6. It's complexity is so reduced to $\mathcal{O}(|Src|)$. The complexity of $Best(R)(t)$ becomes $\mathcal{O}(NS)$ where $N = |R(t)|$ and $S = |Best(R)(t)|$.

The main algorithm used to compute *KBest* from the *BTG* is a Kahn-topological sort limited to k results. See Algorithm 5.

Its complexity is majored by the complexity of the Kahn algorithm which is $\mathcal{O}(N + |Next|)$. The limitation to k introduces a global factor $\frac{k}{N}$. Leading to $\mathcal{O}(k + kD))$, where D is the average degree of each node (majored by $\mathcal{O}(N)$). The complexity of KBest is therefore $\mathcal{O}(N^2)$. Moreover, as k is usually small compared to N, and D is more likely to be very small compared to N (many tuples are not comparable) then the major complexity factor comes from the construction of the BTG.

5.3 Incremental Evaluation of BTG

This section introduces the obtention of BTG in an incremental way. This is motivated by queries over data streams where preferences are evaluated on sequences of windows. The BTG is required for the tuple-set contained in the *current* window. As two successive windows may overlap, then the new BTG can be constructed by incremental updates of the *current* one. The implementation of Astral's window sequences makes available two delta sets wrt the current window and the next one: δ_R^- are the tuples that "exit" from the window and,

---

Algorithm 5. Calculate KBest$(R)(t)$

Data: The BTG structure, k the number of required tuples
Res $\leftarrow$ **new TreeSet**()   /* Ordered set */

if $k < |Src|$ then   /* Src contains more than k best items */
  | $N \leftarrow |Src| - k$ /* The positional order in Src is used */
  | **foreach** $s \in Src$ **do**   /* to keep the k more "recent" items */
  | | **if** $N = 0$ **then** Res.add(s)
  | | **else** $N \leftarrow N - 1$
  ∟ **return** Res
NextLvl $\leftarrow$ Src; id $\leftarrow$ 0; PrecCount $\leftarrow$ **new HashMap**()
while $id < k$ **and** $id < |Src| + |Prec.keys()|$ **do**
  | **if** $Buffer = \emptyset$ **then** /* Buffer contains tuples with the same level */
  | | **foreach** $t \in NextLvl$ **do**
  | | ∟ Buffer.push(t)
  | ∟ NextLvl.clear()
  | $t \leftarrow$ Buffer.pop()
  | **foreach** $s \in Next.get(t)$ **do**   /* For each node dominated by t */
  | | $n \leftarrow$ PrecCount.get(s)
  | | **if** $n = null$ **then** $n =$ Prec.get(s).size()
  | | **if** $n = 1$ **then** NextLvl.add(s) /* s is part of the next level */
  | ∟ **else** PrecCount.put$(s, n - 1)$ /* There are more nodes to browse */
  ∟ Res.add$(t.copy(id\!+\!+))$ /* Update the positional order */

return Res

δ_R^+ are the ones arriving for the new window. There is no intersection between these sets. These delta sets are used to obtain the BTG of the new window based on the preceding one as shown in Algorithm 7.

The complexity of updating the BTG is $\mathcal{O}(|\delta_R^-|.D + |\delta_R^+|.N)$, where D is the average degree of a node in the graph. If we consider that the size of the δ_R are similar and that D is ruled by $\mathcal{O}(N)$ then the complexity becomes $\mathcal{O}(|\delta_R|.N)$.

Table. 1 hereafter summarizes the complexity of the preference operators for the two BTG construction approaches. It is worth noting that the incremental approach is really interesting if the delta sets are small compared to the total number of nodes. A large portion of the current BTG can be reused for the new one. If it is not the case the BTG creation "from scratch" performs better.

Learning Inferred Preferences
The proposed implementation applies the mathematical definition of the preference order and does not keep trace of inferred preferences. For instance, for tuples s_1, s_2, s_3, if $s_1 <_\Gamma s_2$ and $s_2 <_\Gamma s_3$ then by transitivity $s_1 <_\Gamma s_3$. Now, if s_2 is no more in the current scope, we would say $s_1 \not<_\Gamma s_3$. However, at some point in time it was known that $s_1 <_\Gamma s_3$ and this knowledge could be reused. A small change in the Graph.Remove function allows us to provide that semantics

Algorithm 6. Create BTG	**Algorithm 7.** Incremental BTG
Input: A tuple-set TS **Data**: The BTG structure **foreach** $s \in TS$ **do** $Graph.\text{Insert}(s)$	**Input**: δ_R^- and δ_R^+ **Data**: The BTG structure **foreach** $s \in \delta_R^-$ **do** $\quad \lfloor$ $Graph.\text{Remove}(s)$ **foreach** $s \in \delta_R^+$ **do** $\quad \lfloor$ $Graph.\text{Insert}(s)$

Table 1. Best/KBest complexity

	BTG	Incremental BTG
Best	$\mathcal{O}(N.S)$	$\mathcal{O}(\Delta.N)$
KBest	$\mathcal{O}(N^2)$	$\mathcal{O}((\Delta + k)N)$

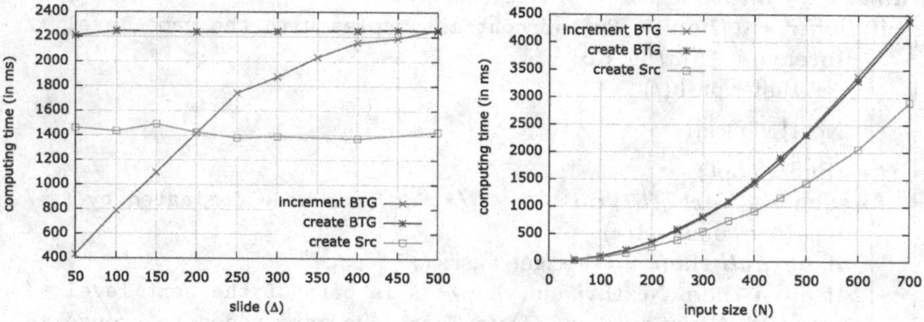

(a) Varying Δ on KBest (for $N = 500$)   (b) Varying N on KBest (for $\Delta = N$)

Fig. 2. Computing time of Incremental BTG, Create BTG and Src maintenance

if desired. When removing a node, the preceding nodes will be linked to the following nodes. The global complexity doesn't change.

6   Experimental Results

The algorithms presented in this paper have been implemented as extensions of the Astral DSMS Prototype[3]. These extensions have been facilitated by its SOA architecture. We performed experiences to study the behaviour of both the **Best** and the **KBest** operators.

Experimentation Setup: A quad-core Intel Xeon 2.6GHz computer with 6GB of RAM is used along with the Sun/Oracle 1.6 JVM, an Apache Felix OSGi platform with Astral. 30,000 tuples have been gathered from real-world quotes[4]. We used the preferences presented in Section 2 and the queries of the running example. Let us focus here on the experiments with queries in the style of **Q3** and **Q4** of Section 2. These are top-k queries over stream.

$$S = Transaction \bowtie (Volatility[StOpName/L] \bowtie StockOption)$$

---

[3] Available at http://astral.googlecode.com under Apache 2 Licence.

[4] Dump provided by Dukascopy's Data Export service Available at http://www.dukascopy.com/swiss/english/data_feed/csv_data_export

The query with the KBest operator uses sliding windows as follows: $\mathbf{KBest}(S[N \text{ slide } \Delta], k)$

Results: Experiments show that the evaluation time of **KBest** is dominated by the construction/update of the BTG. We also observed the evolution of the structure of the BTG from one window to the next one: the maximum level varies from 2 to 6 and the number of non-dominated tuples varies from 1 to N. Big changes in the structure of the graph are *bad cases* for the incremental BTG algorithm (Algorithm 7) of the **KBest** operator.

Figure 2(a) shows the computing time of the two algorithms of BTG: create (Algorithm 6) and incremental (Algorithm 7). It also shows the time for the algorithm reduced to the *Src* maintenance. This can be used for the Best operator. In the experiments we used several window sizes (N) and rates (Δ). We noticed that changes in the rate do not impact the time for create BTG, whereas the incremental algorithm performs 6 times better for a N/Δ ratio = 10. Surprisingly, the two BTG algorithms behave similarly when $\Delta \sim N$ which correspond to few or no intersection between successive windows. This means that in the incremental version, the deletions in the BTG do not take long time compared to insertions. This may be not true when the BTG is a strongly connected graph with nodes with high level (though very unlikely in practice).

The variation of the size of the window (figure 2(b) with $\Delta = N$) shows that the behavior is not impacted by the number of tuples involved. As expected, the evolution is N quadratic and the incremental algorithm strictly follows the performance of the create algorithm.

7 Related Work

The problem of enhancing well-known query languages with preference features has been tackled in several recent and important work in the area. For a comprehensive survey on preference modeling, languages and algorithms see [10]. In this section we present some related work concerning contextual preference support in traditional databases and preference support in stream data.

Contextual Preference Support. In the database field, several proposals for incorporating context in query languages exist in the literature. In [11] preferences are expressed in a *quantitative* format, that is, by means of scores associated to attribute-value clauses. A *contextual query* is a standard query enhanced with a user context. The main problem tackled in these papers is identifying the preferences that are most relevant to a contextual query and presenting an algorithm to locate them. The approach we adopt in this paper follows a *qualitative* model to express preferences: preferences are expressed by a (small) set of rules from which is inferred a strict *partial* order on tuples. Moreover, we assume that the contextual preferences are given and incorporated into the query language syntax. Qualitative approaches has many advantages when compared to quantitative ones due to their conciseness and deduction capability.

Top-k Preference Queries. In [6] the *top-k queries* have been introduced in a *quantitative* preference model setting, that is, where preference between tuples is expressed by a score function defined over the dataset. The *top-k dominating queries* have been introduced in [14] as an extension of the skyline queries of [3] which were originally designed to return the most preferred tuples, without any user control on the size of the result. A *top-k dominating query* returns the k tuples which dominated the maximum amount of tuples in the database. This concept is orthogonal to the skyline and pareto queries, as well as to the approach CPrefSQL we adopt in this paper.

Preference Support on Stream Data. Most work on preference queries in data streams [12,13,9] concern methods for the continuous evaluation of skyline queries, top-k queries and top-k dominating queries under the sliding window model. In these proposals, a preference operator is coupled with two forms of the sliding window operator over data streams: the *count-based* and the *time-based* ones. In the count-based sliding window the last N tuples of a stream are returned and for each arriving tuple, the oldest one expires. In the time-based sliding window, the active tuples are those arrived during the last T time instants. The preference operators are applied to the set of tuples returned after a sliding window execution over the stream. To the best of our knowledge no previous work exists that proposes a stream algebra incorporating both stream and preference operators. A comprehensive survey on continuous processing of skyline, top-k and top-k dominating queries can be found in [8].

Contextual Preferences on Data Streams. A recent work treating contextual preferences in data streams (coming from sensors) is [2]. The authors propose a preliminary and informal methodology described through a real-world example that tries to combine the research topics of context-awareness, data mining and preferences. The paper does not tackles the problem of incorporating the discovered preferences into a query language on sensor data.

8   Conclusion and Future Work

This paper proposes an integrated solution to support user personalized queries in rich data environments involving real time data streams and persistent data and providing powerful querying capabilities. Instantaneous and continuous preference queries are supported. They can benefit of the whole expressivity of Astral and particularly of the large variety of window support (positional, temporal and cross-domain windows) to manage data streams. The contributions of this work include the definition and implementation of preference operators as an extension of Astral. Our experiments allowed to identify patterns of queries (based on the window characteristics) that can be used to decide the best strategy to optimize the query evaluation. Our future research will focus on new optimization approaches and on the distributed evaluation of the preference queries.

Acknowledgment. This work is partially supported by CAPES, the STIC-AmSud ALAP project, CNPq and FAPEMIG and the project BQR Arteco of the Grenoble Institute of Technology. We also would like to thank the Sigma team and M. Echenim of the LIG laboratory for their support.

References

1. Arasu, A., Babcock, B., Babu, S., Cieslewicz, J., Datar, M.: STREAM: The Stanford Data Stream Management System. In: Data Stream Management: Processing High-Speed Data Streams (January 2004)
2. Beretta, D., Quintarelli, E., Rabosio, E.: Mining context-aware preferences on relational and sensor data. In: 6th International Workshop on Flexible Database and Information System Technology (FlexDBIST 2011)in Conjunction with the 22nd International Conference on Database and Expert Systems Applications (DEXA), pp. 116–120 (2011)
3. Borzsonyi, S., Kossmann, D., Stocker, K.: The skyline operator. In: Proc. 17th International Conference on Data Engineering (ICDE 2001), Germany, pp. 412–430 (2001)
4. de Amo, S., Pereira, F.: Evaluation of conditional preference queries. In: Proceedings of the 25th Brazilian Symposium on Databases, Belo Horizonte, Brazil (October 2010); Journal of Information and Data Management (JIDM) 1(3), 521–536 (2010)
5. de Amo, S., Pereira, F.: A context-aware preference query language: Theory and implementation. Technical report, Universidade Federal de Uberlândia, School of Computing (2011)
6. Hristidis, V., Koudas, N., Papakonstantinou, Y.: Prefer: A system for the efficient execution of multi-parametric ranked queries. In: Proceedings of ACM SIGMOD International Conference on Management of Data, Santa Barbara, CA, USA, pp. 259–270 (2001)
7. Kießling, W., Köstler, G.: Preference sql - design, implementation, experiences. In: Proceedings of the Int. Conf. on Very Large Databases, pp. 990–1001 (2002)
8. Kontaki, M., Papadopoulos, A.N., Manolopoulos, Y.: Continuous Processing of Preference Queries in Data Streams. In: van Leeuwen, J., Muscholl, A., Peleg, D., Pokorný, J., Rumpe, B. (eds.) SOFSEM 2010. LNCS, vol. 5901, pp. 47–60. Springer, Heidelberg (2010)
9. Kontaki, M., Papadopoulos, A.N., Manopoulos, Y.: Continuous top k-dominating queries. Technical report, Aristotle University of Thessaloniki (2009)
10. Koutrika, G., Pitoura, E., Stefanidis, K.: Representation, composition and application of preferences in databases. In: International Conference on Data Engineering (ICDE), pp. 1214–1215 (2010)
11. Stefanidis, K., Pitoura, E.: Fast contextual preference scoring of database tuples. In: Proceedings of the International Conference on Extending Database Technology (EDBT), pp. 344–355 (2008)
12. Morse, M., Patel, J.M., Grosky, W.: Efficient continuous skyline computation. Information Sciences 177, 3411–3437 (2007)
13. Mouratidis, K., Bakiras, S., Papadias, D.: Continuous monitoring of top-k queries over sliding windows. In: Proceedings of SIGMOD, pp. 635–646 (2006)
14. Papadias, D., Tao, Y., Fu, G., Seeger, B.: Progressive skyline computation in database systems. ACM Transactions on Database Systems 30, 41–82 (2005)
15. Petit, L., Labbé, C., Roncancio, C.L.: An Algebric Window Model for Data Stream Management. In: Proceedings of the 9th International ACM Workshop on Data Engineering for Wireless and Mobile Access, pp. 17–24. ACM (2010)
16. Petit, L., Labbé, C., Roncancio, C.L.: Revisiting Formal Ordering in Data Stream Querying. In: Proceedings of the 2012 ACM Symposium on Applied Computing. ACM, New York (2012)
17. Wilson, N.: Extending cp-nets with stronger conditional preference statements. In: AAAI, pp. 735–741 (2004)

Fast Block-Compressed Inverted Lists

Giovanni M. Sacco

Università di Torino, Dipartimento di Informatica, Corso Svizzera 185,
10149 Torino, Italy
giovanni.sacco@unito.it

Abstract. New techniques for compressing and storing inverted lists are presented. Differently from previous research, these techniques are especially designed for volatile inverted lists and combine different types of compression (including prefix compression) with block segmentation to allow easy insertion/deletion of pointers and, most importantly, to significantly reduce execution times while keeping storage requirements close to a baseline monolithic inverted list implementation based on Elias's δ codes. Inverted lists for information retrieval are addressed and experiments are reported. The best method uses an optimized block-oriented evaluation that is able to efficiently skip irrelevant pointers and that has an observed average execution time which is less than 65% of the baseline implementation.

1 Introduction

An inverted list $<K_j, \{ p_0, ..., p_i, ..., p_{N-1} \}>$ is a data structure that stores a mapping from a key K_j (such as, for instance, a database key or a query term) to a list of one or more pointers p_i to objects in a structured, semi-structured or unstructured database. Inverted lists are used in many information technology applications. In addition to structured databases, an important application area is information retrieval [1]. In this context, the key represents a term (or, usually, a unique term identifier assigned through a lexicon) and the list enumerates all the documents which contain that term. An important, emerging area is represented by dynamic taxonomies [9, 10], where the deep extension of concepts (i.e. the set of objects classified under a concept or one of its descendants) can be represented by inverted lists.

The key in an inverted list may be of variable size. The list is usually kept ordered because this allows to perform list operations such as intersection, union, subtraction, etc. by merging, in linear time. Inverted lists are accessed through inverted indices that allow quick access to the inverted list corresponding to a search key. Inverted indices are usually stored in secondary storage, and are often organized as B-trees or variations [4]. Indices are usually organized into fixed-size pages, a page being the minimum access unit.

Usually, the inverted list itself is stored in a different secondary-storage area, managed as a heap. This requires an additional access for accessing the inverted list from the index, and is not efficient for volatile environments where the inverted list can grow or shrink. In this present paper, we assume that the inverted lists are stored in

S.W. Liddle et al. (Eds.): DEXA 2012, Part I, LNCS 7446, pp. 412–421, 2012.

the inverted index, i.e., the records inserted in the index are the inverted lists themselves, and the index keys are the inverted list keys. In this context, several problems arise. First, for large information bases or information bases with a non-uniform distribution of keys, many inverted lists may contain a very large number of pointers, and, in general, span several pages. It is quite difficult and expensive to perform insertions and deletions in these cases, if the pointers in the inverted list are to be maintained ordered so that list operations can be performed in a linear time. A main memory space equal to the size of the list is usually required, and lists that cannot be entirely stored in a single page require complex allocation strategies.

Even if list operations on sorted lists can be performed in linear time, they can be too expensive in very large applications, so that more efficient strategies are desirable. At the same time, these strategies must use effective compression techniques because, in naïve implementations, the overhead of inverted lists and indices can be so large as to be impractical.

The present paper introduces a number of different strategies for the representation of inverted lists. Its main contributions are:

- better execution times for inverted list operations. The focus is on the optimization of intersections, because they are the most frequent operations in practice. The block-oriented evaluation that we propose is generally sublinear with respect to the number of pointers in the lists;
- easy support for insertions/deletions on volatile index files, while exhibiting an acceptable overhead over standard methods.

The first strategy we introduce is the *normalized* strategy, in which each inverted list for a key K is "exploded" into its constituent pairs $<K, p_i>$, where p_i is the i-th pointer in the list and each pair is a record in the index file. This architecture is the simplest one to implement, because insertions and deletions are performed through the inverted index insertion and deletion primitives. In order to reduce the storage overhead that is implied by the replication of the key for each pointer, a prefix compression is used [12], in which the prefix in common with the previous record is not stored, but represented by its length. Prefix compression, also called front compression and explained in more detail in the following, is one of the important parts of our overall strategy.

The normalized strategy can be improved by the *interval normalized* strategy which represents an interval L of contiguous values by a single record containing the first pointer in L and the coded length of L minus 1. This strategy is especially effective for longer and consequently "denser" inverted lists. The interval normalized strategy represents a block of values through a single record, and, as we will show, exploits blocking to optimize evaluation.

Blocking is carried over to non-contiguous intervals, and *block-compressed* strategies are presented for sparse intervals. These strategies represent a sparse interval of the inverted list by a single record, which contains a triple $<K, p, s>$, where K is the key of the inverted list, p is the first pointer in the sparse interval, and s is the sparse interval coded by Elias's δ codes [6]. Block-oriented evaluation can be extended to sparse interval in order to improve evaluation. Two variants of this strategy are discussed. Finally, a *hybrid* strategy, which combines interval normalized and block-compressed strategies is presented. These strategies can easily manage insertions and deletions by a B-tree like implementation, discussed in the following.

These strategies are compared to an efficient baseline simple δ implementation which consists of monolithic inverted lists, where pointers in the list are represented by their gaps and these gaps are coded by Elias's δ codes. This method is discussed in the following.

2   Previous Research

Although inverted lists were proposed several decades ago, there has recently been a renewed research interest on their effective implementation. Storage schemes are investigated in [13, 15]. The baseline method we will use for comparison is one of the best global compression methods proposed in [13]. The method stores, for each key, a monolithic inverted list, where pointers in the list are represented by their gaps and these gaps are coded by Elias's δ codes. As an example, consider the list { 1, 13, 27 }. This list is represented by the difference (gap) between pointer p_i and the previous one p_{i-1}. For p_0, this difference is the value of the pointer itself (i.e. $p_{-1}=0$). Thus the list becomes { 1, 12, 14 } and is coded by Elias's δ codes: variable-length universal codes which use a number of bits L_x to code the integer x, where $L_x = 1+2 \lfloor \log \log 2x \rfloor + \lfloor \log x \rfloor$. The compression obtained by this method is shown to be quite good and close to the best known performance for global models, with a reasonable time required for compression and decompression. Global compression models can be outperformed by local models [13], that take into account the observed frequencies in each inverted list. These methods are not considered here, because they are difficult to apply to volatile environments.

Efficient execution schemes that use auxiliary indices are discussed in [2]. Hybrid storage strategies, that use different storage representations for inverted lists in the same index, are discussed in [13] and in [5]. Compression improvements can be obtained by document reordering [11, 14] or by using pointers to blocks of text [8].

All these proposals are based on monolithic architectures that represent the inverted list as an atomic record. Consequently, they suffer from the problems we have discussed before regarding the management of long inverted lists, and, in general, of volatile indices. Also, and most importantly, the execution of operations on lists, e. g. intersections, require that all the pointers in the lists be processed, and this can require high processing times.

3   Inverted List Representation

3.1   Normalized Inverted Lists

A normalized inverted list structure represents the inverted list <Kj, { p0, ..., pi, ..., pN-1 }> through N records containing the constituent pairs of the inverted list, i.e., <Kj, pk>, $0 \leq k \leq N-1$. This type of representation is called normalized by analogy with the First Normal Form in relational databases. A normalized representation makes insertions and deletions in an ordered inverted list trivial, because such operations are performed by the normal record insertion/deletion operations on the inverted index with no additional implementation required. In addition, very large inverted lists can be managed effortlessly through standard index operations. As we noted before, inverted lists are difficult to implement in the standard single-record architecture because they usually require a main memory space equal to the size of the list, and complex allocation strategies if the list cannot be entirely stored into a single page.

The single but quite significant disadvantage of this architecture is that the key of the inverted list is replicated for each pointer, leading to a potentially large waste of storage space. In order to avoid this overhead, a prefix compression can be used. In prefix compression, the records in each page are stored sequentially according to the key logical order, and the prefix of a record which is the same as the prefix of the previous record is not explicitly stored but factored out. Since prefix compression usually requires a larger computational effort to search for a specific key, efficient methods for locating the required key within a page were devised, but are not described here.

In order to access the entire normalized inverted list, a prefix scan is implemented. The prefix scan for a search key K' uses the index to access the first record <Kj, p0>, with Kj =K' (K' is the prefix of the record to be found). Subsequently, it performs a sequential scan fetching all the subsequent records with Kj =K', until a record with Kl >K' or the end of file is encountered. In order to simplify prefix scans and to improve compression, normalized list data (including the list key) are usually stored in such a way as to maintain lexicographic order in byte-by-byte comparisons. In particular, numeric data and pointers are usually stored in big-endian order, i.e., the most significant byte is stored first.

By compressing the common prefix of consecutive records, the overhead of the key in each pair is eliminated at the slight expense of storing the encoded common prefix. In addition, the prefix compression will usually compress a significant part of the pointer, which is stored in big-endian byte order. Finally, prefix compression will also compress the common prefix of two different consecutive keys.

Table 1. Intersection by merging

```
void MergeIntersect(Result& r)
{
  int i, j;
  int cA, cB;
  cA=m_cursors[0]->NextPtr(); cB=m_cursors[1]->NextPtr();
  while(cA>=0 && cB>=0)
  {
    if(cA==cB)
    {
      r.Add(cA);
      cA=m_cursors[0]->NextPtr(); cB=m_cursors[1]->NextPtr();
    }
    else
    if(cA<cB) cA=m_cursors[0]->NextPtr();
    else cB=m_cursors[1]->NextPtr();
  }
}
```

The implementation of list operations is by merging (see Table 1 for intersection) and it is analogous to normal inverted list operations. Only two inverted lists are considered, for simplicity; the extension to n-ary operations is straightforward.

The algorithm for intersection is written in terms of cursors. A cursor is an object used to sequentially scan a (normalized) inverted list. The cursor maintains the current position cp in the list (initially, the cursor is positioned before the first pointer, i.e. cp=-1), and implements two operations: the Next operation and the MoveTo operation. The Next operation advances the current position and returns the corresponding pointer ptr[cp] or an end-of-scan condition. The MoveTo(X) operation advances the current position until the corresponding pointer is equal to or larger than X (or an end-of-scan occurs). If X is equal to or smaller than ptr[cp], the current position is not changed.

3.2   Interval Normalized Inverted Lists

In some cases, such as higher-level concepts in dynamic taxonomies and inverted lists for very frequent terms, inverted lists are very dense and consequently there is a high probability of sequences of contiguous pointers. Although contiguous pointers are efficiently compressed by prefix compression, a different representation can achieve a better compression and, at the same time, a more efficient evaluation of list operations.

We focus on the representation of intervals of contiguous values. An interval normalized record is no longer a pair $<K_j, p_i>$ but a triple $<K_j, p_i, delta_i>$, where delta is the length of a run of contiguous pointers and consequently $delta_i=0$ if the next pointer p_{i+1} is not contiguous to p_i. Since non-uniform key distributions make one-element lists frequent, their space requirements should be minimized. By using a variable-length record and discarding the null most significant bytes in delta, we minimize the overhead of delta. The actual value of delta can be trivially reconstructed, and delta=0 requires no storage, so that normalized inverted list records are just a special case of interval list records.

In addition to a more efficient data compression, interval representation allows list operations that potentially have a sublinear complexity in the number of pointers in the list. Although the same merge implementation used for normalized lists could be used, we can derive a more efficient algorithm by exploiting the organization by intervals.

First, we extend cursors to account for intervals. Each list i has a current cursor C_i whose component $C_i.pos$ identifies the current position in the ordered list of records, and whose component $C_i.cont$ represents a range of contiguous pointers in the form <ps, pe>, where ps identifies the first pointer in the range and pe the last pointer in the range. $C_i.cont$ is set when the current record (identified by $C_i.pos$) is read and it is derived from the current record <K, p, delta>, where p is the first pointer in the range, and delta is the length of the interval of contiguous pointers, by setting ps=p and pe=ps+delta.

The Next operation can be easily supported, but it is not required by the algorithm. Instead, we focus on the MoveTo(x) operation. When the cursor is to be advanced to a value x, the following cases occur. If $x \leq C_i.cont.pe$, the cursor is not advanced, and $C_i.cont.ps$ is set to x if $C_i.cont.ps \leq x$ ($C_i.cont.ps$ is not changed otherwise).

If $x>C_i.cont.pe$, the cursor is advanced to the next record until $C_i.cont.ps\leq x\leq C_i.cont.pe$ (if $C_i.cont.ps<x$, then Ci.cont.ps is set to x). If such record does not exist, an end-of-scan condition is raised and the processing for the list is terminated.

Now, we can process intersection by focusing on contiguous pointer intervals rather than on pointer values. Consider two lists i and j and compute maxp as $max(ps_i, ps_j)$ and minp as $min(pe_i, pe_j)$.

If maxp>minp, the two intervals are disjoint. Both cursors are moved to maxp (the cursor k whose $ps_k=maxp$ is not moved by definition). No pointer value smaller than maxp needs to be considered.

If maxp $\leq$ minp, the intersection of the two lists is given by [maxp, minp]; all the values in the interval or (more efficiently) the interval itself can be output. The cursors are advanced to minp+1, i.e. to the value immediately larger than the end of intersection interval.

Table 2. Block-oriented intersection

```
void BlockIntersect(Result& r)
{
  int i, j; int minp, maxp;
  Cursor *c0=m_cursors[0], *c1=m_cursors[1];
  while(!c0->isEof() && !c1->isEof())
  {
    int end0=c0->getEnd(), start0=c0->getStart();
    int end1=c1->getEnd(), start1=c1->getStart();
    maxp=max(start0, start1);
    if(end0<end1) { minp=end0; j=0; }
    else { minp=end1; j=1; }
    if(minp>=maxp)
    {
      MergeIntersectRecords(r, maxp, minp);
      c0->MoveTo(minp+1); c1->MoveTo(minp+1);
    }
    else m_cursors[j]->MoveTo(maxp);
}

void MergeIntersectRecords(Result& r, int start, int end)
  { for(int i=start; i<=end; i++) r.Add(i); }
```

Table 2 shows the code for the block-oriented intersection of two interval lists. With respect to the algorithm for normalized lists, the complexity of the algorithm is proportional to the number of intervals (i.e. records) rather than to the number of pointers. Thus, in the worst case, the algorithm reported is as expensive as the normal merging algorithm. The worst case occurs when all intervals consist of a single value. As the number of different intervals decreases (or, equivalently, the size of each interval increases), the cost of the new algorithm improves because less irrelevant information is processed. On the one side, disjoint intervals are quickly identified in the main loop of the algorithm. On the other side, the MoveTo operation can quickly skip records that do not contain candidates for intersection. The best case occurs when each list consists of a single record (i.e., a single contiguous interval) and the intersection can be computed by processing two records, regardless of the number of pointers involved.

As regards volatile inverted lists, the interval normalized representation requires a very simple implementation. On insertion of a new pointer in a list, if the new pointer is contiguous to the previous record or to the next record, the appropriate record is updated. Otherwise, a normalized record is inserted. On deletion of a pointer, if the pointer belongs to an interval record, the interval record is updated. Unless the pointer is the first value or the last value in the interval, the interval record is split into two disjoint interval records.

3.3   Block-Compressed Inverted Lists

Interval normalized lists compress an interval of contiguous pointers into a single record, providing both storage savings and improvements in execution time. However, their effectiveness depends on the density of the inverted list they represent and can be expected to be not quite significant in text retrieval applications, as the vast majority of pointers will very likely be non-contiguous and hence represented by single-value intervals.

However, the same idea of a block representation can be used to represent an interval of non-contiguous (i.e., sparse) pointers, which is the general case for intervals. The first strategy we present, called BWO, represents a set of ordered non-contiguous pointers by encoding the gaps between them through Elias's δ codes (other codes such as Golomb codes [7] can also be used). Basically, we use here the same strategy as the baseline simple δ strategy, applying it to a single interval instead of applying it to the entire inverted list.

The use of sparse blocks requires minor changes in the algorithm reported in Table 2. When an intersection interval [maxp, minp] is found, the interval does not represent the actual result but just a candidate interval on which the intersection on actual values is to be subsequently computed. If i and j are two cursors, then

- if both cursors represent a contiguous interval, then [maxp, minp] is the result
- if one of the two cursors, say j, represents a sparse interval, then the result is given by the pointers of j in the interval [maxp, minp]
- if both cursors represent a sparse interval, then the result is given by their intersection in the interval [maxp, minp]. Such intersection can be computed by intersection by merge on the lists on the interval.

A critical problem is the determination of the highest value in the interval for records representing a sparse interval. If the interval is compressed by codes such as Elias's δ codes, in order to find the highest value in the interval one has to decompress all the pointers in the record. This is precisely what we want to avoid, because we are using a block evaluation to quickly skip over irrelevant blocks without decompressing their pointers. If indeed we do decompress them, we will have no performance improvement.

One strategy, called BW, stores the last pointer in the interval immediately after the initial pointer in the interval. This last pointer can be encoded by an Elias's δ code, but it has, by definition, a larger gap and consequently BW is slightly more expensive from the storage point of view than BWO (BWO stands for Blocked With-Out the highest pointer, BW for Blocked With the highest pointer).

In order to avoid this storage overhead without having to decompress the entire block of pointers, BWO estimates the highest pointer in the block by accessing the next record in the scan. If this record exists, then the highest pointer value is estimated as the initial pointer in the next record minus 1. Otherwise, the highest pointer value is taken to be the maximum integer value. Because of estimation errors in the intersection interval, we expect a less effective filtering of useless data and consequently, a potentially slower evaluation.

Volatile inverted lists can be easily managed by using a Btree-like strategy. We define a maximum blocksize for records storing sparse intervals. If the insertion of a pointer in a block causes an overflow, the record is split into two records, each with roughly half of the pointers, and the pointer is inserted into the appropriate record. On pointer deletion, if the resulting record is less than half of the maximum size, the record is either recombined or rebalanced with the next (or the previous) record, if they exist. If they do not exists, the entire list is smaller than the minimum blocksize, and no action is taken: the use of a variable record avoids wasting storage space.

3.4   Hybrid Inverted Lists

Interval and block representations can be used together. When inserting or updating an inverted list, a block representation will generally be used if it is more efficient, storage-wise, than a normalized or an interval normalized representation. It is easy to compare the storage requirements of these two representations with respect to block representation, and choose the best representation among the three. Performance of list operations may also be taken into account in the selection of the representation, thereby biasing the choice towards interval representations.

In this paper we use a rather simplistic hybrid representation (HYBR) that switches from a BW representation to an interval representation whenever an interval of contiguous values larger than θ is found.

4   Experiments

The strategies introduced above have obvious advantages over conventional approaches as far as volatility and the management of large lists are concerned. In order to characterize space and time behavior with respect to the baseline simple δ implementation, we conducted preliminary experiments on TEXTSET, a set of inverted lists that represents text postings for news texts in Italian. Texts are in lowercase with no stopword removal and no stemming/normalization. TEXTSET has 519,347 unique terms contained in 193,259 documents for a total of 50,115,616 postings and an average 96.5 postings per term. This is a worst case for text retrieval since Italian is a heavily inflected language, and the types of compression proposed are more effective for denser inverted lists. This set also approximates the behavior of a large secondary index in database applications.

The index files are prefix-compressed prefix B+-trees [3], organized in 2KB pages managed by a buffer manager with LRU replacement. All the experiments were run on an Intel Q9550 with 4 Gb RAM, running Microsoft XP. As in [5], the experiments were run with files resident in memory, so the comparison does not include any I/O and studies the behavior independently of available memory. The experiments are strongly biased in favor of the baseline simple δ, because it implements a direct access to the appropriate inverted list and completely bypasses the buffer manager, which contributes with a significant overhead to the other methods.

The experiments compared space and time requirements for normalized lists, interval lists, BWO, BW and HYBR. For the last three methods, a maximum blocksize of 22 was used, and, for HYBR, an interval representation was used for all those intervals whose delta was equal to or larger than 24. Results are reported in Table 3.

Table 3. Stats for TEXTSET

	Bpp (bits per posting)	Bpp over simple δ bpp	Number of records	Time over simple δ
Normalized	18.582	2.250	50,115,616	13.926
Interval	15.771	1.910	39,374,581	3.783
BWO	9.204	1.115	2,626,364	0.924
BW	9.532	1.154	2,700,444	0.674
HYBR	9.525	1.154	2,741,231	0.648
Simple δ	8.256	1.000	519,347	1.000

Our experiments show that normalized lists and interval lists are inefficient with respect to space and time, with a ratio over the baseline simple δ of 2.25 and 1.91 respectively (for space) and of 13.93 and 3.78 (for time).

The results are much more interesting for BW and HYBR. The time ratio over the baseline simple δ ranges from .65 (HYBR) to .67 (BW), showing a significant speedup over the baseline. Both methods show roughly the same storage 15.4% overhead over the baseline method, which seems to be acceptable considering the significant speedup, and the other advantages over the baseline. BWO shows the lowest storage overhead with respect to the baseline (11.5%). Although its execution time is lower than the baseline, its improvements are very limited when compared to HYBR, whose speedup is substantial.

5   Conclusions

We have proposed a number of strategies that solve the problems arising in the management of inverted lists, especially for volatile indices and for very large lists. The experiments conducted show that the best solutions are the HYBR and BW strategies. BW segments the inverted list into blocks whose maximum size is fixed, and relatively small (30-byte records were used in the experiments). Each block consists of the inverted list key, the first pointer in the block in big-endian, and a sequence of pointers whose gaps are coded by Elias's δ codes. The first pointer in the sequence is the last pointer in the interval represented by the block. Relatively small blocks make the management of even extremely large lists quite simple and independent of the available main memory. The storage requirements of this representation are considerably reduced by prefix compression, so that the overhead with respect to the baseline monolithic simple δ implementation is acceptable. At the same time, the exact knowledge of the smallest and highest pointer values in the interval, allows to optimize execution and to quickly skip over useless records and pointers within, and consequently provides a significant performance improvement over the baseline strategy. HYBR, the best strategy, refines BW by switching to an interval list representation when long chunks of contiguous values are detected.

These strategies appear to be robust, efficient and universal strategies for the management of volatile and static inverted indices. Although they were derived for volatile files, their significant performance improvements make them the strategies of choice for static files as well.

References

1. Baeza-Yates, R.A., Ribeiro-Neto, B.: Modern Information Retrieval. Addison-Wesley Longman Publishing Co., Inc., Boston (1999)
2. Baeza-Yates, R.A.: A Fast Set Intersection Algorithm for Sorted Sequences. In: Sahinalp, S.C., Muthukrishnan, S.M., Dogrusoz, U. (eds.) CPM 2004. LNCS, vol. 3109, pp. 400–408. Springer, Heidelberg (2004)
3. Bayer, R., Unterauer, K.: Prefix B-trees. ACM Trans. Database Syst. 2(1), 11–26 (1977)
4. Comer, D.: The Ubiquitous B-Tree. ACM Comput. Surv. 11(2), 121–137 (1979)
5. Culpepper, J.S., Moffat, A.: Efficient set intersection for inverted indexing. ACM Trans. Inf. 29(1) (2010)
6. Elias, P.: Universal codeword sets and representations of the integers. IEEE Trans. on Information Theory IT-21(2), 194–203 (1975)
7. Golomb, S.W.: Run-length encodings. IEEE Trans. Info Theory 12(3), 399–401 (1966)
8. Navarro, G., de Moura, S.E., Neubert, M., Ziviani, N., Baeza-Yates, R.: Adding Compression to Block Addressing Inverted Indexes. Information Retrieval 3(1), 49–77 (2000)
9. Sacco, G.M.: Dynamic Taxonomies: A Model for Large Information Bases. IEEE Trans. on Knowl. and Data Eng. 12(3), 468–479 (2000)
10. Sacco, G.M., Tzitzikas, Y. (eds.): Dynamic Taxonomies and Faceted Search: Theory, Practice, and Experience. The Information Retrieval Series, vol. 25. Springer (2009)
11. Scholer, F., Williams, H.E., Yiannis, J., Zobel, J.: Compression of inverted indexes for fast query evaluation. In: Proc. ACM SIGIR Conf. (SIGIR 2002), pp. 222–229 (2002)
12. Wagner, R.: Indexing design considerations. IBM Syst. J., 351-367 (1973)
13. Witten, I.H., Moffat, A., Bell, T.C.: Managing Gigabytes: Compressing and Indexing Documents and Images. Morgan Kaufmann Publishers Inc., San Francisco (1999)
14. Yan, H., Ding, S., Suel, T.: Inverted index compression and query processing with optimized document ordering. In: Proc. Conf. on World Wide Web (WWW 2009), pp. 401–410 (2009)
15. Zobel, J., Moffat, A.: Inverted files for text search engines. ACM Comp. Surv. 38(2) (2006)

Positional Data Organization
and Compression in Web Inverted Indexes

Leonidas Akritidis and Panayiotis Bozanis

Department of Computer & Communication Engineering,
University of Thessaly, Volos, Greece

Abstract. To sustain the tremendous workloads they suffer on a daily
basis, Web search engines employ highly compressed data structures
known as inverted indexes. Previous works demonstrated that organizing the inverted lists of the index in individual blocks of postings leads
to significant efficiency improvements. Moreover, the recent literature
has shown that the current state-of-the-art compression strategies such
as PForDelta and VSEncoding perform well when used to encode the
lists docIDs. In this paper we examine their performance when used to
compress the positional values. We expose their drawbacks and we introduce PFBC, a simple yet efficient encoding scheme, which encodes
the positional data of an inverted list block by using a fixed number of
bits. PFBC allows direct access to the required data by avoiding costly
look-ups and unnecessary information decoding, achieving several times
faster positions decompression than the state-of-the-art approaches.

1 Introduction

Due to the critical importance of the inverted index organization in the overall efficiency of a search engine, a significant part of IR research is conducted
towards the determination of an effective index setup strategy. In particular,
several works proposed methodologies for storing the index data in a special
manner which allows us to skip large portions of the lists during query processing. These approaches suggest partitioning the inverted lists of the index in a
number of adjacent blocks which can be individually accessed and decompressed.
Undoubtedly, the omission of the unnecessary information stored within the index significantly accelerates the evaluation of a query, since the lists traversal is
faster and we also decompress less data.

The benefits of these methods are magnified in the case where we store positional data within the index. This is due to the fact that the size of the positions
is several times larger than that of docIDs and frequencies and the indexes containing positional values are about 3 to 5 times larger than the non-positional
ones. Therefore, it is extremely important to devise an effective mechanism to
organize and compress the positional data, since a naive solution could lead to
prohibitively large indexes and reduced query throughput.

In this work we demonstrate that although the current block compression
methods are both effective and efficient when applied at docIDs and frequencies,
they do not perform equally well when they operate upon the positional data of

S.W. Liddle et al. (Eds.): DEXA 2012, Part I, LNCS 7446, pp. 422–429, 2012.

an inverted list. We introduce PFBC, a scheme which encodes the positions of an inverted list block by using a fixed number of bits allowing us to a) access the required data almost instantly and b) decode only the data actually needed, without touching any unnecessary information. We demonstrate that with a small cost in space, PFBC outperforms all the adversary compression methods in terms of speed, when applied on the positional data of the index.

2   Background and Related Work

In this Section we provide some elementary information about the inverted index organization and compression and we present the most significant related work.

Due to the huge volumes of text and the great length of the inverted lists, Web search engines store their inverted indexes in highly compressed forms either in main memory or disk. There is a multitudinous family of compression algorithms which can be used to encode the index data. The interested reader can refer to [2], [9] and [11] for overviews and performance benchmarks. Some of the recently proposed state-of-the-art schemes such as PForDelta [7] and VSEncoding [9] are capable of encoding entire bundles of integers achieving both satisfying compression effectiveness and very high decompression speeds.

One of the most important goals when designing query processing algorithms is to *skip* any unnecessary information stored within the inverted lists. For this reason, [8] introduced the block-based index organization, which suggests partitioning each list in blocks of fixed or variable sizes. Within each block, the data is organized in chunks which respectively accommodate the docIDs, the frequency values, and the positional data. Other optimized block-based organizations were introduced in [1], [3], [5].

Although the issue of inverted list partitioning is well-studied, the matter of the organization of the positional data is still open. The research that has been conducted towards this problem resulted into two basic approaches: (a) interleaving, i.e. the positional data belonging to a particular block is stored sequentially after the docIDs and the frequency values, and (b) creating a completely separate structure for positions with its own lookup mechanism. For instance, [12] describe a tree-like look-up structure which operates on interleaved positional data. On the other hand, [10] organize the positions by employing a separate structure, namely indexed list. The problems of these methods are (a) they require increased storage, (b) they decelerate query processing due to the look-up operations and (c) they decode redundant information.

3   The Positions Fixed-Bit Compression (PFBC)

In this Section we describe PFBC, a simple, yet efficient approach for encoding and organizing the positional data of an inverted list.

Our analysis begins by considering the block-based list organization of Figure 1. Suppose that the inverted list I_t of a term t is partitioned into B_{I_t} blocks and each block $B_i \in B_{I_t}$ is comprised of S_{B_i} postings. In the sequel, we identify

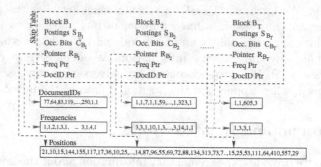

Fig. 1. Organizing an inverted list into blocks according to PFBC

the highest positional value $|p_{B_i}|_{max}$ for each block B_i of the inverted list and we allocate a number of $C_{B_i} = \lceil log_2(|p_{B_i}|_{max} - 1) \rceil$ bits to produce a binary representation of each occurrence in that block. A pseudocode demonstrating how PFBC encodes a bundle of K positional values is presented in Algorithm 1.

The fixed bit compression methodology of PFBC is expected to introduce some compression loss in comparison to PForDelta. Actually, the latter encodes the largest integers of a list as exceptions and the rest of them by using a fixed-bit scheme, similar to the one we described. This operation is proved to be very effective in the case of docIDs, because in the docIDs blocks the number of large integers is small. However, when P4D is applied at blocks of positional data, the benefits are diminished because in such blocks the number of large integers cannot be predicted. Indeed, as we demonstrate by our experiments, PFBC is outperformed by P4D in terms of compressed sizes by only a small margin.

3.1   Accessing and Decompressing the Positional Data with PFBC

To achieve direct access to the compressed positional data, it is required that we store two values for each block B_i of the list: (a) the aforementioned C_{B_i} value which denotes the number of bits we used to encode the positions of the block B_i, and (b) a pointer R_{B_i} pointing at the beginning of the positional data of B_i. Exploiting this limited amount of information, we are able to calculate the location of the positional data for any posting belonging to B_i. The following equation provides the exact bit $\mathcal{S}_j$ where the positions of a posting j start from:

$$\mathcal{S}_j = R_{B_i} + C_{B_i} \sum_{x=0}^{j-1} f_{x,B_i} \tag{1}$$

where f_{x,B_i} is the xth frequency value stored within B_i. Consequently, to locate the positional data for an arbitrary posting j we first need to dereference the corresponding R_{B_i} pointer value. Then, we need to sum up all the frequency values of the previous postings of the block; this sum reveals the number of the positional values stored between the beginning of the block and the location of the desired data. Since the compressed positions are stored by using a fixed

Algorithm 1. Encoding a bundle of K positional values with PFBC. After the identification of the highest positional value (steps 3-8), we calculate C, which is the number of bits required to encode all K integers (step 9). The function $write()$ in step 13 is used to store each p_i value into a storage $\mathcal{P}$ by using C bits.

---

byte $PFBC - Encode(K, p[K])$

1. **int** $i \leftarrow 0, p_{max} \leftarrow 0, C \leftarrow 0$
2. **byte** $\mathcal{P}$
3. **while** $(i < K)$ {
4.   **if** $(p_i > P_{max})$ {
5. $p_{max} = p_i$
6.   }
7. $i++$
8. }
9. $C \leftarrow \lceil \log_2(p_{max} - 1) \rceil$
10. $\mathcal{P} \leftarrow$ **allocate** $\lceil KC/8 \rceil$ bytes
11. $i \leftarrow 0$
12. **while** $(i < K)$ {
13. $write(p_i, \mathcal{P}, C)$
14. $i++$
15. }
16. **return** $\mathcal{P}$

---

number of bits, we just need to multiply the sum by C_{B_i} to locate the first compressed position of the posting. The operation ends by decoding the next $f_{j,B_i} C_{B_i}$ bits and the positions are retrieved.

PFBC exhibits a wide range of advantages over the adversary approaches:

- It facilitates direct access to the positional data by using equation 1. No expensive look-ups for positions in tree-like structures are required. Consequently, query processing is accelerated;
- It saves the space cost of maintaining a separate look-up structure [12], since the involved pointers can be stored within the skip table;
- It uses fewer pointers than the indexed lists of Transier and Sanders [10];
- It enables decoding of the information actually needed, without the need to decompress entire blocks or sub-blocks of integers.

The R_{B_i} and C_{B_i} values are stored within the the skip structure (upper part of Figure 1); for each entry of the skip table, we also record these two values. This strategy is both effective and efficient; no extra space is required, but only the room occupied by the values themselves. Furthermore, in case the query processor decides that a posting belonging to a particular block should be exhaustively evaluated by decoding its corresponding positional data, we are able to immediately access R_{B_i} and C_{B_i}.

Algorithm 2 includes a pseudocode which demonstrates how PFBC is used to decompress the positional data for a specific posting. To access and decode the positions for the jth posting of the block B_i, we initially accumulate all the

---

Algorithm 2. Decoding the positional data of the jth posting of the block B_i.

int $PFBC - Decode(j, B_i, \mathcal{P})$

1. **int** $x \leftarrow 0, s \leftarrow 0$
2. **while** $(x < j)$ {
3. $s \leftarrow s + f_{x,B_i}$
4. $x + +$
5. }
6. **int** $start \leftarrow R_{B_i} + sC_{B_i}$
7. $x \leftarrow 0$
8. **while** $(x < f_{j,B_i})$ {
9. $p[x] \leftarrow read(\mathcal{P}, C_{B_i}, start)$
10. $start \leftarrow start + C_{B_i}$
11. $x + +$
12. }
13. **return** p

---

frequency values of the previous $j - 1$ postings of the block (steps 2–5). In the sequel, we read the R_{B_i} and C_{B_i} values from the skip table and we locate the required data as indicated by Equation 1. If f_{j,B_i} is the associated frequency value of the jth posting, we sequentially read f_{j,B_i} groups of C_{B_i} bits from the compressed sequence; each group represents a positional value of this posting.

4   Experiments

In this Section we compare PFBC against the state-of-the-art compression methods. More specifically, we created three inverted indexes having their docIDs and frequencies organized in the same manner. Each inverted list was split into blocks of 128 postings and P4D was used to encode the DocIDs and the frequencies. In the first index we applied the strategy proposed by [12], i.e. the positions are compressed with OptP4D and accessed by using a separate look-up structure. In the second case, the positions are encoded according to VSEncoding [9] and accessed in a way identical to the one we apply at the OptP4D case (that is, we set pointers every 128 compressed positions).

The sample document collection we employed in our experiments is the Clueweb09-T09B data set, which consists of about 50 million pages. To simulate a real-world search engine environment, the document collection was split up into ten separate segments (called *shards*) consisting of about 5 million documents and each segment was indexed separately [6].

4.1   Compressed Index Sizes

Now let us evaluate the performance of PFBC against the adversary state-of-the-art approaches in terms of compression effectiveness. Apart from the size of the inverted file, we also measure the space occupied by the accompanying data structures (i.e. skip table, pointers to positions, and position look-up structure).

Table 1. Overall space requirements (in GB) of our experimental index setups

Data Structure	OptP4D	VSEncoding	PFBC
Inverted Index	90.8	90.2	92.0
Skip Table	1.7	1.7	1.7
Pointers to positions	-	-	2.1
Positions look-up	4.1	4.1	-
Total	96.6	96.0	95.8

In Table 1 we record the overall space requirements of each index setup that we examine. Notice that each organization approach does not make use of all data structures. For instance, our PFBC approach does not require the existence of a positions look-up structure, whereas all strategies employ a skip table. The absence of a data structure is denoted by using a dash symbol.

Among our examined encoding algorithms, VSEncoding achieved the best compression performance; the ten inverted files of all shards occupy in total roughly 90.2 GB. On the other hand, OptP4D performed imperceptibly worse resulting in an inverted file which occupied less than 1% more space. As we anticipated, the usage of PFBC introduced some slight losses; Compared to VSEncoding, the inverted files of PFBC occupied in total about 2% more space.

In Table 1 we report the sizes of the auxiliary data structures. The skip table which includes the positional pointers (recall the upper part of Figure 1) is much more economic than the look-up structures themselves. As a matter of fact this data structure occupies approximately 65% of the space occupied by the data structures of the OptP4D and VSEncoding approaches (skip table plus the positions look-up structure, 5.8 GB). In the last row of Table 3 we present the overall index sizes (inverted file plus auxiliary data structures) for each of the examined schemes. In conclusion, we notice that the superiority of OptP4D and VSEncoding over PFBC in terms of compressed sizes is compensated by the significantly smaller data structures that accompany our proposed index scheme. As a result, PFBC presents marginal savings of 0.02-0.08%.

4.2   Query Throughput

In this Subsection we examine the performance of PFBC against the adversary approaches in terms of speed during query processing. To perform this experiment, we submitted a set of 50 conjunctive queries drawn from the Web Adhoc Task of the TREC-2009 Web Track. For each query we measure several statistics such as the decompression times and the size of the accessed data.

The submitted queries were answered by employing a two-stage processing method: During the first phase we traverse the inverted lists of the query terms by employing DAAT, and we quickly identify the most relevant results by accessing docIDs and frequency values only. In the second phase we apply more complex ranking schemes such as BM25TP [4] to the K best results determined in the previous stage by retrieving the positional values. We experimented with two values of K; the first one is $K = 200$ and was selected because in [12] the

Table 2. Access and decode times per query and per posting for different values of K

K		OptP4D	VSEncoding	PFBC
	Decompressed positions	2,756,128	2,756,128	374,251
	Decompressed positions/query	55,123	55,123	7,485
$K = 200$	Total access time (msec)	0.11	0.11	0
	Total decompression time (msec)	5.56	5.14	1.01
	Average time per query (msec)	0.11	0.10	0.02
	Decompressed positions	11,192,608	11,192,608	1,044,234
	Decompressed positions/query	223,852	223,852	20,885
$K = 1000$	Total access time (msec)	0.31	0.31	0
	Total decompression time (msec)	23.94	21.10	4.02
	Average time per query (msec)	0.48	0.42	0.08

authors prove that higher values do not lead to any further precision gains. Furthermore, since the major Web search engines return at most 1000 results, we also choose to set $K = 1000$. Since the inverted indexes we constructed were comprised of ten shards, we repeated our experiments ten times; each time the query processor was assigned a different index shard. The results of our experiments are illustrated in Table 2.

Table 2 is divided in two parts; the upper part contains the results we recorded for $K = 200$, whereas the lower one includes the results for $K = 1000$. The first line represents the total number of positional values accessed by each method for all the ten index shards, whereas the second line shows the number of the decompressed positions per query. PFBC outperforms the adversary approaches by a significant margin, since the fixed-bit compression scheme allows us to locate exactly the data we need to access and we do not have to decode entire blocks of integers. In total, the organization method with the look-up structure employed by OptP4D and VSEncoding touched 7.4 times more data than the one applied by PFBC for $K = 200$. In case we set $K = 1000$, PFBC is even more efficient, since the other methods decode about 10.7 times more data.

The next line reveals the average position look-up times consumed by each method per query. The algorithm of Table 2 allows PFBC to calculate the location of the positional data without searching for it, consequently, the latency is nullified in this case. Regarding the other two approaches which employ the aforementioned look-up structure, they introduce a latency of about 0.11 msec per query for $K = 200$ and 0.31 msec for $K = 1000$.

Now let us examine the decompression rates achieved by each method. The lines 3 and 7 of Table 2 include the total amount of time required to decode the positional values for all the 50 queries of our experiment. Furthermore, lines 4 and 8 reveal the average decompression time per query. On average, VSEncoding outperformed the OptP4D approach by a margin ranging between 1% and 2% for different values of K. PFBC was the fastest among the evaluated schemes, since it achieved about 5 times faster decompression compared to VSEncoding for both settings of K.

5 Conclusion

In this paper we introduced PFBC, a method especially designed for organizing and compressing the positional data in Web inverted indexes. PFBC operates by employing a fixed number of bits to encode the positions of each inverted list block, and stores a limited number of pointers which enable the direct retrieval of the positional data of a particular posting. Compared to the current state-of-the-art compression techniques, PFBC offers improved efficiency allowing direct access without look-ups, and very fast decompression. The experiments we have performed on a 50 million document collection demonstrated that in contrast to OptP4D and VSEncoding, the proposed approach touches much fewer data and allows about 5 times faster positions decompression.

References

1. Anh, V.N., Moffat, A.: Structured Index Organizations for High-Throughput Text Querying. In: Crestani, F., Ferragina, P., Sanderson, M. (eds.) SPIRE 2006. LNCS, vol. 4209, pp. 304–315. Springer, Heidelberg (2006)
2. Anh, V., Moffat, A.: Index compression using 64-bit words. Software: Practice and Experience 40(2), 131–147 (2010)
3. Boldi, P., Vigna, S.: Compressed Perfect Embedded Skip Lists for Quick Inverted-Index Lookups. In: Consens, M.P., Navarro, G. (eds.) SPIRE 2005. LNCS, vol. 3772, pp. 25–28. Springer, Heidelberg (2005)
4. Buttcher, S., Clarke, C., Lushman, B.: Term proximity scoring for ad-hoc retrieval on very large text collections. In: Proceedings of the 29th Annual International ACM SIGIR Conference on Research and Development in Information Retrieval, pp. 621–622 (2006)
5. Chierichetti, F., Kumar, R., Raghavan, P.: Compressed web indexes. In: Proceedings of the 18th International Conference on World Wide Web, pp. 451–460 (2009)
6. Dean, J.: Challenges in building large-scale information retrieval systems: invited talk. In: Proceedings of the Second ACM International Conference on Web Search and Data Mining, p. 1 (2009)
7. Heman, S.: Super-Scalar Database Compression between RAM and CPU Cache. Master's Thesis. University of Amsterdam. Amsterdam, The Netherlands (2005)
8. Moffat, A., Zobel, J.: Self-indexing inverted files for fast text retrieval. ACM Transactions on Information Systems (TOIS) 14(4), 349–379 (1996)
9. Silvestri, F., Venturini, R.: Vsencoding: efficient coding and fast decoding of integer lists via dynamic programming. In: Proceedings of the 19th ACM International Conference on Information and Knowledge Management, pp. 1219–1228 (2010)
10. Transier, F., Sanders, P.: Engineering basic algorithms of an in-memory text search engine. ACM Transactions on Information Systems (TOIS) 29(1), 2 (2010)
11. Witten, I., Moffat, A., Bell, T.: Managing Gigabytes: Compressing and Indexing Documents and Images (1999)
12. Yan, H., Ding, S., Suel, T.: Compressing term positions in web indexes. In: Proceedings of the 32nd International ACM SIGIR Conference on Research and Development in Information Retrieval, pp. 147–154 (2009)

Decreasing Memory Footprints
for Better Enterprise Java Application Performance

Stoyan Garbatov and João Cachopo

INESC-id,
Rua Alves Redol, 9,
1000-029 Lisboa,
Portugal
`stoyangarbatov@gmail.com`, `joao.cachopo@ist.utl.pt`

Abstract. In this paper, we present a work for reducing the memory footprint of enterprise Java applications. The work relies on the predictions provided by stochastic models of the applications' data-access patterns. The models, built during the execution of the application, are used both at compile-time, to control the in-memory representation of data, and, at runtime, to decide which portions of the data to load. The combined effect of these two approaches allows for an effective reduction in the memory used by the application, leading to a significant performance improvement. We evaluate the newly developed approaches on the TPC-W benchmark, with different database sizes, and show that our solution increases the benchmark throughput by 10.78% on average, with a maximum of 35.43% when operating over larger databases.

Keywords: heap management, in-memory object representation, persistence.

1 Introduction

The memory footprint of an application may exert significant influence over its performance. This is particularly true for above-than-average sized footprints (e.g., enterprise applications), where performance can suffer severely due to the extra overhead caused by memory management.

In execution environments with automatic garbage collection, such as the Java VM, the garbage collection (GC) mechanism is responsible for identifying data that is no longer accessible and for de-allocating the associated memory resources. Modern generational collectors do this very efficiently for objects that become garbage soon after they are created. On the other hand, objects that survive repeated GC cycles are moved into the tenured region, where they remain for potentially long periods of time.

This GC behaviour is well known by enterprise application and framework programmers, who rely on mechanisms such as SoftReferences to influence the behaviour of the GC to their advantage. The key idea is that objects representing the application's domain data should be kept around, for as long as possible, to avoid the costly operation of loading them. Therefore, these objects should be tenured, whereas other transient objects should be GCed as soon as possible.

S.W. Liddle et al. (Eds.): DEXA 2012, Part I, LNCS 7446, pp. 430–437, 2012.

A common approach for doing this is to keep domain objects within softly-referenced caches. This approach often leads to a very large tenured region that, when full, needs to be GCed, introducing significant overhead into the application. In part, this is due to the need to scan large volumes of data to identify reachable objects. When an application's frequently used objects do not fit in memory, this can become a problem, since the tenured region has to be GCed frequently. In that case, the application is forced to be constantly rotating objects in and out of memory, spending time not only in memory management, but also in the loading of previously GCed objects.

To avoid, or at least minimize this problem, we propose to reduce the memory footprint of an enterprise application's domain objects, so that more instances may fit in the available space. We present two complementary approaches for doing this.

The first approach consists in identifying the most compact and efficient in-memory representation of domain objects. The second approach consists in delaying the loading of data that is not likely to be accessed by the application soon. Both approaches make informed decisions based on stochastic models of the applications' data access patterns that are built during the execution of the application.

To the best of our knowledge, for the purpose of improving an application's performance by keeping its heap smaller and by avoiding and delaying situations demanding intense GC activity, the work we present here is unique. The novelty is both on how we identify the sub-sets of domain objects that have the highest influence on a given application's execution, and on how we use that information to achieve the memory footprint reduction. There are other research publications with similar aims, but their techniques are distinct from those presented in this paper. The contribution of our work consists in the development and evaluation of these two new approaches.

The proper management of an application's memory is essential if any sort of acceptable performance is to be expected. Some research concentrates on analysing the conditions under which data is allocated and manipulated in the heap. The goal is to devise ways for improving the memory management of the data.

Jones and Ryder [8] study the lifetime of Java objects. They demonstrate that the lifetimes of allocated objects belong to a relatively low number of small ranges along with the fact that allocation sites can be strongly grouped according to the length of the lifetime of their allocated objects. Objects associated with identically grouped allocation sites tend to live only in particular phases of an application's operation. The authors conclude that to predict with precision an object lifetime's distribution, one additional stack level should be considered, apart from the allocation site itself.

Chis et al. [2] present a solution for identifying problems leading to unreasonably sized heaps in large applications. They use an innovative ContainerOrContained relation to identify patterns within the heap caused by problems with high impact on memory consumption. Chis et al. establish a relatively small set of patterns that need to be examined to achieve significant memory consumption improvements.

Bhattacharya et al. [1] introduce a new approach for minimizing the performance loss due to memory bloat caused by excessive generation of temporary objects within a loop. The authors identify objects that can be reused and apply a source-code transformation for making the necessary modifications to reuse the objects in a more efficient manner. The practical results of this approach indicate it is possible to decrease an application's memory footprint along with improving its performance.

Based on the present analysis, a few relevant points need to be emphasised. Overly large heaps are capable of causing serious application performance issues. Two main approach styles for dealing with this issue have been identified. One of these consists in attempting to keep the memory footprints small by preventing the loading of unnecessary data or by delaying its loading to the moment when it is effectively needed. The second style includes techniques for keeping the heap trimmed and compact by de-allocating no-longer-necessary data and recycling its memory. Several aspects, essential for the efficient operation of both styles, include performing automatic analysis of target behaviour and its precise prediction to allow the dynamic and adaptive identification of the most appropriate approached to be employed.

2   System Description

In this paper, we present two approaches for reducing the memory footprint of a Java application's domain data, while, at the same time, improving performance when the application is placed under sub-optimal memory availability conditions.

The first technique uses a carefully selected in-memory layout for instances of domain classes. This is done to achieve a compact representation with minimal memory overhead. The second approach consists in delaying the loading of domain data until the moment when it is effectively needed, as opposed to eager loading. This strategy seeks to decrease the upper limit of the effectively used heap, by avoiding the loading of domain data that is never to be accessed during the lifetime of the application.

Both techniques make informed decisions regarding the actions that are to be taken. These decisions are based on predictions made by stochastic models of the application's data access patterns. A comprehensive description and discussion of these models, which have been developed previously, can be found in [7, 5, 4, 6]. To make the article self-contained, the key ideas of these works shall be presented next.

The access pattern analysis approaches were developed for providing high precision answers to questions along the lines of "What (domain) data is accessed in execution context X and Y?", "What is the likelihood of accessing data A in context Z?" and "What is the access probability for all domain data types in the application?". The concept of "execution context" designates the scope within which data accesses occur (for instance during the execution of a method or a service).

To collect a representative volume of statistical data of a target application's access patterns, a short training period is necessary. During it, the application runs with additional code that has been injected automatically. This code is responsible for gathering the data necessary for the access pattern analysis. The overhead incurred by the execution of the injected code has been demonstrated to be sufficiently low to allow the measurements to be performed while the target system is operating normally.

The stochastic models employed for the behaviour analysis are Bayesian Updating, Markov Chains and Importance Analysis [7, 5, 4, 6]. Their implementations have been evaluated and demonstrated to generate correct and highly precise predictions about the effectively accessed domain data throughout the execution contexts of a target application. In the context of the work presented here, these methods are responsible for supplying the access probabilities of all domain classes and their fields.

The work was performed in the context of the Fénix Framework [3]. The Fénix Framework allows the development of Java applications with a transactional and persistent domain model. Programmers using the framework describe the application's domain model structure by using a new domain-specific language, the Domain Modelling Language (DML). After this, they can develop the rest of the application in plain Java, without any further considerations. The transactional and persistence features provided by the framework are orthogonal. For the current work, the transactional functionality shall be disregarded, since it is not relevant to the goals at hand.

A source-to-source compiler is responsible for transforming the DML specification into the domain classes' Java definitions. This compiler selects the particular layout that domain objects have at runtime.

Before describing the new domain object layout, the two previously existing schemas will be considered. The new approach is referred to as DynamicL, while the two previous layouts are labelled as OneBoxPerSlotL and OneBoxPerObjectL.

An important concept when discussing object layouts is that of a "Box". A Box is responsible, among other things, for holding the persistent state of domain objects, as well as for loading it from persistence, whenever it is needed.

A sample of a domain class with the OneBoxPerSlotL can be seen in Listing 1 (left). Every object attribute is contained by a single Box. At runtime, when a domain object is referenced for the first time, it is initialized as a thin wrapper containing the ObjectId (a unique system-wide identification of that instance). The first time that any of an object's attributes are accessed, all Boxes are initialized, followed by a loading of the associated state, performed in a single round-trip to the persistence layer.

```
public class domainObject{              public class domainObject {
  private box<FieldType1> field1;          private ObjState objState;
  ...                                       protected static class ObjState
  private box<FieldTypeN> fieldN;             extends Box{
  public FieldType1 getField1(){              FieldType1 field1;
    return field1.get(); }                    ...
  public void setField1(FieldType1 val){      FieldTypeN fieldN; }
    field1.set(val); }                     public FieldType1 getField1(){
  ...                                         return objState.field1; }
  public FieldType1 getFieldN(){           public void setField1(FieldType1 val){
    return fieldN.get(); }                   objState.field1=val; }
  public void setFieldN(FieldTypeN val){   ...
    fieldN.set(val); }}                  }
```

Listing 1: Domain class with OneBoxPerSlotL (left), OneBoxPerObjectL (right)

This leads to the following properties. The application has a large memory footprint, due to the memory overhead caused by the existence of one Box per object attribute. The eager loading of objects from persistence shortens the warm-up phase of the application because, most, if not all, of the existing domain data ends up in memory after a relatively small portion of it has been accessed. This leads to situations where large volumes of data are unnecessarily loaded and kept in memory.

A sample of a domain object with the OneBoxPerObjectL layout can be seen in Listing 1 (right). All of the persistent state of any domain object is contained by a single Box. The behaviour when referencing an object or accessing any of its attributes for the first time is identical to that of the OneBoxPerSlotL. The short warm-up phases that load most of the existing domain data are present as well. However, even though unnecessary data ends up being kept in memory, the memory footprints of applications employing this layout are significantly smaller due to the lack of memory overhead caused by multiple Boxes per object instance. In terms of the compactness of fully loaded domain objects, this approach corresponds to the optimal solution.

By analysing these two object layout styles, a few pertinent points may be identified. One such aspect is that if a compact object memory representation is to be achieved, then each domain class should have the lowest possible number of Boxes to minimize the induced memory overhead. Another issue is that both approaches load significant volumes of persistent domain data into memory, even though most of it might not end up being necessary for the proper operation of the application.

It is with these considerations that the new DynamicL layout was created. The approach is adaptive inasmuch as decisions about the actual layout configuration of domain objects are made at compile time, when taking into account the data access pattern information provided by the stochastic analysis modules. As such, it is possible that domain classes have different layouts, from one deployment of the application to another, due to changes in the access patterns performed at runtime.

The code generation with the DynamicL layout proceeds as follows. For every domain class, its attributes are split among two sets, based on their access probabilities. High access probability fields are placed in one set (HighP), while low access probability fields are placed in the second set (LowP). Once this has been accomplished, the code generator outputs different code, depending on the set to which a given object field belongs to. HighP attributes are given an individual Box to wrap them and to take care of their persistent loading. On the other hand, all LowP attributes, of a given domain class, are placed in a single Box.

Even though HighP fields are assigned individual Boxes, the overall memory overhead caused by these boxes, when compared against the OneBoxPerObjectL approach is negligible. This is because, in most applications, only a small fraction of data is responsible for the great majority of accesses performed. It is rare to have a domain class with more than a couple of HighP attributes, making this approach close to the optimal solution, with regards to the number of Boxes per instance.

For the DynamicL approach, when a domain class instance is referenced for the first time, the procedure is identical to the other approaches. The difference is when persistent fields whose values have not been loaded yet are accessed. Whenever this occurs, the associated Box is initialized, followed by a roundtrip to the persistence layer to fetch the data needed to load the proper values of the wrapped attributes. Only the attribute(s) contained by the Box that has been accessed are loaded from persistence, regardless of the state in which the remainder of instance Boxes might be. This particular measure was taken to guarantee that only data that is effectively needed by the application is loaded into memory, as opposed to loading a piece of unnecessary data simply because it belongs to an object being accessed.

The delaying of persistent data loading leads to a smaller memory footprint, because most of the data that is never accessed at runtime does not end up in memory. It is still possible for unnecessary data to be loaded, but this is rather unlikely to happen.

It should be noted that a lazy version of the OneBoxPerSlotL is not a viable solution. The overhead of a single box per field cannot be offset by delaying/avoiding the loading of low access probability data. The resulting footprint would be between that of the eager OneBoxPerSlotL and OneBoxPerObjectL.

3   Results

The TPC-W benchmark [9] was used for the validation of the work presented here. It specifies an e-commerce workload that simulates the activities of a retail store website, where emulated users can browse and order products from the website.

The main evaluation metric is the WIPS – web interactions per second that can be sustained by the system under test. The benchmark execution is characterised by a series of input parameters. The first of these indicates the type of workload, which varies the percentage of read and write operations, that is to be simulated by the emulated browser (EB) clients. Three types of workload are considered here, namely - Read-Only (Mix0) - 100% read operations; Browsing (Mix1) - 95% read and 5% write operations and Shopping (Mix2) - 80% read and 20% write operations.

The training of the system was performed with the benchmark executing in Mix2 mode. The same profiling results (from Mix2) were employed for all 3 workload configurations (Mix0, Mix1, Mix2) in the performance testing phase.

The remaining input parameters are as follows: number of EBs - 10; ramp-up time - 300 sec; measurement - 1200 sec; ramp-down time - 120 sec; number of book items in the database - 1k, 10k and 100k; think time - 0, ensuring that the EBs do not wait before making a new request. All results were obtained as the average of 4 independent executions of the benchmark, with the same configurations. The EBs and the benchmark server were always run on the same physical machine.

Table 1. Performance (throughput) comparison, Alpha

	512MB	640MB	768MB	896MB	1024MB	2048MB	Average
mix0_1k	1.28%	0.63%	1.27%	1.74%	1.09%	1.24%	**1.21%**
mix0_10k	-4.22%	-3.50%	-5.27%	-5.78%	-5.63%	-5.47%	**-4.98%**
mix0_100k	34.52%	32.90%	31.92%	33.62%	35.12%	32.93%	**33.50%**
mix1_1k	6.86%	7.55%	0.39%	5.59%	0.70%	5.33%	**4.40%**
mix1_10k	-7.98%	-16.15%	-11.09%	-15.32%	-15.88%	-15.42%	**-13.64%**
mix1_100k	24.06%	18.55%	19.91%	25.73%	24.84%	33.55%	**24.44%**
mix2_1k	21.81%	19.46%	17.13%	8.41%	14.87%	4.42%	**14.35%**
mix2_10k	4.34%	16.32%	9.44%	9.34%	11.62%	4.05%	**9.19%**
mix2_100k	29.35%	26.53%	19.07%	30.88%	29.55%	30.32%	**27.62%**
Average	**12.23%**	**11.36%**	**9.20%**	**10.47%**	**10.70%**	**10.11%**	**10.68%**

Performance measurements were made with the benchmark running on two different machines. The first machine (Alpha) is equipped with 2x Intel Xeon E5520 (a total of 8 physical cores with hyper-threading running at 2.26 GHz) and 24 GB of RAM. Alpha's operating system is Ubuntu 10.04.3, while the JVM is Java(TM) SE Runtime Environment (build 1.6.0 22-b04), Java HotSpot(TM) 64-Bit ServerVM (build 17.1-b03, mixed mode). The second machine (Beta) has 4x AMD Opteron 6168 (a total of 48 physical cores running at 1900 MHz) and 128 GB of RAM. Beta's operating system is Red Hat Enterprise Linux 6.2, while the JVM is Java(TM) SE Runtime Environment (build 1.6.0 24-b07), Java HotSpot(TM) 64-Bit ServerVM (build 19.1-b02, mixed mode). On both machines, the benchmark server was run on top of Apache Tomcat 6.0.24, with the options "-Xshare:off -Xms64m -Xmx${heapSize}m -server -XX:+UseConcMarkSweepGC -XX:+AggressiveOpts".

Table 2. Performance (throughput) comparison, Beta

	512MB	640MB	768MB	896MB	1024MB	2048MB	Average
mix0_1k	-0.26%	-0.50%	0.72%	1.66%	0.97%	0.32%	**0.49%**
mix0_10k	3.69%	2.09%	2.87%	3.36%	4.02%	4.41%	**3.41%**
mix0_100k	31.43%	32.79%	33.45%	32.49%	32.68%	33.19%	**32.67%**
mix1_1k	0.76%	1.61%	3.48%	2.24%	1.69%	0.97%	**1.79%**
mix1_10k	-4.39%	-2.24%	-2.69%	-2.03%	-4.23%	-3.46%	**-3.18%**
mix1_100k	29.76%	25.74%	24.17%	24.74%	28.90%	35.43%	**28.12%**
mix2_1k	-3.43%	-3.72%	14.42%	33.75%	0.22%	7.60%	**8.14%**
mix2_10k	-7.58%	4.68%	9.40%	11.12%	3.06%	7.79%	**4.74%**
mix2_100k	24.51%	24.30%	9.46%	22.94%	25.34%	23.22%	**21.63%**
Average	**8.28%**	**9.42%**	**10.59%**	**14.47%**	**10.29%**	**12.16%**	**10.87%**

The performance of DynamicL and OneBoxPerObjectL is compared for six different configurations of the JVM maximum heap size (Xmx) - 512MB, 640MB, 768MB, 896MB, 1024MB and 2048MB. OneBoxPerSlotL results are omitted because they are worse than the ones provided by any of the other two alternatives. Tables 1 and 2 contain the throughput comparison of DynamicL against OneBoxPerObjectL.

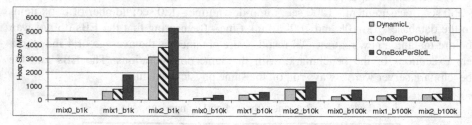

Fig. 1. Effectively used heap sizes, Alpha, with max heap of 16 GB

There are two main reasons for DynamicL displaying consistently better performance metrics than OneBoxPerObjectL. The first relates to situations when Java applications have effectively used heaps that are close to the maximum heap size. Such "low memory availability" conditions force the garbage collection mechanism to be operating without interruption in an attempt to release enough memory so that the effective heap size is no longer close to the upper limit. Garbage collection is well known to be computationally intensive, in particular when it is hard to identify data to be released from memory. As such, DynamicL alleviates these issues by allowing the application to operate with a smaller memory footprint (see Fig. 1), delaying the moment when GC is needed and by keeping the heap smaller for longer periods of time, after it has been GCed. Furthermore, applications with smaller memory footprints allow for improved data locality, leading to better overall system performance.

4   Conclusions

We presented two new approaches for reducing the memory footprint of enterprise Java applications, leading to significantly improved performance. The first uses an adaptive memory representation of domain objects, chosen at compile time. The second one performs, at runtime, a selective loading of domain data necessary for the operation of the application. Both approaches make decisions based on data access pattern stochastic analysis, which provides the access probabilities of domain data through all application execution contexts. The work was evaluated on the TPC-W benchmark against existing solutions. The results demonstrated significant performance improvements. Throughput is increased by 10.78%, on average, and up to a maximum of 35.43%, for larger databases.

Acknowledgments. This work was partially supported by FCT (INESC-ID multiannual funding) through the PIDDAC Program funds and by the Specific Targeted Research Project (STReP) Cloud-TM, which is co-financed by the European Commission through the contract no. 257784. The first author has been funded by the Portuguese FCT under contract SFRH/BD/64379/2009.

References

1. Bhattacharya, S., Nanda, M.G., Gopinath, K., Gupta, M.: Reuse, Recycle to De-bloat Software. In: Mezini, M. (ed.) ECOOP 2011. LNCS, vol. 6813, pp. 408–432. Springer, Heidelberg (2011)
2. Chis, A.E., Mitchell, N., Schonberg, E., Sevitsky, G., O'Sullivan, P., Parsons, T., Murphy, J.: Patterns of Memory Inefficiency. In: Mezini, M. (ed.) ECOOP 2011. LNCS, vol. 6813, pp. 383–407. Springer, Heidelberg (2011)
3. Fernandes, S., Cachopo, J.: Strict serializability is harmless: a new architecture for enterprise applications. In: Proceedings of the ACM International Conference on Object-Oriented Programming Systems, Languages and Applications, pp. 257–276. ACM (2011)
4. Garbatov, S., Cachopo, J.: Importance Analysis for Predicting Data Access Behaviour in Object-Oriented Applications. Journal of Computer Science and Technologies 14(1), 37–43 (2010)
5. Garbatov, S., Cachopo, J.: Predicting Data Access Patterns in Object-Oriented Applications Based on Markov Chains. In: Proceedings of the Fifth International Conference on Software Engineering Advances (ICSEA 2010), Nice, France, pp. 465–470 (2010)
6. Garbatov, S., Cachopo, J.: Data Access Pattern Analysis and Prediction for Object-Oriented Applications. INFOCOMP Journal of Computer Science 10(4), 1–14 (2011)
7. Garbatov, S., Cachopo, J., Pereira, J.: Data Access Pattern Analysis based on Bayesian Updating. In: Proceedings of INForum, Lisbon, Paper 23 (2009)
8. Jones, R.E., Ryder, C.: A study of Java object demographics. In: Proceedings of the 7th International Symposium on Memory Management, Tucson, AZ, pp. 121–130. ACM (2008)
9. Smith, W.: TPC-W: Benchmarking An Ecommerce Solution. Intel Corporation (2000)

Knowledge-Driven Syntactic Structuring: The Case of Multidimensional Space of Music Information

Wladyslaw Homenda[1] and Mariusz Rybnik[2]

[1] Faculty of Mathematics and Information Science, Warsaw University of Technology,
Plac Politechniki 1, 00-660 Warsaw, Poland
[2] Faculty of Mathematics and Computer Science, University of Bialystok,
ul. Sosnowa 64, 15-887 Bialystok, Poland

Abstract. In this paper we study syntactic data structuring as a tool of automatic knowledge discovery. The discussion is focused on domain-nested syntactic processing where paginated music notation is the case of domain. Paginated music notation is a language describing multi dimensional concepts of domain knowledge. Syntactic structuring is based on context-free methods. We propose constructions of context-free grammars driven by concepts of multidimensional knowledge space. Furnishing grammars with attributes allows for information flow between separated knowledge concepts. The study reveals potential and strength of context-free methods in automatic knowledge discovery.

1 Introduction

In this paper we study syntactic data structuring as a tool of automatic knowledge discovery. The discussion is focused on domain-nested syntactic processing with paginated music notation as the case of domain. Paginated music notation is a language describing multi dimensional concepts of domain knowledge. Syntactic structuring is based on context-free methods. We propose constructions of context-free grammars driven by concepts of multidimensional knowledge space. Furnishing grammars with attributes allow for information flow between separated knowledge concepts. The discussion reveals potential of context-free methods in automatic knowledge discovery. The paper is structured as follows. Section 2 outlines multidimensional character of music information and deliberate employment of context-free grammars as a tool for describing such an information. In section 3 different aspects of syntactic structuring are examined. In section 4 an expansion of context-free grammars to attribute grammars is proposed to communicate between different structures of the multidimensional space of music information. Finally, conclusions and directions for future studies are delineated.

2 Grammars as Tools of Syntactical Structuring

As it was shown in several studies, syntactical structuring can be based on context-free grammars covering the language under discussion. The term 'covering the language' means generating all valid and invalid language constructions.

S.W. Liddle et al. (Eds.): DEXA 2012, Part I, LNCS 7446, pp. 438–452, 2012.

This rough approach to syntax makes possible grammar construction for such languages as natural languages or music notation, which are languages of natural communication. Then, integration of syntactical structuring and semantic analysis allows for identification of information structures. These structures can be used for purposes as knowledge discovery or automatic data understanding in man-machine communication, c.f. [3,8,9].

The paginated music notation, the subject of discussion in former studies, is studied in this paper also. In this study we deliberate utilization of grammars in rough description of paginated music notation. Grammars used in syntactic structuring are intended to cover paginated music notation in the sense of paragraph above. Such attempt to syntactic structuring of paginated music notation is endorsed in practice. The grammar will be applied in structuring constructions, which are assumed to be well grounded pieces of paginated music notation. Therefore, incorrect applications of the grammar is not admitted. Of course, such a grammar can neither be applied in checking correctness of constructions of paginated music notation, nor in generation of such constructions, c.f. [4].

The former studies did not discussed details with regard to multi-dimensional nature of paginated music notation. The duration-pitch-voice structuring, the natural awareness in human's perception of the score, have to be solved for automatic processing of paginated music notation. It is especially important in context of automatic understanding of music information.

In Figure 1 we can see an excerpt of a piece of music and its logical split to three dimensions: duration/time, pitch/frequency and voice/instrument. Due to intrinsic properties of music all dimensions are discrete and finite. The duration/time dimension can be measured by the smallest item in terms of notes, rests and rhythmic dots. It is expressed as a part of whole note split to 2^n equal sections. The pitch/frequency is split according to halftones of the music scale. Discreteness allows for representation of music description in two dimensional paginated music notation, where voices are depicted in the form of staves, as shown in upper part of Figure 1. The voices can be interpreted as cross sections of three dimensional space presented in the lower part of this Figure.

2.1   Grammar Driven Structuring

In this section we discuss the grammar previously presented in [4]. The beginning productions of the grammar create the topmost levels of a hierarchy defining *score parts* of the subjected piece of music and split parts to fit frames of pages. Score parts of the piece of music (which are also called *movements*) are arranged in time, hence they can also be placed in duration/time dimension. On the same hand, pages are units created by the necessity to fit printing geometrical constraints. This necessity forces splitting the duration/time axis into intervals of length determined by pages width. Consequently, this geometrical structure may be interpreted in terms of another dimension or rather as a sequence of consecutive time intervals of the duration/time dimension. The latter interpretation, assumed in this paper, defines *systems* as consecutive fragments of duration/time zones of the Duration-Pitch-Voice space:

Fig. 1. Dimensionality of paginated music notation: and excerpt of four parts music (for two pianos) and its decomposition in three dimensional space Duration-Pitch-Voice

$<score> \rightarrow <score\_part> <score>$
$\qquad \rightarrow <score\_part>$
$<score\_part> \rightarrow <page> <score\_part>$
$\qquad \rightarrow <page>$
$<page> \rightarrow <system> <page> \mid <system>$

The following part of the grammar corresponds to the *Voice* dimension. This part of the grammar generates consecutive staves, which correspond to voices of the generated score. Generated staves are included in a system, which is a zone of the Duration-Pitch-Voice space, as explained above.

$<system> \rightarrow <part\_name> <stave> <system>$
$\phantom{<system>} \rightarrow <part\_name> <stave>$
$\phantom{<system>} \rightarrow <stave> <system> \mid <stave>$
$<part\_name> \rightarrow$ Flute $\mid$ Piano $\mid etc.$

The next part of the grammar generates events of the *Duration/Time* dimension. These events are ordered according to their beginning time and they may create sequences (as, for instance, signatures) and hierarchies (e.g. measures and vertical events).

$<stave> \rightarrow$ beg-barline $<bl\_stave>$
$\phantom{<stave>} \rightarrow <bl\_stave>$
$<bl\_stave> \rightarrow <clef> <cl\_stave>$
$\phantom{<bl\_stave>} \rightarrow <cl\_stave>$
$<cl\_stave> \rightarrow <key\_signature> <ks\_stave>$
$\phantom{<cl\_stave>} \rightarrow <ks\_stave>$
$<ks\_stave> \rightarrow <time\_signature> <ts\_stave>$
$\phantom{<ks\_stave>} \rightarrow <ts\_stave>$
$<ts\_stave> \rightarrow <measure>$ barline $<ts\_stave>$
$\phantom{<ts\_stave>} \rightarrow <measure>$ barline
$\phantom{<ts\_stave>} \rightarrow <measure>$
$<measure> \rightarrow <change\_of\_k\_sign.> <ks\_measure>$
$\phantom{<measure>} \rightarrow <ks\_measure>$
$<ks\_measure> \rightarrow <change\_of\_t\_sign.> <ts\_measure>$
$\phantom{<ks\_measure>} \rightarrow <ts\_measure>$
$<ts\_measure> \rightarrow <vertical\ event> <ts\_measure>$
$\phantom{<ts\_measure>} \rightarrow <vertical\ event>$

The last part of the grammar corresponds to the *Pitch/Frequency* dimension. For a given pitch/time event, which is called *vertical event*, the following productions generate notes of different pitch value. These notes are either gathered in chords as having the same duration, or create separated chords of different duration. The general rule is that notes of the same duration are joined to the same stem and create a chord unless they are not separated by other notes of different duration. Of course, this rule has obvious exceptions, i.e. often notes of the same duration belong to different chords. Such exceptions have deep roots in theory of music, which is not a subject of this discussion. Here, for the sake of clarity of presentation, we assume this simple rule of chords' creation.

$<vertical\ event> \rightarrow <stem> <vertical\ event>$
$\phantom{<vertical\ event>} \rightarrow <stem>$
$<stem> \rightarrow <beams> <note\_stem>$
$\phantom{<stem>} \rightarrow <flags> <note\_stem>$
$\phantom{<stem>} \rightarrow <note\_stem>$
$<beams> \rightarrow$ left-beam $<beams> \mid$ right-beam $<beams>$
$<beams> \rightarrow$ left-beam $\mid$ right-beam $\mid$ two-ways

THE MUSIC OF THE NIGHT

Music by ANDREW LLOYD WEBBER
Lyrics by CHARLES HART
Additional lyrics by RICHARD STILGOE

Fig. 2. The subject of discussion: the excerpt of Weber's *The music of the night*, the song from the musical *The phantom of the opera*

$<flags>$ → flag $<flags>$ | flag
$<note\_stem>$ → note-head $<note\_stem>$ | note-head

And finally some productions closing the grammar:

$<clef>$ → treble-clef | bass-clef | $\cdots$
$<key\_signature>$ → | # | b | ## | bb | $\cdots$
$<time\_signature>$ → ¢ | C | $\frac{2}{2}$ | $\frac{4}{4}$ | $\frac{3}{4}$ | $\cdots$

2.2 Derivation Trees and the Lexicon

Syntactic structuring is best identified with derivation trees of language constructions in given grammar. Therefore, we will consider derivation trees of paginated music notation excerpts. The discussion is focused on the excerpt shown in the top part of Figure 2, that comes from Weber's *The music of the night*, the song from the musical *The phantom of the opera*.

Figure 3 presents a part of a derivation tree of the excerpt from Figure 2. The derivation tree covers the second stave. The derivation is developed to the level of single note heads, flags and beams. Derivation of signatures, barlines and symbols of the second measure of this stave is outlined in details. In the derivation tree elements of these groups are included in bounding boxes. On the bottom stave these elements of notation are solid black. Derivations of the first and the third measures are reduced to points of ellipsis. The corresponding elements are greyed out on the bottom stave.

The discussion in next sections is firmly based on the concept of lexicon. The concept of lexicon was introduced in [2] and then developed in [4]. Let us recall that - roughly speaking - lexicon elements includes phrases generated in a grammar; note that a score (a full piece of music) is the phrase as well. Phrases with parts of derivation tree build on them create the entire elements of lexicon. In Figure 3 we can identify many lexicon's elements. For instance, notes

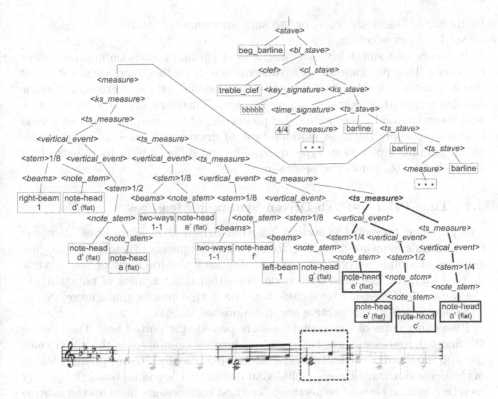

Fig. 3. The second stave of the excerpt from Figure 2: a part of the derivation tree in the grammar shown in section 2.1.

of the stave included in dotted box and corresponding part of the derivation tree, which are in bold, i.e. the subtree with the root <*ts_measure*>, create a lexicon's element. The entire second measure and the left bottom part of the tree, i.e. the subtree with the root in the <*stave*> node, create a lexicon's element. Also, the whole derivation tree together with the whole stave create a lexicon's element.

3 Time versus Pitch Driven Syntax in Measures

Processing information requires accomplishing structural operations. We understand structural operations in terms of selecting, searching, copying, pasting, replacing, transposing, converting etc. Structural operations performed on the space of music information require identification of corresponding areas of the space of information. On the other hand, identification of structures in the space of information is an element of automatic data understanding. Identification of structures of information can be realized based on integrated syntactical structuring and semantical analysis, c.f. [3].

We focus our attention on selection, which is a mother structural operation in spaces of music information. Other operation either are based on selected fragments of the entire score, or use such structures of information, which can be subjected to selection.

It occurs that suitability of syntactic structuring depends on grammar construction. This phenomenon is a consequence of multi dimensional nature of music notation, in this case this is two dimensional nature of cross section for the given voice (the right hand of piano part). Therefore, for the sake of better understanding and easier automatic processing of information structures, it is desirable to apply the most suitable type of grammar. Please notice, that different grammars of various type are similar and differ only in fragments. This feature facilitates automatic processing.

3.1   Time-Prior-to-Pitch-Driven Syntax in Measures

The grammar presented in section 2 is time-driven at measure level. It means that lexicon's elements respective to horizontal fragments of a measure are just subtrees or simple parts of subtrees. In Figure 3 the fragment of the stave is derived from the $<ts\_measure>$ node. Note that this fragment of notation, i.e. the second half measure, is included in the dotted box in this Figure. Node, edges and leaves of this subtree are distinguished in bold.

Remaining notes of the second measure precede the dotted box. They create the first half measure. These notes are four beamed eight notes in the upper voice line and the chord of two half notes in the bottom voice line. Corresponding leaves of the derivation fragment are included in not-bolded bounding boxes. It is easily seen that lexicon element respective to the first half measure includes the subtree with the root surrounded by the ellipse. Of course, the bolded part of this subtree, i.e. the subtree respective to the second part of this measure, is cut off.

3.2   Pitch-Prior-to-Time-Driven Syntax in Measures

Let us consider time-driven structures in the second measure. Let us analyze voice lines, which are the most representative time structures. In Figure 4 the second stave of Weber's piece from Figure 3 is analyzed. There are two voice lines in this stave. Notes of the lower voice line in the second measure are distinguished, all other elements of the stave are greyed out. In the upper part of Figure 4 the derivation subtree of the second measure is shown. This subtree is the part of the tree presented in Figure 3. In this subtree elements corresponding to lower voice line in the second measure are bolded. Notice that these elements do not create a compact structure in frames of the derivation tree. This split makes obvious problems in automatic processing of structured information. Therefore, it would be desirable to find such a structuring, that will keep compactness of lexicon elements for naturally structured pieces of information. We do not intend to develop discussion on compactness and natural structuring terms, we leave it as intuitive notions.

This inconvenience is a result of contradictory nature of the grammar and of the selected construction of music notation. Let us recall that at the measure level the grammar is pitch oriented while the construction is time oriented. In

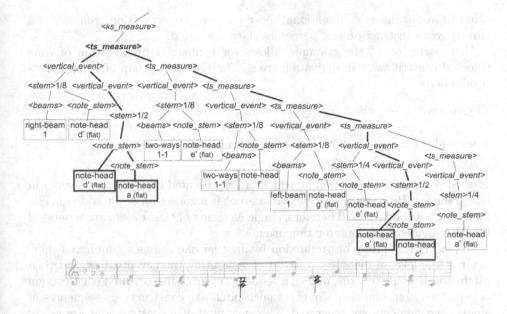

Fig. 4. Selection of the bottom voice line in the second measure and its corresponding derivation tree fragments in pitch-driven grammar

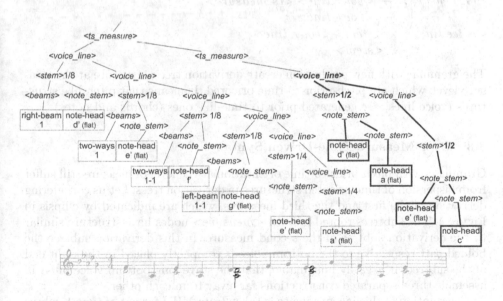

Fig. 5. Selection of the bottom voice line in the second measure and its corresponding fragments of derivation trees in time-driven grammar

other words, in this two dimensional space natures of cross section grammar and time-driven construction are 'perpendicularly' oriented.

Reconstruction of the grammar allows for compact representation of time oriented constructions in derivation trees. The following group of *<measure>* productions:

$$<ts\_measure> \rightarrow \ <vertical\ event> <ts\_measure>$$
$$\rightarrow \ <vertical\ event>$$
$$<vertical\ event> \rightarrow \ <stem> <vertical\ event>$$
$$\rightarrow \ <stem>$$

generate consecutive time events, which are represented by *<vertical event>* in the grammar. However, every such time event is a collection of pitch-driven elements: chords and notes. Therefore, single elements of time events are separated by other elements of the same time events.

Replacing this group of production by another one changes grammar's character at the level of measures. Instead of generating time events first, and then pitch-like elements (chords, notes), a sequence of pitch-like oriented construction is generated first, and then - in every such pitch-like construction - sequences of single time elements are generated. This new pitch-like constructions are represented by *<voice_line>* group of production:

$$<ts\_measure> \rightarrow \ <voice\_line> <ts\_measure>$$
$$\rightarrow \ <voice\_line>$$
$$<voice\_line> \rightarrow \ <stem> <voice\_line>$$
$$\rightarrow \ <stem>$$

The grammar with new production create derivation tree's structure at the measure level, which are pitch-prior-to-time oriented. It means that pitch-like structures (voice lines) are generated prior to time-like ones (chords and notes).

3.3 Over-Measure Time-Driven Syntax

Grammars with pitch-prior-to-time orientation at the level of measure still suffer from dispersion of time-oriented structures in derivation trees. Let us considering derivation of the first and the third measures, which are indicated by ellipsis in Figure 4. The subtrees rooted in both *<measure>* nodes have structure similar to the derivation subtree of the second measure. In this derivation subtree the bolded part respective to the bottom voice is compactly placed in the right end of the subtree. It is easily imaginable, that the stave long bottom voice line will assemble three separated constructions far away from each other.

The solution of this inconvenience is taken from MIDI format of type 1, where - roughly speaking - we have an *empty* track number 0 for representation of signatures while notes are stored in the next track(s).

The following fragment of the initial grammar is responsible for time-prior-to-pitch orientation:

$<stave> \rightarrow$ beginning-barline $<bl\_stave>$
$\rightarrow <bl\_stave>$
$<bl\_stave> \rightarrow$ treble-clef $<cl\_stave>$
$\rightarrow$ bass-clef $<cl\_stave>$
$\rightarrow <cl\_stave>$
$<cl\_stave> \rightarrow <key\_signature> <ks\_stave>$
$\rightarrow <ks\_stave>$
$<ks\_stave> \rightarrow <time\_signature> <ts\_stave>$
$\rightarrow <ts\_stave>$
$<ts\_stave> \rightarrow <measure> <barline> <ts\_stave>$
$\rightarrow <measure> <barline>$
$<ts\_measure> \rightarrow <vertical\ event> <ts\_measure>$
$\rightarrow <vertical\ event>$
$<vertical\ event> \rightarrow <stem> <vertical\ event>$
$\rightarrow <stem>$

The following grammar realizes pitch-prior-to-time orientation of the grammar
at the level of stave:

$<stave> \rightarrow <sign\_stave> <vl\_stave>$
$<sign\_stave> \rightarrow$ beginning-barline $<bl\_stave>$
$\rightarrow <bl\_stave>$
$<bl\_stave> \rightarrow <clef> <cl\_stave>$
$\rightarrow <cl\_stave>$
$<cl\_stave> \rightarrow <key\_signature> <ks\_stave>$
$\rightarrow <ks\_stave>$
$<ks\_stave> \rightarrow <time\_signature> <ts\_stave>$
$\rightarrow <ts\_stave>$
$<ts\_stave> \rightarrow$ dummy-measure barline $<ts\_stave>$
$\rightarrow$ dummy-measure barline
$\rightarrow$ dummy-measure
$<vl\_stave> \rightarrow <voice\_line> <vl\_stave>$
$\rightarrow <voice\_line>$
$<voice\_line> \rightarrow <stem> <voice\_line>$
$\rightarrow <stem>$

The main point of pitch-prior-to-time orientation of the grammar is explained in
Figure 6. Signatures and barlines of the score are derivable from the nonterminal
symbol $<sign\_stave>$. These symbols are displayed in the upper stave. For the
sake of clarity all other symbols are removed from this stave. The top voice line,
derivable from the first nonterminal symbol $<voice\_line>$, is displayed in the
middle stave. Derivation of this voice line is shrunken to ellipsis. And, finally,
the bottom voice line is presented in the bottom stave. Derivation of this voice
line is developed in details.

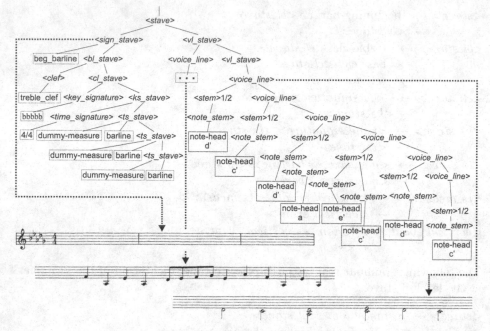

Fig. 6. Selection of the bottom voice line in the second measure and its corresponding fragment of derivation tree in pitch-driven grammar

4 Synchronization

Having pitch-prior-to-time grammar orientation could result in losing of important time-like data. For instance, in the middle and bottom staves *key signature* and *time signature* are unknown. However, signatures are essential for correct reading of notes information. Therefore, we have to provide such data in the grammar or - more precisely - in the derivation tree. To transfer needed data we can use attribute tools in the grammar, i.e. we use attribute grammars. Such a grammar is obtained by furnishing productions with inherited and synthesized attributes. This kind of grammars was introduced by D. Knuth, c.f. [6].

4.1 Attribute Grammars

Let us recall that we deliberate context-free grammars, which are defined as systems $G = (V, T, P, S)$, where V is a finite set of nonterminal symbols, T is a finite set of terminal symbols, P is a finite set of productions and $S \in V$ is the initial symbol of the grammar. Productions are of the form $X_0 \to X_1 X_2 \cdots X_n$, where $X_0 \in V$, $X_1 X_2 \cdots X_n \in V \cup T$, c.f. [4,5].

Nonterminal symbols of a production $X_0 \to X_1 X_2 \cdots X_n$, where $X_0 \in V$, $X_1, X_2, \cdots, X_n \in V \cup T$ have attached sets of attributes in attribute grammars, i.e. the symbol X_I has attached the set $A(X_i)$ of attributes. Attributes are divided into two groups: inherited attributes and synthesized attributes, i.e. $A(X_i) = I(X_i) \cup S(X_i)$, where $I(X_i) \cap S(X_i) = \emptyset$.

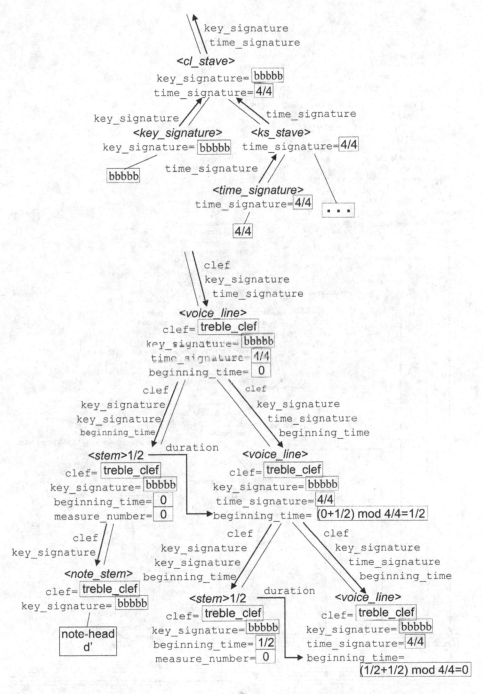

Fig. 7. Synthesized attributes (upper part of the Figure) and inherited attributes (lower part of the Figure)

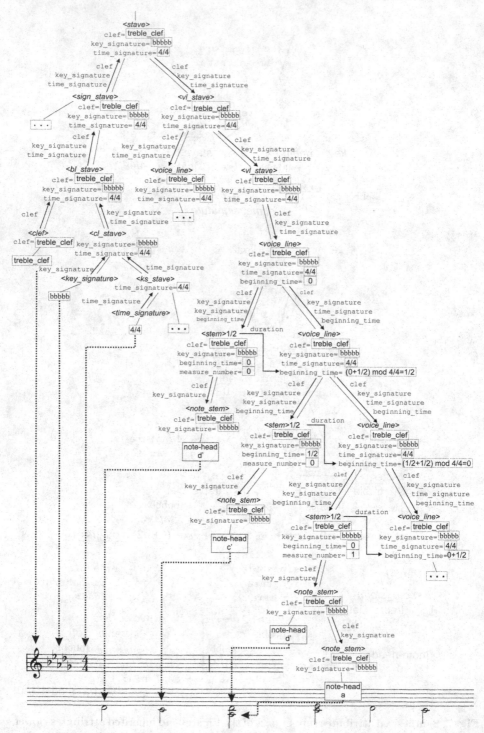

Fig. 8. Information flow in a derivation tree of an attribute grammar

4.2   Information Flow in Attribute Grammars for Music Notation

Attributes are attached to the grammars considered in this paper. They are responsible for transferring important data between items of the information space. We do not attempt to expand formally the grammars with attributes and rules of their evaluation. Instead we outline the idea of using attributes.

The idea of synthesized attributes is shown in the upper part of the Figure 7 comes up by the example of transferring signatures from leaves (terminals) of the derivation tree towards its root. Attributes of given nonterminal and their values are listed below nonterminal's name. Arrows with attributes names show direction of flow of values of attributes. For instance, the nonterminal $<cl\_stave>$ has two attributes: $key\_siganture$ and $time\_signature$. These attributes get values from attributes of $<key\_signature>$ and $<ks\_stave>$, which in turn get values directly from generated elements of music notation: $AbMajor/fMinor$ and $4/4$.

Bottom part of Figure 7 illustrates inherited attributes. For instance, the nonterminal $<cl\_stave>$ has attributes $clef$ and $key\_signature$. Values $treble_clef$ and $AbMajor/fMinor$ of these attributes are inherited from the parent in the derivation tree. On the other hand, the attribute $beginning\_time$ of the nonterminal $<ks\_stave>$ in the lowest node inherits its value from its parent and sibling. The formula shown in this Figure outlines $time$ relative to $measure$.

In Figure 8 attributes' data flow is presented for bigger part of derivation tree. Some obvious attributes synthesized from terminals are skipped for the sake of simplification of the schema.

5   Conclusion

In this paper we study syntactic data structuring driven by information structures. The study is immersed in spaces of music information. Paginated music notation, employed as a language describing music information, subjected to context-free processing, is a powerful tool in the process of automation of information processing. Concepts and ideas studied in the paper stem from practical applications of music processing methods, c.f. [1,7]. We suggest two directions of development of concepts presented in this study (among many possibilities). The first is constructions of grammars generating multidimensional languages and studying properties of such grammars is a general theoretical topic independent on application domain. The second is research in the domain of music information focused on attributed grammars as an inner vehicle of information is a practically oriented direction.

Acknowledgment. The research is supported by The National Centre for Research and Development, grant No N R02 0019 06/2009.

References

1. BrailleScore, grant of The National Centre for Research and Development (2009-2012)
2. Bargiela, A., Homenda, W.: Information structuring in natural language communication: Syntactical approach. Journal of Intelligent & Fuzzy Systems 17(6), 575–581 (2006)
3. Homenda, W.: Integrated syntactic and semantic data structuring as an abstraction of intelligent man-machine communication. In: ICAART - International Conference on Agents and Artificial Intelligence, Porto, Portugal, pp. 324–330 (2009)
4. Homenda, W., Rybnik, M.: Querying in Spaces of Music Information. In: Tang, Y., Huynh, V.-N., Lawry, J. (eds.) IUKM 2011. LNCS (LNAI), vol. 7027, pp. 243–255. Springer, Heidelberg (2011)
5. Hopcroft, J.E., Ullman, J.D.: Introduction to Automata Theory, Languages and Computation. Addison-Wesley Publishing Company (1979, 2001)
6. Knuth, D.E.: The Genesis of Attribute Grammars. In: Deransart, P., Jourdan, M. (eds.) Attribute Grammars and their Applications. LNCS, vol. 461, pp. 1–12. Springer, Heidelberg (1990)
7. SmartScore (2007), http://www.musitek.com
8. Tadeusiewicz, R., Ogiela, M.R.: Automatic Image Understanding - A New Paradigm for Intelligent Medical Image Analysis. Bioalgorithms and Med-Ssystems 2(3), 5–11 (2006)
9. Tadeusiewicz, R., Ogiela, M.R.: Why Automatic Understanding? In: Beliczynski, B., Dzielinski, A., Iwanowski, M., Ribeiro, B. (eds.) ICANNGA 2007. LNCS, vol. 4432, pp. 477–491. Springer, Heidelberg (2007)

Mining Frequent Itemsets
Using Node-Sets of a Prefix-Tree

Jun-Feng Qu[1] and Mengchi Liu[2]

[1] State Key Lab. of Software Engineering, School of Computer, Wuhan University,
Wuhan 430072, China
cocoqjf@gmail.com
[2] School of Computer Science, Carleton University, Ottawa K1S 5B6, Canada
mengchi@scs.carleton.ca

Abstract. Frequent itemsets are important information about databases, and efficiently mining frequent itemsets is a core problem in data mining area. The divide-and-conquer strategy is very applicable to the problem. Most algorithms adopting the strategy construct a very large number of conditional databases when mining frequent itemsets. Representations of conditional databases and methods of constructing them greatly influence the performance of such algorithms. In this study, we propose a **node-set** structure for representing a conditional database, and develop a novel node-set-based algorithm, NS, for mining frequent itemsets. During a mining process, all the node-sets derive from a prefix-tree storing the complete frequent itemset information about the mined database. Compared with previous conditional database representations, node-sets are compact and contiguous on which NS can be performed fast. Constructing conditional databases involves counting for items. In NS, the counting procedure and the construction procedure are blended, which saves the time for scanning conditional databases, and further, the major operations of constructing conditional databases are very simple comparisons. Experimental data show that NS outperforms several famous algorithms including FPgrowth* and LCM, ones of the fastest algorithms, for various databases.

Keywords: Data mining, frequent itemset, node-set.

1 Introduction

Frequent itemsets mined from databases not only are used in many data mining tasks such as association rule mining [4], classification [6], and clustering [20] but also play an important role in real-life applications, e.g., face detection [17]. Therefore, efficient mining of frequent itemsets has received considerable attention since the introduction of the problem by Agrawal et al. [1].

1.1 Problem Definition

Let $\mathcal{I}=\{i_1, i_2, i_3,\ldots, i_n\}$ be a set of items and DB be a transaction database in which each transaction is a subset of $\mathcal{I}$ and has a unique identifier (TID).

S.W. Liddle et al. (Eds.): DEXA 2012, Part I, LNCS 7446, pp. 453–467, 2012.

Each subset of $\mathcal{I}$ is an itemset, and an itemset composed of k items is called a k-itemset. A transaction *satisfies* an itemset if it includes all the items of the itemset. The *support* of an itemset is the number of transactions in DB satisfying the itemset. The support of an itemset can also be defined as the percentage of transactions satisfying the itemset. An itemset is *frequent* if its support exceeds a user-specified minimum support threshold. Given a database and a minimum support, the frequent itemset mining problem is to find out all the frequent itemsets with their supports. For a database with n items, there are 2^n itemsets that have to be checked, and hence the problem is intractable.

1.2   Previous Solutions

An important property of frequent itemsets is that all the supersets of an infrequent itemset are infrequent and all the subsets of a frequent itemset are frequent. The property was first proposed and used in the Apriori algorithm [2] for frequent itemset mining. Apriori first scans a database to find out all the frequent 1-itemsets from which it generates the candidate 2-itemsets. Afterwards, Apriori scans iteratively the database to find out all the frequent k-itemsets ($k >= 2$) from which it generates the candidate $(k + 1)$-itemsets until there is neither any frequent itemset found out nor any candidate itemset generated. The qualification as a candidate $(k + 1)$-itemset is that all of its subsets containing k items, namely $k+1$ k-itemsets, are frequent. Apriori and its variants are called candidate generation-and-test approaches.

More efficient approaches to the frequent itemset mining problem employ the divide-and-conquer strategy, such as the FP-Growth algorithm [11] and the Eclat algorithm [22]. Given a database DB and a minimum support, the approaches first scan DB to find out all the frequent items, namely the frequent 1-itemsets. Subsequently, the frequent items are sorted in some order such as in frequency-descending order, and suppose they are: $fi_1 < fi_2 < fi_3 < \cdots < fi_n$. After eliminating the infrequent items in all transactions, the approaches divide DB into n disjoint sub-databases: $DB_1, DB_2, DB_3, \ldots, DB_n$. DB_k ($1 \le k \le n$) is composed of all the transactions containing fi_k and the frequent items before fi_k but no any frequent item after fi_k, or in the reversed way. DB_k is called the conditional database of fi_k. When each frequent item in DB_k is appended to fi_k, a frequent 2-itemset is generated. In this way, the approaches recursively process DB_k to mine frequent itemsets.

FP-Growth employs prefix-trees to represent (conditional) databases. After identifying all the frequent items in a (conditional) database, FP-Growth constructs a (conditional) prefix-tree by processing each transaction as follows: (1) pick out the frequent items from the transaction; (2) sort the items in frequency-descending order to generate a branch; (3) merge the branch into the prefix-tree. Fig. 1 shows a transaction database and Fig. 2(a) depicts a corresponding prefix-tree. Each node in a prefix-tree contains an *item* and a *counter*. For each item of a prefix-tree, FP-Growth will construct a conditional prefix-tree representing its conditional database. For example, by traversing the sub-trees rooted at the nodes numbered 2 and 8 at their upper left corners in Fig. 2(a), FP-Growth

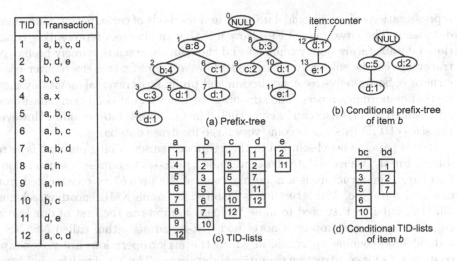

Fig. 1. Database   **Fig. 2.** Representations of (conditional) database
(*minimum support*=2)

counts for the items in the conditional database of item b, and their supports
are $\{c: 5, d: 3, e: 1\}$. Because the minimum support is 2, infrequent item e is
no longer considered when FP-Growth constructs the conditional prefix-tree of
item b. The conditional prefix-tree of item b depicted in Fig. 2(b) is constructed
when FP-Growth traverses the sub-trees again. (This is a modification version
of FP-Growth in which some details are different from those in the original
FP-Growth. They do not matter in this paper.) Eclat uses a vertical database
layout and represents (conditional) databases by TID-lists. In Eclat, each itemset holds a TID-list in which the corresponding transactions satisfy the itemset.
The length of the TID-list is the support of the itemset. Eclat intersects the
TID-lists of two frequent itemsets and thereby obtains the resultant TID-list of
a new itemset composed of the two frequent itemsets. Fig. 2(c) and (d) respectively show the TID-lists representing the database in Fig. 1 and the conditional
TID-lists representing the conditional database of item b.

There are a number of other algorithms for frequent itemset mining. The FP-
growth* algorithm [10] is an efficient variant of FP-Growth, and it reduces half
the traversal cost of FP-Growth. The dEclat algorithm [21] incorporating the
"diffset" technique significantly improves Eclat's performance. Tiling [9] makes
the best of CPU cache to speed previous algorithms up; CFP-growth [15] consumes less memory than other algorithms; LCM [19] and AFOPT [12] adopt
multiple optimization strategies and achieve good performance; BISC [5] and
FIUT [18] employ new ideas to mine frequent itemsets.

1.3   Motivation and Contribution

For many mining tasks, most divide-and-conquer mining algorithms have to construct a very large number of conditional databases during their mining processes
[13]. Therefore, the key factors relevant to the performance of these algorithms are

representations of conditional databases and methods of constructing conditional databases. The advantage of FP-Growth is that prefix-trees representing conditional databases are highly compact and thus database scans, namely prefix-tree traversals, can be efficiently performed. However, a prefix-tree is a pointer-based structure, both prefix-tree construction and prefix-tree traversal inevitably incur costly pointer dereferences. The advantage of Eclat is that conditional databases represented by TID-lists can be constructed fast by simple intersections. However, the sizes of TID-lists will become very large for dense databases.

In this paper, a novel solution to the frequent itemset mining problem is introduced. Firstly, we use a data structure called *node-set* to represent a conditional database. The structure is a node mapping of the prefix-tree constructed from a mined database. The structure is inspired by some XML encoding schemes [3], [14], but it is first used to mine frequent itemsets to the best of our knowledge. Secondly, we propose a novel node-set-based algorithm called NS. NS is a divide-and-conquer approach, in which the major operations are very simple comparisons for constructing conditional databases. The NS algorithm holds not only the advantage of FP-Growth, namely a compact representation of conditional databases, but also that of Eclat, namely a fast method of constructing conditional databases. Thirdly, extensive experimental data show that NS outperforms several significant algorithms including FPgrowth* and LCM, ones of the fastest algorithms, for various databases. The remainder of this paper is organized according to the three aforementioned points that are in Section 2, 3, and 4. The paper ends in the conclusion of Section 5.

2   Preliminaries

This section will introduce the key concepts and fundamental principles of the NS algorithm and illustrate the node-set structure. In the following, we suppose that (1) all the items of an itemset are ordered; (2) P is a prefix itemset (can be empty); (3) x and y are items and x is before y; (4) Px, Py, and Pxy are the itemsets composed of P, x, and y.

2.1   Conditional Node

Given a prefix-tree representing a database, the support of an itemset relates only to a number of nodes rather than all nodes.

Definition 1 (Conditional node). *In a prefix-tree, a conditional node of an itemset is such a node that it contains the last item of the itemset and that all the items of the itemset are in the path from the node to the root.*

In Fig. 2(a), the conditional nodes of itemset bc are the nodes numbered 3 and 9 at their upper left corners respectively. The path from a conditional node of an itemset to the root includes all the items of the itemset and corresponds to the transactions in the database satisfying the itemset, and hence the counter of the conditional node registers the partial support of the itemset.

Lemma 1. *In a prefix-tree, the sum of the counters of all the conditional nodes of an itemset is the support of the itemset.*

Proof. In the process of constructing a prefix-tree, for any transaction satisfying an itemset, the construction algorithm is bound to either take it through or create a conditional node of the itemset when it is merged into the prefix-tree.□

For example, the support of itemset bc is 5, namely the sum of the counters of its conditional nodes (numbered 3 and 9 respectively in Fig. 2(a)). The sum of the counters of itemset ac's conditional nodes (numbered 3 and 6 respectively) is its support 4.

Lemma 2. *In a prefix-tree, a conditional node of itemset Py is a conditional node of itemset Pxy, as long as x is in the path from the node to the root.*

Proof. If x is in the path from a Py's conditional node to the root, the Py's conditional node meets the two requirements of being a Pxy's conditional node according to Definition 1.   □

The nodes numbered 4, 5, and 7 respectively in Fig. 2(a) are the conditional nodes of itemset ad, and the nodes numbered 4 and 7 respectively are also the conditional nodes of itemset acd because the paths from them to the root include item c.

Lemma 3. *In a prefix-tree, the set of Pxy's conditional nodes is a subset of the set of Py's conditional nodes.*

Proof. Any node in the set of Pxy's conditional nodes meets the two requirements of being a Py's conditional node.   □

2.2 Topology Number

If the set of Py's conditional nodes is available, based on Lemma 3, the set of Pxy's conditional nodes can be evaluated by picking out the eligible nodes from Py's conditional nodes according to Lemma 2. A number of Py's conditional nodes are Pxy's conditional nodes and the other nodes are not.

Definition 2 (Candidate node). *All of Py's conditional nodes are called the candidate nodes for Pxy's conditional nodes.*

In Fig. 2(a), the conditional nodes of itemset ad are the nodes numbered 4, 5, and 7 respectively, and they are also the candidate nodes for itemset abd when abd's conditional nodes are evaluated. After they are checked, the nodes numbered 4 and 5 respectively are proved to be abd's conditional nodes. To evaluate the conditional nodes of Pxy, namely to judge whether or not x exists in the paths from Py's conditional nodes to the root, a primitive method is to traverse these paths. However, the longer the distance between x and y is, the more the traversal cost is. For example, the search for item a from the nodes containing item c may go through the nodes containing only item b in Fig. 2(a), but the same search from the nodes containing item d may go through the nodes containing either item b or item c. Further, traversals incur costly pointer dereferences.

Can one directly pick out Pxy's conditional nodes from Py's conditional nodes without traversals for related paths?

Definition 3 (Topology number). *In a prefix-tree, the topology number of a node is i if it is the ith node to be visited in depth-first order.*

The topology number of each node in the prefix-tree in Fig. 2(a) is at the node's upper left corner. By such a numbering way, the topology numbers of all the descendant nodes of a node constitute an integer-consecutive region that can be defined by a pair of boundaries. For example, the topology numbers of all the descendant nodes of the node numbered 1 are in $[2, 7]$.

Definition 4 (Descendant-node-topology-number region). *The integer-consecutive region composed of the topology numbers of all the descendant nodes of a node is the descendant-node-topology-number (abbr. dntn) region of the node.*

Lemma 4. *If the topology number of a Py's conditional node is in the dntn region of a Px's conditional node, the Py's conditional node is a Pxy's conditional node.*

Proof. The topology number of a Py's conditional node is in the dntn region of a Px's conditional node, which means that the former node is a descendant node of the latter and that x must be in the path from the Py's conditional node to the root, and thereby it can be deduced from Lemma 2 that the Py's conditional node is also a Pxy's conditional node.   □

For a depth-first mining algorithm, Px's conditional nodes and Py's have been available when Pxy's conditional nodes are evaluated. Therefore, Lemma 4 gives a solution to the aforementioned problem provided the information about both topology number and dntn region is available. For example, for the prefix-tree in Fig. 2(a), after the conditional node (numbered 2) of itemset ab and those (numbered 4, 5, and 7 respectively) of itemset ad have been evaluated, the conditional nodes of itemset abd, namely the nodes numbered 4 and 5 respectively, can be picked out from the conditional nodes of itemset ad because they are in the dntn region of the node numbered 2, namely $[3, 5]$.

2.3 Node-Set Structure

In the NS algorithm, a (conditional) database is represented by a set of nodes in a prefix-tree constructed from a mined database, and the set with related information is stored in a node-set structure.

A node-set contains two tables: one called *mapping* and the other called *item-list*. The mapping table registers three pieces of information about each node in a (conditional) database, and they are a node's topology number denoted as tn, the upbound of the node's dntn region denoted as *upbound*, and the node's

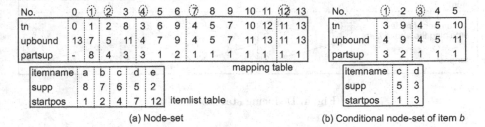

No.	0	①	②	3	④	5	6	⑦	8	9	10	11	⑫	13
tn	0	1	2	8	3	6	9	4	5	7	10	12	11	13
upbound	13	7	5	11	4	7	9	4	5	7	11	13	11	13
partsup	-	8	4	3	3	1	2	1	1	1	1	1	1	1

No.		①	2	③	4	5
tn		3	9	4	5	10
upbound		4	9	4	5	11
partsup		3	2	1	1	1

mapping table

itemname	a	b	c	d	e
supp	8	7	6	5	2
startpos	1	2	4	7	12

itemlist table

itemname	c	d
supp	5	3
startpos	1	3

(a) Node-set   (b) Conditional node-set of item *b*

Fig. 3. Core data structure

counter denoted as *partsup*. The dntn region of a node can be defined by (tn, upbound]. All the nodes containing an item are mapped onto a contiguous part of the mapping table in topology number ascending order. The itemlist table registers the information about all the items in a (conditional) database. For each item, its name denoted as *itemname*, its support denoted as *supp*, and the position of the first node containing the item in the mapping table denoted as *startpos* are stored in the itemlist table.

The node-set constructed from the prefix-tree in Fig. 2(a) is shown in Fig. 3(a). Fig. 3(b) depicts the conditional node-set representing the conditional database of item *b*.

3 NS Algorithm

Given a transaction database and a minimum support, the NS algorithm first constructs a prefix-tree representing the database. After mapping all the nodes of the tree onto a initial node-set, NS mines frequent itemsets by recursively constructing conditional node-sets from the initial node-set. Prefix-tree construction has been discussed in previous literature [11], and the mapping and mining procedures of NS will be introduced in this section.

3.1 Mapping All the Nodes of a Prefix-Tree

In the process of constructing a prefix-tree from a database, the following information can be gathered: the names of the items, the supports of the items, and the numbers of nodes containing an item. The itemlist table of the initial node-set is built based on the information. Except that the first two vectors of the itemlist table are directly available, the startpos vector can be deduced from the third piece of information. For example, for the prefix-tree in Fig. 2(a), the numbers of nodes containing items *a*, *b*, *c*, *d*, and *e* are respectively 1, 2, 3, 5, and 2. The evaluation of the startpos vector is demonstrated in Fig. 4.

After the itemlist table of the initial node-set is built, all the nodes of the prefix-tree rooted at *root* can be mapped onto the mapping table of the node-set by calling **Mapping**(*root, mapping, startpos*, 0). Algorithm 1 is its pseudo-code, in which the *startpos* vector is indexed by the names of the items.

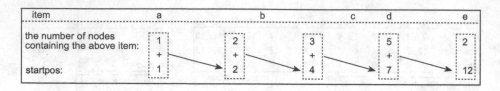

item	a	b	c	d	e
the number of nodes containing the above item:	1	2	3	5	2
	+	+	+	+	
startpos:	1	2	4	7	12

Fig. 4. Deducing startpos vector

For node N, $startpos[N.item]$ indicates its position in *mapping*. Firstly, parameter $curTN$ as N's topology number is assigned to the tn field (line 1). Subsequently, N's counter is mapped onto the *partsup* field (line 2). Before mapping the next node, the algorithm provides the node with its topology number stored in $nextTN$ (line 3). After all the child nodes of N have been recursively processed (lines 4-6), please note that $nextTN$ is updated and stores the topology number of the node that will be processed immediately after the sub-tree rooted at N is processed. If $nextTN$ remains unchanged, which means that N has no child node, the *upbound* of N's dntn region is assigned as $curTN$ (line 8), and otherwise the *upbound* of N's dntn region is $nextTN$-1 (line 10).

Once N is mapped, $startpos[N.item]$ is increased by 1 and indicates the position of the next node containing $N.item$ in *mapping* (line 12). In the mining phase, the prefix-tree is no longer used and hence N is deleted immediately after being mapped (line 13). At last, the topology number for the next node that will be mapped is returned (line 14).

Algorithm 1. *Mapping* algorithm

Input: N is a node in the prefix-tree representing a database;
　　　　mapping is the mapping table of a node-set;
　　　　startpos is the startpos vector of the itemlist table of the node-set;
　　　　$curTN$ is the current topology number.
Output: the initial node-set representing the database.
1　$mapping[startpos[N.item]].tn = curTN$;
2　$mapping[startpos[N.item]].partsup = N.counter$;
3　$nextTN = curTN + 1$;
4　**foreach** child c of N **do**
5　　| $nextTN = \mathtt{Mapping}(c, mapping, startpos, nextTN)$;
6　**end**
7　**if** $nextTN==(curTN+1)$ **then**
8　　| $mapping[startpos[N.item]].upbound = curTN$;
9　**else**
10　| $mapping[startpos[N.item]].upbound = nextTN - 1$;
11　**end**
12　$startpos[N.item] = startpos[N.item] + 1$;
13　delete N;
14　**return** $nextTN$;

3.2   Mining the Frequent Itemsets from a Node-Set

After the initial node-set is constructed from a prefix-tree, all the frequent item-sets can be identified by calling NS's mining procedure, which is showed in Algorithm 2.

All the items in node-set S (the second parameter) are processed successively. For item x in $S.itemlist$, the combination of prefix itemset P (the first parameter) and x, denoted as Px, is the prefix itemset of the next level of recursion (line 2). After Px with its support is outputted (line 3), the main task of the algorithm is to identify all the frequent items in the conditional database of Px and to construct the conditional node-set of Px denoted as $subS$ (line 4). Each item y after x will be checked (lines 6-17). Whether y is a frequent extension of Px or not, namely whether Pxy is frequent or not, depends on the support of Pxy. From another perspective, $S.mapping$ stores the conditional nodes of both Px and Py, and then Pxy's conditional nodes can be picked out from Py's conditional nodes (candidate nodes) according to Lemma 4. When all the conditional nodes of Pxy are found out, the support of Pxy can be figured out according to Lemma 1.

Algorithm 2. *Mining algorithm*

Input: P is a prefix itemset, initially empty;
　　　　 S is the conditional node-set of P;
　　　　 $minsup$ is the minimum support threshold.
Output: all the frequent itemsets with P as prefix.

1 **foreach** item x in $S.itemlist$ **do**
2 　│ $Px = P \cup x$;
3 　│ output Px with the support of x in $S.itemlist$;
4 　│ $subS = NULL$;
5 　│ **foreach** item y after x in $S.itemlist$ **do**
6 　│ 　│ $support = 0$;
7 　│ 　│ **foreach** candidate node n containing y in $S.mapping$ **do**
8 　│ 　│ 　│ **if** $\exists m \in S.mapping$ **and** m contains x **and** $n.tn \in (m.tn, \ m.upbound]$
　　　　　　　　　then
9 　│ 　│ 　│ 　│ append $<n.tn, \ n.partsup, \ n.upbound>$ to $subS.mapping$;
10 　│ 　│ 　│ 　│ $support = support + n.partsup$;
11 　│ 　│ 　│ **end**
12 　│ 　│ **end**
13 　│ 　│ **if** $support \geq minsup$ **then**
14 　│ 　│ 　│ append $<y, \ support, \ y$'s starting position in $subS.mapping>$ to
　　　　　　　　　$subS.itemlist$;
15 　│ 　│ **else**
16 　│ 　│ 　│ delete all the nodes containing y from $subS.mapping$;
17 　│ 　│ **end**
18 　│ **end**
19 　│ Mining(Px, $subS$, $minsup$);
20 **end**

Item y is checked as follows. Firstly, variable *support* storing the support of Pxy is initialized (line 6). Secondly, for each candidate node n, namely for each conditional node of Py, as long as there is node m containing item x in $S.mapping$ and the topology number of n is in the dntn region of m (line 8), it can be deduced from Lemma 4 that n is a conditional node of Pxy. All the conditional nodes of Pxy belong to the conditional database of Px and are appended to $subS.mapping$, and their *partsup*s are added to *support* (lines 9-10). If the support of Pxy exceeds *minsup*, y is a frequent extension of Px. In this case, <item y, its support, and the position of the first node containing y in $subS.mapping$> is appended to $subS.itemlist$ (line 14). Otherwise, all the nodes containing y are deleted from $subS.mapping$ (line 16).

After constructing the conditional node-set of Px, the algorithm recursively processes Px with its conditional node-set (line 19).

3.3   An Example of Mining Algorithm

The following will illustrate how Algorithm 2 constructs the conditional node-set of item b in Fig. 3(b) from the initial node-set in Fig. 3(a). P is empty and *minsup* is 2 now.

Px with its support, namely itemset b with 7, is first outputted. Subsequently, the algorithm starts to construct the conditional node-set of item b. There are the two conditional nodes of b whose dntn regions are (2, 5] and (8, 11] respectively. The items after b are checked one by one:

- Item c has the two conditional nodes whose tns are $3 \in (2, 5]$ and $9 \in (8, 11]$ respectively. The accumulated support is 5 larger than 2. Therefore, c is a frequent extension of b.
- Item d has the three conditional nodes whose tns are $4 \in (2, 5]$, $5 \in (2, 5]$, and $10 \in (8, 11]$ respectively. The accumulated support is 3 larger than 2. Therefore, d is a frequent extension of b.
- Item e has the only conditional node whose tn is $11 \in (8, 11]$. The accumulated support is 1 smaller than 2. Therefore, e is not a frequent extension of b and the node containing e in the mapping table of the conditional node-set of item b is deleted.

3.4   Atom Operation

There are the three steps in the NS algorithm: constructing a prefix-tree, mapping nodes, and mining frequent itemsets. In general, the time complexities of the first two steps are linear, namely $\mathcal{O}(t)$ and $\mathcal{O}(n)$ respectively (t is the number of transactions in a mined database and n is the number of nodes in the corresponding prefix-tree), but that of the last step is $\mathcal{O}(2^i)$ (i is the number of frequent items). Therefore, the mining time dominates NS's running time in most cases.

The main task of the mining algorithm is to construct conditional node-sets, and the atom operation is to judge whether a Py's conditional node is a Pxy's

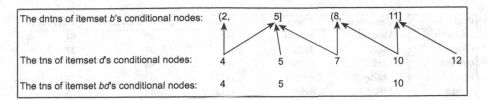

Fig. 5. Evaluation of conditional node

conditional node or not (line 8 in Algorithm 2). Suppose the number of Px's conditional nodes is m and that of Py's conditional nodes is n. For each Py's conditional node, the judgment of line 8 in Algorithm 2 will use $2 \times m$ comparisons in the worst case. Therefore, it is possible to perform $2 \times m \times n$ comparisons for judging whether Pxy is frequent or not.

However, in the mapping table of a node-set, the nodes containing an item are mapped onto a contiguous part in topology number ascending order and are processed in the same order, and hence the nodes in the mapping table of any conditional node-set keep the same order. Then, $2 \times m + n$ comparisons at most are enough for the evaluation of Pxy's conditional nodes from Py's. For example, the conditional nodes of itemset b are the nodes whose dntn regions are $(2, 5]$ and $(8, 11]$ respectively, and those of itemset d are the nodes whose tns are 4, 5, 7, 10, and 12 respectively in Fig. 3(a). The process of evaluating the conditional nodes of itemset bd is demonstrated in Fig. 5. We can observe that this process is actually a 2-way comparison between the dntn regions' boundaries of itemset b's conditional nodes and the topology numbers of itemset d's conditional nodes. In this way, the atom operation of the NS algorithm can be performed fast.

4 Experiments

To test the performance of the NS algorithm, we have done extensive experiments that are reported in the section.

4.1 Experimental Setup

We have implemented the NS algorithm, and it was compared with the FP-Growth algorithm [11], the FPgrowth* algorithm [10], the AFOPT algorithm [12], the dEclat algorithm [21], and the LCM algorithm [19]. These algorithms have been proven to be superior to the Apriori algorithm, and thus Apriori was no longer tested. To avoid implementation bias, the implementation of FP-Growth was downloaded from [8], and the implementations of FPgrowth*, AFOPT, dEclat, and LCM were downloaded from [7]. FPgrowth* is the fastest algorithm in IEEE ICDM Workshop on frequent itemset mining implementations (FIMI'03), and LCM is the fastest algorithm in IEEE ICDM Workshop on FIMI'04. Except for FP-Growth (coded by Bart Goethals) and dEclat (coded by Lars Schmidt-Thieme [16], the fastest dEclat implementation that we have

Database	Size(bytes)	#Trans	#Items	AvgTransLen	MaxTransLen
accidents	35509823	340183	468	33	51
chess	342294	3196	75	37	37
connect	9255309	67557	129	43	43
kosarak	32029467	990002	41270	8	2498
pumsb	16689761	49046	2113	74	74
webdocs	1481890176	1692082	5267656	177	71472

Fig. 6. Statistical information about experimental databases

tested), the other algorithms were coded by their authors respectively. All of the codes were written in C/C++, used the same libraries, and were compiled using gcc (version 4.3.2).

The six dense/sparse databases from [7] were used. The statistical information about the databases is showed in Fig. 6, including the size (bytes), the number of transactions, the number of distinct items, the average transaction length, and the maximal transaction length. The experiments were performed on a 2.83GHz PC machine (Intel Core2 Q9500) with 4×10^9 bytes memory, running on a Debian (Linux 2.6.26) operating system. Running time was recorded by "time" command, and it contains input time, CPU time, and output time. Output was directed to "/dev/null".

4.2   Experimental Results

The experimental results are depicted in Fig. 7. The running time of NS includes the times of constructing a prefix-tree, mapping nodes, and mining frequent itemsets. Note that we did not plot when an implementation terminated abnormally due to either segmentation faults or memory allocation failures.

For almost all the databases and minimum supports, NS performs the best. For example, in Fig. 7(a), the running times of these algorithms are respectively: NS(17.553 seconds), FPgrowth*(91.959s), FP-Growth(449.568s), AFOPT(126.469s), dEclat(212.283s), and LCM(67.498s) when the minimum support is 6% for dense database *accidents*. NS is over an order of magnitude faster than FP-Growth and dEclat, and it is several times faster than FPgrowth*, AFOPT, and LCM. For sparse database *kosarak* in Fig. 7(d), their running times are respectively: NS(5.099 seconds), FPgrowth*(20.672s), FP-Growth(117.257s), AFOPT(17.228s), dEclat(terminated abnormally), and LCM(9.179s) when the minimum support is 0.085%. NS is still the fastest.

4.3   Discussion

The NS algorithm distinctly outperforms the previous algorithms for the reasons below.

There are a very large number of conditional databases constructed by divide-and-conquer mining algorithms in most cases [13], and thereby the representations

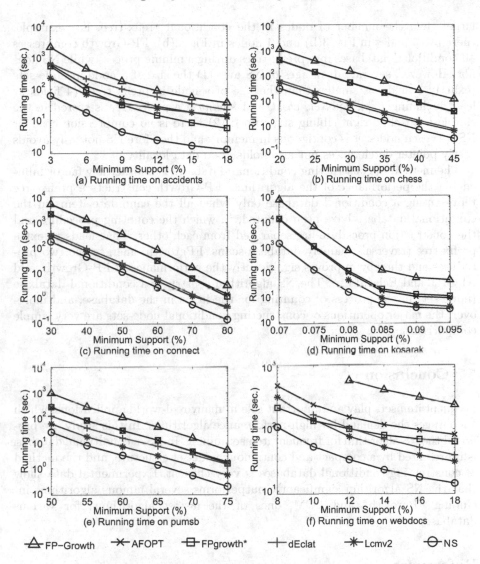

Fig. 7. Performance comparison

of conditional databases have an important impact on the performance of such algorithms. NS uses a node-set structure to represent a conditional database. For a mining task, all the node-sets derive from a prefix-tree representing the mined database. A prefix-tree is a high compact structure [11], and thus a node-set is a compact conditional database representation. For example, when all the transactions in the database in Fig. 1 are replicated 100 times, the size of the prefix-tree in Fig. 2(a) and the size of the node-set in Fig. 3(a) remain unchanged while the size of TID-lists in Fig. 2(c) increases by 100 times.

Please note that except for the initial node-set, for a conditional database, the number of nodes in the mapping table of the conditional node-set is generally larger than the number of nodes in the conditional prefix-tree. For example, there are 5 nodes in Fig. 3(b) and 4 nodes in Fig. 2(b). FP-Growth compresses all conditional databases into prefix-trees during a mining process while NS does not. However, several advantages of NS are: (1) the size of a node in node-sets (exactly 3 fields) is smaller than the size of a node in prefix-trees (4 fields at least including *item, counter, child,* and *sibling* if a prefix-tree is stored in the simplest left-child-right-sibling structure); (2) there is no compression cost for NS; (3) each node-set is contiguous in memory and therefore NS not only avoids costly pointer dereferences but also holds good data locality [9].

The method of constructing conditional databases is another key factor influencing the performance of the algorithms. FP-Growth constructs a prefix-tree representing a conditional database only when all the candidate items in the conditional database have been counted, in which the counting procedure and the construction procedure are separated from each other. That leads to extra prefix-tree traversals, namely database scans. FPgrowth* merges the two procedures and thereby improves significantly the performance of FP-Growth, and AFOPT and Eclat do so. The NS algorithm constructs a conditional database (node-set) in the process of counting for the items in the database, and moreover, the major operations of constructing conditional node-sets are very simple comparisons.

5   Conclusion

Frequent itemsets play an important role in many real-world applications, which encourages the demand for high-performance algorithms. In this study, we proposed the NS algorithm for frequent itemset mining. In NS, a conditional database is represented by a compact and contiguous node-set structure, and the method of constructing conditional databases is very efficient. Experimental data show that the NS algorithm significantly outperforms several famous algorithms including FPgrowth* and LCM, ones of the fastest algorithms, for various databases.

References

1. Agrawal, R., Imieliński, T., Swami, A.: Mining Association Rules between Sets of Items in Large Databases. In: Proc. ACM SIGMOD, pp. 207–216 (1993)
2. Agrawal, R., Srikant, R.: Fast Algorithms for Mining Association Rules in Large Databases. In: Proc. VLDB, pp. 487–499 (1994)
3. Bruno, N., Koudas, N., Srivastava, D.: Holistic Twig Joins: Optimal XML Pattern Matching. In: Proc. ACM SIGMOD, pp. 310–321 (2002)
4. Ceglar, A., Roddick, J.F.: Association Mining. ACM Comput. Surv. 38(2), 1–42 (2006)
5. Chen, J., Xiao, K.: Bisc: A Bitmap Itemset Support Counting Approach for Efficient Frequent Itemset Mining. ACM Trans. Knowl. Disc. Data 4(3), 12:1–12:37 (2010)

6. Cheng, H., Yan, X., Han, J., Yu, P.S.: Direct Discriminative Pattern Mining for Effective Classification. In: Proc. ICDE, pp. 169–178 (2008)
7. Frequent Itemset Mining Implementations Repository, http://fimi.ua.ac.be/
8. Frequent Pattern Mining Implementations, http://adrem.ua.ac.be/~goethals/software/
9. Ghoting, A., Buehrer, G., Parthasarathy, S., Kim, D., Nguyen, A., Chen, Y.K., Dubey, P.: Cache-Conscious Frequent Pattern Mining on Modern and Emerging Processors. The VLDB Journal 16(1), 77–96 (2007)
10. Grahne, G., Zhu, J.: Fast Algorithms for Frequent Itemset Mining Using FP-Trees. IEEE Trans. Knowl. Data Eng. 17(10), 1347–1362 (2005)
11. Han, J., Pei, J., Yin, Y., Mao, R.: Mining Frequent Patterns without Candidate Generation: A Frequent-Pattern Tree Approach*. Data Min. Knowl. Disc. 8(1), 53–87 (2004)
12. Liu, G., Lu, H., Lou, W., Xu, Y., Yu, J.X.: Efficient Mining of Frequent Patterns Using Ascending Frequency Ordered Prefix-Tree. Data Min. Knowl. Disc. 9(3), 249–274 (2004)
13. Liu, G., Lu, H., Yu, J.X., Wang, W., Xiao, X.: Afopt: An Efficient Implementation of Pattern Growth Approach. In: Proc. IEEE ICDM Workshop FIMI (2003)
14. Lu, J., Ling, T.W., Chan, C.Y., Chen, T.: From Region Encoding to Extended Dewey: on Efficient Processing of XML Twig Pattern Matching. In: Proc. VLDB, pp. 193–204 (2005)
15. Schlegel, B., Gemulla, R., Lehner, W.: Memory-Efficient Frequent-Itemset Mining. In: Proc. EDBT, pp. 461–472 (2011)
16. Schmidt-thieme, L.: Algorithmic Features of Eclat. In: Proc. IEEE ICDM Workshop FIMI (2004)
17. Tsao, W.K., Lee, A.J., Liu, Y.H., Chang, T.W., Lin, H.H.: A Data Mining Approach to Face Detection. Pattern Recogn. 43(3), 1039–1049 (2010)
18. Tsay, Y.J., Hsu, T.J., Yu, J.R.: FIUT: A New Method for Mining Frequent Itemsets. Inf. Sci. 179(11), 1724–1737 (2009)
19. Uno, T., Kiyomi, M., Arimura, H.: Lcm ver. 2: Efficient Mining Algorithms for Frequent/Closed/Maximal Itemsets. In: Proc. IEEE ICDM Workshop FIMI (2004)
20. Wang, H., Wang, W., Yang, J., Yu, P.S.: Clustering by Pattern Similarity in Large Data Sets. In: Proc. ACM SIGMOD, pp. 394–405 (2002)
21. Zaki, M.J., Gouda, K.: Fast Vertical Mining Using Diffsets. In: Proc. ACM SIGKDD, pp. 326–335 (2003)
22. Zaki, M.J.: Scalable Algorithms for Association Mining. IEEE Trans. Knowl. Data Eng. 12(3), 372–390 (2000)

MAX-FLMin: An Approach for Mining Maximal Frequent Links and Generating Semantical Structures from Social Networks

Erick Stattner and Martine Collard

LAMIA Laboratory
University of the French West Indies and Guiana, France
{estattne,mcollard}@univ-ag.fr

Abstract. The paper proposes a new knowledge discovery method called *MAX-FLMin* for extracting frequent patterns in social networks. Unlike traditional approaches that mainly focus on the network topological structure, the originality of our solution is its ability to exploit information both on the network structure and the attributes of nodes in order to elicit specific regularities that we call "Frequent Links". This kind of patterns provides relevant knowledge about the groups of nodes most connected within the network. First, we detail the method proposed to extract maximal frequent links from social networks. Second, we show how the extracted patterns are used to generate aggregated networks that represent the initial social network with more semantics. Qualitative and quantitative studies are conducted to evaluate the performances of our algorithm in various configurations.

1 Introduction

In last decade, social network analysis has become an active research area called *"network science"* [1], an emerging discipline that focuses on relationships maintained between entities. While traditional network analysis methods, popularized with numerous studies in sociology [2], have mainly exploited the works conducted in the domain of graph theory, the current tendency, known as *"social network mining"* or more simply *"link mining"*, have attempted to apply the concepts of data mining on networks [3].

One of the classical task of social network mining consists into searching for patterns in social networks. Many existing pattern extraction methods mainly focus on the network topological structure for extracting patterns such as communities or subgraphs, but ignore the node attributes, which does not allow to take full advantage of the whole network information. Indeed, in many real situations both network structure and node features may be relevant. Especially now with the explosion of social networks, in which nodes are often represented with a set of heterogeneous attributes.

In this paper, we address the problem of the search for frequent patterns in social networks, by proposing a new and original approach that combines both

S.W. Liddle et al. (Eds.): DEXA 2012, Part I, LNCS 7446, pp. 468–483, 2012.

network structure and node features. One of the first issues of our work was to propose a definition of *"a pattern"* that combines these two aspects. Thus in our context, we search for regularities among links that connect groups of nodes. Here, a group is a set of nodes that share common characteristics. We call such patterns *"frequent links"*.

Thus, this article presents *MAX-FLMin* (Maximal Frequent Link Mining), a new knowledge discovery algorithm for extracting maximal frequent links from social networks. The algorithm works without any a priori knowledge on the social network and performs a bottom-up research by reducing the search space at each iteration. The patterns extracted are then synthesized by generating a semantic network that summarizes the whole knowledge elicited. To demonstrate the efficiency of our solution, we conduct several experiments for understanding how our algorithm behaves according to various parameters and we compare the results to those obtained by a naive approach.

This papers is organized as follows. Section 2 reviews the traditional pattern extracting methods in social networks. Section 3 formally defines the concepts of frequent links and aggregated networks. In Section 4 we detail *MAX-FLMin* and discuss its complexity. Section 5 is devoted to the aggregated networks and their generation. Section 6 demonstrates the efficiency of our solution through experimental results. Finally, Section 7 concludes and presents our future directions.

2   Related Work

Recent social network analysis methods, known as being *link mining* or *social network mining* areas, refers to *"data mining techniques that explicitly consider links when building predictive or descriptive models of linked data"* [3]. As for traditional data mining area, social network mining includes several categories of methods that address classical tasks such as node classification, group detection, link prediction, node clustering and the search for subgraphs.

The search for subgraphs is certainly the family of methods that is the most similar to a task of searching for frequent patterns in social networks. Indeed, as explained in [3,4], the most natural and widely used definition of a pattern in the context of social networks is that of *"connected subgraph"*.

Thus in the context of social networks, the problem of frequent pattern discovery consists on searching for subnetworks found in a collection of networks [5] or a single large network [6] according to a minimum support threshold. The traditional approach is to use labels associated with nodes and links. Afterwards, by using such a network representation, the problem consist on searching for sets of connected labels occurring frequently enough. A classical example is the collection of networks obtained from baskets of items. Nodes correspond to items and all items in a same basket are connected. Once such a network is created for each basket, subgraphs occurring frequently form frequent patterns in the traditional sense.

The main frequent subgraph discovery algorithms can be classified according to two basic approaches [7]: (1) the Apriori-based approach and (2) the pattern-growth approach.

(1) Apriori-based frequent subgraph discovery algorithms refer to techniques that exploit the properties of the *Apriori* algorithm [8] for finding substructures through a mining process that performs two main phases. (i) A candidate generation stage for generating candidate subgraphs and (ii) an evaluation phase that evaluates how much frequent the candidates are, by using the properties of graph isomorphism. Typical Apriori-based approaches are for instance *AGM*, proposed by Inokuchi et al. [5] for minimizing both storage and computation, or *FSG* by Kuramochi and Karypis [9] that is fitted to large network databases.

(2) Pattern-growth techniques are approaches that extend a frequent structure by adding a new edge in every possible directions [10]. The main problem of this approach is that the same structure can be generated at several iterations. For instance, *gSpan* [4] attempts to avoid the discovery of duplicate structures.

3   "Frequent Links" and "Aggregated Network" Concepts

Unlike traditional methods that only focus on structural regularities, we propose in this paper a new vision of the frequent pattern discovery in social networks by redefining the notion of *"pattern"*. Indeed, rather than defining a pattern as a subgraph, we propose a definition that combines structure and attributes by defining a pattern as a *"set of links between two groups of nodes, where nodes in each group share common characteristics"*. When these patterns are found frequently enough in the overall network, they are frequent patterns in the traditional meaning and we call them *"frequent links"*.

More formally, let $G = (V, E)$ be a network, where V is the set of nodes (vertexes) and E the set of links (edges) with $E \subseteq V \times V$.

V is defined as a relation $R(A_1, ..., A_p)$ where each A_i is an attribute. Thus, each vertex $v \in V$ is defined by a tuple $(a_1, ..., a_p)$ where $\forall k \in [1..p], v[A_k] = a_k$, the value of the attribute A_k in v.

An item is a logical expression $A = x$ where A is an attribute and x a value. The empty item is denoted $\emptyset$. An itemset is a conjunction of items for instance $A_1 = x$ *and* $A_2 = y$ *and* $A_3 = z$. An itemset which is a conjunction of p items is called a p-itemsets.

Let us note m_1 and m_2 two itemsets and V_{m_1}, V_{m_2}, respectively the sets of nodes in V that satisfy m_1 and m_2.

We denote E_{m_1} the set of links that start from nodes satisfying m_1, namely nodes in V_{m_1}: $E_{m_1} = \{e \in E \; ; \; e = (a, b) \quad a \in V_{m_1}\}$

Similarly, we note E_{m_2} the set of links that arrive to nodes in V_{m_2}:
$E_{m_2} = \{e \in E \; ; \; e = (a, b) \quad b \in V_{m_2}\}$

Thus, we define $E_{(m_1, m_2)}$ the set of links connecting nodes in V_{m_1} to nodes in V_{m_2}:

$$E_{(m_1, m_2)} = E_{m_1} \cap E_{m_2}$$
$$= \{e \in E \; ; \; e = (a, b) \quad a \in V_{m_1} \text{ and } b \in V_{m_2}\}$$

Definition 1. We call *support* of $E_{(m_1,m_2)}$, the proportion of links in E that belong to $E_{(m_1,m_2)}$, i.e.

$$supp(E_{(m_1,m_2)}) = \frac{|E_{(m_1,m_2)}|}{|E|}$$

Definition 2: We say there is a frequent link between m_1 and m_2, and we note (m_1, m_2), if the support of $E_{(m_1,m_2)}$ is greater than a minimum support threshold β,

$$supp(E_{(m_1,m_2)}) > \beta$$

Notation. Let I be the set of all itemsets built with V, we denote FL the set of frequent links among these itemsets in I.

$$FL = \bigcup_{m_1 \in I, m_2 \in I} \{ (m_1, m_2) \in I^2 ; \frac{|E_{(m_1,m_2)}|}{|E|} > \beta \}$$

Property 1: Thus, according to definition 2, if link (m_1, m_2) is frequent then the sets E_{m_1} and E_{m_2} satisfy the following condition:

$$|E_{m_1}| > \beta \times |E| \quad and \quad |E_{m_2}| > \beta \times |E| \tag{1}$$

Proof. Indeed, we have stated that a link (m_1, m_2) is frequent if

$$\frac{|E_{(m_1,m_2)}|}{|E|} > \beta$$

$$\Rightarrow \frac{|E_{m_1} \cap E_{m_2}|}{|E|} > \beta$$

$$\Rightarrow |E_{m_1} \cap E_{m_2}| > \beta \times |E|$$

$$\Rightarrow |E_{m_1}| > \beta \times |E| \quad and \quad |E_{m_2}| > \beta \times |E|$$

As for the traditional research of frequent patterns in the data mining area, the extraction of all frequent patterns is marred by the extraction of the sub-patterns that are also frequent. We thus define the *"maximal frequent links"*. For this purpose, let us first introduce the notion of *"sub-link"*.

Property 2. If sm_1 (resp. sm_2) is a sub-itemset of m_1 (resp. m_2), (for instance $m_1 = xyz$ and $sm_1 = xy$) then $|E_{(m_1,m_2)}| \leq |E_{(sm_1,m_2)}|$ and $|E_{(m_1,m_2)}| \leq |E_{(m_1,sm_2)}|$.

Proof. Let sm_1 and sm_2 be respectively sub-itemsets of m_1 and m_2. We have $V_{m_1} \subseteq V_{sm_1}$ and therefore $\forall m_2 \in I, \quad E_{(m_1,m_2)} \subseteq E_{(sm_1,m_2)}$. Similarly, $V_{m_2} \subseteq V_{sm_2}$ and $\forall m_1 \in I, \quad E_{(m_1,m_2)} \subseteq E_{(m_1,sm_2)}$
So $|E_{(m_1,m_2)}| \leq |E_{(sm_1,m_2)}|$ and $|E_{(m_1,m_2)}| \leq |E_{(m_1,sm_2)}|$

Definition 3. Let sm_1 and sm_2 be respectively sub-itemsets of m_1 and m_2. We call each link (sm_1, sm_2) *sub-links* of (m_1, m_2). Similarly, (m_1, m_2) is called *super-link* of (sm_1, sm_2).

Notation. If (sm_1, sm_2) is a sub-link of (m_1, m_2) we note $(sm_1, sm_2) \subseteq (m_1, m_2)$

Property 3. Let sm_1 and sm_2 be two itemsets. If (sm_1, sm_2) is not frequent, then any super-link (m_1, m_2) of (sm_1, sm_2) is not frequent.

Similarly, if (m_1, m_2) is frequent, then any sub-link (sm_1, sm_2) of (m_1, m_2) is frequent.

Definition 4. Let β be a minimum support threshold, we call *maximal frequent link*, any frequent link (sm_1, sm_2) such as, there exists no super-link (m_1, m_2) of (sm_1, sm_2) that is also frequent.
More formally, $\nexists (m_1, m_2) \in FL$ such as $(sm_1, sm_2) \subset (m_1, m_2)$.

Notation. We denote FL_{max} the set of maximal frequent links in FL

$$FL_{max} = \bigcup_{m_1 \in I, m_2 \in I} \{ (m_1, m_2) \in FL \; ; \; (m_1, m_2) \text{ maximal} \}$$

The usefulness of such an approach is quite obvious since the extracted patterns provide relevant knowledge about groups of nodes the most connected into the network. This knowledge can be synthesized by generating a reduced network that summarizes all these connections. We call this network an *"aggregate network"*.

Definition 5. Let $G = (V, E)$ be a social network, I the set of itemsets in V and β the minimum support threshold.
We call aggregate β-network of G, the network $AggG_\beta = (V_\beta, E_\beta)$ defined as follows.
V_β is the set of meta-nodes x such that $x = V_m$ if $\exists m' \in I$ such as $(m, m') \in FL_{max}$ or $(m', m) \in FL_{max}$.
E_β is the set of links $(x, y) \in V_\beta \times V_\beta$ such that $x = V_{m_1}$, $y = V_{m_2}$ and $(m_1, m_2) \in FL_{max}$.

Thus a node in V may belongs to several meta-nodes in V_β. A link in E_β represents a semantic relationship between two groups of V-nodes that are each described by a set of features. A link (x, y) of E_β means that there is a maximal frequent link between nodes in x and nodes in y. But some nodes in x or in y may not participate in this frequent links.

4   Maximal Frequent Link Mining

Searching all maximal frequent links into a given network may be time consuming if the search space is wide. In this work, we propose a *bottom-up* research that gradually reduces the search space. Section 4.1 details our approach for the classical cases of unimodal and oriented networks and Section 4.2 discusses flexibility, optimizations and complexity of our solution.

4.1   MAX-FLMin Algorithm

Designing an algorithm that searches for maximal frequent links is particularly challenging and computationally intensive since, in the network analysis area, it is admitted that the number of links play a key role throughout the computation phases in networks. A naive approach would be to generate all possible itemsets from attributes of nodes and then evaluate the frequency of each itemset pair.

The algorithm we propose performs a bottom-up research and exploits properties 1 and 3 for gradually reducing the search space to super-itemsets potentially involved in frequent links. Without loss of generality, the search for frequent links involving t-itemsets, can be reduced to super-link involving $(t\text{-}1)$-itemsets. *MAX-FLMin* is detailed in algorithm 1.

Algorithm 1. *MAX-FLMin* Algorithm

Require: $G = (V, E)$: **Network, and** $\beta \in [0..1]$: **Minimum support threshold**
 1. FL_{max}: **Set of all maximal frequent links** $\leftarrow \emptyset$
 2. C_{m_1}: **Stack of** m_1 **candidates itemsets** $\leftarrow \emptyset$
 3. C_{m_2}: **Stack of** m_2 **candidates itemsets** $\leftarrow \emptyset$
 4. L: **Lists of frequent links** $\leftarrow \emptyset$
 5. t: **Iteration** $\leftarrow 1$
    {*Generation of all frequent links between 1-itemsets*}
 6. $I_l \leftarrow$ Generate all 1-itemsets m_1 from V such as $|E_{m_1}| > \beta \times |E|$
 7. $I_r \leftarrow$ Generate all 1-itemsets m_2 from V such as $|E_{m_2}| > \beta \times |E|$
 8. **for all** itemset $m_1 \in I_l$ **do**
 9.   **for all** itemset $m_2 \in I_r$ **do**
10.   **if** $|E_{(m_1,m_2)}| > \beta \times |E|$ **then**
11.   add m_1 to C_{m_1}
12.   add m_2 to C_{m_2}
13.   add (m_1, m_2) to L
14.   add (m_1, m_2) to FL_{max}
15.   **end if**
16.   **end for**
17. **end for**
    {*Generation of the other frequent links*}
18. $t \leftarrow t + 1$
19. **while** $L \neq \emptyset$ **and** *allCombinations()* = *false* **do**
20. $C_{m_1} \leftarrow$ {joint of all distinct $(t\text{-}1)$-itemsets m_1 of L sharing $(t\text{-}2)$ items, such as $|E_{m_1}| > \beta \times |E|$ } $\bigcup C_{m_1}$
21. $C_{m_2} \leftarrow$ {joint of all distinct $(t\text{-}1)$-itemsets m_2 of L sharing $(t\text{-}2)$ items, such as $|E_{m_2}| > \beta \times |E|$ } $\bigcup C_{m_2}$
22. $L \leftarrow \emptyset$
23.   **for all** itemset $m_1 \in C_{m_1}$ **do**
24.   **for all** itemset $m_2 \in C_{m_2}$ **do**
25.   **if** $((|m_1| = t$ or $|m_2| = t)$ and $\nexists l \in L$ such as $(m_1, m_2) \subset l$ and $\frac{|E_{(m_1,m_2)}|}{|E|} >$ $\beta)$ **then**
26.   add (m_1, m_2) to L
27.   remove all $q \in FL_{max}$ such as $q \subset (m_1, m_2)$
28.   add (m_1, m_2) to FL_{max}
29.   **end if**
30.   **end for**
31.   **end for**
32. $t \leftarrow t + 1$
33. **end while**
34. **return** FL_{max}

More precisely, at iteration $t = 1$, *MAX-FLMin* starts by constructing the sets I_l and I_r that are respectively the set of m_1 and m_2 1-itemsets that verify property 1 (see lines 6-7 algorithm 1). Then, frequent links are searched among these itemsets and stored in a temporary list L, that will store at each iteration t all frequent links involving t-itemsets. During this process, m_1 and m_2 itemsets involved in frequent links are also stored in separate stacks for further generations. At this level, all frequent links are considered as maximal ones and stored in FL_{max} (see lines 8-17).

At iteration $t+1$, m_1 (resp. m_2) candidate itemsets are generated and stored in C_{m_1} (resp. C_{m_2}) (see lines 20-21). These candidates are the union of (i) the super-itemsets generated from the t-itemsets of L, i.e. the $(t+1)$-itemsets potentially involved in frequent links according to properties 1 and 3, and (ii) the previous m_1 (or m_2) itemsets already involved in frequent links since maximal frequent links do not necessarily imply maximal itemsets. Note that sets C_{m_1} and C_{m_2} are sorted from the largest ones to smallest ones (in terms of number of items).

Once candidates are generated, the list L is cleared and frequent links can be extracted (see lines 22-31). The comparison is performed only if at least one of the candidate itemsets has a size $t+1$ (in order to not compare sub-links already processed) and if a frequent link is not already in L. Indeed, as C_{m_1} and C_{m_2} are sorted, the first (m_1, m_2) frequent links identified are necessarily the maximal ones regarding iteration $t+1$. The comparison is thus done to check if a super-link has not already been added to L, i.e. if (m_1, m_2) is maximal. If (m_1, m_2) is a maximal frequent link regarding iteration $t+1$, it is added to L and FL_{max} and all its sub-links are removed from FL_{max}.

These operations are repeated until no more maximal frequent links is detected or all the combinations are performed (lines 19-33).

4.2 Discussion

Real-world networks have various features: directed, undirected, unipartite or multipartite. Thus, a valuable characteristic of any algorithm that aims to analyze networks is the ability to adapt to all kinds of networks. This is the case of *MAX-FLMin*.

Since it is common to represent undirected networks as directed ones, in which links are stored in both directions, *MAX-FLMin* can be directly applied. However, an interesting property of the frequent link definition (see definition 2), in the case of undirected networks, is that if the link (m_1, m_2) is frequent, the link (m_2, m_1) is frequent too. Thus, as shown in algorithm 2, only one set can be used for storing the 1-itemsets and symmetrical comparisons can be avoided.

Regarding multipartite networks, the algorithm can also be directly applied, but performs unnecessary comparisons since the initial 1-itemsets are calculated on the overall set of nodes (see lines 6-7 of algorithm 1). Thus, if we have a knowledge on the nodes involved on both sides of the links, generation of the sets I_l and I_r in algorithm 1 can be performed as follows:

- $I_l \leftarrow$ Generate all 1-itemsets m_1 from $v \in V$ involved to the left of the links such as $E_{m_1} > \beta \times |E|$

Algorithm 2. Adaptation of lines 6-17 of algorithm 1 for undirected networks

---

Require: $G = (V, E)$: **Network**, and $\beta \in [0..1]$: **Minimum support threshold**
1. i, j: **integer**
2. $I \leftarrow$ Generate all 1-itemsets from V such as $|E_{m_1}| > \beta \times |E|$
3. **for all** i from 0 to $I.size$ **do**
4.   **for all** j from i to $I.size$ **do**
5.   {*Lines 10-15 of algorithm 1 remain unchanged*}
6.   **end for**
7. **end for**

- $I_r \leftarrow$ Generate all 1-itemsets m_2 from $v \in V$ involved to the right of the links such as $E_{m_2} > \beta \times |E|$

Regarding complexity, the key computation step is the generation of the sets E_{m_1}, E_{m_2} and $E_{(m_1, m_2)}$ at each iteration. A straightforward and efficient way to implement this task and speed up the process is to use a node structure that stores its input and output neighbors. Thus, rather than iterate over all the network links, the search is reduced to nodes as detailed on algorithm 3.

Algorithm 3. Optimization of the generation of the E sets

Require: $G = (V, E)$: **Network**, m_1: **Itemset**, m_2: **Itemset**
1. E_{m_1}: **set of links** $\leftarrow \emptyset$
2. E_{m_1}: **set of links** $\leftarrow \emptyset$
3. $E_{(m_1, m_2)}$: **set of links** $\leftarrow \emptyset$
4. **for all** node $v \in V$ **do**
5.   **if** v matches with m_1 **then**
6.   add all output links of v to E_{m_1}
7.   **end if**
8.   **if** v matches with m_2 **then**
9.   add all input links of v to E_{m_2}
10.   **end if**
11. **end for**
12. $E_{(m_1, m_2)} \leftarrow E_{m_1} \cap E_{m_2}$

More formally, two parameters are involved in the amount of computations: (i) the number of attributes $|R|$ and (ii) the size of the network ($|V|$ and $|E|$). As explained previously, in a naive approach, $2^{|R|} \times 2^{|R|} \times |E|$ computations are required to extract maximal frequent links.

For studying the complexity of *MAX-FLMin*, let us first consider the empirical case of a complete network (configuration 1). In such a configuration, all links are frequent and therefore the entire itemsets lattice has to be explored for extracting the maximal frequent links. Our solution performs $|R| \times |R| + \sum_{k=1}^{|R|} C_{|R|}^k \times C_{|R|}^k \times |V|$ computations.

Let us now consider the other extreme case in which no link is frequent (configuration 2). When searching for the 1−frequent links (see lines 8-17 of algorithm 1), we detect that no link is frequent. In this configuration $|R| \times |R|$ computations are required.

For more clarity, consider a complete network $G = (V, E)$ with $|V| = 10000$ and $|E| = |V| \times (|V| - 1)$. Figure 1 shows, according to the number of attributes, (a) the estimation of the number of computations (log), and (b) the gain with respect to a naive approach.

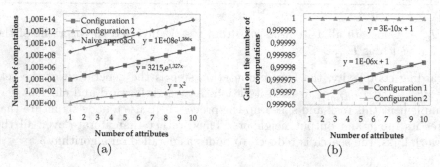

(a)   (b)

Fig. 1. Estimation of (a) the number of computations (log) and (b) the gain compared to a naive approach for configurations 1 and 2

As shown on Figure 1(a), the number of computations in configuration 1 increases exponentially since it can be approximated by $y = 3215.1e^{1.3275 \times |R|}$. However, for configurations 1 and 2, we can observe the good performances of *MAX-FLMin* since the gain on the number of computation is systematically above 99% (see Figure 1(b)).

5   Aggregate Network Generation

Once the maximal frequent links have been extracted, the aggregate network can be generated. It is important to understand that, for a given threshold β, the overall set of maximal frequent links allows obtaining the β maximal aggregate network. The generation process performs in two steps:

1. The maximal frequent links are extracted from G
2. Then, the aggregate network is generated from FL_{max}

In the aggregated network, nodes correspond to itemsets while links represent maximal frequent links. Note that the aggregated network conserves the same properties. If G is directed (resp. undirected), *AggNet* is also directed (resp. undirected). Similarly, if G is a multipartite network, *AggNet* is multipartite too. The algorithm for generating the maximal aggregated network, called *AggNet-MFL* (aggregated network based on maximal frequent links), is detailed in algorithm 4.

Let us specify that aggregated network is not a direct or lighter representation of the initial network. Indeed, there is no simple mapping between the aggregate

---

Algorithm 4. *AggNet-MFL*: Aggregated Network Generation

Require: Fl_{max}: **Set of maximal frequent links**
1. $V_\beta \leftarrow \emptyset$: **Set of meta-nodes**
2. $E_\beta \leftarrow \emptyset$: **Set of couple of meta-nodes (links)**
3. $AggG_\beta \leftarrow (V_\beta, E_\beta)$: **Aggregated network**
4. **for all** maximal frequent link $l=(m_1, m_2) \in FL_{max}$ **do**
5.   **if** $m_1 \notin V_\beta$ **then**
6.   add m_1 to V_β
7.   **end if**
8.   **if** $m_2 \notin V_\beta$ **then**
9.   add m_2 to V_β
10.   **end if**
11.   add (m_1, m_2) to E_β
12. **end for**
13. **return** $AggG_\beta$

---

and the initial network since nodes in the initial network may be represented by several nodes in the aggregate one and inversely, nodes in the aggregate network correspond to groups of nodes in the initial network. Thus, the resulting structure is a much more semantic network, since it may be viewed as a form of knowledge representation acquired on the connections between groups of nodes in the initial network.

6   Experimental Results

Various sets of experiments have been conducted to evaluate the performances of *MAX-FLMin*. First, we describe in Section 6.1 the dataset used and the test environment. Afterwards we study in Section 6.2 how evolve maximal frequent links according to qualitative and quantitative point of views. Finally, Section 6.3 focus on the associated aggregated networks and their evolution.

6.1   Testbed

The dataset used is a geographical proximity contact network obtained with Episims [11], a simulation tool that reproduces the daily movements of individuals in the city of Portland. In the network, two individuals are connected when they were geographically close during the simulation. The main features of the network are described in Figure 2.

Data have been processed so that each node is identified by (1) age class, i.e. $\lfloor \frac{age}{10} \rfloor$ (2) gender (1-male, 2-female), (3) worker (1-has a job, 2-has no job), (4) relationship to the head of household (1-spouse, partner, or head of household, 2-child, 3-adult relative, 4-other) and (5) contact class, i.e. $\lfloor \frac{degree}{2} \rfloor$ (6) sociability (i.e. 1-$cc > 0.5$, 2-$else$).

The application of our approach on this network makes sense since our belief is that underlying patterns can be highlighted between attributes of individuals and proximity contacts they maintain.

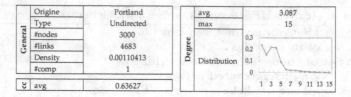

General	Origine	Portland
	Type	Undirected
	#nodes	3000
	#links	4683
	Density	0.00110413
	#comp	1
cc	avg	0.63627

Fig. 2. Main features of the contact network used (#comp is the number of components and cc is the clustering coefficient)

MAX-FLMin has been developed in JAVA and included into the graphical tool GT-FLMin [12]. All experiments have been averaged on 100 runs and conducted on a Intel Core 2 Duo P8600, 2.4Ghz, 3Go Ram, Linux Ubuntu 10.10 with Java JDK 1.6. In experiments, the size of the network is varied by extracting subgraphs into the overall network, which allows us to make varying both nodes and links. Combinations of network size used are $(|V|, |E|) = \{(500 , 806),$ $(1000,1750), (1500,2685), (2000 , 3304), (2500,3988), (3000,4683)\}$. For simplicity in the rest of this paper, we will discuss the size of the network by simply referring to $|V|$. Similarly, the number of attributes $|R|$ evolves by removing, starting with the last, attributes from node information.

6.2   MAX-FLMin: Qualitative and Quantitative Results

In a first approach, we focus on the quality of the extracted patterns. As shown in Figure 3, which shows the maximal frequent links obtained with configuration $|V| = 3000$, $|R| = 4$ and $\beta = 0.10$, the patterns extracted are relevant since they provide a direct knowledge about the connected groups of nodes in the network ('*' means that the attribute can take any value). For example, the first row indicates that 10.7% of the links of the network connect 40 years old individuals who have a job to individuals who do not have a job.

In a quantitative point of view, we compare the incidence of different support thresholds ((a) $\beta = 0.11$, (b) $\beta = 0.15$ and (c) $\beta = 0.2$) on (1) the number of maximal frequent links, (2) the runtime (sec) and (3) the gain on the runtime compared to a naive approach. Figure 4 describes these results for different $|R|$ values and according to the network size.

Several observations can be made regarding the evolution of the number of extracted patterns (see Figures 4(1)).

Max. Frequent Link	Support
((4;*;1;*),(*;*;2;*))	0.107
((2;*;*;2),(*;*;2;2))	0.105
((*;1;1;*),(*;*;1;*))	0.113
((1;*;2;2),(*;1;*;*))	0.102
((*;1;1;1),(*;2;*;*))	0.133

Fig. 3. A sample of maximal frequent links obtained with configuration $|V| = 3000$, $|R| = 4$ and $\beta = 0.10$

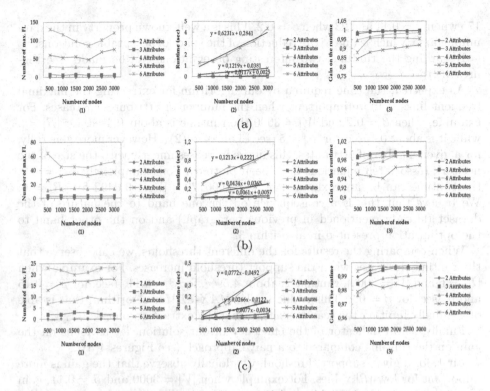

Fig. 4. Effects of different support thresholds ((a) $\beta = 0.11$, (b) $\beta = 0.15$ and (c) $\beta = 0.2$) on (1) the number of maximal frequent links, (2) the runtime (sec) and (3) the gain on the runtime compared to a naive approach

First of all, we observe that the number of maximal frequent links is more important when the number of attributes is high as expected. For example, for $\beta = 0.11$ and $|V| = 3000$, the number of extracted patterns is approximately 125 when $|R| = 6$, while it is about 75 when $|R| = 5$ (see Figure 4(a)(1)). This can be explained by the fact that when the $|R|$ value is increased, the amount of itemsets potentially involved in frequent links is statistically increased too.

However, we were surprised to observe that, whatever the β threshold is, the number of patterns remain relatively stable when the network size increases. This is a very interesting result that we explain by two factors. (i) The nature of the attributes. Indeed many attributes are binary and therefore when you focus on a subset of the dataset (a subgraph in our context), the probability of generating the same itemsets as for the overall dataset is strong. (ii) The human behaviors in general, since the underlying factors that generate or influence the behaviors can often be found at smaller scales. In other words, if we focus on a subset relevant enough, the data distribution is such that it is likely to extract a large majority of patterns.

Finally as expected, when comparing the results obtained for the different thresholds, we observe that the number of extracted patterns decreases when the β increases. For example, for $|R| = 6$ and $|V| = 3000$, the number of patterns is about

175 when $\beta = 0.11$ and 23 when $\beta = 0.2$. This is a well-known property in the data mining area, which is due to the reduction of the space of acceptable solutions.

Regarding the runtime, two interesting observations can be made (see Figures 4(2)).

As expected, the time required by our algorithm for extracting the maximal frequent links is more important when the number of attributes increases. For example, when $\beta = 0.2$ and $|V| = 3000$ the runtime is about 0.4sec. for $|R| = 6$ while it is about 0.15sec. for $|R| = 5$ (see Figure 4(c)(2)). However more generally, for a given number of attributes, this figure increases linearly with the size of the network (associated equations have been plotted). For example when $\beta = 0.2$, the runtime can be approximated by $y = 0.00772 \times |V| + 0.0492$ for $|R| = 6$. We believe that this is a consequence on the one hand to the nature of the dataset (already mentioned in previous paragraph) and on the other hand to the optimization presented in algorithm 3.

When comparing the results for the different thresholds, we can observe that the runtime decreases when the support threshold increases. For example when $|R| = 6$ and $|V = 3000|$, runtime is about 4.5sec. for $\beta = 0.11$, 1sec. for $\beta = 0.15$ and 0.4sec. for $\beta = 0.2$. This is due to the *MAX-FLMin* algorithm, that is able of gradually limiting the search space during the extraction phase.

Finally, as an indicator of the efficiency of our solution, let us focus on the gain on the runtime compared to a naive approach (see Figures 4(3)).

First, for a given support threshold, we globally observe that the gain is more important for low $|R|$ values. For example when $|V| = 3000$ and $\beta = 0.11$, gain is about 90% for $|R| = 5$ while it is about 99% for $|R| = 4$ (see Figure 4(a)(3)).

Moreover, we observe that the gain increases with the β threshold. For example for $|V| = 3000$ and $|R| = 5$, the gain is about 96% for $\beta = 0.15$ while 98% for $\beta = 0.2$.

Thus, these results confirm the study conducted on the complexity in Section 4 and demonstrate both good performances and efficiency of *MAX-FLMin* for extracting maximum frequent links. Indeed whatever is the β threshold used, the gain on the runtime is always greater than 80%.

As we have observed that the runtime increases more or less linearly with the size of the network (see Figures 4(2)), we have studied how evolves the slope of these curves according to the number of attributes. Figure 5 shows the *logarithm* of the slope for different β value. As a reference, the result is also plotted for a naive approach.

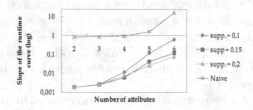

Fig. 5. Log of the slope of the runtime curve according to $|R|$

We can observe that the curves are more or less straight, which suggests that this slope is growing exponentially with the number of attributes; a result that the study conducted on complexity implied. However, we also note that, compared to the naive approach, the difference of the evolution of this slope is significant. This confirms the good performances of our algorithm.

6.3 Aggregated Networks: Examples and Evolution

To conclude this section on experiments, we focus on the aggregated networks by studying how they evolve according to the minimum support threshold β.

As a first approach, Figure 6 shows some examples of aggregated networks. (a) is the initial network presented in Figure 2 with $|V| = 3000$ and $|R| = 6$, and (b), (c), (d) and (e) are respectively the aggregated networks obtained with $\beta = 0.05$, $\beta = 0.1$, $\beta = 0.15$ and $\beta = 0.2$.

As you can see, our approach significantly reduces the network size. However, it is important to keep in mind that the resulting network is a much more semantic network that aims to represent the groups of nodes the most connected in the initial network. We insist on the fact that it must not be seen as a direct or lighter representation of the initial network (see Section 5).

In Figure 7, we describe how evolve the main features of the aggregated network according to the support threshold: (a) presents the evolution of the number of nodes and links, (b) describes the evolution of the network clustering coefficient and (c) details the degree distribution for some β values ($\beta = 0.11$, $\beta = 0.15$ and $\beta = 0.2$).

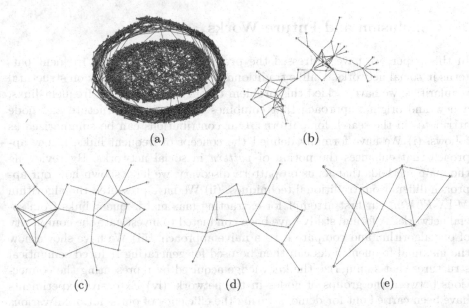

(a)   (b)

(c)   (d)   (e)

Fig. 6. Aggregated networks obtained from network (a) (see also Figure 2) with $\beta = 0.05$ (b), $\beta = 0.10$ (c), $\beta = 0.15$ (d) and $\beta = 0.2$ (e)

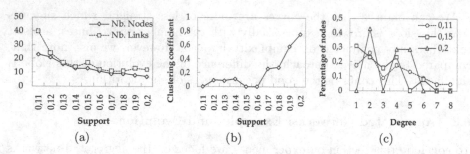

Fig. 7. Evolution of some features of the aggregated networks: (a) network size, (b) clustering coefficient and (c) degree distribution

As expected, the size of the aggregated network (number of nodes and links) decreases when the support increases (see Figure 7(a)). Indeed, increasing the support threshold leads to a reduction of the acceptability space of solutions. Thus, fewer frequent links are extracted, which explains the observed phenomenon.

It is difficult to explain the general shape of the curve obtained for the clustering coefficient (see Figure 7(b)). Our belief is that when the support is increased, only groups of nodes that have very strong connections between them are preserved. This may explain the growth of the network clustering coefficient.

Regarding the network degree distribution (see Figure 7(c)), we globally observe that the aggregated networks have a structure often observed, in which a high proportion of nodes have a very low number of connections, while the highly connected nodes are rare.

7   Conclusion and Future Works

In this paper, we have addressed the problem of the search for frequent patterns in social networks. Unlike traditional works that focus solely on structural regularities, we have tackled this problem through the search for frequent links, a new and original approach that combines both network structure and node attributes in the search for patterns. Our contributions can be summarized as follows. (i) We have formally defined the concept of frequent links, a new approach that enhances the notion of *pattern* in social networks. By reviewing the main methods that focus on pattern discovery, we have shown how our approach differs from traditional techniques. (ii) We have presented the algorithm *MAX-FLMin*, our first attempt for extracting maximal frequent links from social networks. A formal study have been conducted to investigate the complexity of our algorithm and compare it to a naive approach. (iii) We have shown how the maximal frequent links can then be used for generating reduced semantical structures that summarize the knowledge acquired by representing the connections between the groups of nodes in the network. (iv) Extensive experiments have been carried out for demonstrating the efficiency of our solution in various configurations. These experiments have highlighted interesting features of human behaviors, that have allowed to observe that the number of attributes is a parameter more involved in the number of patterns extracted than the network size. (v) Our solution has been implemented into the graphical tool *GT-FLMin*.

As perspectives in a short term, we want to improve the performances of our algorithm by reducing the combinations phases. As a first attempt, some tracks have already been presented in the article.

In a long term, the proposed approach and especially the aggregate network, raises a variety of new interesting research issues that we plan to address. One first issue is the definition commonly attributed to a community. Indeed in some extent, our approach highlights different communities and the link they maintain. Nevertheless, these communities are far from the traditionally accepted definition, namely a set of nodes densely connected, since in our approach nodes in the same community are not necessarily connected. Thus this work raises a fundamental question on the notion of *community* in social networks.

Similarly, another interesting track would be to use the aggregated network as a predictive model for addressing the link prediction problem. Indeed, our belief is that the patterns extracted by *MAX-FLMin* could be used to predict with great accuracy, the occurrence of new links in social networks. Thus, it would be very interesting to compare such a solution to traditional methods.

References

1. Barabasi, A., Crandall, R.: Linked: The new science of networks. American Journal of Physics 71, 409 (2003)
2. Milgram, S.: The small world problem. Psychology Today 1, 61–67 (1967)
3. Getoor, L., Diehl, C.P.: Link mining: a survey. SIGKDD Explor. 7, 3–12 (2005)
4. Yan, X., Han, J.: gspan: Graph-based substructure pattern mining. In: Proceedings of the 2002 IEEE International Conference on Data Mining (2002)
5. Inokuchi, A., Washio, T., Motoda, H.: An Apriori-Based Algorithm for Mining Frequent Substructures from Graph Data. In: Zighed, D.A., Komorowski, J., Żytkow, J.M. (eds.) PKDD 2000. LNCS (LNAI), vol. 1910, pp. 13–23. Springer, Heidelberg (2000)
6. Kuramochi, M., Karypis, G.: Finding frequent patterns in a large sparse graph. Data Min. Knowl. Discov. 11, 243–271 (2005)
7. Cheng, H., Yan, X., Han, J.: Mining graph patterns. In: Managing and Mining Graph Data, pp. 365–392 (2010)
8. Agrawal, R., Srikant, R.: Fast algorithms for mining association rules in large databases. In: Proceedings of the 20th International Conference on Very Large Data Bases, pp. 487–499 (1994)
9. Kuramochi, M., Karypis, G.: Frequent subgraph discovery. In: Proceedings of the 2001 IEEE International Conference on Data Mining, pp. 313–320 (2001)
10. Nijssen, S., Kok, J.N.: The gaston tool for frequent subgraph mining. Electr. Notes Theor. Comput. Sci. 127(1), 77–87 (2005)
11. Barrett, C.L., Bisset, K.R., Eubank, S.G., Feng, X., Marathe, M.V.: Episimdemics: an efficient algorithm for simulating the spread of infectious disease over large realistic social networks. In: Conference on Supercomputing, pp. 1–12 (2008)
12. Stattner, E., Collard, M.: Gt-flmin: Un outil graphique pour lextraction de liens frquents dans les rseaux sociaux. In: 12e Conference Internationale Francophone sur l'Extraction et la Gestion de Connaissance, EGC (2012)

Sequenced Route Query
in Road Network Distance
Based on Incremental Euclidean Restriction

Yutaka Ohsawa[1], Htoo Htoo[1], Noboru Sonehara[2], and Masao Sakauchi[2]

[1] Graduate School of Science and Engineering, Saitama University
[2] National Institute of Informatics

Abstract. This paper proposes a fast trip planning query method in the road network distance. The current position, the final destination, and some number of point of interest (POI) categories visited during the trip are specified in advance. Then, the query searches the shortest route from the current position with stops at one of each specified POI category from the visiting sequence before reaching the final destination. Several such types of trip planning methods have been proposed. Among them, this paper deals with the optimal sequenced route (OSR) which is the simplest query because it has a strongest restriction on the visiting order. This paper proposes a fast incremental algorithm to find OSR candidates in the Euclidean space. Furthermore, it provides an efficient verification method for the road network distance.

1 Introduction

In recent years, several types of trip planning query methods have been proposed for location based services (LBS). In the typical trip planning, some point of interest (POI) categories are given as stopovers before arriving at a final destination. Li et al.[1] proposed a trip planning query (TPQ) that does not specify a visiting order for the POI categories. For example, a restaurant, a department store, and a movie theater may be visited before reaching the final destination; however, the visiting order is not specified. Sharifzadeh et al. [2] proposed an optimal sequenced route (OSR) in which a unique visiting order is given. For example, the department store should be visited first, next the restaurant, and finally the movie theater. Multi-rule partial sequenced route (MRPSR) query by Chen et al.[3] was a generalization approach from TPQ and OSR.

The framework employed in this paper is based on an incremental Euclidean restriction (IER)[4] approach, which searches candidates in the Euclidean distance first, and then verifies the results of the road network distance. This approach is versatile and has been applied to several types of queries based on the road network distance. However, few attempts have been made to apply the approach to trip planning queries.

For the Euclidean distance search, several efficient incremental search algorithms for range queries, k-NN queries, and ANN queries, that use the minimum

S.W. Liddle et al. (Eds.): DEXA 2012, Part I, LNCS 7446, pp. 484–491, 2012.
© Springer-Verlag Berlin Heidelberg 2012

bounding rectangle (MBR) in the R-tree index have already been presented. An incremental search reports the results one at a time, starting with the best. This characteristic is essential for the IER framework because all possible routes in the Euclidean distance that are shorter than the shortest route in the road network should be searched. However, the road network distance cannot be known before verification. This paper proposes efficient algorithms for both steps of the IER framework.

The main contributions of this paper are as follows:

- to present a novel incremental OSR search algorithm in the Euclidean distance. This algorithm determines the OSR by a best-first search in R-trees;
- to present an efficient algorithm for verifying the road network distance that reduces the number of pair-wise distance calculations.

2   Incremental Queries in the Euclidean Distance

2.1   OSR Queries in the IER Framework

We define the OSR query as:

Definition 1 (OSR query). *Given a current point s, a final trip destination d, and a visiting order of POI category sets $C_i (1 \leq i \leq m)$, the OSR query finds the minimum distance route starting from s, selecting one POI from each C_i according to the visiting sequence, and finally arriving at d.*

The IER framework generates candidates for OSRs in the Euclidean space, and then verifies those candidates in the road network distance. Let the shortest OSR given by searches in Euclidean space be Sr and its verified length in the road network be $L_N(Sr)$. The shortest OSR in the Euclidean space is not always the shortest OSR in the road network distance. Therefore, all OSRs whose length are less than $L_N(Sr)$ also have the potential to be the shortest route in the road network. Therefore, all OSRs less than $L_N(Sr)$ must be searched in the Euclidean space, and then the results must be verified in the road network. Finally, the shortest OSR in the road network is returned as the result. These are the essential steps of an OSR query based on the IER framework.

In this paper, when two points a and b are given, $d_E(a, b)$ denotes the Euclidean distance between a and b, and $d_N(a, b)$ denotes the road network distance between a and b. IER depends on the relationship $d_E(a, b) \leq d_N(a, b)$. Therefore, if an OSR with the length $L_N(Sr)$ is obtained, the OSR candidates in Euclidean distance longer than $L_N(Sr)$ can be safely discarded.

All OSRs whose lengths are less than $L_N(Sr)$ can be determined by an incremental search. In an incremental search, OSR candidates are searched from the shortest up to k OSRs. Therefore, all OSRs shorter than $L_N(Sr)$ can be determined by repeating the incremental search while the length of the determined OSR is shorter than $L_N(Sr)$.

2.2   Simple Trip Route Query in Euclidean Distance

Before describing general OSR queries in which multiple POI categories to be visited are specified, we discuss the simplest trip planning query case, a simple trip route (STR) query. An STR query finds the shortest route from a starting point (s) to a destination (d) via a POI belonging to a specified category.

This section presents an incremental search algorithm for an STR query in the Euclidean distance. In general, the number of the POIs belonging to the specified category is large, therefore, we assume that the POIs are indexed by an R-tree [5]. The basic strategy to find an STR is a best-first search by calculating the lower bound route length (LBRL) to the MBRs in the R-tree.

Fig. 1 shows typical examples of positional relationships among s, d, and three MBRs ($mbr1, mbr2, mbr3$) in an R-tree. The dotted lines show the lower bound routes for each MBR. All possible arrangements for two points and an MBR can be categorized into these three cases to evaluate the LBRL.

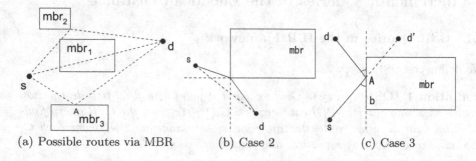

(a) Possible routes via MBR   (b) Case 2   (c) Case 3

Fig. 1. Lower bound route of STR

The LBRL calculation method can be summarized below. Let the line segment whose end points are s and d be $\ell^{s,d}$, the objective MBR to calculate the LBRL be mbr, and the four vertices of the MBR be $v_1 - v_4$.

Case1: Where $\ell^{s,d}$ intersects mbr, the LBRL is the length of $\ell^{s,d}$, i.e., $|\ell^{s,d}|$. This case corresponds to $mbr1$ in Fig. 1(a).

Case2: Where $\ell^{s,d}$ intersects both extended lines of the horizontal and vertical sides of mbr, the LBRL is the minimum length through a vertex of mbr (Fig. 1(b)), i.e., $\min(|\ell^{s,v_i}| + |\ell^{v_i,d}|) : \{i = 1, \ldots, 4\}$.

Case3: $\ell^{s,d}$ is located on one side of an edge (b) of mbr (Fig. 1(c)). In this case, the point d' which is symmetrical with respect to d across an edge of the MBR b, is obtained. Then the intersection point A of b and $|\ell^{s,d'}|$ is calculated. When point A is located in the extent of the edge b, the LBRL is $|\ell^{s,d'}|(=|\ell^{s,A}| + |\ell^{A,d}|)$. Otherwise, the LBRL is calculated by the same method as that of Case 2.

Hereafter, the LBRL obtained from the method described above is denoted as $L_E^{s,d}(e)$, where e is either an MBR in the R-tree or a POI. When e is an MBR, the value of $L_E^{s,d}(e)$ shows the LBRL against the MBR. when e is a POI, the value shows the trip route length in the Euclidean distance via the POI. The R-tree is traversed by a best-first search using a PQ. Here, the PQ manages the following records.

$$< L_E^{s,d}(e), e >$$ (1)

Fig. 2 illustrates the process of finding the trip route on the R-tree. Fig.2(a) shows an R-tree; Fig.2(b) and (c) show the arrangement of the MBRs (rectangles) and the POIs (black dots). In Fig.2(b) and (c), the dashed rectangles show the MBR of the root node, the dotted lines illustrate trip routes, and the accompanying numbers show the length of the trip routes.

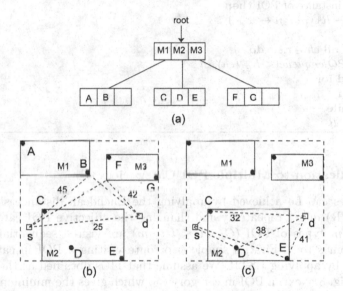

Fig. 2. Example of R-tree

Initially, the LBRL is calculated for each MBR in the root node, the record of Eq.(1) is composed and it is enqueued into the PQ. At this point, the content of the PQ is as follows.

$$< 25, M2 >, < 42, M3 >, < 45, M1 >$$

By dequeuing, $< 25, M2 >$ is obtained from the PQ; hence, the child node of $M2$ is descended one level and reaches the leaf node that contains POIs C, D, and E. The LBRL is calculated for each POI, and the corresponding records are enqueued. At this point, the PQ contains the following records.

$$< 32, C >, < 38, D >, < 41, E >, < 42, M3 >, < 45, M1 >$$

Dequeuing the PQ again, we obtain record $< 32, C >$, and e of the record is a POI. Thus, the shortest trip route via C is determined. If we continue the search until we get the shortest trip routes for k number, we can find k shortest routes in the ascending order of length. Algorithm 1 shows a pseudo-code for the STR search based on the Euclidean distance.

Algorithm 1. Euclidean distance simple trip route query (ESTR)

Input: $s,d,root,k$
Output: kSTR
 1: $n \leftarrow 0, R \leftarrow \emptyset$
 2: $PQ.enqueue(< d_E(s,d), root >)$
 3: **while** $PQ.size() > 0$ and $n < k$ **do**
 4:    $r \leftarrow PQ.dequeue()$
 5:    **if** $r.e$ instance of POI **then**
 6:       $R \leftarrow R \cup r.e, \ n \leftarrow n + 1$
 7:    **else**
 8:       **for all** $ch \in r.e.c$ **do**
 9:          $PQ.enqueue(< L_E^{s,d}(ch), ch >)$
10:       **end for**
11:    **end if**
12: **end while**
13: **return** R

2.3   Application to Multiple POI Categories

OSR queries can be achieved by applying the Euclidean distance simple trip route (ESTR) query repeatedly and changing the objective POI category. Assume that m types of POI ($C_i : 1 \le i \le m$) are visited sequentially during the trip from s to d. First, a simple trip route visiting a POI in category C_1 is searched by applying ESTR. We assume that p^1 is obtained as the result as shown in Fig. 3. Next, a POI in category C_2, which gives the minimum distance during the trip from p^1 to d, is searched by applying the ESTR again. Repeating this search, we can obtain a route by visiting a number of m POIs sequentially during the trip from s to d.

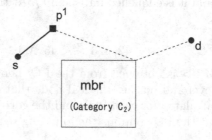

Fig. 3. OSR query using ESTR

The entire search is controlled by a PQ. The records in the PQ are ordered by the distance of the route from s to d by visiting already determined POIs and an MBR, which is searched next. For example, in Fig. 3, the cost value is $d_E(s, p^1) + L_E^{p^1,d}(m)$. The PQ contains records whose categories of targets are different. The PQ record has the following format.

$$< Cost, prev, dfs, tgt, e, PSR > \qquad (2)$$

Here, $prev$ is the POI that belongs to the category preceding the current target tgt category, and its initial value is s. Furthermore, dfs is the partial route length from s to $prev$. tgt is the target POI category number next to be searched, e is a node in the R-tree managing the POIs in the category C_{tgt}. and PSR is a sequenced POI set determined up to this point. The PQ returns records in the ascending order of the $Cost$ value. For example, in Fig. 3, $prev$ is p^1, dfs is $d_E(s, p^1)$, tgt is 2, e is mbr, and PSR is $\{s, p^1\}$.

Let the record dequeued from the PQ be r. When e of r ($r.e$) is an MBR, new records are composed for all child nodes of $r.e$, and then the records are enqueued into the PQ. Otherwise, when $r.e$ is a POI, it is the POI to be visited next. Therefore, the POI category is advanced by one, and then the next target category is changed to $C_{e.tgt+1}$. When the category $C_{e.tgt+1}$ is the final destination d, a complete route is found, and it is the shortest OSR. Therefore, the result route is returned.

This algorithm can generate OSRs incrementally from the shortest to the next shortest if the function retains the contents of the PQ after the shortest OSR is found. The verification on road network distance requires all OSR candidates whose route lengths are less than L^{min}. This search can be achieved by iterating the algorithm while the route length is less than L^{min}. Algorithm 2 shows the pseudo-code of the OSR search in the Euclidean distance.

The verification on road network distance can be achieved several ways including pair-wise A* algorithm and several materializing methods of shortest path distance on road network.

3   Experimental Result

We implemented the algorithms described in the previous section in Java and conducted experimental evaluations. The hardware used in the experiments was an Intel Core i7 CPU (3.2GHz) with 9 GB memory. The road map data used in the experiments covers a 200-km^2 area including urban and suburban areas, and consists of 25,586 road segments. The POIs locations were generated by a pseudo-random sequence generator with a specified probability ($Prob$). For example, $Prob = 10^{-3}$ indicates a POI on one thousand road segments.

Fig. 4 compares the referred R-tree node numbers of PNE [2] and the EOSR in OSR queries in the Euclidean distance. In the experiments, the number of visiting POI categories (m) is set at 3. The horizontal axis shows POI density and the vertical axis shows the number of referred nodes in R-trees. The size of the R-tree nodes was set to 64 slots (size of a node was 2KB).

Algorithm 2. Euclidean Optimal Sequenced Route (EOSR)

Input: $s,d,m,T(i : i \le i \le m)$
Output: Euclidean OSR

```
 1: PQ.enqueue(< d_E(s,d), s, 0, 1, T(1).root, {s} >)
 2: while PQ.size() > 0 do
 3:    r ← PQ.dequeue()
 4:    if r.tgt > m then
 5:       return r.PSR
 6:    end if
 7:    if r.e instance of POI then
 8:       i ← r.i + 1
 9:       d ← r.dfs + d_E(r.prev, r.e)
10:       PQ.enqueue(< d + d_E(r.e, d), r.e, d, i, T(i).root, r.PSR ∪ r.e >)
11:    else
12:       for all ch ∈ r.e.c do
13:          PQ.enqueue(< r.dfs + L_E^{r.prev,d}(ch.e), r.prev, r.dfs, r.tgt, ch, r.PSR >)
14:       end for
15:    end if
16: end while
```

In Fig.4, PNE-1st and EOSR-1st show the number of visited R-tree nodes when the first (the shortest) result was obtained. PNE-10th and EOSR10-th show the number of visited R-tree nodes when the tenth shortest result was obtained. As shown in this figure, the referred node number in PNE increases rapidly according to the POI density. In contrast, the increase is lower in the EOSR. For example, the ratio of the visited R-tree node number between two methods reaches 100 times when the POI density is 0.02.

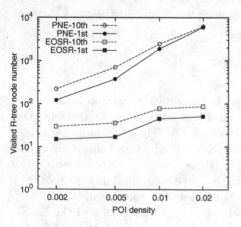

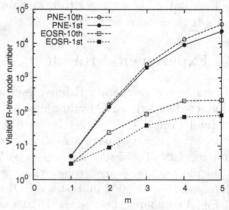

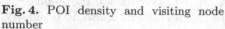

Fig. 4. POI density and visiting node number

Fig. 5. Relationship between m and visiting node number

Fig. 5 shows the relationship between the referred R-tree node number and the number of the POI categories to be visited (m) during the trip. In this experiment, the POI density was set to 0.01 for all POI categories. The number of nodes increases in accordance with the increase in m in PNE. The ratio of PNE and the EOSR reaches more than 300 times when $m = 5$.

4   Conclusion

This paper proposed an efficient trip planning method for the road network distance based on IER framework. First, an incremental search algorithm, the EOSR, for the Euclidean distance is presented. Compared with PNE, which is the only existing incremental algorithm applicable to the OSR in the Euclidean distance, experimental results demonstrate that the EOSR query significantly outperforms PNE, particularly when POIs are densely distributed or the number of POI categories to be visited during the trip is large.

This paper proposed an algorithm to determine only one shortest route; however, the top k shortest routes are sometimes required to facilitate users' choices. The algorithm proposed in this paper can be easily adopted for this requirement because the EOSR generates candidates incrementally and the algorithm for verifying the road network distance can be easily applied to k OSR queries. Furthermore, the algorithm and the methodology in this paper can also be directly adapted to the TPQ and the MRPSR.

Acknowledgments. The present study was partially supported by the Japanese Ministry of Education, Science, Sports and Culture (Grant-in-Aid Scientific Research (C) 21500093 and (B) 2300337).

References

1. Li, F., Cheng, D., Hadjieleftheriou, M., Kollios, G., Teng, S.H.: On Trip Planning Queries in Spatial Databases. In: Medeiros, C.B., Egenhofer, M., Bertino, E. (eds.) SSTD 2005. LNCS, vol. 3633, pp. 273–290. Springer, Heidelberg (2005)
2. Sharifzadeh, M., Kolahdouzan, M., Shahabi, C.: The optimal sequenced route query. The VLDB Journal 17, 765–787 (2008)
3. Chen, H., Ku, W.S., Sun, M.T., Zimmermann, R.: The multi-rule partial sequenced route query. In: ACM GIS 2008, pp. 65–74 (2008)
4. Papadias, D., Zhang, J., Mamoulis, N., Tao, Y.: Query processing in spatial network databases. In: Proc. 29th VLDB, pp. 790–801 (2003)
5. Guttman, A.: R-Trees: a dynamic index structure for spatial searching. In: Proc. ACM SIGMOD Conference on Management of Data, pp. 47–57 (1984)

Path-Based Constrained Nearest Neighbor Search in a Road Network

Yingyuan Xiao, Yan Shen, Tao Jiang, and Heng Wang

Tianjin Key Laboratory of Intelligence Computing and Novel Software Technology,
Tianjin University of Technology, 300384, Tianjin, China
{yingyuanxiao,tjutshenyan,jiangtaoxyy,hengwang}@gmail.com

Abstract. Nearest Neighbor (NN) queries are frequently used for location-dependent information services. In this paper, we study a new NN query called Path-based Constrained Nearest Neighbor (PCNN) query, which involves the additional constraints on non-spatial attribute values of data objects on processing a continuous NN search along a path. For PCNN query processing, we propose an efficient PCNN query method based on transformation idea. The proposed method transforms a continuous NN search into static NN queries at discrete intersection nodes. We further leverage peer-to-peer sharing to improve the proposed method. Extensive experiments are conducted, and the results demonstrate the effectiveness of our methods.

Keywords: Location-based services, path-based constrained nearest neighbor query, peer-to-peer sharing.

1   Introduction

Location-Based Services (LBS) [1] enable mobile clients to search for facilities such as restaurants, shops, and car-parks close to their route. In general, mobile clients send *location-dependent queries* to an LBS server from where the corresponding location-related information are returned as query results. However, conventional *location-dependent queries* (e.g., range query and NN query) purely focus on the proximity of objects while neglect the additional constraints on non-spatial attribute values of data objects. This paper addresses a new kind of NN query called Path-based Constrained Nearest Neighbor (PCNN) query, which involves the specified constraints on non-spatial attribute values of data objects on processing a continuous NN search along a path. Specifically, a PCNN query is defined between a query path P and a set of interest objects S, and retrieves the set of the nearest interest object of every point on P and meanwhile satisfies the specified constraints on non-spatial attribute values of data objects. The following is a typical example about the PCNN query.

Example: A car is approaching a path and the driver intends to find a hotel nearby the path. Then, he uses the on-board computer to issue the query "let me know the set of the nearest hotel of every point in the path, whose average price is ranged from \$40 to \$60."

S.W. Liddle et al. (Eds.): DEXA 2012, Part I, LNCS 7446, pp. 492–501, 2012.

In this paper, we explore the problem of efficient PCNN query processing in a road network and present the proposed processing methods. The remainder of this paper is organized as follows. We review the related work in Section 2. In Section 3, we formally define the PCNN query and describe the reference infrastructure for supporting PCNN queries. In Section 4, we first establish the theoretical foundation for efficiently answering PCNN queries, and then present the proposed processing approaches. We evaluate the proposed approaches through comprehensive experiments in Section 5. Finally, Section 6 concludes this paper.

2   Related Work

Location-dependent queries in spatial networks have been investigated in recent years. Papadias et al. [2] firstly address the problem of *location-dependent query* processing in spatial networks, and develop a Euclidean restriction and a network expansion framework to answer the popular spatial queries (e.g., NN query, range query, etc.). Different from our work, [2] neither considers continuous spatial queries nor involves the additional constraints on non-spatial attribute values of data objects.

Continuous Nearest Neighbor (CNN) query, as an extension of NN query, has been studied in the Euclidean space [3-8]. A CNN query retrieves the nearest neighbor of every point in the specified line segment. In particular, the result, which is different from the PCNN query, contains a set of <R, T> tuples, where R is an interest object, and T is the interval during which R is the nearest neighbor of each point on T. Due to many real-life objects moving on pre-defined spatial networks, several CNN algorithms have been developed for spatial networks. In [9], Feng et al. adopt heuristics to generate computation points and the search region for accelerating the CNN search process. Kolahdouzan et al. [10] present a solution called UBA based on VN^3 for CNN queries in spatial network databases. In [11], Cho et al. present UNICONS which incorporates the use of precomputed NN lists into Dijkstra's algorithm for CNN queries. All the algorithms mentioned above need to find split points to gain the set of <R, T> tuples while PCNN queries only retrieve the set of the nearest neighbor of every point on the path. In addition, CNN query does not involve the additional constraints on non-spatial attribute values. In [12], Jun et al. explore the problem of generalized spatial query processing in the wireless data broadcasting system. The generalized spatial queries are constructed by adding the additional constraints on non-spatial attribute values in conventional *location-dependent queries*. Ku et al. [13] present a novel approach for reducing *location-dependent query* access latency by leveraging results from nearby peers in wireless data broadcasting environments. Different from our work, [12, 13] aim at the wireless data broadcasting setting and only consider static spatial queries in the Euclidean space.

3   Preliminary

In this section, we first formally define the PCNN query and related concepts, and then describe the reference infrastructure for supporting PCNN queries.

3.1   Notations and Definitions

A road network can be modeled as a graph $G = (E, V)$, where V is a set of nodes corresponding to road junctions and E is a set of edges between two nodes in V corresponding to road segments. A *path* is a sequence of successively neighboring edges. Usually, we use the sequence of successively neighboring nodes on a path to denote the path. The start and end nodes of a path are called *terminating nodes* of the path, and all nodes except terminating nodes on a path are called *intermediate nodes* of the path. A *subpath* of a given path P is a part of P between any two nodes of P. Table 1 summarizes the symbolic notations used throughout this paper.

Table 1. Symbolic notations

Symbol	Meaning
S	A set of interest objects
P	A given path in a road network
n_k	A node corresponding to a road junction
n_s	The start node of a given path P
n_e	The end node of a given path P
o	An interest object in S
$N(n_k)$	The nearest interest object of n_k
$R_{path}(P)$	The set of the nearest interest object of every point on the query path P
$O_{path}(P)$	The set of interest objects on the query path P
MC	A mobile client issuing a location-dependent query
BS	A mobile support base station
$peer_i$	A single-hop mobile client of the MC
$q<P, C>$	a PCNN query where P denotes the query path and C is the constraints on non-spatial attribute values of data objects

Consider a road network with a set of interest objects S. We formally define an intersection node, an intersection node sequence and a PCNN query.

Definition 1. Intersection node: A node where three or more edges meet is called an *intersection node*. Otherwise, it is called a *non-intersection node*.

Definition 2. Intersection node sequence: For a given path P, the intersection node sequence of P is defined as the node sequence that is constructed by all *intersection nodes* from the start node n_s to the end node n_e of P except n_s and n_e.

Definition 3. PCNN query: Let C be the specified constraints on non-spatial attribute values of interest objects. For a given path P and a set of interest objects S, a PCNN query retrieves the set $R = \{o \in S | \exists t \in P \ (o=N(t)) \wedge (o \ s.t. \ C)\}$, where $o=N(t)$ represents o is the nearest interest object of t, and $(o \ s.t. \ C)$ denotes o satisfies the constraints C.

3.2   Reference Infrastructure

Fig. 1 depicts the reference infrastructure for supporting PCNN queries in a road network, which is a LBS system based on Personal Communication Systems (PCS) or

Global System for Mobile Communications (GSM). A set of general purpose computers is interconnected through a high-speed wired network, which are categorized into Fixed Host (e.g., LBS) and mobile support Base Stations (BSs). One or more BSs are connected with a BS Controller (BSC), which coordinates the operations of BSs using its own software program when commanded by the Mobile Switching Center (MSC). Unrestricted mobility in PCS and GSM is supported by wireless link between BS and mobile clients. Mobile clients refer to mobile intelligent terminals such as PDA, on-board computer, etc., which equip with GPSs and can communicate with BSs using wireless channels. The power of a BS defines its communication region, which we refer to as a cell. A mobile client (MC) can freely move from one cell to another and transparently accesses the spatial database residing at the fixed network.

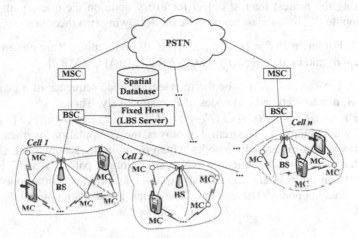

Fig. 1. A reference infrastructure

Due to the increasing deployment of new peer-to-peer (P2P) wireless communication technologies, mobile clients are now being equipped with wireless P2P capabilities. This enables mobile clients to become parts of self-organizing, wireless mobile ad hoc networks (MANETs) that allow mobile clients to communicate with neighboring peers in an ad hoc manner for data sharing.

4 PCNN Query Processing Approaches

We start with a baseline approach and two basic theorems in subsection 4.1, and then an efficient PCNN query algorithm is proposed in subsection 4.2. Finally, we improve the proposed algorithm by utilizing peer-to-peer sharing in subsection 4.3.

4.1 Basic Ideas

Lemma 2 presented in [11] provides an insight into how to compute the set of the k nearest interest objects of every point on a given query path. We consider the case of $k=1$ and rephrase Lemma 2 in terms of our notations in the following Lemma 1.

Lemma 1. For any path $P = \{n_1, n_2, ..., n_k\}$, $R_{path}(P) = O_{path}(P) \cup \{N(n_1)\} \cup \{N(n_2)\}... \cup \{N(n_k)\}$.

On the basis of Lemma1, a straightforward algorithm to compute PCNN query consists of four steps: 1) compute the nearest interest object for every node on the given query path P using an existing NN algorithm, and use S_1 to denote the set of these nearest interest objects; 2) search all interest objects along P, and use S_2 to denote the set of these interest objects; 3) union S_1 and S_2 into S_3 and 4) filter out from S_3 those objects that do not qualify the specified constraints on non-spatial attribute values and return the final result.

We refer to this algorithm as *ABA* (*a baseline approach*). Although *ABA* is correct, as can be proved easily by Lemma 1, it generates a great deal of processing overhead on computing the nearest interest object for every node on the query path. To efficiently compute PCNN queries, we propose the following two theorems.

Theorem 1. For any path $P = \{n_1, n_2, ..., n_k\}$, if all nodes along P are non-intersection nodes except n_1 and n_k, then $R_{path}(P) = O_{path}(P) \cup \{N(n_1)\} \cup \{N(n_k)\}$.

Theorem 2. Let $<n_1, n_2, ..., n_i>$ be the intersection node sequence of a given path P and n_s and n_e denote start and end nodes of P, respectively. Then,

$$R_{path}(P) = O_{path}(P) \cup \{N(n_s)\} \cup \{N(n_1)\} \cup \{N(n_2)\} \cup ... \cup \{N(n_i)\} \cup \{N(n_e)\}$$

Compared with Lemma 1, Theorem 1 removes the computation overhead running static queries at those intermediate nodes which are non-intersection nodes. Theorem 2 proves that to perform a continuous NN search along a path, it is sufficient to retrieve objects on the path and to run static queries at its intersection node sequence and two terminating nodes. The proofs of Theorems 1 and 2 are omitted due to space limitations.

4.2   Intersection Node-Based Method

In this subsection, we propose an efficient PCNN query method, called *INBM* (*intersection node-based method*), which leverages Theorem 2 to erase the processing overhead of running NN queries at all non-intersection nodes of the query path.

Algorithm 1: *INBM* (P, C)

Input: P is a query path and C denotes the specified constraints on non-spatial attribute values
Output: *Result*, i.e., the result of the PCNN query
1: *Result* := $\varnothing$; *Sequence* := $\varnothing$; *ObjectSet* := $\varnothing$;
2: *Sequence* := GetINS(*P*, *adjacency-list*);
3: *ObjectSet* := GetObject(*P*);
4: **for** $\forall\, t \in$ *Sequence* $\cup \{n_s, n_e\}$ **do**
5:      *Result* := *Result* $\cup \{N(t)\}$;
6: *Result* := *Result* $\cup$ *ObjectSet*;
7: **for** $\forall\, o \in$ *Result* **do**
8:      **if** o does not qualify the specified constraints C
9:          *Result* := *Result*\ $\{o\}$;
10: **return** *Result*;

When a BS receives a PCNN query request issued by a mobile client located at its cell, it transmits the request to the LBS server. The LBS server is responsible for employing *INBM* to process the request and returns the query result to the mobile client. Specifically, *INBM* contains four steps: 1) compute the nearest interest object for every intersection node on P using an existing NN algorithm, and use S_1 to denote the set of these nearest interest objects; 2) search all interest objects along P, and use S_2 to denote the set of these interest objects; 3) union S_1 and S_2 into S_3 and 4) filter out from S_3 those objects that do not qualify the specified constraints and return the final result. We formalize *INBM* method in the following Algorithm 1.

In Algorithm 1, $N(t)$ is computed by means of the existing NN processing algorithm, like IER or INE [2], *adjacency-list* represents the adjacency list of the road network, the operation GetINS(P, *adjacency-list*) is responsible for gaining the intersection node sequence of P, and the operation GetObject(P) returns the set of interest objects located at P. The formalized descriptions of GetINS(P, *adjacency-list*) and GetObject(P) are omitted due to space limitations. As the nearest interest object of a node is fixed unless the updates to the interest object set S happen, we can precompute $N(t)$ for each intersection node t to improve the performance of Algorithm 1.

4.3   Sharing-Based Method

MANETs enable mobile clients to leverage query results cached in their neighboring peers. Based on peer-to-peer sharing, we propose *SBM* (*sharing-based method*) to improve the performance of *INBM*. *SBM* needs each mobile client to cache its last query path and the NN set of the query path for sharing. When a PCNN query request is issued by a mobile client *MC*, *MC* first broadcasts the PCNN query request to all its single-hop mobile clients (peers) for data sharing instead of fetching data solely from the remote LBS server. Only when these data cached in single-hop peers do not match the PCNN query, is the PCNN query request transmitted to the LBS server through the neighboring BS.

Let $q<P, C>$ denote a PCNN query issued by a mobile client *MC*, $peer_i$ ($1 \le i \le k$) denote a single-hop mobile client of *MC*, and P_i denote the last query path cached by $peer_i$. The processing strategy of *SBM* is described as follows: *MC* first broadcasts the request $q<P, C>$ to all its single-hop mobile clients $peer_i$ for $i=1, 2, \ldots, k$. For each $peer_i$ ($1 \le i \le k$), once receiving the request $q<P, C>$ from *MC*, it checks whether P and P_i are the same path. If P and P_i are the same path, $peer_i$ sends the message "Yes" to *MC* and selects from the cached objects those object that qualify the specified constraint C; otherwise it sends the message "No" to *MC*. If *MC* receives "Yes" from a $peer_i$ within the specified deadline T, it sends the message "Acknowledgement" to the $peer_i$ and waits the result from the $peer_i$. The specified deadline T is calculated by the formula: $T = t_{issue} + 2 \times t_{delay} \times slack$, where t_{issue} is the time at which the request $q<P, C>$ is broadcast by *MC*, t_{delay} denotes the largest network delay between two neighboring mobile clients based on MANETs, and *slack* represents the slack factor which is a random variable uniformly chosen from a slack range. Once receiving "Acknowledgement" from *MC*, the $peer_i$ sends those qualified objects to *MC*. If all $peer_i$ send "No" or *MC* does not receive any "Yes" within the specified deadline, *MC* transmits the query path P to its BS and the BS is responsible for forwarding P to the LBS Server. The LBS Server retrieves the set H of the nearest interest object of every point in P and sends H to MC. MC caches H and returns from H those object qualifying the specified constraint C.

Based on the above processing strategy, we formalize SBM in the following Algorithm 2-4. Algorithm 2 is executed at mobile clients, Algorithm 3 is executed at these single-hop $peer_i$ and Algorithm 4 is executed at the LBS Server.

Algorithm 2: SBM -MC (P, C)

Input: P is a query path and C denotes the specified constraints on non-spatial attribute values
Output: the result of the PCNN query
1:   $Result := \varnothing$;
2:   Broadcast the request $q<P, C>$ to all its single-hop mobile clients $peer_i$ for $i=1, 2, ..., k$ and wait response messages from them;
3:   **if** receiving the message "Yes" from a $peer_i$ within the specified deadline   **then**
4:      Send the message "Acknowledgement" to the $peer_i$ and wait the result from the $peer_i$;
5:      Receive the result from the $peer_i$;
6:      **return** the result;
7:   **else**
      / * all $peer_i$ send "No" or MC does not receive any "Yes" within the specified deadline */
8:      Transmit the query path P to the LBS server through its BS and wait the result H from the LBS server;
9:      Cache H which is received from the LBS server;
10:    **for** $\forall o \in H$ **do**
11:      **if** o qualify the specified constraint C **then**
12:        $Result := Result \cup \{\ o\ \}$;
13:    **return** $Result$;

Algorithm 3: SBM -$Peer$ (P, C)

Input: P and C, which are sent by MC
Output: the result of the PCNN query or message "No"
1:   $Result := \varnothing$;
2:   **if** $P = P_i$;
      /* P_i denote the last query path cached by $peer_i$ */
3:      Send "Yes" to MC;
4:   **else**
5:      Send "No" to MC and **return**;
6:   **for** $\forall o \in CacheData$ **do**
/*$CacheData$ is the set of the nearest interest object of every point in P_i, cached in $peer_i$ */
7:      **if** o qualifies the specified constraints C **then**
8:        $Result := Result \cup \{\ o\ \}$;
9:      **if** receiving the message "Acknowledgement" from MC **then**
10:    Send $Result$ to MC;

Algorithm 4: *SBM -LBS* (*P*)

Input: a query path *P*
Output: *H*, i.e., the set of the nearest interest object of every point in *P*
1: *H* := ∅; *Sequence* := ∅; *ObjectSet* := ∅;
2: *Sequence* := GetINS(*P*, *adjacency-list*);
3: *ObjectSet* := GetObject(*P*);
4: **for** ∀ *t* ∈ *Sequence* ∪ {n_s, n_e} **do**
5:     *H* := *H* ∪ {*N*(*t*)};
6: *H* := *H* ∪ *ObjectSet*;
7: **return** *H*;

---

5   Experimental Evaluation

This section experimentally assesses the performance of the proposed *INBM* and *SBM*. We mainly compare the proposed methods with *ABA* in terms of I/O cost (page accesses), network delay and execution time by simulation experiments. In our simulation experiments, the LBS server runs MS Windows XP with a 2.4GHz CPU and 1GB main memory. Mobile clients include lap-tops and smart PDAs running Windows CE with 500MHz CPU and 128M main memory. All algorithms of Section 4 are developed in C++. Our evaluations are based on a real road network. The road network data description is as follows: $|V|=2104$ and $|E|=21692$, where *V* is the set of nodes and *E* is the set of edges. In order to control the density of the interest objects, we use synthetic spatial objects as interest objects which are uniformly generated on the network edges. Each interest object has a spatial attribute denoted by its 2D coordinates, together with eight non-spatial attributes. In order to reduce the randomness effect we average the results of the algorithms over 20 PCNN queries. The key simulation parameters contain N_n, the number of nodes in a given query path, and *Pr*, the probability of *MC* receiving "Yes" from a $peer_i$ within the specified deadline.

Fig. 2 shows the performance of the three methods in terms of I/O cost, as a function of N_n. As shown in Fig. 2, *SBM* gets a distinct advantage over *INBM* and *ABA* in term of I/O cost. This is because *SBM* enables the great majority of PCNN queries to be computed by utilizing peer-to-peer sharing and thus avoids a large number of disk I/O accesses which are required by *INBM* and *ABA*. Moreover, we can also see from Fig. 2 that *INBM* consistently outperforms *ABA*. The reason is that *INBM* leverages Theorem 2 to erase the processing overhead of running NN queries at non-intersection nodes. Fig. 3 depicts the network delay of the three methods. The network delay refers to the time required to complete all the network transmission for computing a PCNN query. We can learn from Fig. 3 that *SBM* has a slightly advantage over *INBM* and *ABA*. This is not surprising because *SBM* answers PCNN queries by directly fetching data cached in the neighboring peers in most case. This enables *SBM* to generate less routing and forwarding messages than that of *INBM* and *ABA*. Fig. 4 demonstrates the performance of the three methods in terms of execution time, as a function of N_n. As shown in Fig. 4, *SBM* evidently outperforms *INBM* and *ABA*.

The reason mainly includes the two aspects: firstly, *SBM* decreases heavily disk I/O accesses by means of peer-to-peer sharing; secondly, *SBM* reduces the network delay compared with *INBM* and *ABA*. We can also see from Fig. 4 that *INBM* has a distinct advantage over *ABA*. The main reason is *INBM* erases the processing overhead of running NN queries at non-intersection nodes, required by *ABA*. Figs. 5, 6 and 7 show how *Pr* influences I/O cost, network delay and execution time. Fig. 5 plots page accesses as a function of *Pr* for various values of *Pr*. As shown in Fig. 5, the page accesses of *SBM* decrease as *Pr* increases. Moreover, *SBM* gets a distinct advantage over *INBM* and *ABA* in terms of I/O cost. The reason is increasing *Pr* enables *SBM* to compute the more PCNN queries by utilizing these data cached in the neighboring mobile clients. Fig. 6 illustrates network delay as a function of *Pr* for various values of *Pr*. We can learn from Fig. 6 that the network delay of *SBM* decreases as *Pr* increases and became smaller than that of other two methods when *Pr* is greater than a fixed value. This is because increasing *Pr* enables *SBM* to answer the more PCNN queries by directly fetching data cached in the neighboring mobile clients instead of requesting data solely from the remote LBS server. Fig. 7 shows execution time as a function of *Pr* for various values of *Pr*. We can see from Fig. 7 that the execution time of *SBM* decreases as *Pr* increases and meanwhile *SBM* gets a distinct advantage over *INBM* and *ABA*. This is not surprising because for *SBM*, the larger *Pr* results in the less I/O cost and network delay.

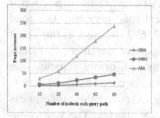

Fig. 2. I/O cost vs. N_n

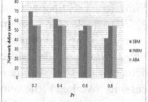

Fig. 3. Network delay vs. N_n

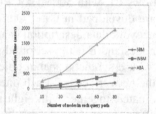

Fig. 4. Execution time vs. N_n

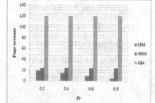

Fig. 5. I/O cost vs. *Pr*

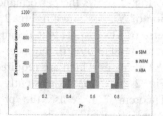

Fig. 6. Network delay vs. *Pr*

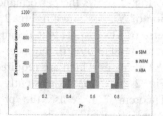

Fig. 7. Execution time vs. *Pr*

6   Conclusion

In this paper, we address a new kind of NN query called Path-based Constrained Nearest Neighbor (PCNN) query, which involves the additional constraints on non-spatial attribute values of data objects on answering a continuous NN search along a path. We propose two important prepositions (i.e., theorems 1 and 2) to decrease computation overhead for PCNN query processing. On the basis of them, we propose an efficient PCNN query method. The proposed method transforms a continuous NN search into static NN queries at discrete intersection nodes. Further, we leverage peer-to-peer sharing to improve the proposed method. Extensive experiments are conducted, and the results demonstrate the effectiveness of our methods.

References

1. Lee, D.L., Lee, W.C., Xu, J., Zhang, B.: Data management in location-dependent information services. IEEE Pervasive Computing 1(3), 65–72 (2002)
2. Papadias, D., Zhang, J., Mamoulis, N., Tao, Y.: Query processing in spatial network databases. In: Proc. of VLDB, pp. 802–813. ACM Press, Berlin (2003)
3. Tao, Y., Papadias, D., Shen, Q.: Continuous Nearest Neighbor Search. In: Proc. of VLDB, pp. 287–298. ACM Press, Hang Kong (2002)
4. Mokbel, M.F., Xiong, X., Aref, W.G.: SINA: Scalable Incremental Processing of Continuous Queries in Spatio-temporal Databases. In: Proc. of ACM SIGMOD, pp. 623–634. ACM Press, Paris (2004)
5. Mouratidis, K., Hadjieleftheriou, M., Papadias, D.: Conceptual Partitioning: An Efficient Method for Continuous Nearest Neighbor Monitoring. In: Proc. of ACM SIGMOD, pp. 634–645. ACM Press, Maryland (2005)
6. Iwerks, G.S., Samet, H., Smith, K.: Continuous K-Nearest Neighbor Queries for Continuously Moving Points with Updates. In: Proc. of VLDB, pp. 287–298. ACM Press, Berlin (2003)
7. Xiong, X., Mokbel, M.F., Aref, W.G.: SEA-CNN: Scalable processing of continuous K-nearest neighbor queries in spatio-temporal databases. In: Proc. of ICDE, pp. 643–654. IEEE Press, Tokyo (2005)
8. Xiao, Y.Y., Wang, H.Y.: An Efficient Algorithm for Continuous Nearest Neighbor Queries Based on the VDTPR-tree. Journal of Computational Information Systems 4(2), 527–534 (2007)
9. Feng, J., Watanabe, T.: Search of Continuous Nearest Target Objects along Route on Large Hierarchical Road Network. In: Proc. of Control and Applications, New York, USA, pp. 33–41 (2004)
10. Kolahdouzan, M., Shahabi, C.: Continuous K-Nearest Neighbor Queries in Spatial Network Databases. In: Proc. of STDBM, Citeseer,Toronto, Canada, pp. 33–40 (2004)
11. Cho, H.J., Chung, C.W.: An Efficient and Scalable Approach to CNN Queries in a Road Network. In: Proc. of VLDB, pp. 865–876. ACM Press, Trondheim (2005)
12. Jun, H., Choi, H., Chung, Y.D.: Generalized Spatial Queries in the Wireless Data Broadcasting System. In: Proc. of MDM, Taipei, Taiwan, pp. 279–284 (2009)
13. Ku, W.S., Zimmermann, R., Wang, H.: Location-Based Spatial Query Processing in Wireless Broadcast Environments. IEEE Transactions on Mobile Computing 7(6), 778–790 (2008)

Efficient Fuzzy Ranking
for Keyword Search on Graphs

Nidhi R. Arora[1], Wookey Lee[2], Carson Kai-Sang Leung[3],
Jinho Kim[4], and Harshit Kumar[1]

[1] University of Suwon, Hwaseong, South Korea
[2] Inha University, Incheon, South Korea
[3] University of Manitoba, Winnipeg, MB, Canada
[4] Kangwon National University, Kangwon, South Korea
trinity@inha.ac.kr, kleung@cs.umanitoba.ca, jhkim@kangwon.ac.kr

Abstract. When compared with the traditional single-node results returned by search engines, keyword search over graphs is a new answering paradigm that brings new challenges to ranking. In this paper, we propose an efficient fuzzy-set theory based ranking measure called FRank. This measure captures the presence and relevance of query keywords and their query-dependent edge weights. It evaluates the query answer based on the distribution of keywords in the query and the structural connectivity between these keywords. Experimental results show that our proposed FRank measure led to superior performance when compared with traditional ranking measures.

Keywords: Fuzzy sets, graph rank, information retrieval (IR), keyword search.

1 Introduction and Related Work

Graph data are available in numerous application domains (e.g., relational databases, Web, semantic Web). To access information from structured data, users usually need to (i) learn complex query languages (e.g., SQL, XPath) and (ii) acquire knowledge about the data schema and organization. Hence, wealth of information that is present in these data may not be easily accessible by non-technical users. This calls for efficient query processing of keyword search (which is a convenient and user-friendly mechanism used by many search engines for information retrieval (IR) [1,10,11]) over graph data. A *keyword search over graph structured data* usually returns an answer as tree structures [4,7,9] or sub-graphs [12,15] due to the linkage of information across multiple database relations or web pages.

As the number of answers (i.e., structures that match query keywords) can be large and not all the answers are equally relevant to the query, it is critical for efficient keyword search to rank and return top-k structures. Existing ranking functions are usually based on (i) a content-relevance based score (e.g., tf-idf), (ii) some structural properties (e.g., PageRank or its variants, path length or reciprocal path length), or (iii) the aggregation of the above two. However, there

S.W. Liddle et al. (Eds.): DEXA 2012, Part I, LNCS 7446, pp. 502–510, 2012.

are limitations associated with these ranking functions. For instance, the IR-style ranking [13] computes the content-relevance score of the result structure by assigning a tf-idf based relevance score to each node and then combining these scores using an aggregate function such as SUM. However, scores are assigned irrespective of the keyword distribution or the structure. Same scores are assigned to graphs having the same structures or structures having the same number of nodes and edges. BLINKS [5] and BANKS-II [7], on the other hand, used a distance-based score in terms of *path length* (or *reciprocal path length*) from the root node to each of the leaf nodes that contain query keywords. However, this score only works for tree structures (having a single root node) but *not* subgraphs (having multiple root nodes). In contrast, the graph ranking score (GRS) [9] ranks the answer structures based on the proximity of nodes containing query keywords by evaluating the eigenvalues of a matrix. However, the creation of a matrix at each stage of the graph exploration process can be computationally expensive.

To overcome the shortcomings of the above approaches, we propose a novel fuzzy-set based ranking function, called *FRank*, for efficient and effective top-k keyword search. Although fuzzy-set theory has been used in (i) modelling uncertain and imprecise data in fuzzy IR [2] or fuzzy XML [14] and (ii) constructing web ontology [18], this is the first attempt—to the best of our knowledge—to consolidate fuzzy sets for efficient and effective ranking of keyword search results. Our **key contributions** of this paper are as follows:

1. We use fuzzy sets to capture the presence and relevance of each individual query keyword discretely and measure the query-specific node-relevance.
2. We propose a new aggregate operator to compute query-dependent content-relevance based edge weights (cf. distance-based edge-weights).
3. The FRank measure aggregates the edge weights capturing their relevance to the keyword query. It automatically favours answers with higher information content than those with only partial information.

This paper is organized as follows. In next section, we propose our relevance scoring function, which helps in processing query and generating top-k results for keyword searches over graphs. Experimental results are reported in Section 3, and conclusions are given in Section 4.

2   Our Proposed Relevance Score

Consider an undirected graph $G=(V,E)$, where V is a set of nodes (e.g., web pages, database tuples, XML documents) and E is a set of edges connecting the nodes (e.g., hyperlinks between web pages, primary key-foreign key relationships between database tuples, parent-child relationships among XML elements). A keyword query Q consists of a set of M keywords—i.e., $Q = \{t_1, t_2, \ldots, t_M\}$, where each t is a general query keyword term. We use V_t to denote a set of nodes containing the query keyword term t; we use $V_Q = \bigcup_{t \in Q} V_t$ to denote the union of all keyword nodes for a particular query Q. In this section, we propose a ranking function to effectively rank the answer structures.

2.1   Fuzzy Node Sets

To independently and discretely accumulate IR-style content-relevance based score for each of the query keywords, we use fuzzy sets which provide natural means to model gradual relevance by means of its membership function. For each $u \in V_Q$ (i.e., for each keyword node for a query Q), we create a fuzzy set f_u such that each element in the set corresponds to a query keyword term t and its membership degree $\mu_{u,t}$ measuring the content relevance of t towards the node u:

$$f_u = \{(t, \mu_{u,t}) \mid t \in Q\}. \tag{1}$$

We define $\mu_{u,t}$ as a product of (i) the normalized inverse document frequency (idf) $nidf_t$ and (ii) the normalized term frequency (tf) $ntf_{u,t}$. In other words,

$$\mu_{u,t} = nidf_t \times ntf_{u,t}, \quad \text{where} \quad nidf_t = \frac{\ln(d_t)}{\ln(N)} \quad \text{and} \quad ntf_{u,t} = \frac{tf_{u,t}}{dl_u}. \tag{2}$$

Here, (i) d_t is the number of nodes containing t (i.e., document frequency), (ii) N is the total number of nodes in the graph G, (iii) $tf_{u,t}$ is the term frequency of t in u, and (iv) dl_u is the document length (i.e., the number of terms) in Node u.

Example 1. Consider (i) a graph G with keyword distribution as shown in Fig. 1 and (ii) query $Q=\{$graph, semantic, web$\}$. Then, as the keyword term "graph" appears in Nodes 2, 4 and 5 (out of $N=10$ nodes), its document frequency $d_{\text{graph}}=3$. Thus, the normalized idf $nidf_{\text{graph}} = \frac{\ln(3)}{\ln(10)} = 0.4771$. As the term "graph" appears once in Node 2 (i.e., $tf_{2,\text{graph}}=2$) and three additional terms—namely, "pattern", "recognition", "data"—are associated with Node 2 (i.e., the document length $dl_2=4$), the normalized tf $ntf_{2,\text{graph}}=\frac{1}{4}=0.25$. Hence, the resulting fuzzy set $f_2 = \{$(graph, $0.4771 \times 0.25 = 0.1193$), (semantic, 0), (web, 0)$\}$. Similarly, $f_6=\{$(graph, 0), (semantic, 0.2007), (web, 0.2007)$\}$ because (i) 4 out of $N=10$ nodes contain "semantic" and 4 out of $N=10$ nodes contain "web" (i.e., $nidf_{\text{semantic}}=nidf_{\text{web}}=\frac{\ln(4)}{\ln(10)}=0.6021$) and (ii) "semantic" and "web" are 2 of $dl_6=3$ terms in Node 6 (i.e., $ntf_{6,\text{semantic}}=ntf_{6,\text{web}}=\frac{1}{3}$). See Table 1.   □

As $\mu_{u,t}$ can be pre-computed for every node u and keyword term t in G, it can be stored offline during the creation of inverted index. Hence, fuzzy sets for keyword nodes can be generated efficiently at runtime. For any *latent Steiner node u'* (i.e., node that appears in the shortest path between two keyword nodes), it does not contain any query keyword. We assign $\mu_{u',t}=1$.

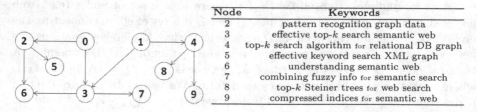

Node	Keywords
2	pattern recognition graph data
3	effective top-k search semantic web
4	top-k search algorithm for relational DB graph
5	effective keyword search XML graph
6	understanding semantic web
7	combining fuzzy info for semantic search
8	top-k Steiner trees for web search
9	compressed indices for semantic web

Fig. 1. A data graph G and text associated with nodes based on a publication DB

Table 1. Tf-idf for $Q=\{$graph, semantic, web$\}$

u	t	$tf_{u,t}$	dl_u	$ntf_{u,t}$	$\mu_{u,t}$
2	graph	1	4	0.25	0.1193
3	semantic	1	5	0.2	0.1204
	web	1	5	0.2	0.1204
4	graph	1	6	0.1667	0.0795
5	graph	1	5	0.2	0.0954
6	semantic	1	3	0.3333	0.2007
	web	1	3	0.3333	0.2007
7	semantic	1	5	0.2	0.1204
8	web	1	5	0.2	0.1204
9	semantic	1	4	0.25	0.1505
	web	1	4	0.25	0.1505

$N=10$ keyword terms		
t	d_t	$nidf_t$
graph	3 (in Nodes 2, 4 & 5)	0.4771
semantic	4 (in Nodes 3, 6, 7 & 9)	0.6021
web	4 (in Nodes 3, 6, 8 & 9)	0.6021

2.2 Content-Based Edge Score

Next, we devise an *aggregate operator* to compute edge weights based on the query-specific information contained in the two nodes connecting the edge. To avoid the problem associated with conventional fuzzy union (which may produce abundant results) or fuzzy intersection (which may be too restrictive), we combine $\mu_{u,t}$ and $\mu_{v,t}$ for the edge connecting u & v. Specifically, to weigh complementary information of an edge heavier than extensions of the same keywords, we define a new aggregate operator called the *complemented weighted average* (*CWA*). Let Δ_v represent the total amount of information available in a node v. It is the sum of query-specific content-relevance scores of query keywords contained in v. Let $\Delta_{v/t}$ represent the sum of query-specific scores of query keywords (except t) contained in v. Then,

$$CWA(e_{u,v}) = \frac{1}{M(M-1)}\left[\sum_{t \in Q}(\mu_{u,t} \times \Delta_{v/t})\right],\qquad(3)$$

where M is the total number of query keywords and $\mu_{u,t}$ defines the membership degree as content relevance of a keyword t towards node u. *CWA* enumerates the total amount of information available in the incident nodes of an edge and computes the content relevance of an edge.

Example 2. Reconsider Example 1, in which edge $e_{2,6}$ contains complete query-specific information but edge $e_{3,6}$ contains only partial information. Recall that $f_2=\{$(graph, 0.1193), (semantic, 0), (web, 0)$\}$ & $f_6=\{$(graph, 0), (semantic, 0.2007), (web, 0.2007)$\}$. For $e_{2,6}$, as Node 2 contains only one keyword "graph" (out of $M=3$ keywords), $\Delta_{6/graph}=0.2007+0.2007=0.4014$. Then, $CWA(e_{2,6}) = \frac{1}{3(3-1)}\left[\sum_{t \in Q}(\mu_{2,t} \times \Delta_{6/t})\right] = \frac{1}{6}[(0.1193 \times 0.4014) + 0 + 0]=0.0080$. Similarly, $CWA(e_{3,6}) = \frac{1}{6}[0 + (0.1204 \times 0.2007) + (0.1204 \times 0.2007)] = 0.0081$ because $e_{3,6}$ contains $m_{e_{3,6}}=2$ out of $M=3$ query keywords. Thus, CWA together with weighing function assigns heavier weight to the complete information and lighter weight to edges containing less relevant or partial information. □

2.3   Ranking of Answer Structures

Given the answer structure A, the relevance score of an answer can be computed by aggregating the edge weights of all the edges in the structure. In contrast to existing approaches (where all the edges are of equal importance), each edge weighs differently in ours. Each edge weight is computed according to the amount of query specific information contained by it (Section 2.2). The final rank (or overall score) of A is computed as follows:

$$rank(A) = \frac{m_e}{M}CWA(e^k) + \left(1 - \frac{c}{n}\right), \tag{4}$$

where (i) m_e is the total number of query keywords present in the incident nodes, (ii) e^k is an edge incident on one or both *keyword nodes*, (iii) c is the total number of edges incident on both latent Steiner nodes, and (iv) n is the total number of edges in the answer structure. The term $\left(1 - \frac{c}{n}\right)$ enforces a higher rank for the result structures containing fewer number of latent Steiner nodes. Note that c enumerates the total number of edges incident on latent Steiner nodes out of the total number of edges present in the result structure. Thus, if the result structure contains many edges incident on latent Steiner nodes, then $\left(1 - \frac{c}{n}\right)$ will be smaller or closer to 0. Conversely, if the result structure contains fewer latent Steiner nodes, then $\left(1 - \frac{c}{n}\right)$ will be larger. Hence, for two or more result structures having the same number of edges, $\left(1 - \frac{c}{n}\right)$ helps to enforce higher ranking of structures with keyword nodes.

3   Experimental Evaluation

To evaluate the effectiveness and search quality of our proposed ranking measure (implemented in Java), experiments were performed on an Intel Core 2 Duo PC with 2.13GHz processor and 2GB of RAM on Windows XP platform. The experiments are designed to check the performance of creating and processing fuzzy sets as an inherent part of the search process. We also evaluated the quality of FRank by comparing it with the state-of-the-art ranking measures.

We first used two popular real-world datasets from the literature for experiments: (i) DBLP dataset was downloaded from dblp.uni-trier.de/xml with the resulting graph consists of N=2M nodes (authors), n=3.7M edges, and M=328K keywords; (ii) Stanford dataset was formed by crawling web pages on www.cs.stanford.edu with the resulting graph consists of N=8K nodes, n=27K edges, and M=70.8K keywords. We picked many queries with length ranging from 2 to 5 keywords.

In the experiments, we compared with BANKS-II [7] (denoted as BANKS) and BLINKS [5] (denoted as IRDB). BANKS-II generated top-k answer trees using graph exploration heuristics. The final rank of an answer tree was computed as $ES \times NP^{0.2}$, where (i) ES is the tree edge score (computed by aggregating the edge scores for all edges on the path from the root to the leaf node containing keyword t) and (ii) NP is tree node prestige (computed by summing the

node weights—a function of the in-degree—of leaf nodes and the answer root).
BLINKS returned only node pairs (instead of tree/graph structure) as answers.
Since it did not return a proper tree structure, we instead used the IR-style
ranking to rank the answer trees in relational databases. The relevance score of
an answer tree was computed by aggregating node relevance scores with the size
of the answer tree. Relevance score of a node u is calculated as the sum of $\mu_{u,t}$.

3.1   Fuzzy Set Performance

Fig. 2 shows the processing time required to create fuzzy sets for
top-10 and top-20 results from a batch of the first 50 keyword nodes for each of
the query keywords. The difference in the processing time of fuzzy sets depends
on the distribution of query keywords in the data graph. For example, consider
$Q4=\{$mobile, web, search$\}$ and $Q4'=\{$personalize, web, search$\}$ on the DBLP
dataset. The query keywords "web" and "search" are simultaneously present in
1,229 nodes, but the numbers of common nodes containing all query keywords
in $Q4$ and $Q4'$ are 0 and 3 respectively. Similar comments apply to the Stanford
dataset. For some queries, the fuzzy set processing time was as low as 15ms. As
shown in Fig. 2, the time for creating fuzzy sets gradually increased when the
value of k (for top-k results) increased. Note that the elapsed time depends on
the distribution of keywords in the query and the connectivity among keyword
nodes.

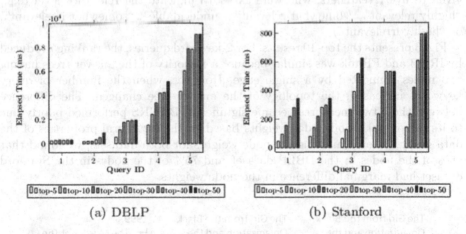

(a) DBLP   (b) Stanford

Fig. 2. Fuzzy set performance: time to process DBLP and Stanford data

3.2   Quality of FRank

We evaluated the effectiveness of FRank in terms of the quality of ranked answer
trees based on their position in the top-k result. To measure the quality of each
ranking mechanism, we used the discounted cumulative gain (DCG) [6] measure,
which is one of the standard measures used in many IR systems [3,17]. As DCG

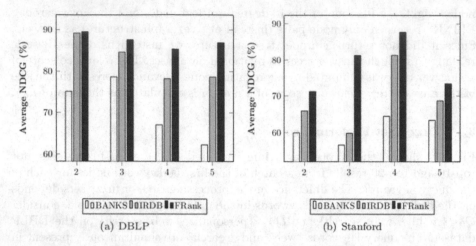

(a) DBLP   (b) Stanford

Fig. 3. Quality of FRank: search accuracy for query with 2 to 5 keywords

is defined as $DCG(p) = rel_1 + \sum_{i=2}^{p} \frac{rel_i}{\log(i)}$ where p is the position of the answer tree in the result set and rel_i is the relevance level of the i-th answer tree in the result set, DCG identifies different levels of ranking and favours the ranking that follows the actual relevance order. The results for each query were then given to five researchers, who were asked to indicate the relevance level (e.g., "highly relevant", "somewhat relevant", "undecidable", "somewhat irrelevant", or "highly irrelevant").

Fig. 3 presents the top-10 results. For 2-keyword queries, the ranking produced by IRDB and FRank was similar because a majority of the answer trees having two nodes connected by a single edge. However, when the number of query keywords increased, the topology of the answer tree changed. The difference between these two measures became significant. BANKS performed poorly due to its use of static node/edge weights based on the structural properties of the data graph. After examining the node weights for both datasets, we found that 65% of the nodes in the DBLP dataset and 75% of the nodes in the Stanford dataset had marginal difference in the node weights.

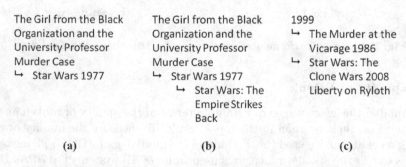

Fig. 4. Quality of FRank: results with different keyword distribution

The reason for effective and improved ranking of FRank over IRDB is because of the calculation of proximity between actual query keywords (instead of keyword nodes). For example, when we experimented with the third dataset (a subset of IMDB data downloaded from www.imdb.com/interfaces for a IMDB graph capturing n=1M movie links), keywords "star" and "wars" appear together in 161 nodes (out of N=285K nodes) in the resulting graph for Q={star, wars, murder}. Being a popular series, many other series refer to old episodes of Star Wars. Thus, we get the answer tree topology as shown in Fig. 4, along with the previous answer tree. Eight different answer trees can be obtained by replacing "star wars: the empire strikes back" in Fig. 4(b). Unlike BANKS and BLINKS, our FRank does not prune out these results. Instead, the results are presented to the user in the order of their relevance. Although the structural compactness score is the same for both Figs. 4(b) & (c), the content-relevance based node weights for Fig. 4(b) are heavier than those for Fig. 4(c). Because IRDB accumulates the total content-relevance scores as node weights, the ranking provided by IRDB was $\langle$(b), (a), (c)$\rangle$. In contrast, the ranking provided by FRank was $\langle$(a), (c), (b)$\rangle$. BANKS and BLINKS do not return Fig. 4(b) as an answer tree. Our FRank, on the other hand, ranks Fig. 4(b) lower in the result set (instead of pruning it out), and thus leads to a much better ranking.

4   Conclusions

In this paper, we proposed a new ranking measure based on fuzzy set theory, called FRank, for effective processing of keyword queries over graph data. FRank helps to acknowledge the presence of multiple query keywords, and computes content-relevance based query specific edge weights. It is effective as it automatically favours results with complete information than those with partial information.

Acknowledgement. This project is partially supported by (i) MKE under ITRC support program supervised by NIPA (S. Korea), (ii) NSERC (Canada), and (iii) University of Manitoba.

References

1. Agrawal, S., Chaudhuri, S., Das, G.: DBXplorer: a system for keyword-based search over relational databases. In: IEEE ICDE 2002, pp. 5–16 (2002)
2. Bruno, N., Wang, W.H.: The threshold algorithm: from middleware systems to the relational engine. IEEE TKDE 19(4), 523–537 (2007)
3. Clarke, C.L.A., et al.: Novelty and diversity in information retrieval evaluation. In: ACM SIGIR 2008, pp. 659–666 (2008)
4. Dalvi, B.B., Kshirsagar, M., Sudarshan, S.: Keyword search on external memory data graphs. In: VLDB 2008, pp. 1189–1204 (2008)
5. He, H., et al.: BLINKS: ranked keyword searches on graphs. In: ACM SIGMOD 2007, pp. 305–316 (2007)

6. Järvelin, K., Kekäläinen, J.: Cumulated gain-based evaluation of IR techniques. ACM TOIS 20(4), 422–446 (2002)
7. Kacholia, V., et al.: Bidirectional expansion for keyword search on graph databases. In: VLDB 2005, pp. 505–516 (2005)
8. Kargar, M., An, A.: Keyword search in graphs: finding r-cliques. In: VLDB 2011, pp. 681–692 (2011)
9. Kim, S., et al.: Retrieving keyworded subgraphs with graph ranking score. ESWA 39(5), 4647–4656 (2012)
10. Lee, W., Leung, C.K.-S.: Structural top-k web navigation with inclusive query. In: IEEE ICIT 2009 (2009), doi:10.1109/ICIT.2009.4939712
11. Lee, W., Leung, C.K.-S., Lee, J.J.H.: Mobile web navigation in digital ecosystems using rooted directed trees. IEEE TIE 58(6), 2154–2162 (2011)
12. Li, G., et al.: EASE: an effective 3-in-1 keyword search method for unstructured, semi-structured and structured data. In: ACM SIGMOD 2008, pp. 903–914 (2008)
13. Liu, F., et al.: Effective keyword search in relational databases. In: ACM SIGMOD 2006, pp. 563–574 (2006)
14. Liu, J., Ma, Z.M., Yan, L.: Efficient processing of twig pattern matching in fuzzy XML. In: CIKM 2009, pp. 117–126 (2009)
15. Qin, L., et al.: Querying communities in relational databases. In: IEEE ICDE 2009, pp. 724–735 (2009)
16. Talukdar, P.P., et al.: Learning to create data-integrating queries. In: VLDB 2008, pp. 785–796 (2008)
17. White, R.W., Bailey, P., Chen, L.: Predicting user interests from contextual information. In: ACM SIGIR 2009, pp. 363–370 (2009)
18. Zhang, F., et al.: Fuzzy semantic web ontology learning from fuzzy UML model. In: CIKM 2009, pp. 1007–1016 (2009)

Author Index